D1420385

E77U6

Computing in Nonlinear Media and Automata Collectives

Computing in Nonlinear Media and Automata Collectives

Andrew Adamatzky

University of the West of England, Bristol

Institute of Physics Publishing
Bristol and Philadelphia

British Library Cataloguing-in-Publication Data

A catalogue record for this book is available from the British Library.

ISBN 0 7503 0751 X

Library of Congress Cataloging-in-Publication Data are available

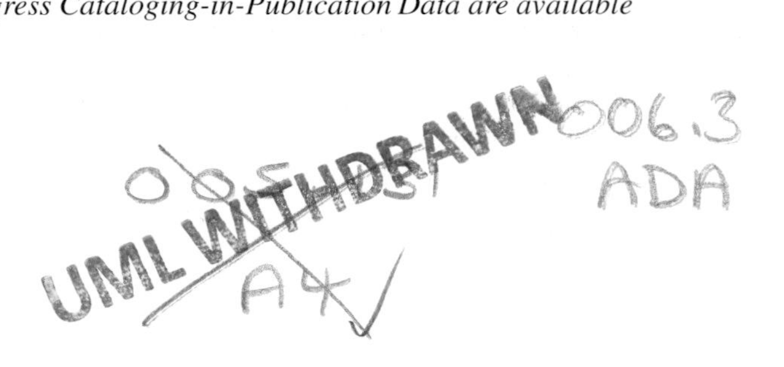

Commissioning Editor: James Revill
Production Editor: Simon Laurenson
Production Control: Sarah Plenty
Cover Design: Frédérique Swist
Marketing Executive: Colin Fenton

Published by Institute of Physics Publishing, wholly owned by The Institute of Physics, London

Institute of Physics Publishing, Dirac House, Temple Back, Bristol BS1 6BE, UK

US Office: Institute of Physics Publishing, The Public Ledger Building, Suite 1035, 150 South Independence Mall West, Philadelphia, PA 19106, USA

Typeset in TEX using the IOP Bookmaker Macros
Printed in the UK by MPG Books Ltd, Bodmin, Cornwall

Contents

Preface

This book aims to show how to design parallel processors from a massive pool of simple locally interacting uniform substances. Computing in excitable lattices, processing in chemical thin liquid layers and construction from amorphous swarms of social insects form the core subject of this volume which guides the reader from mathematical models of reaction–diffusion and excitable media to automata implementation of massively parallel processors to fabrication of real-life computing devices. Numerous amusing examples aim to entertain those already bored with dry theories or grounded applications.

The structure of this book is fairly logical: introduction, intuitive examples from computational geometry and graph theory, universal computing and the emergence of computation. In the first chapter we introduce the basics of unconventional computing. This class of computation deals with either new computational paradigms (e.g. evolutionary computing and ant colony optimization); advanced architectures (e.g. cellular automata); or novel materials (e.g. DNA molecules and microtubules). Computing in reaction–diffusion and excitable media amalgamates all three approaches. We paradigmatically represent the structure of the information by chemicals, waves or localizations. We architecturally implement chemistry- or wave-based algorithms in cellular-automata machines. We design laboratory prototypes of reaction–diffusion and excitable processors from natural components.

The spreading and interaction of substances are the main tools of reaction–diffusion computing. Various types of propagation fronts are the essential attributes of all natural systems from heat transfer to population dynamics to neural activity. Most of these phenomena can be adequately described in terms of reaction–diffusion systems, where reactants are converted into products during the propagation of a wavefront.

Most imagined, simulated and real-life chemical processors obey a very simple scheme of action. A medium is at rest at the beginning of a computation. Data are transformed into the spatial configuration of a geometrical object. This configuration is cast onto the medium. After projection, the elements of the data configuration form an ornament of local disturbances in the medium's characteristics, e.g. drops of a reagent or an excitation pattern. The local disturbances generate waves. The waves spread in the medium. Eventually the waves, originating from different data sources, meet each other. They somehow interact and produce either a concentration profile of a precipitate or a stationary excitation structure. The emerging pattern represents the result of the computation. This scenario is exemplified in the second chapter.

From the first chapter we obtain a clue as to how a reaction–diffusion medium can compute. We are ready now to apply wave phenomenology to the solution of real problems. Reasonably we start with a problem that has an underlying spatial structure and is open for intuitive solutions. We picked out three problems, which can be straightforwardly solved in either a reaction–diffusion or excitable medium. So, the second chapter deals with the construction of Voronoi diagrams, skeletons and convex hulls. To construct a Voronoi diagram or a skeleton we employ distance-to-time transformations implemented in the spread of diffusive or phase waves. Thus, for example, in the Voronoi diagram of a planar set, waves are generated at given sites of data space, they spread out, interact with each other and form a stationary structure as a result of the interaction. This structure represents bisectors, separating the given site. Computation of a convex hull is also simple. Waves are generated at the given sites, spread out and merge into a single travelling pattern. Eventually the excitation pattern stops its growth at the boundaries of the requested convex hull. The chapter familiarizes readers with all the essential steps of reaction–diffusion processor design: from the mathematics of computation to cellular-automata models to laboratory prototypes of chemical processors. The text is spiced up with the computation of a Voronoi diagram in a swarm of mobile entities inhabiting a lattice and even more wild examples of the subdivision of the British Isles.

'From electricity to ant families through reaction–diffusion and excitation' could be a slogan for the third chapter. In contrast to the previous chapter, this part of the book deals with graphs. Here we consider the computation of various types of graphs, where selected sites of a space are connected, usually by graph edges, as well as computation on graphs, when graph edges and nodes update their states in parallel. The shortest-path problem is a good test for a new computing device, particularly unconventional ones. We show how to find the shortest path in cellular-automata models of excitable media and in real-life Belousov–Zhabotinsky reactors. In automata models and in laboratory prototypes of excitable chemical processors the shortest path is calculated from the wave velocity field, generated by waves propagating through computational space. Several types of proximity graphs are constructed in nonlinear media. Computation of a spanning tree is discussed in full detail. The mathematical background for the reaction–diffusion construction of spanning trees is based on three propositions: trees are approximated by random walks, random walks are similar to electrical flows and electrical flows are similar to diffusion. Various designs of analog devices for shortest-path computation and spanning-tree construction are discussed thereafter.

Inspired by ideas of field computing we express some biological findings related to the motility-linked behaviour of amoeboids and the growth of neural terminals in an algorithm for the distributed construction of a minimum spanning tree. The algorithm is then transformed to computation of spanning trees in cellular-automata models of a reaction–diffusion medium. Reaction and diffusion at a macro-scale are exploited in our algorithms of swarm-based construction of spanning trees, where 'diffusive ant families' play the role of an active nonlinear medium.

Chapters 2 and 3 give us examples of 'specialized' processors. A processor is specialized if it is designed to solve some particular problem; for example, we could not use a reaction–diffusion processor that constructs the skeleton of a planar shape to do arithmetical calculations. Most real-life computers are universal. A device is called computation universal if it computes any logical function. Are reaction–diffusion and excitable media universal computers? The positive answer and proofs of media universality are given in the fourth chapter. There we give an overview of various types of computation universality, namely architecture-based and collision-based universality, and offer practical realization of universal processors in active nonlinear media.

Architecture-based computation assumes that a Boolean circuit is embedded into a system in such manner that all elements of the circuit are represented by the system's stationary states. This architecture is static. Three examples of architecture-based universality—sand-piles, mass transfer gates and wave gates—are discussed in detail. A dynamic, or collision-based, computation employs finite patterns, mobile self-localized excitations, travelling in space and executing computation when they collide with each other. The lion's part of the chapter is devoted to cellular-automata models of excitable media that exhibit a rich variety of mobile self-localizations. A remarkable spectrum of logic gates implemented in self-localization collisions is constructed. Most phenomena found in cellular-automata models are substantiated by real-life examples; i.e. we analyse the behaviour of solitons, light bullets, breathers, gaussons, excitons in monomolecular arrays and defects in tubulin microtubules in the context of collision-based computing.

While studying the construction of tesselations, proximity graphs and Boolean circuits in reaction–diffusion and excitable media we became proficient in the subject. We learn how to simulate nonlinear media, how to interpret a mathematical problem in terms of reaction and diffusion, how to design an algorithm and how to build a real-life chemical processor. We understand how problems with an underlying parallelism, like image-processing, graph theory and computation geometry, can be solved in cellular-automata models and natural media. Now we can assess, at least roughly, which problems can be solved in what types of nonlinear media. Should we fabricate these media from scratch or could we instead search for already existing specimens in nature? Chapter 5 offers an answer. We show which parameters of behaviour of medium's local elements can be taken into account when the computational properties of the medium are under scrutiny. Morphological, dynamical and computational classifications of excitable lattices are provided in this chapter together with several applications of emerging computation properties to image-processing and the control of robot navigation.

Most concepts, algorithms and designs discussed in this book are original unless otherwise stated. Several collaborative results must be acknowledged. The laboratory prototypes of the reaction–diffusion processors for the computation of the Voronoi diagram and skeleton were fabricated together with Dmitrij Tolmachev. Early results on lattice swarm construction of Voronoi diagrams and the reaction–diffusion load balancing of communication networks were published jointly with Owen Holland. The basic designs of excitable-lattice controllers for mobile robots were pioneered in partnership with Chris Melhuish. Much thanks, guys, for these contributions!

The book may be used in many different ways. You can read it from cover to cover, you can select certain topics, you can look through the illustrative examples or you can just consult the bibliography; the recipes for the chemical processors are very handy indeed. If you are in a hurry follow one of the fast tracks shown in the diagram.

Andrew Adamatzky
Bristol 2001

Fast tracks for busy readers

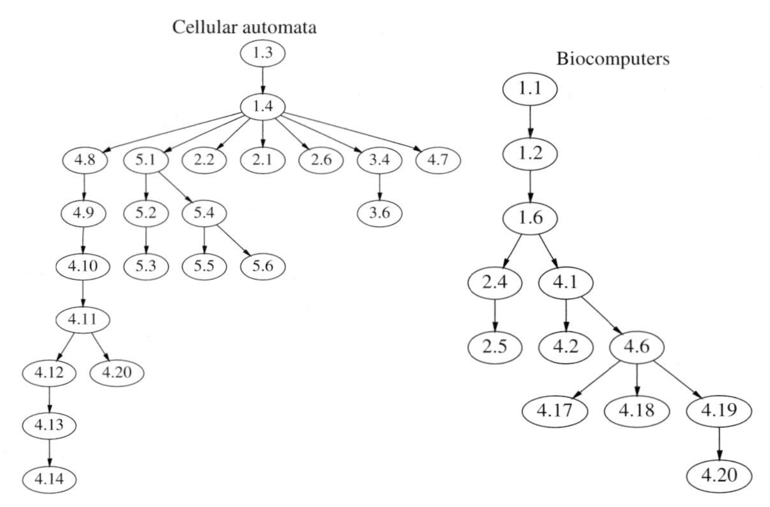

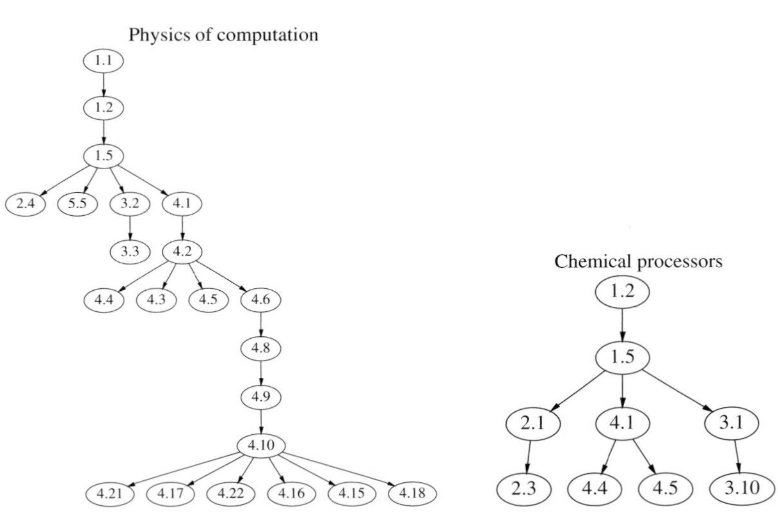

Chapter 1

Reaction–diffusion, excitation and computation

This book is built on case studies. So is this chapter. Despite its mosaic structure this chapter nevertheless shows a logical interrelation of sections. We are mainly concerned with computations in reaction–diffusion and excitable media. Such computations are unconventional. This is why we start with a brief discussion of unconventional computing. Most computational techniques discussed in this book rely on wave dynamics in nonlinear media. Therefore we devote one of the chapter's sections to this subject. Various methods for modelling reaction–diffusion and excitable media are discussed here both to ease understanding of the theoretical algorithms and their practical implementations. We have allocated a special section to cellular automata because they represent an intuitive link between real nonlinear media and mathematical models of massively parallel computing architectures. At the end of the chapter we provide several examples of computational problems solved in nonlinear active media to give readers a sense of the subject.

1.1 Unconventional computing

What is unconventional computing? This is a matter of time scales, personal preferences and orthodoxy in education. As Toffoli writes:

> *a computing scheme that today is viewed as unconventional may well be so because its time hasn't come yet—or is already gone.* [593]

We would possibly associate unconventional with the advanced, however this is not customary at the moment. Unconventional computing, as is commonly agreed, includes:

- computing in stirred chemical reactors (mass-transfer and kinetic-based computing) and thin-layer chemical reactors (this is discussed widely in this book, so we do not give any references at the moment);

- assembly-based computing, particularly DNA computing (see, e.g. [147, 35, 495, 381, 544, 342, 255, 335, 249, 149, 173]) and, partially, microtubule-based-computing [287, 110];
- computing in neural networks (see e.g. [285, 328, 391, 154, 45]);
- computing in cellular automata (the present book certainly deals with this subject, however see also [591, 566, 274, 176, 254]);
- quantum computing (see e.g. [399, 109, 642, 380]);
- genetic algorithms and evolutionary programming (see e.g. [260, 364, 462, 153, 237, 455]); and
- swarm optimization, the solution of optimization problems using the principles of social insect behaviour (see e.g. [97, 552, 553, 96, 139, 191, 252, 272])

Some of the entries are sometimes classified as the subjects of *natural computing*; others are considered to belong to the *physics of computation*.

> *We first build a computer that hides the underlying spatial structure of Nature, and then we try to find the best way to contort our spatial computation to fit into that mold!... and so our most efficient computers and algorithms will eventually evolve toward the underlying spatial 'hardware' of Nature.* [424]

These words by Margolus [424] sufficiently characterize the basic ideas of unconventional, particularly spatial computing.

Molecular computing and chemical computing are the two most popular plots in the field of natural computing (see e.g. [142, 143, 144, 145, 146, 150, 256, 542, 149, 339]). DNA computing [35] gives us an example of recent successes in molecular computing: DNA computing has become a synonym for molecular computing [74], thus replacing the more well-known term 'molecular electronics'. A DNA computer solves the Hamiltonian path problem by producing a molecule, which encodes the path, if such a path exists. A Hamiltonian path is not the only problem solvable in a DNA computer [495, 381, 544, 342, 255, 249]. Ideally, any problem can be solved in this molecular computer. Shortly after pioneer Adleman published his work [35] it was demonstrated that a universal Turing machine can be built from DNA processors (see also [75]), i.e. DNA processors are simulation universal.

Berry and Boudul's chemical abstract machine [82, 83, 319] is a brilliant example of how a chemical paradigm of the interactions between molecules can be utilized in concurrent computations, e.g. in algebraic process calculi. The state of a concurrent system is seen as a stirred solution in which the molecules of reagents interact with each other according to certain rules [83]. This idea has evolved from an abstract machine. The machine employs a gamma formalism, which is based on a set of transformations. The transformations consume the elements of a multiset and produce new elements by specified rules, developed for parallel programming [62]. The process is parallel because all pairs of molecules can interact at once. Molecules are represented by multisets and the reaction rules are rewritings of multisets. Thus, a specific machine is designed by defining the molecular algebra and the rules of interaction between the molecules. The basic elements of the machine are ions in which the interactions are determined by their valences. Therefore, we can talk about a reaction rule when a molecule is formed from ions: heating when a molecule is split into several ions and cooling when a compound molecule is assembled from several other molecules [83].

The ideas of a chemical abstract machine are also reflected in the abstract chemistry developed from the theory of the self-maintenance of systems, autopoiesis [620] and λ-calculus [137]. In abstract chemistry

> *molecules are viewed as computational processes supplemented with a minimal reaction kinetics.* [238]

Fontata and Buss [137] derived several levels of λ-calculus models of chemistry to build a link between abstraction and dynamics [238]. In the models a physical molecule corresponds to a symbolic representation of an operator. The operator's action represents the molecules' behaviour. A chemical reaction is seen as evaluation of this functional application. So, a collision between molecules is interpreted as a production rule or a relation between the abstract molecules [238]; in the typed calculi chemical model we can additionally represent the specifics of an interaction by a type discipline [238].

When did physics-based natural computing start? Computing via analog means was known long before digital computing. It can be dated back to 1876 when Lord Kelvin published his first results on analog-related computers (see [601]; cited in [512]).

> *It may be possible to conceive that nature generates a computable function of a real variable directly and not necessarily by approximation as in the traditional approach.* [512]

The differential analyser is a typical example of a general purpose analog computer. It generates functions of time; the functions are measured in volts. The analyser consists of the following blocks [512].

- An integrator (one input and one output): for the input $u(t)$ and the initial condition $e(a)$ at time step $t = a$ we have $\int_a^t u(t)\,\mathrm{d}t + e(a)$ on the output.
- An adder: two inputs $u(t)$ and $v(t)$ and one output $u(t) + v(t)$.
- A constant multiplier, which produces $k \cdot u(t)$.
- A variable multiplier, which produces $u(t) \cdot v(t)$.

The ideas of analog computing evolved into field computing, i.e. computation over a continuum of continuously variable quantities [408], where the basic operations are field transformations rather than discrete operators [407, 408]. Neural networks, in general, and retina, in particular, may be classified as subclasses of field computers, for which silicon chip realizations are now widespread (see e.g. [456]).

In general, any physical system can play the role of a computer, if the system's behaviour and reaction to perturbations are appropriately interpreted. Thus, for example,

> *any physical system whose dynamics in phase space is dominated by a substantial number of locally stable states to which it is attracted can therefore be regarded as a general content-addressable memory.* [314]

Amongst other things the physics of computation deals with

- the physical meaning of information:
  physical action measures amount of computation [595];
- the thermodynamics of computation:
  the digital computer may be thought of as an engine that dissipates energy in order to perform mathematical work [78];
- the relation of the minimal action principle to computation abilities; do you think that *a system's natural trajectory is the one that will hog most computational resources* [301]?

However, one of the most attractive principles of physics-based computing, which is utilized in this book, is highlighted by Margolus [424]:

> *Because physical information can travel at only finite velocity, portions of our computation that need to communicate quickly must be physically located close together.*

1.2 Waves in reaction–diffusion and excitable media

The majority of physical, chemical, biological and sociological processes can be described in terms of propagating fronts of one nature or another. The spread of genes in population dynamics, heat transfer, the growth of microbial, plant and animal colonies (and even cities) and the dynamics of excitation in nervous systems are just a few arbitrarily picked examples of phenomena with propagating fronts. Calcium waves in a cytoplasm of frog oocytes, waves in a mixture supporting RNA replication and aggregation in colonies of *Dictyostelium discoideum* are instances of propagating fronts in reaction–diffusion and excitable media [72, 612, 580]. Let us briefly discuss three examples of spatial wave dynamics from different fields of science.

- In a point-wide stimulation of a cerebral cortex a spreading EEG depression is observed. The suppression of electrical activity propagates from the stimulation point as a concentric wave called Leão's wave. The wave propagates on the cerebral cortex, due to K^+ movements between intra- and extra-cellular spaces [432]. Similar waves are observed in laboratory experiments with isolated retina [114, 432].

- Auto-waves are phenomena typically observed in morphogenesis. They include proximal–distal waves of cell re-orientations and peristaltic annulations in *Hydrozoa*, cortical peristaltic waves in *Nematoda*, waves of cell contraction in embryos, contact cell polarizations, calcium waves accompanying cell division in *Paramecium*, mitotic activity waves in retinal differentiation of crayfish [77, 107, 547, 503, 283, 481].
- Circular patterns, wave-like in nature, are generated in bacterial cultures infected by phage [323]. Spiral spatial waves are very typical for host–parasite systems [91]. Almost all population systems, especially those with predator–prey relations exhibit wave patterns in their development [37, 465, 351, 379, 86].

Most types of propagating fronts, including those propagating in excitable media, can be analysed in terms of reaction–diffusion systems [212].

The first works on wave phenomena in excitable and chemical media were published at the beginning of this century by Mayer [437] and Luther (see [558, 559] for details). Their publications dealt with phenomenological results on the arhythmic properties of jellyfish muscle and the propagation of a wavefront in a tube containing a chemical liquid [559]. Robert Luther produced a propagating front in a permanganate oxidation of oxalic acid, where a purple solution becomes uncoloured at the moving boundary [212]. However, it was some time before the ideas were developed further, particularly in theoretical works. Thus, the 1946 paper by Wiener and Rosenblueth [640] is usually cited as the basic reference for mathematical models of excitable media, but in fact the first mathematical work, published in 1937, on the subject belongs to Kolmogorov, Petrovsky and Piscounoff [360], who analysed wave motion in excitable media and obtained a reaction–diffusion equation. Fisher independently published a similar equation in the same year. Then a quantitative mathematical model of wave propagation in a nerve was presented in 1952 by Hodgkin and Huxley [308]. Hodgkin and Huxley's analytical studies of a travelling excitation in nerve membranes [308] were developed further by FitzHugh [231]. Actually, most of the basic features of wave motion in excitable media had already been discovered mathematically by the late 1930s. These features include the notion of a threshold or a necessary condition for the formation of a travelling wave: a sufficiently large region of a medium must be excited to start the excitation wave dynamic [545, 360] (see also [573]).

The simplest kind of a chemical wave emerges when

> *reactants are converted into products as the front propagates through the reaction mixture.* [212]

Background information on reaction–diffusion and pattern formation in chemical, physical and biological systems can be found in [441, 160, 212, 357]. We consider just a few examples for illustrative purposes.

Turing's paper [607] showed that stable patterns are produced in reaction–diffusion systems, where two reacting substances diffuse with different rates. This idea has been successfully applied to analyses of pattern formation on mollusc shells [441, 443] that gives us a bright example of a biological application of reaction–diffusion theory. Meinhardt and Klinger [442, 443, 444] developed a model of pattern formation in a mollusc shell where pigment deposits oscillate and travelling waves of pigment production are formed, when pigment-producing cells trigger their neighbours. It is assumed that the patterns develop during the growth of the shell, which is a $(1+1)$-dimensional process. The one-dimensional pulses of pigment production may collide with one another, annihilate, reflect or produce new travelling waves. As a result a wave dynamic which is typical for mollusc-shell patterns is generated. Patterns similar to those of *Strigilla* and *Palmadusta diluculum* can be produced [442, 443, 444]. The reaction–diffusion model itself is based on the interaction between an autocatalytic activator a and its inhibitor h, concentrations of which evolve in a one-dimensional space x as follows:

$$\frac{\partial a}{\partial t} = \frac{\rho a^2}{h} - \mu a + D_a \frac{\partial^2 a}{\partial x^2} + \rho_0$$

and

$$\frac{\partial h}{\partial t} = \rho a^2 - \nu h + D_h \frac{\partial^2 h}{\partial x^2} + \rho_1$$

where ρ is a source density, μ is the activator lifetime, ν is the inhibitor lifetime, and ρ_0 and ρ_1 are source densities for the activator and inhibitor. Sometimes, we talk about the depletion of a substrate, instead of the production of an inhibitor, which may be consumed in the autocatalytic production of an activator [444]. The parameters for reaction and diffusion determine the exact structure of the developing pattern. A discrete cellular-automaton-based model of mollusc shell-pattern formation is developed in [500]. Various patterns that arise from collisions of classical waves, chemical waves and mobile localization are discussed there.

The nontrivial behaviour of nonlinear chemical systems can be explained by their bistability. Bistability means that two stable steady states coexist. For example, in a stirred tank a chemical system undergoes a transition from one stable state to another as a result of perturbations [212]. In reaction–diffusion systems

> *fronts typically propagate with a constant velocity and wave form, converting reactants that lie ahead into products which are left behind. The bulk of the chemical reaction occurs within a narrow reaction zone.* [212]

The Belousov–Zhabotinsky-reaction-based medium [664, 669] is the most well-known example of an active nonlinear medium [644]. The classical reaction incorporates a one-electron redox catalyst, an organic substrate (that can be brominated and oxidized) and a bromate ion in either the form $NaBrO_3$ or $KbrO_3$. The reagents are dissolved in sulphuric or nitric acid. Ferroin is the most typical catalyst in classical reactions, and ruthenium is a standard catalyst in light-sensitive reactions. Malonic acid is used as an organic substrate; Belousov himself used citric acid. Oxidized and reduced forms of the catalysts have different colours, for example, oxidized ferroin has a blue colour and reduced ferroin has a red colour.

A very typical recipe for a Belousov–Zhabotinsky reaction in a stirred reactor includes 0.3 M malonic acid, 0.1 M $NaBrO_3$, chloride-free ferroin (or $MnSO_4$), diluted in 1 M sulphuric acid. Namely, we dilute, in sulphuric acid, malonic acid, the catalyst and the bromate. Another, more detailed recipe, is as follows: 0.2–0.36 M BrO_3^-, 0.03–0.2 M malonic acid, 0.21–0.51 M H_2SO_4, 0.001–0.1 M catalyst (we can choose a catalyst from $Fe(phen)_3SO_4$, $Ru(bipy)_3SO_4$, $Ce(SO_4)_2$ and $Mn(SO_4) \cdot 4H_20$) and 0.09 M bromomalonic acid [475].

What happens in the reaction mixture liquid? In the standard version of the reaction the following sequence of events is repeated [226, 47]. BrO_3^- is transformed into bromic acid $HBrO_2$. This transformation process is coupled with oxidizing ferroin to give ferriin. At this stage the colour of the mixture changes because ferroin is red and ferriin is blue. Ferroin can be recovered from ferriin with the help of malonic acid. Bromide ions are produced in the ferriin-to-ferroin transformation. These bromide ions inhibit the production of the bromic acid. More generally we can state that the essentials of the Belousov–Zhabotinsky reaction are:

- the oxidation of BrO_3^- leads to the production of HOBr and Br_2, which brominate the organic substrate
- the autocatalytic generation of $HBrO_2$ with parallel oxidation of metal catalyst
- the oxidation of the organic substrate by the catalyst to regenerate Br^-.

These three stages are cyclically repeated [212].

The following schematic description [501] of the Belousov–Zhabotinsky reaction is adopted for the well-known Oregonator model [350] (see also [670]):

$$A + W \xrightarrow{k_1} U + P$$

$$U + W \xrightarrow{k_2} 2P$$

$$A + U \xrightarrow{k_3} 2U + 2V$$

$$2U \xrightarrow{k_4} A + P$$

$$B + V \xrightarrow{k_5} hW$$

where A is the concentration of BrO_3^-, B the concentration of bromomalonic acid, P the concentration of HOBr, and U the concentration of $HBrO_2$. V is the concentration of the catalyst (for example, Ce^{4+} or $Ru(bpy)_3^{2+}$), W is the concentration of Br^-. The rate constants range from k_1 to k_5 and h is a stoichiometric factor, which counts the number of moles of Br^- released in the reduction of 1 mole of a catalyst [501]. The Oregonator model for light-sensitive modification of the Belousov–Zhabotinsky reaction may be formulated in the following system of two partial differential equations [579, 531]:

$$\frac{\partial u}{\partial t} = \epsilon^{-1}\left(u - u^2 - (fv + h)\frac{u-q}{u+q}\right) + D_u \nabla^2 u$$

$$\frac{\partial v}{\partial t} = u - v + D_v \nabla^2 v$$

where u is the concentration of $HBrO_2$, v is the concentration of the catalyst, h is used to show the contribution of the catalyst produced as a result of illumination, ϵ, f and g are small real constants, and D_u and D_v are diffusive coefficients.

The Brusselator is yet another well-accepted model for reaction–diffusion systems:

$$\frac{\partial u}{\partial t} = a - (b+1)u + u^2 v + D_u \nabla^2 u$$

$$\frac{\partial v}{\partial t} = bu - u^2 v + D_v \nabla^2 v.$$

The concentrations of the two chemical species are represented by the variables u and v. The control parameters are a and b, and D_u and D_v are the diffusive coefficients. Various dissipative structures including spiral waves can be formed as the system evolves according to the initial parameters.

Experiments with the Belousov–Zhabotinsky reaction can be done either in stirred or non-stirred, thin-layer reactors. In the stirred reactor periodic changes between oxidized and reduced states happen concurrently in all sites in the medium. Therefore the whole solution changes colour from blue to red and back to blue if ferroin is the catalyst [227, 543]. If cerium is used as the catalyst the solution oscillates between colourless and yellow. It oscillates between colourless and red if we use manganese ions. The oscillations start after some induction interval. A typical oscillation period is approximately 1 min; however, it strongly depends on the temperature and other factors.

To make an unstirred reactor prepared solutions are poured into a Petri dish to make a layer 1–3 mm thick. To generate circular waves one can immerse Ag wire (100 μm diameter) in the solution [475]. The wave is detectable by its blue front, i.e. a blue colour indicates ferriin, the oxidized state of the catalyst. It spreads into a resting, red coloured solution where the catalyst is resting in its reduced ferroin form [472]. The wavelength is usually around 1 mm. The waves spread with the velocity 1 mm min^{-1}.

What is going on in a non-stirred Belousov–Zhabotinsky reactor? We spread the solution as a thin layer and disturb it with a silver wire, if the catalyst is ferroin or just illuminate part of the solution, if ruthenium is used as a catalyst. The disturbance leads to the generation of a blue oxidation front that propagates on the red background of reduced ferroin.

Two types of waves are observed, and sometimes they coexist, in Belousov–Zhabotinsky solutions [47]:

- trigger waves (they have constant shape and low speed, 1–4 mm min^{-1})
- phase waves (waves with a velocity much higher than that of the trigger waves); they may be similar to phase-shifted oscillations in chains of weakly coupled pendula.

We could possibly consider a continuum where trigger waves occupy one end and phase waves the other [47]. Diffusion phase waves are somewhere between the poles.

Fronts in a Belousov–Zhabotinsky medium are described by a quadratic reaction–diffusion equation, which is similar to the Fisher–Kolmogorov equation [212]. As previously discussed, the medium supports trigger waves—the propagation pulses that move through the resting reaction solution; and the system returns to the resting state behind the wave. So, the Belousov–Zhabotinsky medium is an excitable medium.

In excitable chemical media wave propagation occurs because of coupling between diffusion and autocatalytic reactions. When autocatalytic species are produced in one micro-volume of the medium they diffuse to the neighbouring micro-volumes and they trigger an autocatalytic reaction there. If an excitable system has spatial extent it will respond to perturbations exceeding the excitation threshold by producing excitation waves [212, 341].

Excitation waves travelling in excitable media are often called auto-waves because they propagate at the expense of energy stored in the medium [368, 322]. Even a superficial comparison of auto-waves with classical waves shows entirely different behaviour for the excitation waves [365, 368]:

- In contrast with classical waves auto-waves do not conserve energy. Rather they propagate at the expense of the energy of the chemical medium; however, they do conserve amplitude and waveform.
- Auto-waves, in general, are not reflected from a medium's boundaries or inhomogeneities; however, they may annihilate when they collide with one another. It has been demonstrated in [671, 213] that waves in a Belousov–Zhabotinsky medium can be reflected as the result of collisions with rectangular inhomogeneities depending on wave speed and angle of incidence.
- Auto-waves do not exhibit interference.
- Both classical and excitation waves obey Huygens' principle, therefore they diffract.

If an excitation wave is broken, for example due to defects in the medium, the broken wave may either contract or be transformed into a spiral wave. The dynamics of two- and three-dimensional spiral waves has been well investigated in [349, 350, 453].

An attractive setup for investigating excitation waves on substrates with a discrete topology may be implemented in a so-called heterogeneous active Belousov–Zhabotinsky medium. The medium consists of a reaction zone with an immobilized catalyst, usually ferroin fixed in silicahydrogel substrate, covered by a feeding solution without a catalyst [396]. One can also immobilize the catalyst on cation exchange resin beads [435]; in such a setup we can also obtain a pseudo-cellular topology in the reactor space.

What is the essence of excitation?

> *A system is excitable if it possesses a stable fixed point and responds to perturbation in such a manner that if a perturbation magnitude exceeds a certain threshold the system moves in a phase space along some closed trajectory until it returns to the fixed point. While moving along its trajectory the system is not sensitive to any perturbations and it is may be called refractory.* [341]

A grass fire, particularly on large time and space scales, is a typical example of an excitable medium. If we set a path of dry grass on fire the fire will spread. Ideally, a fire front is circular, hence it may lose its regularity due to inhomogeneity of grass cover and landscape. The burnt zones cannot be repeatedly fired. However, over the years the grass grows again and the territory is ready to be set on fire again. The following analogies are remarkable.

- *Setting on fire* is a *perturbation* action.
- *A fire-front* is an *excitation front*.
- *A burnt zone* is the part of medium in a *refractory state*.
- *Growth of a grass* is the *recovery of a substrate*.

The grass-fire example demonstrates four essential features of every excitable medium:

(i) There is a spatially homogeneous resting state which is stable to small perturbations.
(ii) When a certain number of a medium's sites are perturbed above a certain threshold self-propagating waves are generated.
(iii) The waves travel in the medium without damping.
(iv) The medium returns to its resting state after propagation of the waves.

In this book we do not deal with reaction and diffusion only in chemical species. Therefore, it is reasonable to mention the Fisher equation, which deals with the spread of a mutant gene in the population:

$$\frac{\partial p}{\partial t} = D\frac{\partial^2 p}{\partial x^2} + kp(1-p)$$

where p is the frequency of the mutant gene. We should also appreciate that not only chemical and biological media are subject to excitation. Many other systems exhibit excitation dynamics. We can demonstrate this using a system studied by Muratov [474]. The example clearly exposes the underlying mechanics of excitation. This is an elastic medium with friction exhibiting stick–slip motion. The Burridge–Knopoff model is under consideration here. The model is represented by a one-dimensional array of blocks resting on a frictional surface. The blocks are connected by springs, with spring constant k_c, one by one. Thus they form a chain. Each block is also connected by a spring, with spring constant k_p, to a loader plate. The blocks are pulled by a loader plate moving with a constant velocity. Each block is linked by springs to its left- and right-hand neighbours and to the loader plate, therefore the following total force

$$F_i = k_c(x_{i+1} + x_{i-1} - 2x_i) - k_p x_i - f_i$$

acts on the ith block, where x_i is a displacement of the block i, and f_i is a friction force [474]. When the force applied to a block exceeds the static friction force the block starts to slip and the friction force drops immediately to some value below the static friction. When the blocks stick the friction returns to its static value. The accumulated stress accelerates the block at the onset of motion. The released potential energy is enough to support travelling waves in the chain of blocks. In this model the static friction plays the role of excitation threshold while the springs connecting the blocks represent local coupling [474].

1.3 Cellular automata

Space, time and state of the space site are the basic constituents of any model. Each element may be either discrete or continuous. Partial differential equations are a classical instance in which all elements are continuous. We have coupled map lattices when space and time are discrete and the state is continuous. Cellular automata give us an example in which all three elements—time, space and state—are discrete.

The following words were recorded just at the beginning of the Second World War:

> *Suppose one has an infinite regular system of lattice points in E^n, each capable of existing in various states $S_1, \ldots, S_k$. Each lattice point has a well defined system of m neighbours, and it is assumed that the state of each point at time $t+1$ is uniquely determined by the states of all its neighbours at time t. Assuming that at time t only a finite set of points are active, one wants to know how the activation will spread.* [616]

This quotation from Stanislaw Ulam's book [616] gives us not only the first evidence of cellular-automata theory but also quite accurately defines a cellular automaton itself.

> *A cellular automaton is a local regular graph, usually an orthogonal lattice, each node of which takes discrete states and updates its states*

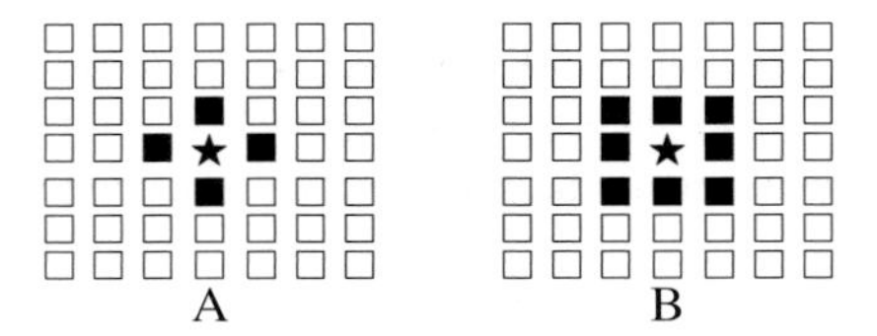

Figure 1.1. Classical neighbourhood of a two-dimensional cellular automaton: (A) von Neumann neighbourhood; (B) Moore neighbourhood. The central cell is shown by ⋆. Other elements of the cell ⋆'s neighbourhood are shown by ■s. Cells not belonging to the neighbourhood of cell ⋆ are shown by □s.

in a discrete time; all nodes of the graph update their states in parallel using the same state transition rule.

The adjective '*cellular*' possibly reflects the original representation of such automata in matrix terms, as Ulam and, at least initially, von Neumann did; or Neuman's early drawing of ruled squares. Biological cells, as is quite often wrongly assumed, never remained behind the cellular-automata concept. Sometimes one can meet cellular automata under such names as tesselations and tesselation structures, homogeneous structures, mosaic automata, uniform automata networks and cellular arrays.

A cellular automaton is a tuple $\langle \mathbf{L}, \mathbf{Q}, u, f \rangle$ *of a many-dimensional array* $\mathbf{L}$ *of uniform finite automata, finite state set* $\mathbf{Q}$*, local connection template, or automaton's neighbourhood,* u*,* $u : \mathbf{L} \rightarrow \mathbf{L}^k$*,* k *is a positive integer, and automaton transition function* f*,* $f : \mathbf{Q}^k \rightarrow \mathbf{Q}$*. Automata of the array update their states simultaneously and in discrete time. Each automaton* x *of the array* $\mathbf{L}$ *calculates its next state by the rule:* $x^{t+1} = f(u(x)^t)$*, where* x^{t+1} *is the state of automaton* x *at time step* $t+1$ *and* $u(x)^t$ *is the state of automaton's neighbourhood* $u(x)$ *at time step* t*.*

Automata constituting the array $\mathbf{L}$ are called cells. Cells are locally interconnected. The exact architectonics of local links between the cells determine the structure of the neighbourhood. In most situations, to define the neighbourhood $u(x)$ of a cell x it is enough to determine the cells, cell x's neighbours, being at a distance no more than r, where r is a positive integer, in a metric M:

$$u(x) = \{y \in \mathbf{L} : |x - y|_M \leq r\}.$$

For $M = L_1$ we obtain the so-called von Neumann (cruciform or four-cell, because $k = 4$) neighbourhood (figure 1.1(A)), and the so-called Moore (complete or eight-cell, because $k = 8$) neighbourhood (figure 1.1(B)) for

$M = L_\infty$. There are no restrictions on the type of neighbourhood, therefore quite a variety of them can be defined. If the neighbourhood is not simply organized it can be defined as a list of neighbours: $u(x) = (y_1, \dots, y_k)$, k is the size of neighbourhood. The state of the neighbourhood at time step t can be written as $u(x)^t = (y_1^t, \dots, y_k^t)$.

The cell state transition rule can be represented in many different ways, for example, as a set of `IF-THEN` rules, a formula or a lookup table.

The configuration of a cellular automaton is a mapping $c : \mathbf{L} \to \mathbf{Q}$, which assigns a state from $\mathbf{Q}$ to each cell of the array $\mathbf{L}$. A global transition function is the mapping $G : \mathbf{Q}^\mathbf{L} \to \mathbf{Q}^\mathbf{L}$, which transforms any configuration c to another configuration c', $c' = G(c)$ by applying local transition function f to the neighbourhood of every cell of the array $\mathbf{L}$. In this book we mostly deal with deterministic automata. A cellular automaton is called deterministic if for any configuration c there is exactly one configuration c' that $G(c) = c'$.

Given initial configuration c^0 the evolution of a cellular automaton is represented by a sequence of the automaton's configurations:

$$c^0 \to c^1 \to \cdots \to c^t \to c^{t+1} \to \cdots \to c^\tau \to \cdots \to c^{t+p}.$$

Eventually the cellular automaton falls into a cycle with period p, $p > 1$, or a fixed point of its evolution, $p = 1$.

The global dynamics of finite cellular automata can be visualized in a graph of global transitions. Every node of the graph corresponds to a unique configuration of the cellular automaton and the arcs connecting the nodes represent global transitions between the configurations.

Here is a small example. Given a one-dimensional cellular automaton with five nodes. The lattice is concaved in a torus, i.e. the automaton has periodic boundaries. Every cell of the automaton takes two states, 0 and 1, updates its state depending on the states of its left- and right-hand neighbours. Indexing the cells of the lattice by i, $i = 1, \dots, 5$, we can write cell x_i's neighbourhood as $u(x_i) = (x_{i-1}, x_{i+1})$; in this particular example a cell is not included in its neighbourhood. Let a cell take the state 1 if only one of its neighbours is in state 1, independently of the state of the cell itself. Otherwise, the cell takes the state 0. The cell state transition rule $x_i^{t+1} = f(x_{i-1}^t, x_{i+1}^t)$ can be represented as $x_i^{t+1} = (x_{i-1}^t + x_{i+1}^t) \bmod 2$, if states are arithmetical, or $x_i^{t+1} = \overline{x_{i-1}^t} \wedge x_{i+1}^t \vee x_{i-1}^t \wedge \overline{x_{i+1}^t}$, if the states have Boolean values. We could also represent the rule as

$$x_i^{t+1} = \begin{cases} 1 & \text{if } x_{i-1}^t = 1 \text{ and } x_{i+1}^t = 0 \text{ or } x_{i-1}^t = 0 \text{ and } x_{i+1}^t = 1 \\ 0 & \text{otherwise.} \end{cases}$$

Finally, we can describe the rule via a lookup table, where every entry has the format $x_{i-1}^t x_{i+1}^t \to x_i^{t+1}$:

$$00 \to 0 \qquad 01 \to 1 \qquad 10 \to 1 \qquad 11 \to 0.$$

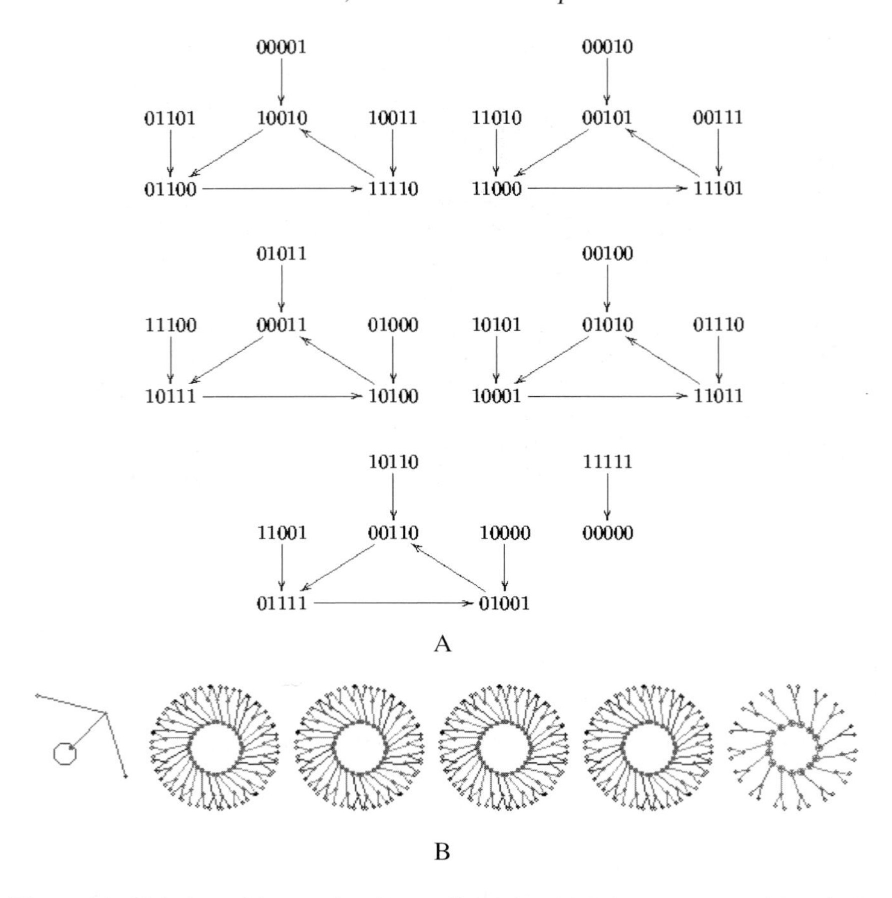

Figure 1.2. Global transition graphs of a one-dimensional cellular automaton with periodic boundaries. (A) The automaton consists of five cells. Nodes of the graph are labelled by explicit configurations of automaton states. (B) The automaton consists of ten cells. Explicit configurations are not shown. This graph was built using `DDLab` [651].

The automaton consists of five cells, and every cell takes two states, therefore the global transition graph has 32 nodes. The graph consists of the six components of connectivity as shown in figure 1.2(A).

Increasing the number of cells in a cellular-automaton lattice one can obtain impressive results in spacetime (see, e.g., figure 1.3) and in the global integral dynamic (see, e.g., figure 1.2(B)) of the cellular automaton.

The development of cellular automata was certainly facilitated by early results from the beginning of computation [611], the formal modelling of neural networks [438] and the widespread application of automata theory in the natural sciences [641]. For the last 30 years cellular automata have grown from

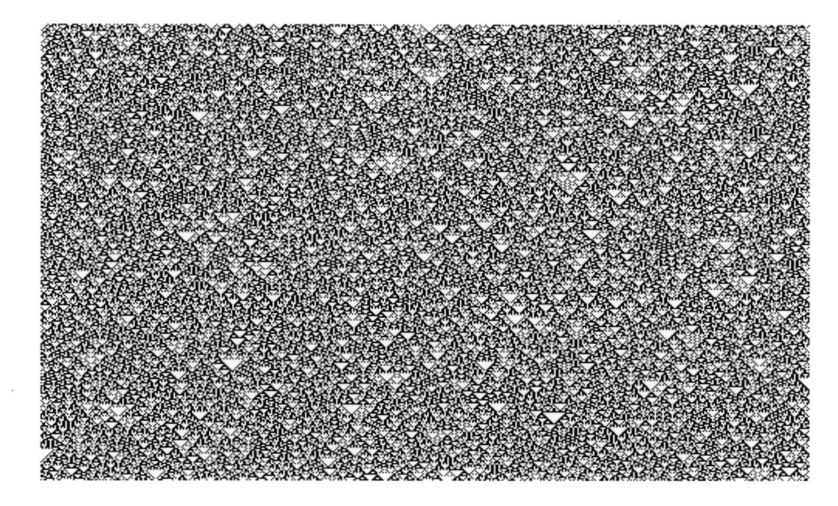

Figure 1.3. Spacetime evolution of one-dimensional cellular automaton, defined in the text.

entertainment toys for intellectuals to an interdisciplinary field that contributes to almost all sciences, from pure mathematics and logic, to physics and chemistry, engineering and biology, psychology and geology. As Toffoli stated:

> *Cellular automata are dynamical systems that play in discrete mathematics a role comparable to that partial differential equations in the mathematics of the continuum.* [595]

This can be paraphrased as

> *cellular automata are parallel models of computation, discrete dynamic systems and models of natural processes that replace traditional mathematics in natural sciences and substitute nature in contemporary mathematics.*

The half-century of struggle it has taken for cellular automata to become a recognized discipline reflects, on a small scale, the time it has taken humanity to recognize the discreteness of the surrounding world. This started with Leeuwenhoek's discoveries of the cellular structure of living matter [388], was continued in the identification of the discreteness of 'grey matter', when Santiago Ramon y Cajal convinced the world that neurons are separated by synaptic spaces and not continuously connected to each other as Camillio Golgi believed, and reached its apogee in Neumann's self-organizing automata [479], Zuse's *discrete universe* [676], Toffoli's *programmable matter* [592, 593, 594] and Minsky's *cellular vacuum* [457, 458]. For a period of time the theory of cellular automata flourished significantly and dozens of quite distinct classes of cellular automata emerged.

The definition of a cellular automaton given earlier concerns deterministic and synchronous cellular automata with a stationary architecture of local connections. There are also such attractive classes as asynchronous cellular automata, stochastic cellular automata and structurally dynamical cellular automata.

In *asynchronous* cellular automata cells do not update their states at the same time, state-switching may be triggered by internal or external events, sometimes stochasticity is involved (see e.g. [477, 116, 8, 152, 403, 551, 278, 499, 574]). We must mention that no proper asynchronous cellular automaton has ever existed. Either such automata are probabilistic or cells use some external input to control the delay of state-switching.

Stochastic cellular automata rely on the probabilistic update of cell state: a cell will choose one of the states with a predetermined conditional probability (see e.g. [189, 411, 277, 392, 9, 338, 84, 398]).

Structurally dynamic cellular automata are a lovely class. Cellular automata and automata networks were the seeds from which a spacetime theory on the Planck scale in [534, 535] emerged. In Requardt constructions the strength of the link between two neighbouring cells of a cellular automaton depends on the intensity of the interaction between the cells, where the interaction is considered to occur via transportation of quanta between the cells. The strength of each link is changed in such a manner to balance the intensity of the energy exchange between the cells. Considering the model in the context of reaction and diffusion we can see the strength of the link as an analogue of dynamical diffusive coupling. These representations have evolved from 'structurally dynamic' cellular automata [318], the most promising class of cellular automata. The notion of structurally dynamic automata implies that the lattices evolve dynamically: the links connecting cells evolve together with the cells. A link, as usual, may take two states: *is* and *is not*. The states of the links are updated depending on the states of adjacent cells and nearby links. Some examples of the identification of these automata can be found in [9]. Advanced discussion of dynamical cellular automata is given in [292], where models assume that the local connectivity of the cells changes as the automaton evolves depending on the cell states. In such an interpretation, a link is analogous to a synapse, in which the strength depends on the 'excitation flow' between axon and dendritic terminals.

Instead of re-telling numerous results from cellular-automata theory and applications I prefer to refer readers to some introductory and advanced texts.

Thus, Egan's *permutation city* [208] explores the idea of compactly encoding human beings and their worlds in cellular-automata non-constructible configurations—Gardens of Eden. This is the only example of good science fiction dealing directly with cellular automata. As to non-fiction readers should start with von Neumann's *self-organization* [479] and Zuse's *discrete universe* [675] as texts that should be necessarily read through.

Finally, classical results on one-dimensional cellular automata can be found in Wolfram's collection of papers [647]. Most results in [647] are still valid and certainly will be for a long time.

The book [648] by Wuensche and Lesser deals mainly with a graph representation of the spacetime dynamics of one-dimensional binary cellular automata. Really wonderful examples of global transition graphs are to be found there. Some sections of the book must be paid attention, particularly those devoted to glider dynamics in cellular automata.

The monographs by Toffoli and Margolus [591] and by Weimar [635] discuss simulation of physical phenomena in cellular automata. They both present useful techniques of cellular-automata modelling together with analysis of the efficiency and correctness of cellular-automata simulators.

Supervised evolution of cellular-automata machines is the main subject of Sipper's monograph [566]. Anyone eager to design parallel processors using genetic algorithms which appeal to evolution should consider this book.

There is one chapter in the book [9] that raises the question of designing molecular computers but it focuses entirely on the identification problem and considers the essence of the problem rather simplistically.

Aladyev's non-standard account of his personal results in cellular-automata theory, particularly the algebras of local transition functions and simulation of biological processes [41] is quite interesting, at least in a historical perspective. Idealogically similar, but entirely different in implementation, is the project realized by Voorhees [627].

Preston and Duff's text [506] deals with cellular-automata algorithms and processors for image-processing. Clearly, image-processing has always been close to cellular-automata architectures. A systolic screen [175] is a very typical example, where a digitized image is represented in a rectangular array of pixels, the array in its turn is projected onto the same size array of simple processors, where there is a one-to-one correspondence between the pixel and the processor. When the processor updates its state the pixel also switches its state.

A mathematical theory of cellular automata, which also includes problems of decidability and mutual simulations between automata classes, is derived in [373].

And last but not least, if you like `Mathematica` you could be happy to analyse the programs from Gaylord and Nishidate's book [257]. Quite useful examples of reaction–diffusion and excitable media are considered there, you can even find implementations of sand-piles and traffic flows.

> *My question is, 'Can physics be simulated by a universal computer?'... I would like to have the elements of this computer locally interconnected, and therefore sort of think about cellular automata as an example... we might change the idea that space is continuous to the idea that space perhaps is a simple lattice and everything is discrete ... and that time jumps discontinously.* [224]

Feynman's words may form a bridge between the current section and the next one.

1.4 Modelling reaction–diffusion and excitable media

Mathematical models of excitable media usually employ partial differential equations of a reaction–diffusion type. Some examples can be found in [366, 679, 401, 214, 489, 678, 324]. Use of a finite difference method of numerical integration is a typical move when we deal with solutions of reaction–diffusion equations. Let us tackle a two-dimensional excitable medium described by a couple of reaction–diffusion equations:

$$\frac{\partial a}{\partial t} = f(a,b) + D_a\left(\frac{\partial^2 a}{\partial x^2} + \frac{\partial^2 a}{\partial y^2}\right)$$
$$\frac{\partial b}{\partial t} = g(a,b) + D_b\left(\frac{\partial^2 b}{\partial x^2} + \frac{\partial^2 b}{\partial y^2}\right)$$

where a and b are the concentrations of the species, D_a and D_b are the diffusion coefficients of the species. The functions $f(\cdot)$ and $g(\cdot)$ describe the local kinetics of the species. When there is no general analytic technique we should appeal to numerical methods. A classical finite difference method of numerical integration is based on a system of difference equations:

$$a_{i,j}^{t+\triangle t} = f(a_{i,j}^t, b_{i,j}^t)\triangle t + D_a\frac{\triangle t}{(\triangle x)^2}(a_{i-1,j}^t + a_{i+1,j}^t + a_{i,j-1}^t + a_{i,j+1}^t - 4a_{i,j}^t)$$
$$b_{i,j}^{t+\triangle t} = g(a_{i,j}^t, b_{i,j}^t)\triangle t + D_b\frac{\triangle t}{(\triangle x)^2}(b_{i-1,j}^t + b_{i+1,j}^t + b_{i,j-1}^t + b_{i,j+1}^t - 4b_{i,j}^t)$$

where indices i and j mean coordinates $(i\triangle x, j\triangle y)$ at the orthogonal lattice with step $\triangle x$. The multipliers $D_a\frac{\triangle t}{(\triangle x)^2}$ and $D_b\frac{\triangle t}{(\triangle x)^2}$ are usually taken to be close to $\frac{1}{4}$. Not surprisingly, most patterns emerging in reaction–diffusion systems relate to time and space scales. Therefore, numerically obtained patterns may be incorrect, compared to analytical solutions, if no numerical diffusion anisotropy is taken into account and no calibrating is provided [188]. The following neighbourhood templates, or discrete representations of Laplacians, seem to be correct when used in a numerical integration [188]:

- One-dimensional system:

$$u_i^{t+\delta t} = u_i^t + \tfrac{1}{6}(u_{i-1}^t + u_{i+1}^t - 2u_i^t) + f(u_i^t)\triangle t$$

- Two-dimensional system:

$$\begin{aligned}u_{i,j}^{t+\delta t} &= u_{i,j}^t + \tfrac{1}{9}(u_{i-1,j}^t + u_{i+1,j}^t + u_{i,j-1}^t + u_{i,j+1}^t - 4u_{i,j}^t)\\ &\quad + \tfrac{1}{36}(u_{i-1,j-1}^t + u_{i+1,j+1}^t + u_{i+1,j-1}^t + u_{i-1,j+1}^t - 4u_{i,j}^t)\\ &\quad + f(u_{i,j}^t)\triangle t\end{aligned}$$

- Three-dimensional system:

$$
\begin{aligned}
u^{t+\delta t}_{i,j,z} = u^t_{i,j,z} &+ \tfrac{1}{18}(u^t_{i-1,j,z} + u^t_{i+1,j,z} + u^t_{i,j-1,z} + u^t_{i,j+1,z} \\
&+ u^t_{i,j,z-1} + u^t_{i,j,z+1} - 6u^t_{i,j,z}) \\
&+ \tfrac{1}{36}(u^t_{i-1,j-1,z} + u^t_{i+1,j-1,z} + u^t_{i-1,j+1,z} + u^t_{i+1,j+1,z} \\
&+ u^t_{i+1,j,z+1} + u^t_{i-1,j,z+1} + u^t_{i+1,j,z} + u^t_{i-1,j,z-1} \\
&+ u^t_{i,j+1,z+1} + u^t_{i,j-1,z+1} + u^t_{i,j-1,z-1} + u^t_{i,j-1,z-1} - 12u^t_{i,j,z}) \\
&+ f(u^t_{i,j,z})\Delta t.
\end{aligned}
$$

The space and time steps should satisfy the scaling relation $\frac{D\Delta t}{(\Delta x)^2} = \frac{1}{6}$ [188].

Several examples of recent results on excitation dynamics include the motion equations of excitation derived in [210], Galilean transformations in reaction–diffusion equations which allow the consideration of auto-wave regimes in two- and three-dimensional excitable media [170] and rigorous results on the three-dimensional simulation of excitation in cardiac tissue [46].

We do not need to have a deep imagination to draw an analogy between the discrete representations of Laplacians and cellular-automata neighbourhoods; and, in fact, between differences and a cell state transition function. How good are cellular-automata models for reaction–diffusion systems?

Historically, cellular-automata models are of the same age as conventional continuous models. Thus, an abstract model of excitation in a continuous medium, namely propagation of impulses in a fibre with refractivity can be traced back to [259, 604]. The models of the excitation dynamics in continuous media [259], designed in the early 1960s, deal with the propagation of the impulses in a ring of refractive fibres, and the pacemaker model where the medium, the elements of which have different periods of spontaneous excitation, becomes excited at the frequency of its fastest element. One of the early two-dimensional models dealt mainly with the synchronization of oscillatory networks [402] and models of heart pathology based on wave propagation in a two-dimensional excitable medium with defects [60]. The first ever cellular-automata model of an excitable medium was developed by Balakhovsky [61] in 1965. This model was promptly elaborated in great detail by Reshod'ko [536] and Zykov [678]. And already by 1966 Krinsky [366, 367] had described the spiral properties of two-dimensional excitation waves.

Recall that the behaviour of an excitable element [275, 640, 341] may be seen to be determined by three intervals: $[-R, \theta]$, $[\theta, U]$ and $[U, 1]$. The element relaxes back to the stable state 0 (and outputs 0 as well) if it is perturbed by a value from the interval $[-R, \theta]$. The element is 'excited', and its output equals 1, if its perturbed by a value from $[\theta, U]$. The value θ is called a threshold of excitability, or excitation threshold. Outside the excitation interval, i.e. in the interval $(U, 1]$, the element takes a refractory state where it is absolutely insensitive to input perturbations. In [341] we can find models which take excitability as originating from the interaction between two chemical species, sometimes called the propagator and controller variables.

Feldman *et al* [221, 222] used cellular automata on a randomized lattice to simulate two-dimensional excitable media. They established a relationship between the excitation time scale, the diffusion constant of the medium and the corresponding characteristics of a cellular-automata model and derived a general equation describing wavefront propagation. Expanding the finite-cell state up to the real interval yields 'cellular neural nets' which can also be used to investigate various regimes of excitation dynamics [329]. An excellent approximation of continuous models by discrete ones has also been demonstrated in a series of papers [494, 643, 48, 159, 428, 429, 427, 261, 262, 263, 264, 550, 273, 554, 215, 377]. It is shown that almost all possible spatial phenomena and effects of curvature and dispersion on wave propagation, e.g. wobbling cores, scroll rings, linked pairs of scroll rings and ripples, obtained in continuous models can also be described, to some degree of accuracy, in far more tractable discrete or cellular-automata models.

Cellular-automata models of reaction–diffusion media exhibit the same spectrum of coherent patterns as numerical simulations of reaction–diffusion equations: travelling waves, spiral waves, stable spots, strips and labyrinths [425, 200].

A strong agreement between the numerical solutions of partial differential equations and cellular-automata models is demonstrated in [130]. The accuracy and performance of three-dimensional cellular automata for the simulation of continuous excitable media is demonstrated in [300].

One realistic cellular-automata model for simulating the spread of excitation during ventricular fibrillation under a wide range of conditions is designed in [344]. The model reproduces 'electrocardiograms' similar to real data in their time series.

In the case of a Belousov–Zhabotinsky reaction, probabilistic cellular automata—in the form of networks of stochastic oscillators—allow the simulation of turbulent stirring and of a stochastic Oregonator [619].

Let us now tackle some peculiarities of cellular-automata-based simulation.

Cellular-automata models of reaction–diffusion systems phenomenologically correspond to the solution of partial differential equations, as do most of our models presented in this book, qualitatively [163, 632] or quantitatively [634, 635].

Here is an example. In [218] a cellular-automata model is used to investigate the stability of vortex rotation. In this model each element of a two-dimensional array takes a finite number of states: 0 (resting state), 1 (excited state), $2, \dots, \rho_a$ (absolutely refractory states), and $\rho_a + 1, \dots, \rho$ (relative refractory state). The values ρ_a and $\rho - \rho_a$ express the duration of the absolute refractory and relative refractory periods. Every cell x of the array updates its state in the following manner:

$$x^{t+1} = \begin{cases} x^t + 1 & \text{if } x^t \in [1, \rho_a] \text{ or } (x^t \in (\rho_a, \rho) \text{ and } \sigma(x)^t \leq \theta(x^t)) \\ 1 & \text{if } (x^t = 0 \text{ or } x^t \in (\rho_a, \rho]) \text{ and } \sigma(x)^t > \theta(x^t) \\ 0 & \text{if } (x^t = 0 \text{ or } x^t = \rho) \text{ and } \sigma(x)^t \leq \theta(x^t) \end{cases}$$

where $\sigma(x)^t = \sum_{y \in u(x)} w_{xy} \chi(y^t)$ is a weighted sum of states of cell x's neighbours. In the model the threshold $\theta(x^t)$ of cell excitation is a function of cell state. We can use any kind of weighting of the neighbourhood. For example, a linear dependence seems to be good enough. If the neighbourhood $u(x)$ looks like $u(x) = \{y : |x - y| \leq r\} - x$ then we can define the weight w_{xy}, for example, as $w_{xy} = r - |x - y| + 1$, where r is the neighbourhood radius. The radius r determines the range of velocities in the dispersion curve; however, large values of r certainly slow down the simulation [218].

In the previous model we noticed that the excitation threshold depends on the current values of the medium site. This is because the threshold may be determined by a decreasing linear function with domain (ρ_a, ρ), the closer cell state is to the resting state the lower the excitation threshold. Using such a model we can investigate most of the properties of excitable media, e.g. the dependence of velocity on wavefront curvature [677], observed in laboratory experiments and in numerical models of partial differential equations (see [218] and [261, 262]).

Anisotropy of wave propagation is visible in lattices with eight-cell neighbourhoods and even more recognizable in lattices with four-cell neighbourhoods. Thus in the eight-connected lattice the wavefront velocity in the direction of the cell diagonal is $\sqrt{2}$ higher than in the direction of the cell edge [426]. How do we cure the anisotropy? The problem of the discretization of a diffusion equation was addressed by the image-processing community long before this happened in physics: various types of local neighbourhoods and time delays were exploited (see [141] for relevant references).

One of the easiest approaches is simply to delete some lattice nodes at random; we need to increase the neighbourhood size in this case. It is shown that a square lattice, where a node is deleted with the probability 0.5, and a hexagonal lattice, where a node is deleted with the probability 0.3, are sufficient to obtain isotropic wave propagation [377].

Another approach to randomization is discussed in [426]. There some type of dynamical randomization of the cell neighbourhood is applied. Some neighbours of every cell are omitted at the stage when the cell counts the number of excited neighbours to calculate its own next state. With large neighbourhoods and several gradations of refractory states we obtain very attractive wavefronts [426].

A systematic investigation of the different kinds of neighbourhood templates, or masks, is provided in [633]. Several neighbourhood templates are under consideration there:

- Efimov–Fast weighted templates, where a neighbour y contributes to the sum of 'excitation' of cell x with weight $w(y, x) = \lfloor c\mathrm{e}^{-\frac{d(x,y)^2}{4b}} \rfloor$, where c and b are constants and $d(x, y)$ is the distance between the cells x and y;

- the classical cruciform template;
- the square template: $w(x_{ij}, x_{lm}) = \frac{1}{(2r+1)^2}$ if $l, m \leq r + \frac{1}{2}$;
- the circular template: $w(x_{ij}, x_{lm}) = \frac{1}{2\pi r^2}$ if $l^2 + m^2 \leq r^2$; and
- the Gaussian template: $w(x_{ij}, x_{lm}) = \frac{1}{2\sqrt{\pi b}}\mathrm{e}^{-\frac{l^2+m^2}{4b}}$

where i, j, l and m are the cell indices of a two-dimensional lattice. Weimar *et al* [633] demonstrate that the velocity–curvature relation $\boldsymbol{n} = v - DK$ between the normal velocity of the wave $\boldsymbol{n}$, wave velocity v, diffusion coefficient D of an excitable variable and wavefront curvature K holds quite well for most types of weighted neighbourhood templates.

In cellular-automata models of reaction–diffusion systems the diffusive part, propagation and scattering, is implemented via modified finite difference schemes and the reactive part is usually realized probabilistically [634]. Moreover, the diffusive part may utilize the technique of a moving average in which the discrete Laplacian is calculated.

Weimar *et al* [632] offer a classification for cellular-automata models of excitable media. Three generations of models are put forward. The first generation includes cellular automata with nearest neighbours, four-cell and eight-cell neighbourhoods, and very few cell states. First-generation models are mostly used in our book. The second generation consists of automata with larger, and usually weighted, neighbourhood templates and a large set of cell states. The models in this class can deal with those aspects of wave dynamics, which are not tackled well by the first-generation models [218, 261, 263, 261, 425, 426]: the curvature effect (velocity depends on wavefront curvature), the dispersion relation (takes into account the gradations of recovery, i.e. relative refractory periods) and the spatial isotropy (smoothness of the wavefront shape—this factor is cured by extending the cell neighbourhood). The curvature effect is achieved by increasing the cell neighbourhood size. The dispersion relation is simulated by including several recovery states in the cell state set; a cell in the recovery state has a higher excitation threshold than a cell in the resting state. And anisotropy is avoided by randomization of either lattice or neighbourhood. The third generation goes back to classical numerical algorithms and combines the accuracy of difference schemes with the computational efficiency of cellular automata [632]. The combination of techniques used in the third-generation models resembles Barkley's technique [67], which will be discussed later.

Coupled map lattice models, in contrast to cellular automata, employ continuous states but still retain discrete time and discrete space [341]. Every node of a coupled map lattice updates its state by the rule:

$$x^{t+1} = f(x^t) + \sum_{y \in u(x)} g(y^t)$$

where $f(\cdot)$ is the function responsible for cell state transition or the local dynamic of a cell, when no coupling with neighbours exists (autonomous evolution); $g(\cdot)$ is a linear or nonlinear function of the states of cell x's neighbours [341]. The spacetime dynamics of coupled map lattice excitation models is very rich and varies from chaotic to expanding spatial domains [341]. However, we could not find anything new in the dynamics of coupled map lattices that has not already been qualitatively described in cellular-automata models.

In 1991 Barkley offered a simple hence efficient hybrid of numerical methods and cellular automata [67]. His technique explicitly employs reaction–diffusion equations and allows continuous adjustment of spacetime resolutions [67]. Now the method is used widely.

Let us discuss the Barkley model [67]. It employs a two-variable system of reaction–diffusion equations, which simulates the dynamics of a excitable medium:

$$\frac{\partial u}{\partial t} = f(u, v) + \nabla^2 u \qquad \text{and} \qquad \frac{\partial v}{\partial t} = g(u, v).$$

The functions $f(\cdot)$ and $g(\cdot)$ represent the dynamics of the variables u and v, respectively. The component v is assumed not to diffuse. This may be appropriate when one of the reagents is immobilized. The functions are described as follows:

$$f(u, v) = \epsilon^{-1} u(1-u)[u - u_{\text{th}}(v)] \qquad \text{and} \qquad g(u, v) = u - v$$

$u_{\text{th}}(v) = \frac{a+b}{a}$, ϵ is very small, a and b are parameters. Sometimes, the variables u and v are called the excitation and recovery variables. In the excitation region, the time scales of the variable u are very fast, therefore either very small or very large time steps should be used to prevent instabilities [67]. Moreover, a simulation must take into account the fact that the recovery region of the system is very simple; we can think of an exponential decay of the variable v. As in lattice gases and cellular-automata models of reaction and diffusion two essential steps are considered: the local dynamic and diffusion. The local dynamic of the variables is simulated as follows:

$$\texttt{if } u^t < \delta \qquad u^{t+1} \leftarrow 0,\ v^{t+1} \leftarrow (1-\Delta t)v^t$$
$$\texttt{else } u_{\text{th}} \leftarrow (v^t + b)/a,\ v^{t+1} \leftarrow v^t + \Delta t(u^t - v^t),\ u^{t+1} \leftarrow F(u^t, u_{\text{th}}).$$

The values of the variables at the tth time step are shown as u^t and v^t; Δt is the time increment. A special region of the (u, v) phase space, where no excitation occurs and a site in the excitable medium simply returns to its resting value after a small perturbation, is bounded by δ. Function F is responsible for the excitation dynamic. Barkley [67] discusses several ways to represent this function. Thus, for example, for large time steps we have

$$\texttt{if } u^t \le u_{\text{th}} \qquad u^{t+1} \leftarrow u^t + (\Delta t/\epsilon)u^{t+1}(1-u^t)(u^t - u_{\text{th}})$$
$$\texttt{else } u^{t+1} \leftarrow u^t + (\Delta t/\epsilon)u^t(1-u^{t+1})(u^t - u_{\text{th}}).$$

For small $\Delta t/\epsilon$ we can use the following explicit form of the function F:

$$F(u^t, u_{\text{th}}) = u^t + (\Delta t/\epsilon)u^t(1 - u^t)(u^t - u_{\text{th}}) + O(\Delta t^2)$$

while for large $\Delta t/\epsilon$, which is quite similar to classical cellular automata, we obtain

$$F(u^t, u_{\text{th}}) = \begin{cases} 0 & \text{if } u^t < u_{\text{th}} \\ u_{\text{th}} & \text{if } u^t = u_{\text{th}} \\ 1 & \text{if } u^t > u_{\text{th}}. \end{cases}$$

If we do not worry about a slight slowdown in computation speed we can use the following approximation of the Laplacian to simulate the diffusion of the excitation component u:

$$h^2\nabla^2 u_{i,j} = u_{i+1,j} + u_{i-1,j} + u_{i,j+1} + u_{i,j-1} - 4u_{i,j}.$$

Several techniques to speed up the simulation are discussed in [67]. The method is tested on three-dimensional excitable media in [195]. In the three-dimensional model a reactive step of the simulation in [195] is similar to that in the two-dimensional model [67] while the diffusive process requires some attention; i.e. the seven-point finite-difference scheme is adopted for modelling three-dimensional diffusion:

$$h^2\nabla^2 u_{i,j,z} = u_{i+1,j,z} + u_{i-1,j,z} + u_{i,j+1,z} + u_{i,j-1,z} + u_{i,j,z+1} + u_{j,j,z-1} - 6u_{i,j,z}.$$

Neither difference schemes nor cellular-automata models nor even a combination of them take into account the behaviour of a single molecule or a particle. Lattice-gas automata do.

Lattice-gas models imply that discrete particles spread on the integer lattice in integer time and undergo discrete collisions with each other, with a finite number of possible outcomes. A lattice-gas model is invented in [290, 246, 247] to simulate fluid dynamics efficiently (see also fluid flow in cellular automata in [646]). Since then lattice-gas automata have been employed in a myriad of simulations from multiphase flow and the Dirac equation [450] to the Schrödinger [92] equation and reaction–diffusion systems [98]. Recently, Hasslacher and Meyer [292] have produced a remarkable example of a dynamically evolving model, where they combine a lattice gas with structurally dynamic cellular automata.

How is the lattice-gas model organized? If there is no reaction between the gas molecules only two instead of three phases, which will be discussed later, must be simulated. The first phase is an advection: every particle is transferred to a lattice node depending on its previous position and momentum. The second phase is scattering: the particles at the lattice nodes are scattered by deterministic, probabilistic or quantum-mechanical rules [292].

The term *lattice chemistry* is derived in [163, 98]. This combines hydrodynamics with chemistry and is expressed in terms of lattice-gas automata.

More generally, lattice chemistry offers a microscopic approach to the dynamics of spatially distributed reacting systems where hydro-dynamics and statistical mechanics are also exploited.

Lattice chemistry obviously operates on two- and three-dimensional lattices, where molecules are seen as point particles, which travel or undergo displacements along the links connecting lattice nodes, in the lattice. The particles travel with discrete velocities; i.e. they hop from one node to a neighbouring one, determined by the particle's velocity vector [163, 98]. The total number of particles at a node is usually restricted by lattice connectivity. This is an exclusion principle. There are also lattice chemistry models without exclusion [98]; they accept that any number of particles can reside at a lattice node, thus particles can simply diffuse on the lattice and react with one another [134].

Given a lattice with particles an evolution is implemented by application of an evolution operator E. The evolution operator is equivalent to the cellular-automaton global transition rule. Typically, the evolution operator E is represented by a composition of three basic operations [163]:

$$E = C \circ R \circ P$$

with propagation operator P, velocity randomization operator (collision operator) R and chemical transformation, or reaction, operator C. The operators P and R are analogues of the advection and scattering operators in classical lattice-gas models [246, 247].

When the propagation operator P is applied each particle jumps from one node to one of its neighbouring nodes in a manner determined by its velocity vector.

The collision operator R randomizes the velocities of the particles at every node. The velocities are randomly shifted. This simulates elastic collisions between the particles and molecules of an implicit solvent. Velocity randomization can be defined via state transition probability matrices.

As a result of applying the chemical transformation operator C the particles react with one another. Particles are created and annihilated at every node: $\alpha X \to \beta X$. There is no conservation of particles and their local momenta. The difference between R and C is that R conserves the number of particles while C does not.

If there is no reaction all particles execute discrete random walks because only operators P and R are applied. This is a diffusion.

Which real computers are suitable for the discussed models? Amongst many possibilities we would like to mention two types of massively parallel computing devices.

The first one is a connection machine [302]. Its architecture resembles a globally connected automata network, where millions of very primitive processors communicate via the spread of diffusive waves of packets. Each packet includes data and a recipient address. The packets are accepted by those processors with the appropriate address while other packets are simply sent away to other processors.

Cellular-automata machines [591] are the second example. A cellular-automata machine is a multiprocessor board with a large graphical memory and a graphic accelerator. Such an architecture does not imitate cellular-automaton architecture but allows a reasonably fast run and, in particular, displays the evolution of large cellular-automata arrays. Recent versions of the cellular-automata machine, as well as recent versions of the connection machine, have become more similar to widespread multiprocessor systems such as, for example, SPARC, for the sole reason that end-users were not prepared to program cellular-automata processors and quite a few problems with straightforward data parallelism were known at that time.

1.5 Computing in nonlinear active media

Most of the problems with natural spatial parallelism can be efficiently solved in nonlinear active media. A problem has a natural parallelism if it consists of many subdomains that can be, at least up to some limit, analysed independently of each other and the results of analyses are combined.

> *A problem is spatially parallel if sites of the problem space can be updated in parallel, i.e. distant sites can be analysed independently at the same time step.* [593]

The so-called programmable matter paradigm gives a commonsense approach which may be used in the design of unconventional computers for problems with natural spatial parallelism. Programmable matter architectures are local and uniform. They are finely grained and massively parallel. The computational tasks that can be solved on programmable matter computers are mainly spatial tasks with nonlinear interactions between a large number of elements, which results in the emergence of macroscopic phenomenology from microscopic mechanisms [593]. So, in fact, the dynamic of the computer simulates, or even directly corresponds to, the dynamic of the simulated problem. We, therefore, have the mapping

$$\textit{spatial parallel problem} \rightarrow \textit{programmable matter architecture}.$$

Both the connection machine [302] and the cellular-automata machine [591] are programmable matters [593]. In this section we consider several examples of wet-ware realizations of programmable matter. When thinking about the reaction–diffusion and excitable computers, as well as all prototypes of real chemical processors discussed in this book we must remember that no competition between wet-ware and conventional silicon hardware can even be imagined at this stage. As Toffoli reasonably notes:

> *programmable matter architectures play a complementary rather than competitive role vis-à-vis traditional architectures; they are tuned, as it were, to a different band of the computational spectrum.* [593]

However, cellular-automata models of reaction–diffusion and excitable computers, if implemented in silicon, may certainly overrun not only classical serial but also existing parallel architectures.

Kuhnert was probably the first to design a laboratory prototype of a reaction–diffusion chemical processor. He developed a recipe for a light-sensitive modification of the thin-layer Belousov–Zhabotinsky reaction and showed how the reaction medium can be used in image-processing [375, 376]. The idea was developed further by Agladze, Krinsky, Rambidi, Aliev and their collaborators [376, 368, 514, 515, 516, 517, 519, 520, 522, 523, 524, 525, 526, 527]. The reaction–diffusion approach to image-processing has also been implemented in mathematical models [50, 674, 507, 508, 155]. Thus, for example, Alvarez and Esclarin [50] use the diffusion component of the reaction–diffusion partial differential equations to reduce the noise in data images; this is an analogue of averaging or homogenization of the image. That is, in the Alvarez and Esclarin [50] model diffusion is used to filter noise and the reaction is employed to enhance contrast.

In general, image-processing in a Belousov–Zhabotinsky medium may be implemented as follows [376, 368, 514, 515, 516, 517, 519, 520, 522, 523, 524, 525, 526, 527]. An image is projected onto a reaction layer. As a result of such a projection a non-uniform oscillation phase distribution is generated in the medium. The distribution represents image features, i.e. encoded data. After that image-processing is implemented because of the interaction between the phase and diffusion waves.

What exactly is happening in excitable chemical media when an image is projected onto a thin-layer reactor? When an image is projected onto a photosensitive excitable chemical medium, the micro-volumes, which become illuminated above some threshold, become excited. Excitation waves start to spread in the medium and interact with each other. By changing the parameters of wave generation and propagation it is possible to implement a wide range of image-processing operations, for example, the restoration of broken contours, edge detection and contrasting [376, 368].

Here we consider several examples of image-processing in Belousov–Zhabotinsky thin-layer chemical reactors. They include the detection of closed curves and the restoration of a broken contour [368, 518] as well as image contouring, enhancement and detection of certain image features [528].

An algorithm for the detection of the closeness of a curve works as follows [368]. We project a curve onto an excitable medium. The sites corresponding to points of the curve become non-excitable or ever-resting. We make the curve sites ever-resting by physical intervention in a medium's characteristics. Then we excite an arbitrary site of the medium far from the projected curve. If the

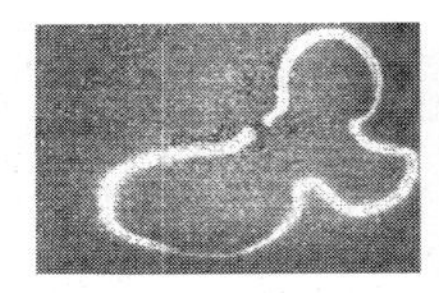

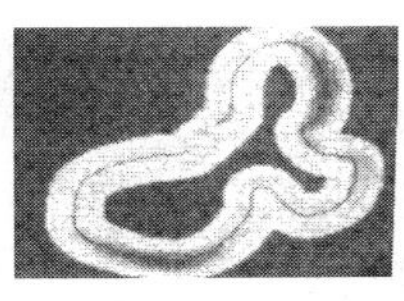

Figure 1.4. Reparation of a broken contour in a Belousov–Zhabotinsky medium [518]. Published with kind permission of Nicholas Rambidi.

curve is not closed the excitation fills the whole lattice. If the curve is closed the sites inside the curve remain at rest. Obviously, we need to 'scan' all medium sites to check whether the whole medium is excited or not. The recognition of closed curves from open curves has also been implemented in a two-dimensional array of cellular neural networks, i.e. resistively coupled Chua circuits, using the propagation of excitation waves [496].

The next solution is even more tricky.

A broken contour is projected onto the excitable medium. We want to restore a contour's continuity. Sites belonging to the contour are excited and an excitation wave propagates in the medium. The broken edges of the contour, represented by excitation waves, expand and finally merge. As a result we have some kind of inflated but restored shape, see figure 1.4 [518]. Its thickness may be later reduced, either by conventional procedures or by back propagating the excitation waves and stopping them at a certain time. Again, you see, the solution is not in the exact form we wanted [518].

The two previous examples are rather playful instances of very simple operations. Could we do something more sophisticated with complex images. A series of works published by Rambidi *et al* [514, 515, 516, 517, 519, 520, 522, 523, 524, 525, 526, 528] gives us a positive answer.

The typical setup of a Belousov–Zhabotinsky thin-layer processor includes a chemical reactor and a computer-controlled video projector. A light-sensitive catalyst $Ru(bpy)_3Cl_2$ is immobilized on a thin, approximately 0.2 mm, layer of a silica gel, placed at the bottom of a reactor. The gel film is covered by a 0.5–1.5 mm thin non-stirred layer of reaction liquid with the following concentration of reagents: $KBr0_3$ (0.3 M), H_2S0_4 (0.3–0.6 M), malonic acid (0.2 M) and KBr (0.05 M). Image data are 'loaded' in the chemical processor by light radiation, i.e. by projecting an image onto the thin layer surface. As indicated in [528] it is very handy to immobilize a ferroin catalyst, in a low concentration, on the gel film surface, after immobilization of the ruthenium catalyst; this would increase the optical density of the processed images.

At the beginning of an experiment a medium is illuminated by a white light to reboot the chemical processor initially. This causes an increase in the concentration of Ru^{3+}. The medium therefore becomes light coloured. Then the data image is projected onto the medium. The concentration of Ru^{3+} in those parts of the medium, which correspond to white fragments of the image, becomes

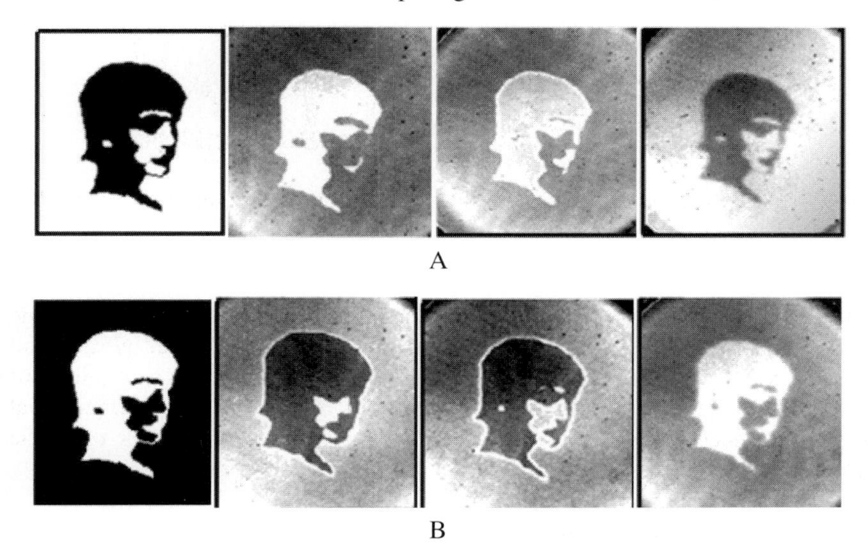

Figure 1.5. Two examples of 'black and white' image-processing in a Belousov–Zhabotinsky medium: (A) processing a positive image; (B) processing a negative image. Published with kind permission of Nicholas Rambidi [528].

higher than in other parts of the medium. At the same time, the concentration of Ru^{2+} increases and the concentration of Ru^{3+} decreases in darkened parts of the medium. Due to the particulars of the reaction an oscillating process starts in the medium. This leads to an increase in Ru^{3+} in previously darkened areas of the medium and a decrease in the reagent in the previously white (illuminated) area. As a result, the image, represented in a reagent concentration, becomes negative. An opposite evolution happens in the medium when we project a negative image. Several snapshots of the evolution of a black and white image in a Belousov–Zhabotinsky medium are shown in figure 1.5.

By varying the contrast of the original image, projected onto a Belousov–Zhabotinsky reactor, we could obtain either the external contour of the image (figure 1.6(A)) or a typical contour of the image (figure 1.6(B)). If a negative image is processed, the basic shape of the image is enhanced (figure 1.6(C)). The projection of a positive image onto the medium accentuates particular features of the image (figure 1.6(D)) [528].

Could we 're-program' Belousov–Zhabotinsky processors or control the computation process? In general, no. However, there are some hints which may allow us to modify the properties of the media during the computation. Thus, for example, we could apply an electromagnetic field.

A spacetime electric field gradient affects the transport of ionic components. Therefore the field gradients may influence the wavefront dynamic. A negative

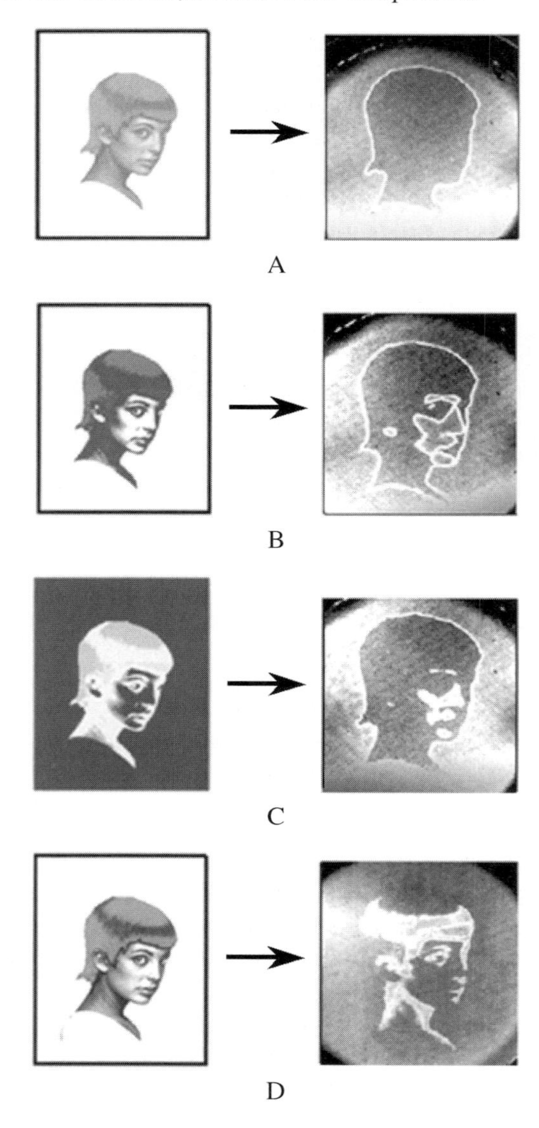

Figure 1.6. Image-processing in a Belousov–Zhabotinsky medium: (A) detection of external boundaries; (B) contour enhancement; (C) shape enhancement; and (D) enhancement of image particulars. Published with kind permission of Nicholas Rambidi [528].

field accelerates the wave. In positive electrical fields the propagating front may slow down or even stopps. Moreover, the wave may start moving backwards if the field is strong enough [421, 422]. When the field is switched off the front recovers and continues its propagation, probably with changed parameters. Varying the electric field applied to the chemical medium we can achieve transitions between several regimes of the reagents' spacetime dynamic [473, 213]. So, the electric field can be used for data inputs, control of computation process or even dynamic re-programming of Belousov–Zhabotinsky reaction-based computing devices.

Reaction–diffusion media can handle image-processing. Do they cope with image recognition? Yes, at least at a very primitive stage. Chemical processors may recognize patterns because they react differently to different input patterns and similarly to similar ones. This is demonstrated in laboratory experiments on patterns recognition by a small network of chemical reactors [384]. Each reactor is a continuous flow stirred tank. Its states are bistable. The reactors are mass-coupled with each other. The iodate-arsenous acid reaction is used (shown for the case when the iodate is in stoichiometric excess) [384]:

$$2IO_3^- + 5H_3AsO_3 + 2H^+ \rightarrow I_2 + 5H_3AsO_4 + H_2O.$$

Each reactor of the network takes binary states of low and high concentrations of reagents. The states of the reactors are visualized by adding starch: the high iodine state is blue and the low iodine state is colourless. The state of a reactor is a pixel of the analysed pattern.

Each reactor in the network is fed with an identical flow of reagents. Such a flow rate is taken to mean that the reactor is in a bistable state of either high or low iodine concentration. The chemical processor is programmed by modifying the network connectivity strength. The reactor network of this chemical processor is programmed by choosing such a flow rate between the reactors that store binary pattern which correspond to the network's stable state attractor [384]. The stored patterns can be evoked by applying similar initial configurations.

In mass-coupling the coupled reactors must be spatially close to each other. Instead of mass-coupling we can also employ electrical coupling between the reactors of the network [309].

Results from pattern recognition have been developed into logical functions via chemical kinetics [306, 307] and chemical implementation of finite automata and neural networks (see, for example, [304, 305]). These works will be discussed in the main body of the book, so we omit detailed discussion here.

The Babloyantz–Sepulcher algorithm [56] for finding the shortest obstacle-free path using an oscillatory two-dimensional medium is another brilliant example of pioneering results on reaction–diffusion computing. The model consists of a two-dimensional lattice of oscillators [56]. The behaviour of each oscillator is represented by the variables x and y and is described by either the kinematic equations of the Brusselator:

$$\dot{x} = a - (b+1)x + x^2y \qquad \text{and} \qquad \dot{y} = bx - x^2y$$

or the Maginu scheme

$$\dot{x} = x - y - \frac{x^3}{3} \qquad \text{and} \qquad \dot{y} = \frac{x - ky}{b}$$

where a and b are abstract quantities and k is the threshold for bifurcation.

It is reasonable to describe the whole oscillator network by the following classical equations:

$$\dot{x}_{i,j} = f(x_{i,j}, y_{i,j}) + c(x_{i+1,j} + x_{i,j+1} + x_{i-1,j} + x_{i,j-1} - 4x_{i,j})$$
$$\dot{y}_{i,j} = g(x_{i,j}, y_{i,j}) + c(y_{i+1,j} + y_{i,j+1} + y_{i-1,j} + y_{i,j-1} - 4y_{i,j})$$

where $f(\cdot)$ and $g(\cdot)$ are functions that describe the local kinetics of the oscillatory elements, and c is a constant coupling coefficient between the oscillators.

Initially, all oscillators are in an unstable rest state. To start the wave we perturb the lattice by changing the state of a few oscillators of the lattice. An oscillation front is generated by the perturbed elements. The front propagates in the medium. A wave propagates behind the front. Incoming, or backpropagating, waves are also formed. This leads to periodical generation of circular waves by the once perturbed region of the lattice. In [56] front propagation in a continuous medium is simulated by approximating the kinetics of the oscillator lattice by the Ginzburg–Landau equation for the complex amplitude w:

$$\frac{\partial w}{\partial t} = (1 + \mathrm{i}\omega_0)w - (1 + \mathrm{i}\beta)|w|^2 w + D\nabla^2 w$$

where ω_0 is the initial frequency, β the dispersion coefficient and D the diffusive coefficient [56].

The main point of the oscillatory network is that if such a system is perturbed the front propagates from the point of perturbation. The front velocities indicate from where the front came and where the initial perturbation is. This idea will be exploited throughout this book. As to the Babloyantz–Sepulcher experiments [56] the following problem is solved: a robot travels in a room with obstacles and an oscillatory network guides the robot toward a target. So the nonlinear medium finds the shortest collision-free path. We associate the physical space of the problem with a network of oscillators, which corresponds, in turn, to the Ginzburg–Landau equation. The equation shows the propagation of waves that travel in the opposite direction to the propagating front. Obstacles are represented by absent oscillators or empty nodes on the grid. The target is assumed to be a signalling centre or pacemaker. Once the target region is perturbed the wavefronts propagate in the lattice. Eventually they reach the robot. The travelling waves start to propagate behind the front and toward the target. As a result the periodic propagation of the waves allows the robot to find a pacemaker centre and, therefore, to hit the target [56].

1.6 Reaction–diffusion media with smart molecules

Chemical media are not the only ones described by the reaction–diffusion equations and thus there are other candidates for the material base of reaction–diffusion algorithms. Sometimes even crowds, especially in extreme situations of panic or excitement, act according to the laws of physics [297], partly due to the absence of self-control, anonymity, suggestibility and the presence of a collective mind [564, 184, 345]. When moving intelligent entities employ mainly self-propulsion, the control of individual motion is by local interaction with other individuals and randomness of environment [625, 297, 298].

Attractive analogies between reaction–diffusion, morphogenesis and pattern formation in social insect societies are provided in Bonabeau *et al* [94, 95]. They show that the distribution of dead ant bodies in ant cemeteries—the distribution itself is governed by pheromone concentrations—is well described by a system of reaction–diffusion equations. The dynamics of opinion formation is another field where nonlinear physics is flourishing [336, 502]. The new generation of physics-based models of opinion formation considers the spacetime dynamics of individual opinions; thus, e.g., Plewczyński [502] describes social change in a collective with a lattice topology, and that of individuals by a nonlinear Schrödinger equation by analogy respectively with a superfluid and a weakly interacting Bose gas in an external potential.

Despite the complex nature of human and animal communication some primitive aspects of their collective behaviour can be described by mathematical models of chemical communications and still bear great predicting power [454]. *Dictyostelium discoideum* forms an impressive example of such primitive modela. There the collective behaviour of myriads of simple organisms is phenomenologically similar to the processes occurring in a Belousov–Zhabotinsky reaction, including the wave motions (see e.g. [420, 194, 471]). Another good example deals with an activation in an ant colony. Ant colonies show global oscillations of activity levels, i.e. changes in the number of active workers. The activity spreads concentrically from the points of a disturbance [571].

In populations of cells or insects individuals may release chemical mediators. These mediators diffuse in the medium. The chemical composition of the mediators represents an activity pattern of the population. In some assumptions, such activity patterns may be considered to be a representation of public opinion or distributed knowledge in the system [454]. Moreover, we can design a reaction-like interaction between doxastic states of each individual in the population and investigate the spacetime dynamics of the population in stirred and non-stirred reaction mixtures [29, 30].

The molecular dynamic applied to societies of social insects may produce quite original computation algorithms in collectives of dumb agents. Computing with social insects forms the new field of *swarm computing* and *ant-colony optimization*. The subject is extensively discussed in [97]. Specific applications in graph optimization and telecommunication networks may be found in [139, 191, 252, 22, 552, 192, 100, 415]. Development of the social insect paradigm in collective robotics, particularly in collective sorting and transport, can be found in [76, 446, 371, 447].

Chapter 2

Subdivision of space

Most imagined, simulated or laboratory prototypes of reaction–diffusion and excitable processors obey one very simple action scheme.

- The medium is at rest at the beginning of computation.
- Data are transformed into the spatial configuration of a geometrical object. The configuration is projected onto the medium.
- Local disturbances of the medium's characteristics, for example, drops of a new reagent or an excitation pattern form on the projected elements of this data configuration.
- The local disturbances generate trigger or phase waves. The waves spread through the medium.
- Eventually waves originating from different data sources meet one another. They somehow interact and produce either the concentration profile of a precipitate or a stationary excitation structure.
- The final configuration of the medium represents the result of the computation.

Almost all points of the scheme are fairly simple except the transformation of the problem into a spatial configuration. Which problem is easy to represent in a chemical mixture poured into a Petri dish: the contouring of an image or computation of a square root of a given number? Actually, both problems are fine; however, the first has a straightforward and intuitive representation. In this chapter we consider several spatial problems that can be straightforwardly represented in the development of nonlinear media. These are Voronoi diagrams, the skeletons of a planar shape and convex hulls. The reaction–diffusion solution of both Voronoi diagram and skeleton problems relies on a distance-to-time transformation implemented by the spread of diffusive or excitation waves. Computation of a convex hull employs instead a curvature–velocity dependence: waves generated at a medium's sites, corresponding to data points, merge into one spreading pattern, the pattern stops its centrifugal motion at the boundaries of the convex hull of the given set.

2.1 Voronoi diagram

Computation of the Voronoi diagram of a finite planar set is one of the most intriguing problems of combinatorial geometry. The problem is

> *to construct a polygon around each point of a given set in such a manner that any point inside the polygon lies nearer to the point that generated the polygon than to any other point of the given set.*

This subdivision of the plane was independently proposed by Dirichlet in 1887, Voronoi in 1908 [628] and Thiessen in 1911. Myriads of papers have been published on the subject and we do not discuss all of them. Any modern book on computational geometry would possibly offer a good overview. Here we simply mention those results which are relevant, to some degree, to our reaction–diffusion algorithms.

A basic knowledge and a detailed description of serial and parallel algorithms for computing Voronoi diagram can be found in [505, 202, 248, 123, 217, 333, 126]. Our original algorithms for computing the planar Voronoi diagram and for inverting the Voronoi diagram were previously published in [5] and [7]. A series of my results on discrete Voronoi diagrams appeared in [10, 15] and [9]. Quite similar ideas were exploited earlier in papers dealing with discrete Voronoi diagram (the first notion may be found in [434]), computation of Voronoi diagram by growth, see e.g. spreading [363], wave-based construction of graphs [570] and thinning algorithms of skeletonization [480].

Some results on the approximation of the Voronoi diagram in collectives of mobile finite automata were reported in [22]. In this chapter we mainly discuss two types of reaction–diffusion algorithms: the $O(n)$-algorithm and the $O(1)$-algorithm. The first algorithm employs at least n different reagents to build the Voronoi diagram of n sites. The second algorithm is oriented towards excitable media and utilizes a constant number of reagents, or cell states, independently of the number of given data sites.

Our algorithms for the solution of discrete Voronoi diagram problems have been rediscovered or extensively utilized in various applications, where the $O(n)$-reagent version of the algorithms is exploited, see, for example, the parallel algorithm for computing a collision-free path in [615, 586], the determination of service areas for power substations [223], the algorithm for the largest empty figure [614] and nearest-neighbour pattern classification [613]. In some cases the algorithm was enriched by VLSI implementation [614], as, e.g., where it was applied for robot motion planning on a two-dimensional cellular-automata lattice [615]. Several useful adjustments to the $O(1)$-algorithms are made by Maniatty and Shimasky in [414]. These improvements made our algorithm correct without increasing its complexity. The Maniatty–Shimansky modification [414] of the Adamatzky algorithm [15] allows the construction of all types of bisectors on a discrete lattice.

Applications of the Voronoi diagram for collision-free motion planning in a room with obstacles are given in [197, 390]. The Voronoi diagrams of objects other than points and metrics other than Euclidean ones are analysed in [36, 658, 584, 385, 386, 355, 356]. Several results on the parallel computing of the Voronoi diagram should also be mentioned here. Most of these do not relate directly to the intuitive construction of a Voronoi diagram but indicate the way to develop the modern field. Thus, the local specification of the sites of a query region sufficient for the computation of the diagram is provided in [372]. The relations of the pattern growing in a lattice with image-processing in general and Voronoi tesselations, in particular, are discussed in [141]. The dynamic computation of the Voronoi diagram of movable planar points and the stability of the Voronoi diagram form another promising bunch of problems where reaction–diffusion algorithms may be applied [629, 42]. The general, hence standard, algorithm for the computation of a multidimensional generalized Voronoi diagram is discussed in [626]. New results on the local computation of the Voronoi diagram in a network of elementary processors are presented in [211, 182]

2.1.1 Planar Voronoi diagram

A planar Voronoi diagram for a set $\mathbf{P}$ *of* n *points is the partition of a plane into* n *polygonal regions, one to each point, each of which is associated with a point* $p \in \mathbf{P}$ *and contains all points of the plane that are closer to* p *than to any other point of* $\mathbf{P}$.

Given $\mathbf{P} = \{p_1, \ldots, p_n\} \subset \mathbf{R}^2$*, the Voronoi cell of point* $p \in \mathbf{P}$ *is the set*

$$\mathrm{Vor}(p) = \{q \in \mathbf{R}^2 | d(p,q) \leq d(z,q) \forall z \in \mathbf{P}\}.$$

A planar Voronoi diagram of $\mathbf{P}$ *is the set of the boundaries of the Voronoi cells:*

$$\mathrm{Vor}(\mathbf{P}) = \cup_{p \in \mathbf{P}} \mathrm{Vor}(p).$$

The following properties are extensively used in the computation of a planar Voronoi diagram:

- A Voronoi cell $\mathrm{Vor}(p)$ is a closed convex polygon if p does not belong to the shell of the convex hull of $\mathbf{P}$; otherwise, $\mathrm{Vor}(p)$ is unbounded.
- Every vertex of the Voronoi diagram is the top point of at least three edges of the diagram.
- Every edge v of Voronoi cell $\mathrm{Vor}(p)$ is determined by the closest neighbour q of point p, i.e. $\forall x \in v : d(x,p) = d(x,q)$.

An example of a planar Voronoi diagram is shown in figure 2.1.

This section mainly deals with discrete analogues of planar Voronoi diagrams. However, to understand the differences between planar and lattices we discuss some intuitive ideas on how to compute a Voronoi diagram in an abstract

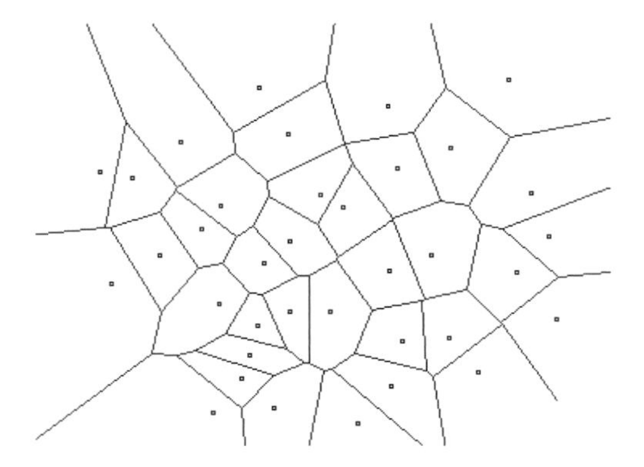

Figure 2.1. Example of a planar Voronoi diagram.

parallel processor with a cellular-automata architecture. Then we consider the reaction–diffusion algorithms in full detail including cellular-automata and laboratory implementations.

Let there be a regular network N of n processors arranged at the nodes of a two-dimensional integer lattice. Every elementary processor P_i of the network N is a RAM-machine that makes arithmetical calculations, reading and writing to the local memory; it also reads the contents of the communication registers of its neighbours. The configuration of points of $\mathbf{P}$ is uploaded into the network N in the following manner.

Let

$$D_x = \max_{p_i, p_j \in \mathbf{P}} |i_x - j_x| \qquad D_y = \max_{p_i, p_j \in \mathbf{P}} |i_y - j_y| \qquad D = \max\{D_x, D_y\}$$

$$\delta_x = \min_{p_i, p_j \in \mathbf{P}} |i_x - j_x| \qquad \delta_y = \min_{p_i, p_j \in \mathbf{P}} |i_y - j_y| \qquad \delta = \min\{\delta_x, \delta_y\}$$

then the network N has at least $M_x \times M_y = M$ elementary processors, where

$$M_x = \left\lceil \frac{D_x}{\delta_x} \right\rceil \qquad M_y = \left\lceil \frac{D_y}{\delta_y} \right\rceil.$$

We will assume that $M_x = M_y = M = \lceil \frac{D}{\delta} \rceil$. In other words, every processor P_i is responsible for the definite square of the plane and keeps in its local memory the coordinates of the points of $\mathbf{T}_i \subset \mathbf{P}$:

$$\mathbf{T}_i = \{p_k \in \mathbf{P} : |k_x - i_x| \leq \delta_x \text{ and } |k_y - i_y| \leq \delta_y\}.$$

At the beginning of a computation an elementary processor of N keeps one of n coordinates of the points of $\mathbf{P}$. As a result of the computation the elementary processor has one of $2n - 5$ coordinates of the vertices of Voronoi diagram VD($\mathbf{P}$) and references on the three closest vertices.

Every processor P_i of the network can be in one of two following situations according to the distribution of data amongst the processors. Either the processor has at least one point of **P** or has no points of **P**. In the first case, the processor scans its neighbours, i.e. it reads the contents of the communication registers of the processors of

$$B_r(P_i) = \{P_j \in N : |i_x - j_x| = |i_y - j_y| = r\} \qquad r = 1, 2, \ldots$$

and generates the set $\mathbf{C}(p_i)$ of the closest neighbours of point p_i. Having computed one closest neighbour of p_i the processor searches for the next such neighbour the addition of which to $\mathbf{C}(p_i)$ does not violate the convexity of $\mathrm{Vor}(p)$. This can be easily done by transferring the centre of the local coordinate system to the currently computed neighbours.

If the processor P_i has no points in **P** then:

- it finds such points $a, b, c \in B_r(p_j)$, $r = 1, 2, \ldots$ that $a \in \mathbf{C}(b) \cap \mathbf{C}(c)$, $b \in \mathbf{C}(a) \cap \mathbf{C}(c)$, $c \in \mathbf{C}(a) \cap \mathbf{C}(b)$;
- it computes a point v of the intersection of the median perpendiculars l_{ab}, l_{bc} and l_{ac} with the segments ab, bc and ac, respectively; point v is one of the vertices of the Voronoi diagram;
- if processor P_i has the coordinates of some vertex v of the diagram, then it finds vertices v_1, v_2 and v_3 in the communication registers of processors from $B_r(p_i)$ such that $v_1 \in l_{ab}$, $v_2 \in l_{ab}$, $v_3 \in l_{ac}$, $v_1 \neq v$, $v_2 \neq v$ and $v_3 \neq v$.

Note that we have provided this algorithm just to show how the problem may be solved in a cellular-automata system using standard approaches.

2.1.2 Discrete Voronoi diagram

Let **L** be a two-dimensional lattice of $m \times m$ nodes with a subset $\mathbf{P} \subset \mathbf{L}$, $|\mathbf{P}| = n$ of labelled nodes; nodes from **P** must be subdivided by a Voronoi diagram.

Let a and b be two different nodes of **L**; then we can define distances between a and b in metrics L_1 and L_∞ as

$$d_\infty = \max\{|x_a - x_b|, |y_a - y_b|\}$$

and

$$d_1 = |x_a - x_b| + |y_a - y_b|$$

respectively. When there is no need to specify a metric we will not do so. The notion of a bisector can be introduced by analogy with the planar case:

$$B_1(a, b) = \{c \in \mathbf{L} : d(a, c) = d(b, c)\}$$

where $B_1(a, b)$ is the bisector of the nodes a and b.

Later results will show that such a definition of the discrete bisector is not sufficient to separate the nodes of **L** in a correct way; sometimes nodes will have no bisector but can be separated in a commonsense way.

To overcome the disadvantage of the standard definition we determine a second sort of bisector:

$$B_2(a, b) = \{c \in \mathbf{L} : |d(a, c) - d(b, c)| \leq 1\}.$$

The discrete Voronoi diagram of a set **P** is the partition of a lattice into sublattices; each of them corresponds to a unique node $p \in \mathbf{P}$ and contains such elements of **L** that are closer to p than to any other node of **P**.

The discrete Voronoi diagram of $\mathbf{P} \subset \mathbf{L}$ *is the set*

$$\mathrm{DVD}(\mathbf{P}) = \cup_{p \in \mathbf{P}} \partial\, \mathrm{DVC}(p)$$

where for any $p \in \mathbf{P}$

$$\mathrm{DVC}(p) = \{x \in \mathbf{L} : d(x, p) < d(x, p)\ \forall z \in \mathbf{P}\}.$$

Here we use two kinds of cell neighbourhood:

$$u_1 = \begin{matrix} & \circ & \\ \circ & \bullet & \circ \\ & \circ & \end{matrix} \qquad u_2 = \begin{matrix} \circ & \circ & \circ \\ \circ & \bullet & \circ \\ \circ & \circ & \circ \end{matrix}$$

where $\bullet$ is the central cell and $\circ$ is a neighbour of the cell in state $\bullet$.

Automaton cells take their states from the set $\mathbf{Q} = \{\bullet, \star\} \cup \mathbf{R}$, where $\bullet$ is a rest state, $\star$ is a precipitate state, and **R** is a set of reagents, or a set of so-called excited states. Elements of the set **R** are the basic constituents of simulated diffusion and reaction. The state $\star$, the so-called precipitate, represents the result of the reaction between different reagents from **R**. When computation is finished, nodes in state $\star$ represent the bisectors, or boundaries, of the Voronoi cells.

Given set $\mathbf{P} \subset \mathbf{L}$ we wish to find the cell state transition function of a cellular automaton A such that for every cell x of the lattice **L** we have $x^0 \in \{\bullet\} \cup \mathbf{R}$ and $\mathbf{P} = \{x \in \mathbf{A} : x^0 \in \mathbf{R}\}$ and there is such a $\tau \in \mathbf{N}$ that for $\forall t \geq \tau$ we have $c^{t+1} = c^t$ and $\forall x \in \mathbf{L} : x \in \{\bullet, \star\}$ and $\mathrm{DVD}(P) = \{x \in \mathbf{L} : x^t = \star\}$.

In words, we wish to find a cellular automaton which starts in configuration c^0 representing set **P** and finishes the evolution $c^0 \to \cdots \to c^\tau \to c^*$ in a fixed point c^* such that $\forall t > \tau, c^t = c^*$. The fixed point configuration c^* of the cellular automaton represents DVD(**P**).

2.1.3 Bisectors

Let $B_{11}(\cdot, \cdot)$, $B_{1\infty}(\cdot, \cdot)$, $B_{21}(\cdot, \cdot)$, $B_{2\infty}(\cdot, \cdot)$ be the bisectors $B_1(\cdot, \cdot)$ and $B_2(\cdot, \cdot)$ computed in L_1 and L_∞ metrics, respectively. We will define the distances

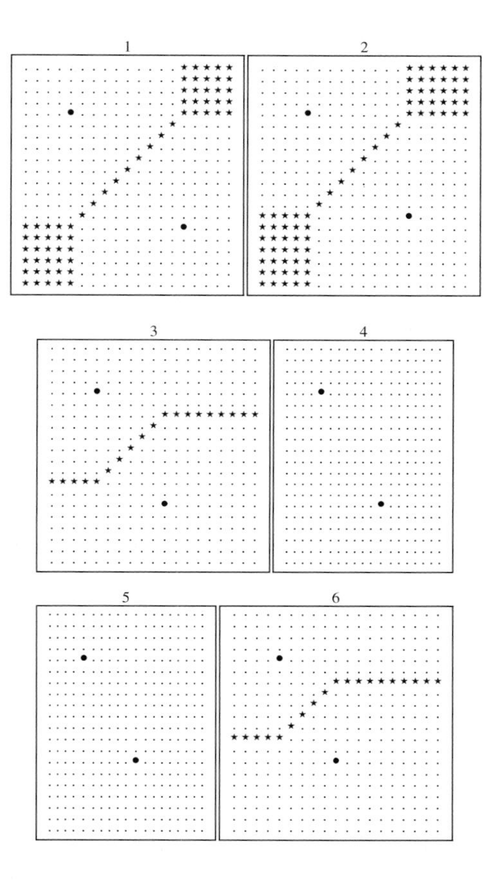

Figure 2.2. The bisectors $B_{11}(a, b)$.

between the nodes $a, b \in \mathbf{L}$ on the x- and y-axes by $s_x = |x_a - x_b|$ and $s_y = |y_a - y_b|$. The bisectors $B_1(a, b)$ and $B_2(a, b)$ calculated for various mutual dispositions of nodes a and b in L_1 and L_∞ metrics are shown in figures 2.2–2.5, where bisector nodes are marked by the symbol $\star$ and cells corresponding to nodes a and b are marked by the symbol $\circ$.

To identify the characteristics of the bisectors we use the predicate $Odd : \mathbf{N} \to \{True, False\}$. We write $Odd(h)$ when h is odd and $\neg Odd(h)$ for even h, where $h \in \mathbf{N}$.

In figures 2.2–2.5 the following correspondence between the patterns (in fact, their indices) and the values of s_x and s_y takes place:

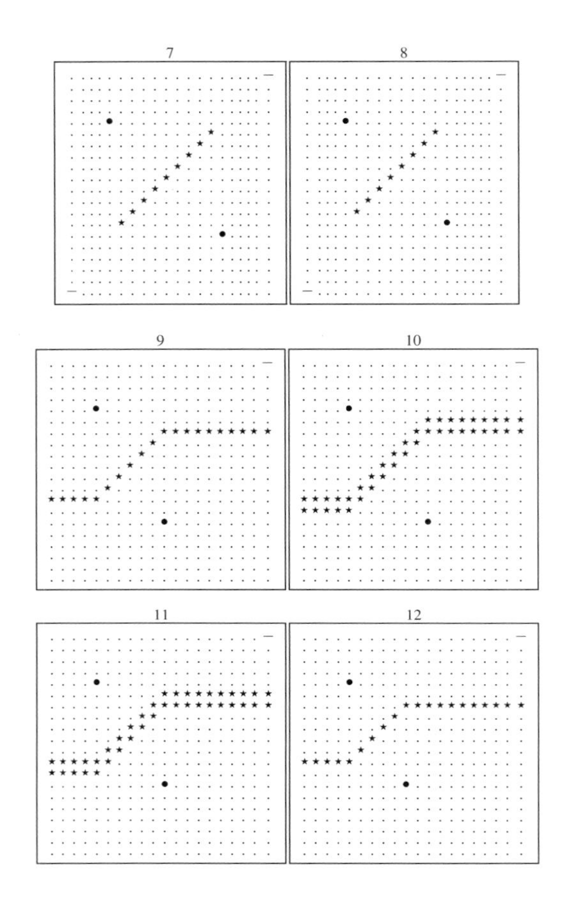

Figure 2.3. The bisectors $B_{1\infty}(a, b)$.

1, 7, 13, 19:	$s_x = s_y$ and $\neg Odd(s_x)$
2, 8, 14, 20:	$s_x = s_y$ and $Odd(s_x)$
3, 9, 15, 21:	$s_y < s_x$ and $\neg Odd(s_x)$ and $\neg Odd(s_y)$
4, 10, 16, 22:	$s_y < s_x$ and $Odd(s_x)$ and $\neg Odd(s_y)$
5, 11, 17, 23:	$s_y < s_x$ and $\neg Odd(s_x)$ and $Odd(s_y)$
6, 12, 18, 24:	$s_y < s_x$ and $Odd(s_x)$ and $Odd(s_y)$.

The correspondence between patterns and types of bisector is as follows:

$B_{11}(a, b)$:	the patterns 1–6 (figure 2.2)
$B_{1\infty}(a, b)$:	the patterns 7–12 (figure 2.3)
$B_{21}(a, b)$:	the patterns 13–18 (figure 2.4)
$B_{2\infty}(a, b)$:	the patterns 19–24 (figure 2.5).

Let one of the following conditions be satisfied:

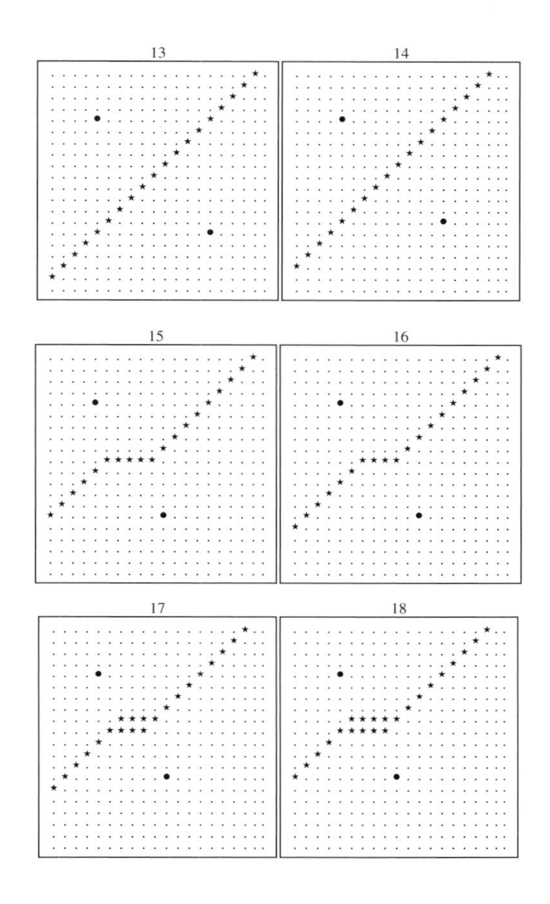

Figure 2.4. The bisectors $B_{21}(a, b)$.

- $x_a \neq x_b$ *and* $y_a = y_b$ *and* $Odd(s_x)$
- $x_a = x_b$ *and* $y_a \neq y_b$ *and* $Odd(s_y)$.

Then $B_{11}(a, b) = \emptyset$.

There is an even number of nodes, different from a and b, of $\mathbf{L}$ on a segment y_a between nodes a and b. Therefore we cannot label any such node $e \in \mathbf{L}$ with coordinates (x_e, y_e), $y_e = y_a$, that $|x_a - x_e| = |x_b - x_e|$. A similar proposition is true when $y_e \neq y_a$. DVC(a) and DVC(b) are the half-lattices, e.g.

$$\text{DVC}(a) = \{e \in \mathbf{L} : x_e \leq x_a + (x_b - x_a)/2\}$$

and

$$\text{DVC}(b) = \{e \in \mathbf{L} : x_e \geq x_b - (x_b - x_a)/2\}$$

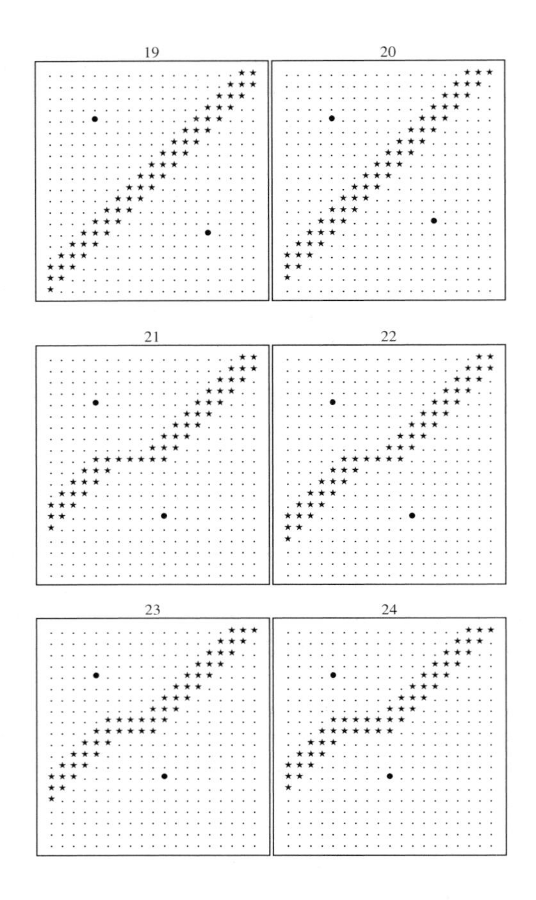

Figure 2.5. The bisectors $B_{2\infty}(a, b)$.

when $x_a > x_b$, $y_a = y_b$ and $Odd(s_x)$. We aim to label nodes which form the bounds of DVD(p), $p \in \mathbf{P}$. Therefore we introduce the modified bisectors $B_{21}(\cdot, \cdot)$ and $B_{2\infty}(\cdot, \cdot)$.

Let one of the following conditions be satisfied:

- $x_a \neq x_b$ *and* $y_a = y_b$ *and* $Odd(s_x)$
- $x_a = x_b$ *and* $y_a \neq y_b$ *and* $Odd(s_y)$.

Then $B_{21}(a, b) \neq \emptyset$ *and* $B_{2\infty}(a, b) \neq \infty$.

Let us assume now that $x_a < x_b$, $y_a < y_b$, and $s_x \neq s_y$. The following findings give the essential features of the bisectors.

- Let $s_x = s_y$. Then

  - $B_{11}(a,b) = B_{21}(a,b) = B_{2\infty}(a,b) \neq \emptyset$; and
  - each of the bisectors consists of nodes lying on the diagonal $((x_a, y_b), (x_b, y_a))$ and two quarter-lattices originating from the top of this diagonal (figure 2.2, patterns 1 and 2; figure 2.4, patterns 13 and 14).

- Let $Odd(s_y) \neq Odd(s_x)$. Then $B_{11}(a,b) = \emptyset$ (figure 2.2, patterns 4 and 5).
- Let $Odd(s_y)$ be true. Then $B_{1\infty}$ is broken. There are no bisector nodes inside the rectangle with opposite vertices a and b (figure 2.3, the patterns 11 and 12).
- For any s_x and s_y we have $B_{21}(a,b) \neq \emptyset$ and $B_{2\infty}(a,b) \neq \emptyset$. Moreover, these bisectors consist of one or more L_1- or L_∞-straight segments separating lattice **L** into two DVCs (figure 2.4, patterns 13 and 14).

The last property indicates that bisectors $B_{21}(\cdot,\cdot)$ and $B_{2\infty}(\cdot,\cdot)$ are more appropriate than $B_{11}(\cdot,\cdot)$ and $B_{1\infty}(\cdot,\cdot)$ because they are connected sets of lattice nodes.

Now we know enough about bisectors to tackle the reaction–diffusion algorithms for computing the bisectors.

The main idea of a reaction–diffusion algorithm is as follows. A configuration of given nodes of **P** is projected onto the lattice **L** which is in a quiescent state in such a way that cells corresponding to the elements of **P** are switched to excited state $+$. Waves of $+$ states spread from the source cells outwards with constant unit speed. The wave moves away to a distance t from source cell within t time steps. Waves originating from two different cells meet each other in cells lying at the same distance from the source cells. Therefore, in order to label bisecting nodes we should assign cells, where waves collide, a special state $\star$. If a cell passes to state $\star$ then it will be in state $\star$ forever.

Here we will discuss two kinds of two-dimensional cellular automata, which are essentially very primitive models of a reaction–diffusion medium.

They both compute DVD in the same time proportional to the lattice size (or diameter of the given set of nodes). The first automaton, A_n, has $O(n)$ cell states, where n is the number of elements of a given set **P**. The second automaton, A_1, has $O(1)$ cell states. Information-processing in cellular automaton A_n is discussed in section 2.1.4 on $O(n)$-algorithms. The computation of DVD in the automaton A_1 is discussed in section 2.1.5 on $O(1)$-algorithms.

2.1.4 $O(n)$-algorithm

Let **P** *be a set of n labelled nodes of a two-dimensional lattice* **L** *of $m \times m$ cells. Then there exists such a two-dimensional cellular automaton A_n of $m \times m$ cells, every cell of which has $n + 3$ states and four (eight) neighbours, which constructs* DVD(**P**) *in the metric L_1 (L_∞) in $O(m - n/m)$ time.*

The rest of this subsection proves this theorem.

To avoid self-interaction of a wave we mark waves originating from different sources by different states. This is the first approach to eliminating self-interaction, the second will be discussed later.

Let $\mathbf{Q} = \{1, 2, \ldots, n, -, \bullet, \star\}$ be a cell state set, where

(i) $1, 2, \ldots, n$ are excited states, the so-called reagent states,
(ii) $\bullet$ is a rest state,
(iii) $-$ is a refractory state,
(iv) $\star$ is a precipitate state or a bisector state.

There are $n + 3$ elements in the set $\mathbf{Q}$.

Given a set $\mathbf{P}$ of nodes we place a unique reagent on every node of $\mathbf{P}$. The reagents $1, 2, \ldots, n$ diffuse on the lattice from their sources, which themselves represent elements of $\mathbf{P}$. We assume that a reagent can diffuse at every time step from its current position, a node of a lattice, to its closest neighbouring sites without a decrease in concentration. Different reagents interact locally with one another. A stationary structure is formed by the nodes in state $\star$ from these results of the reactions.

The pseudo-chemical reactions are as follows:

$$\begin{gathered}\text{for } \alpha, \beta \in \{1, 2, \ldots, n\} \qquad \alpha \neq \beta \\ \alpha + \bullet \rightarrow \alpha \\ \alpha \rightarrow - \\ \alpha + \beta \rightarrow \star \\ - \rightarrow \bullet.\end{gathered}$$

To define the cell state transition rules we use an additional set

$$I(x)^t = \{h \in \mathbf{Q} | \exists y \in u(x) : y^t = h\} \cap \{1, 2, \ldots, n\}.$$

Thus, any cell x of A_n calculates its state x^{t+1} from state x^t and neighbourhood state $u(x)^t$ (we assume that $x \notin u(x)$) by the following rule:

$$x^{t+1} = \begin{cases} \star & x^t \in \{1, \ldots, n\} \text{ and } |I(x)^t| \geq 2 \text{ or } x^t = \star \\ + & x^t = \bullet \text{ and } I(x)^t = \{+\} \\ - & x^t = + \text{ and } (I(x)^t = \{+\} \text{ or } I(x)^t = \emptyset) \\ \bullet & x^t = - \text{ or } (I(x)^t = \emptyset \text{ and } x^t = \bullet) \end{cases}$$

where $+ \in \{1, \ldots, n\}$.

The state transition diagram shown in figure 2.6 explains the behaviour of each individual cell. We see that state $\star$ is an absorbing state. Once in state $\star$ a cell will never move to another state. The transition $- \rightarrow \bullet$ is unconditional.

Other transitions take place in the following conditions:

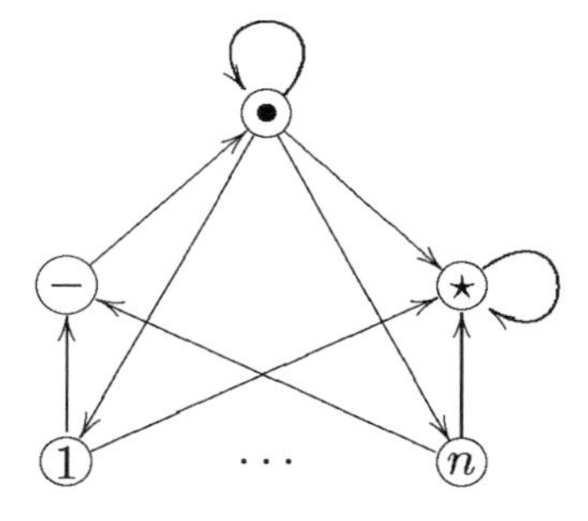

Figure 2.6. A state transition diagram for a cell of the cellular automaton A_n.

$+ \to -$ (exhaustion): $|I(x)^t| \in \{0, 1\}$
$\bullet \to \bullet$ (resting): $I(x)^t = \emptyset$
$\bullet \to +$ (diffusion): $|I(x)^t| = 1$
$\bullet \to \star$ (reaction): $|I(x)^t| \geq 2$
$+ \to \star$ (reaction): $|I(x)^t| \geq 2$.

In figures 2.7 and 2.8 we produce some snapshots of the evolution of a cellular automaton which computes the bisector of the point a, with coordinates (5, 5), and the point b, with coordinates (15, 11). Given set $\mathbf{P} = \{a, b\}$ has two elements, we therefore need only two reagents (call them α and β) to compute the bisector that separates the points. In figure 2.7 the evolution of cellular automaton A_n with a cruciform, four-cell, neighbourhood is presented. The eight-cell neighbourhood is specific to the automaton, the evolution of which is shown in figure 2.8. Every cell of the cellular automaton can take five states: two reagent states, one precipitate state, one rest state and one refractory state. Two cells, corresponding to the elements of **P**, are put into states α and β at the beginning of the computation. They are excited at step $t = 0$. The waves of α and β states are generated around these cells and spread out toward the generators ($t = 3, 4$). At some step of the evolution the waves collide with one another ($t = 5$–14) and the precipitate states are formed. The computation ends when the waves of α and β states reach the edges of the lattice ($t = 15$); the boundary conditions are assumed to be absorbing.

If the given set **P** is a singleton all lattice cells return to the rest state after at most m evolution steps. Assume $\mathbf{P} = \{a, b\}$. If a and b are on the main diagonal of the lattice then the bisector can be detected in the time $m/4$, when the distance between a and b is $m/2$ and the distance between one of the nodes and the near corners of the lattice equals $m/4$. However, the computation time is increased up to m steps when the given nodes are neighbours of each other and lie on the boundaries of the lattice. All this gives us the $O(m - n/m)$ upper bound for the computation time.

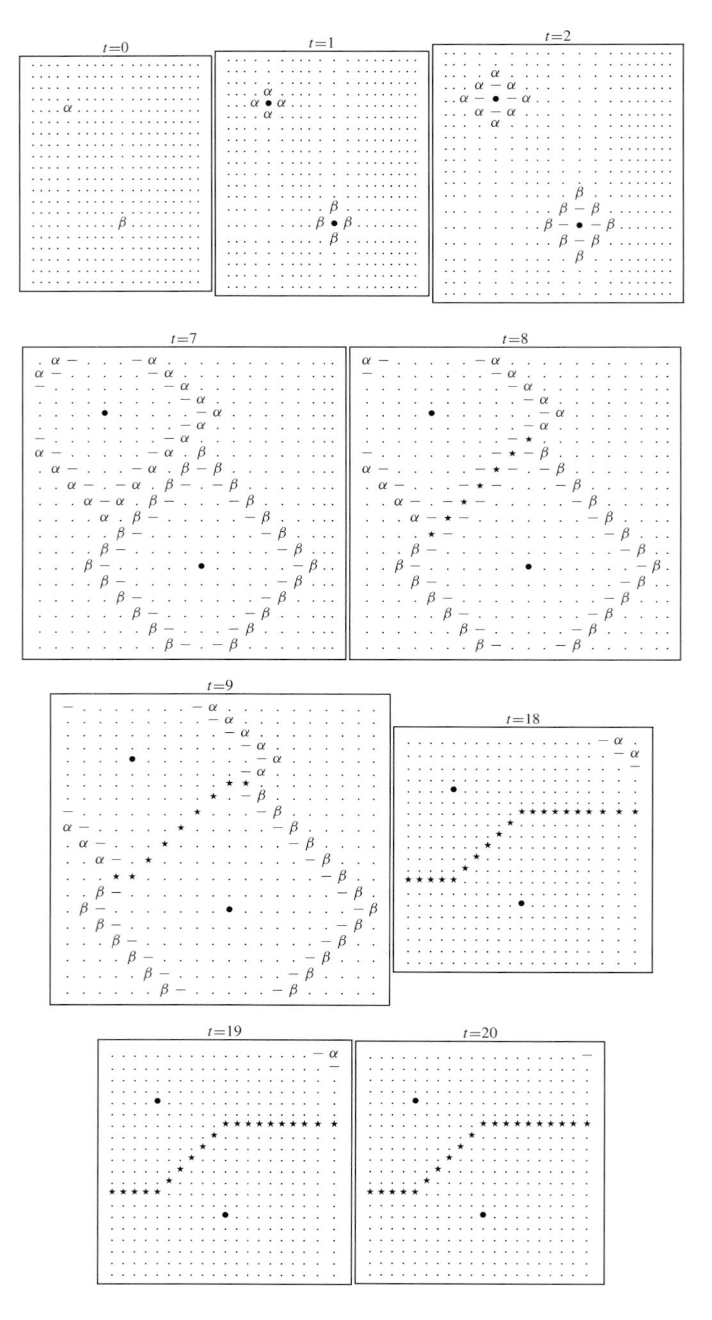

Figure 2.7. Computation of a bisector in the cellular automaton A_n with cell neighbourhood u_1.

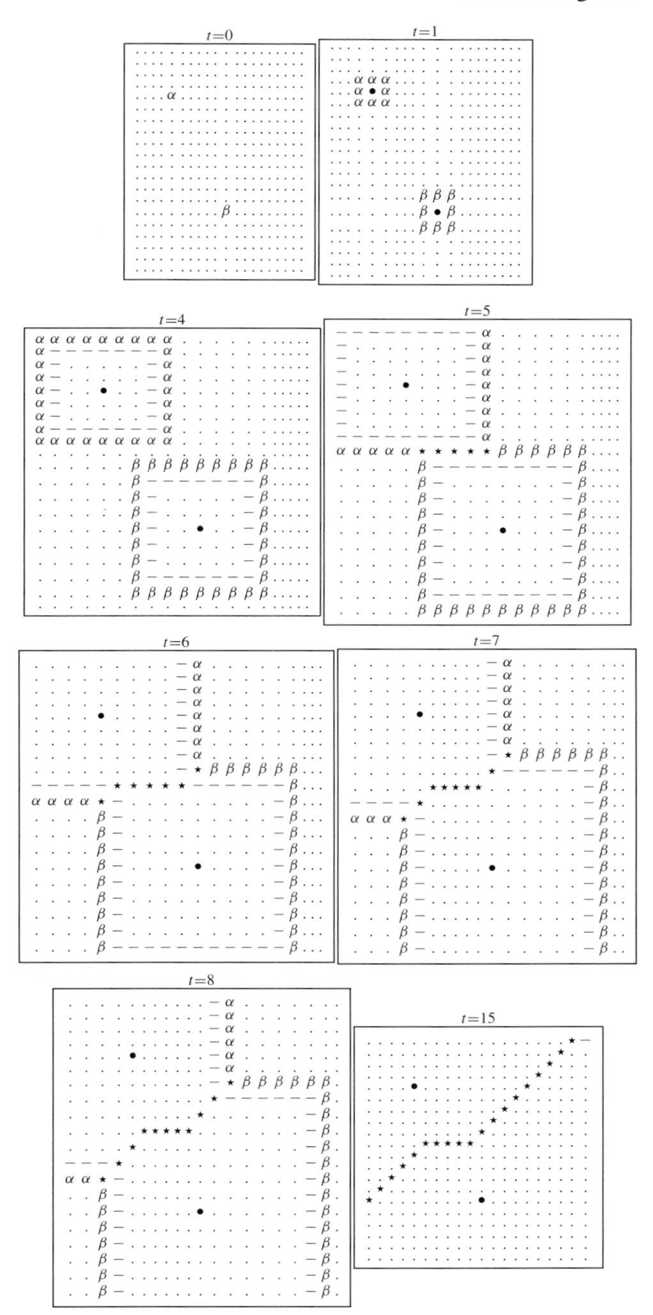

Figure 2.8. Computation of a bisector in the cellular automaton A_n with cell neighbourhood u_2.

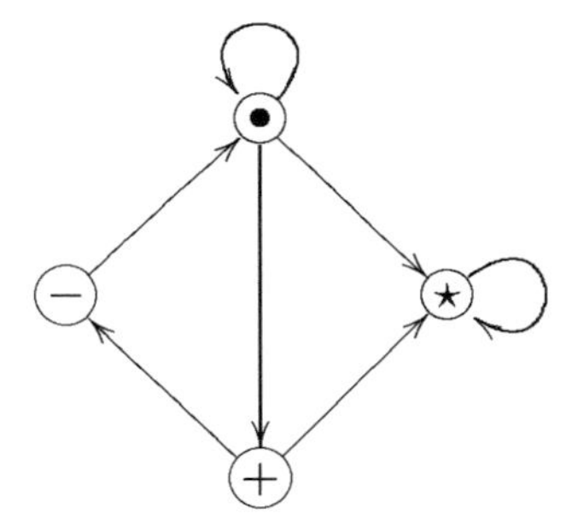

Figure 2.9. A cell state transition diagram of the cellular automaton A_1.

2.1.5 $O(1)$-algorithm

Let **P** *be a set of n labelled nodes of a two-dimensional lattice* **L** *of $m \times m$ nodes. Then there exists a two-dimensional cellular automaton A_1 of $m \times m$ cells, every cell of which has four states and four (eight) neighbours, which constructs* DVD(**P**) *in the metric L_1 (L_∞) in the $O(m - n/m)$ upper time.*

At first, we consider the cellular automaton A_1 which computes DVD(**P**) in the metric L_1. Any cell x of the cellular automaton A_1 has neighbourhood $u_1(x)$ of four neighbours. Let the cells of the automaton take states from the set $\mathbf{Q} = \{\bullet, +, -, \star\}$ and change their states in the order shown in figure 2.9. The cell state diagram of automaton A_1 can be produced from the diagram of automaton A_n (figure 2.6) by merging nodes $1, \ldots, n$ of the diagram into the single node $+$ and the following reduction of duplicated links.

Again we have to define some auxiliary set. Let it be

$$S(x)^t = \sum_{y \in u_1(x)} \xi(y^t, +)$$

where $\xi(y^t, +) = 1$ if $y^t = +$ and 0, otherwise.

We also virtually subdivide the neighbourhood $u_1(x)$ of a cell x into two subneighbourhoods $u_v(x)$ and $u_h(x)$:

$$u_v(x) = \begin{matrix} \otimes \\ \odot \\ \otimes \end{matrix} \qquad \text{and} \qquad u_h(x) = \otimes \odot \otimes.$$

That is, the neighbourhood $u_h(x)$ consists of the north and south neighbours of cell x; the neighbourhood $u_v(x)$ is formed of the east and west neighbours of cell x.

The sets $S_v(x)^t$ and $S_h(x)^t$ are defined by analogy with $S(x)^t$ over the elements of neighbourhoods u_h and u_v.

Any cell x changes its rest state $\bullet$ to the excited state $+$ if at least one of its neighbours is in state $+$. At the same time, cell x changes the state $\bullet$ to state $\star$ when

$$S_v(x)^t = 2 \qquad \text{or} \qquad S_h(x)^t = 2.$$

This rule takes into account situations when two or more excitation waves meet each other at the lattice cells:

$$f\left(\begin{smallmatrix} & + & \\ \alpha & \bullet & \alpha \\ & + & \end{smallmatrix}\right) = f\left(\begin{smallmatrix} & \alpha & \\ + & \bullet & + \\ & \alpha & \end{smallmatrix}\right) = \star$$

for $\alpha \in \mathbf{Q}$.

Simple analysis of the waves spreading from a single source (figure 2.10, $t = 1, 2, 3$) shows that for a central cell in the *rest* state the following states of the neighbourhood never appear in the evolution:

$$\left(\begin{smallmatrix} & + & \\ \alpha & \bullet & \alpha \\ & + & \end{smallmatrix}\right) \qquad \text{and} \qquad \left(\begin{smallmatrix} & \alpha & \\ + & \bullet & + \\ & \alpha & \end{smallmatrix}\right).$$

Therefore a wave originating from a single source does not generate $\star$-states as the result of self-interaction. Unfortunately, the rule becomes more complicated when cell x is in the state $+$ at time t.

Assume that a cell x implements transition $+ \rightarrow \star$ when $S(x)^t \geq 1$. Inspection, at time $t \geq 2$, of the configurations of the waves originating from the single sources demonstrates that the central (relative to its neighbourhood) cell being in an excited state $+$ has no excited neighbours (figure 2.10). That is, the state $\star$ cannot be generated during the spread of a single wave. At sites where at least two different waves meet, every excited cell has at least one excited neighbour.

The cell state transition rule is as follows:

$$x^{t+1} = \begin{cases} \star & ((x^t = \bullet) \wedge (S_v(x)^t = 2) \vee (S_h(x)^t = 2)) \\ & \quad \vee((x^t = +) \wedge (S(x)^t \geq 1)) \vee (x^t = \star) \\ + & (x^t = \bullet) \wedge (S(x)^t \geq 1) \wedge (S_v(x)^t \neq 2) \wedge (S_h(x)^t \neq 2) \\ - & (x^t = +) \wedge (S(x)^t = 0) \\ \bullet & (x^t = -) \vee [(x^t = \bullet) \wedge (S_v(x)^t < 2) \wedge (S_h(x)^t < 2)]. \end{cases}$$

The following conditions determine the cell state transitions:

$+ \rightarrow -$ (exhaustion/refracterness):	$S(x)^t = 0$
$\bullet \rightarrow \bullet$ (resting):	$S(x)^t = 0$
$\bullet \rightarrow +$ (excitation):	$(S(x)^t \geq 1) \wedge (S_v(x)^t < 2) \wedge (S_h(x)^t < 2)$
$\bullet \rightarrow \star$ (precipitation):	$(S_v(x)^t = 2) \vee (S_h(x)^t = 2)$
$+ \rightarrow \star$ (precipitation):	$S(x)^t \geq 1$.

The transition $- \rightarrow \bullet$ is unconditional, i.e. it happens in any condition.

By analogy we can design a cell state transition diagram for the cellular automaton A_1, every cell x of which has $u_2(x)$ neighbourhood of eight cells. The automaton acts in the metric L_∞.

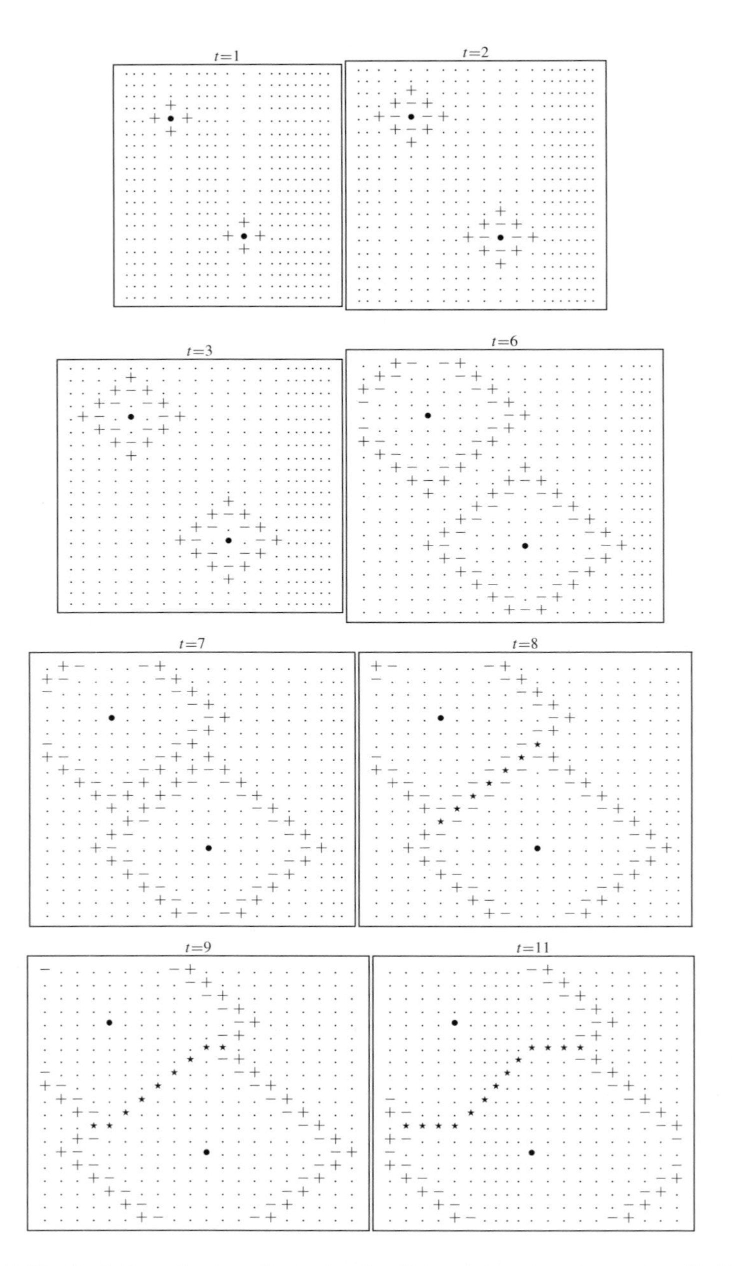

Figure 2.10. Evolution of a two-dimensional cellular automaton A_1, each cell of which has a four-cell cruciform neighbourhood and takes four states. The problem is to separate two elements of given set **P**, marked by • at $t = 0$, by a bisector.

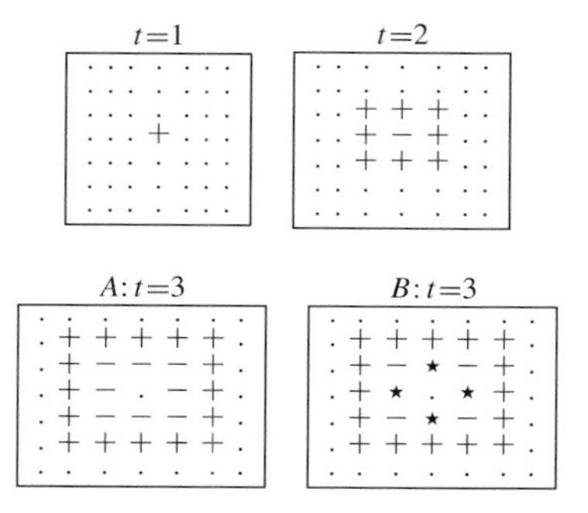

Figure 2.11. Self-interaction of an excitation wave in the initial steps of its development.

Let the predicate $P(x)^t$ be defined as shown here:

$$P(x)^t = ((x^t_{\mathrm{N}} = + \wedge x^t_{\mathrm{S}} = +) \vee (x^t_{\mathrm{W}} = + \wedge x^t_{\mathrm{E}} = +) \\ \vee (x^t_{\mathrm{NW}} = + \wedge x^t_{\mathrm{SE}} = +) \vee (x^t_{\mathrm{NE}} = + \wedge x^t_{\mathrm{SW}} = +))$$

where subindices mean north, south, west, east, north-west, south-east, north-east, north-east and south-west neighbours of cell x, respectively. Then we have the following conditions for cell state transitions:

$+ \rightarrow -$ (exhaustion/refracterness):	$S(x)^t < 4$
$\bullet \rightarrow \bullet$ (resting):	$S(x)^t = 0$
$\bullet \rightarrow +$ (excitation):	$(4 > S(x)^t \geq 1) \wedge \neg P(x)^t$
$\bullet \rightarrow \star$ (precipitation):	$(S_v(x)^t \geq 4) \wedge P(x)^t$
$+ \rightarrow \star$ (precipitation):	$S(x)^t \geq 4$.

Note that the transition $+ \rightarrow \star$ takes place in the condition $S(x)^t \geq 4$ only for $t > 2$. However, at the first steps we have

$$x^1 = + \rightarrow x^2 = \star \qquad \text{if and only if } S(x)^1 \geq 5.$$

Consider the first steps of a spreading wave originating from a single source (figure 2.11). Being in the global state c^2 (figure 2.11, $t = 2$) the cellular automaton has, in general, two opportunities for its next action. That is the next global state, c^3, may be in two forms as shown in figure 2.11(A) and (B). The first alternative, figure 2.11(A), can be realized in the following cell state transition rule:

$$x^t = + \rightarrow x^{t+1} = \star$$

if

$$((S(x)^t \geq 5) \wedge (t \leq 2) \vee ((S(x)^t \geq 4) \wedge (t \leq 2)).$$

The second alternative, figure 2.11(B), is implemented if cells update their states by the rule

$$x^t = + \rightarrow x^{t+1} = \star \qquad \text{if } S(x)^t \geq 4.$$

If the automaton follows branch B, four cells receive the precipitate state $\star$ incorrectly (figure 2.11(B)).

An illustrative example of the separation of the nodes (5, 5) and (15, 11) in the cellular automaton A_1 with cell neighbourhood u_2 is shown in figure 2.12.

2.1.6 Quality of the bisectors

Now that we know how bisectors are constructed in cellular-automata models of reaction–diffusion media, what can we say about how good they are? For the same disposition of nodes a and b of given set $\mathbf{P}$ used in figures 2.2–2.5, cellular automata A_1 and A_n computed the bisectors and the configurations of the automata were fixed. The final configurations of the automata, or configurations of the computed bisectors, are shown in figures 2.13–2.16. The bisectors constructed in automaton A_1 are shown in figures 2.13 (the u_2 neighbourhood case) and 2.14 (the u_1 neighbourhood case). Figures 2.15 and 2.16 represent bisector configurations computed in cellular automaton A_n with the cell neighbourhoods u_2 and u_1, respectively.

In these figures we see that cellular automata produce discrete bisectors, which are good enough to be used in practice. Moreover, the automata sensibly do not fill the quarter lattices and, for the majority of the dispositions of the elements of $\mathbf{P}$, do not generate incomplete bisectors. Compare, for example, patterns 1 and 2 in figure 2.2 and 13 and 14 in figure 2.4 with patterns 1 and 2 in figure 2.13, 13 and 14 in figure 2.15, respectively.

There are some disadvantages however. Studying patterns 7 and 8 in figure 2.14 we find that the following situation takes place for certain dispositions of the given nodes a and b.

Let $|x_a - x_b| = |y_a - y_b|$. Assume that we construct the bisector of the nodes a and b in the cellular automaton A_1 with cell neighbourhood u_1. Then the bisecting nodes are detected only inside a box bounded by columns x_a and x_b and rows y_a and y_b.

Why does this happen?

Look at figure 2.17. Let $|x_a - x_b| = |y_a - y_b| = 4$. At time step $t = 4$, the cells belonging to the diagonal between cells marked by $\clubsuit$ take excitation state $+$, not the precipitate state $\star$. This is because their current neighbourhood states are not included in the conditions of the transition $\bullet \to \star$. Therefore, the waves originating from these two source cells merge into the one wave that continues to spread on the lattice (figure 2.17).

To avoid the situations shown in figure 2.17 we may assume that

$$f\left(\begin{matrix} & + & \\ + & \cdot & \cdot \\ & \cdot & \end{matrix}\right) = f\left(\begin{matrix} & + & \\ \cdot & \cdot & + \\ & \cdot & \end{matrix}\right) = f\left(\begin{matrix} & \cdot & \\ + & \cdot & \cdot \\ & + & \end{matrix}\right) = f\left(\begin{matrix} & \cdot & \\ \cdot & \cdot & + \\ & + & \end{matrix}\right) = \star.$$

However, this leads to the formation of falsely detected bisecting nodes because the resting cells of the automaton will consider the single wavefront as the interaction of two waves (figure 2.18).

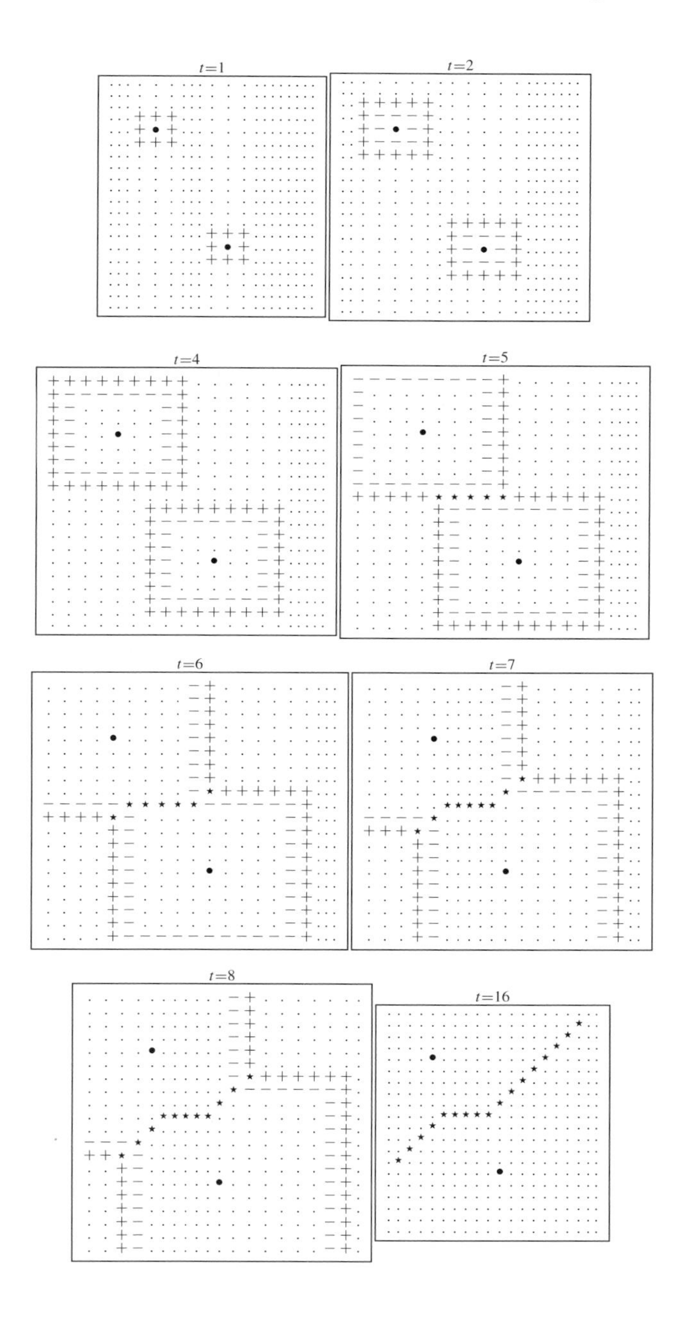

Figure 2.12. Computation of the bisector of two nodes in metric L_∞ in the cellular automaton A_1 with a cell neighbourhood u_2 of eight cells.

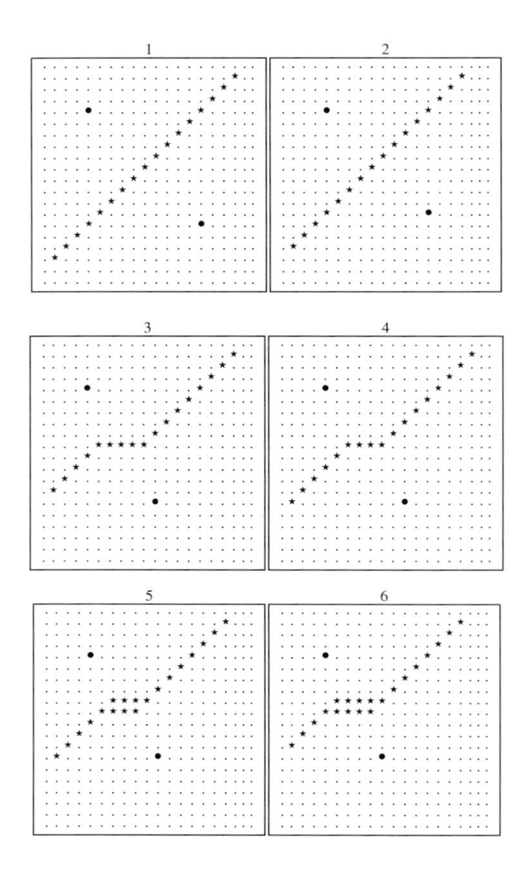

Figure 2.13. The bisectors detected in the cellular automaton A_1 with cell neighbourhood u_2.

The so-called triple bisectors (bisectors consisting of chains of three neighbouring cells in the state $\star$) may be thinned down to one-cell bisectors by applying the following erosion rule:

$$x^{t+1} = \begin{cases} \star & (x^t = \star) \wedge (\forall y \in u(x) : y^t = \star) \\ \bullet & \text{otherwise.} \end{cases}$$

2.1.7 Generalized Voronoi diagram

One of the generalizations of the planar Voronoi diagram includes diagrams of objects other than points and an abstract Voronoi diagram based on a system of bisecting curves as primary objects.

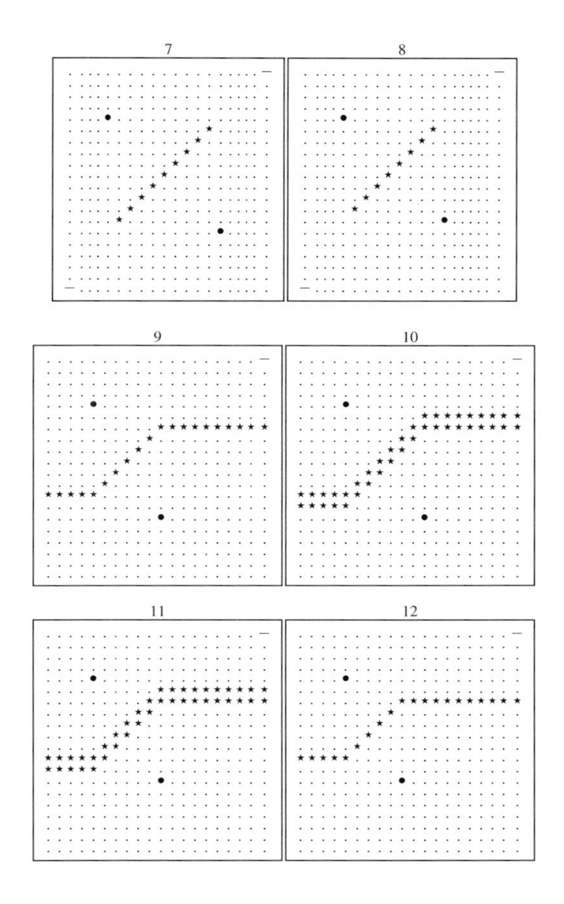

Figure 2.14. The bisectors detected in the cellular automaton A_1 with cell neighbourhood u_1.

Let $\mathbf{P} = \{\mathbf{S}_1, \ldots, \mathbf{S}_m\}$ *be a set of m disjointed subsets of the nodes of* **L***, where every* $\mathbf{S} \subset \mathbf{P}$ *is a connected set. Then a generalized discrete Voronoi diagram (GDVD) of the set* **P** *is defined as*

$$\text{GDVD}(\mathbf{P}) = \bigcup_{i=1,\ldots,k} \partial D \operatorname{Vor}(\mathbf{S}_i)$$

where

$$D \operatorname{Vor}(\mathbf{S}_i) = \{z \in \mathbf{L} | \forall a \in \mathbf{S}_i \forall b \in \mathbf{P} \setminus \mathbf{S}_i : d(a, z) \leq d(b, z)\}$$

Every bound $\partial D \operatorname{Vor}(\mathbf{S}_i)$ of a Voronoi cell $D \operatorname{Vor}(\mathbf{S}_i)$ of a subset $\mathbf{S}_i$ is formed by the nodes belonging to bisectors $\mathbf{B}(\mathbf{S}_i)$ that separate elements of $D \operatorname{Vor}(\mathbf{S}_i)$

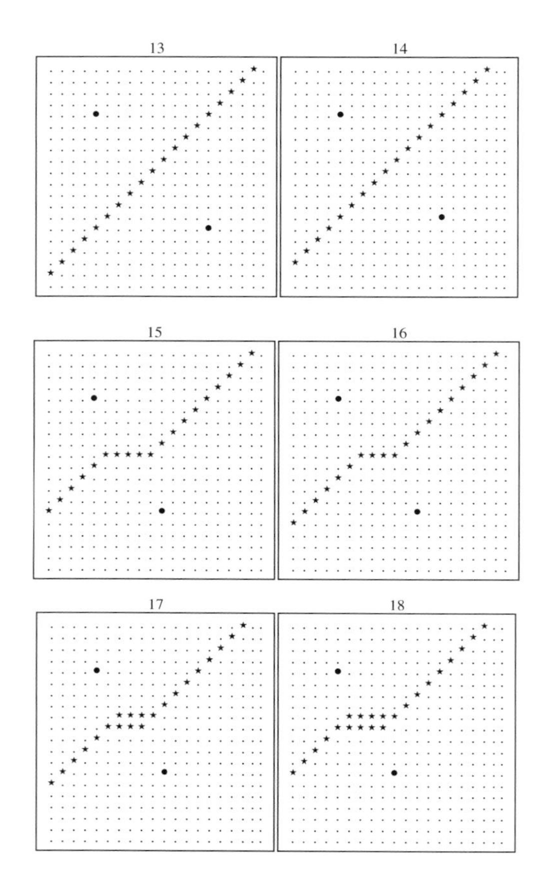

Figure 2.15. The bisectors detected in the cellular automaton A_n with cell neighbourhood u_2.

from elements of D Vor($\mathbf{S}_j$). If $\mathbf{S}_i$ and $\mathbf{S}_j$ are neighbours we have

$$\mathbf{B}(\mathbf{S}_i, \mathbf{S}_j) = \{z \in \mathbf{L} | \forall a \in \mathbf{S}_i \forall b \in \mathbf{P} \setminus \mathbf{S}_i : d(a, z) = d(b, z)\}.$$

So, we have the following problem: given set $\mathbf{P} = \{\mathbf{S}_1, \ldots, \mathbf{S}_m\}$ of n disjoint subsets of lattice nodes, we wish to construct a minimal reaction–diffusion model (minimal cellular automaton) that detects nodes of GDVD(**P**).

> *The* GDVD(**P**) *in the metric L_1 (L_∞) is constructed in a two-dimensional cellular automaton with a four (eight)-cell neighbourhood and $n + 3$ cell states in $O(n)$ time.*

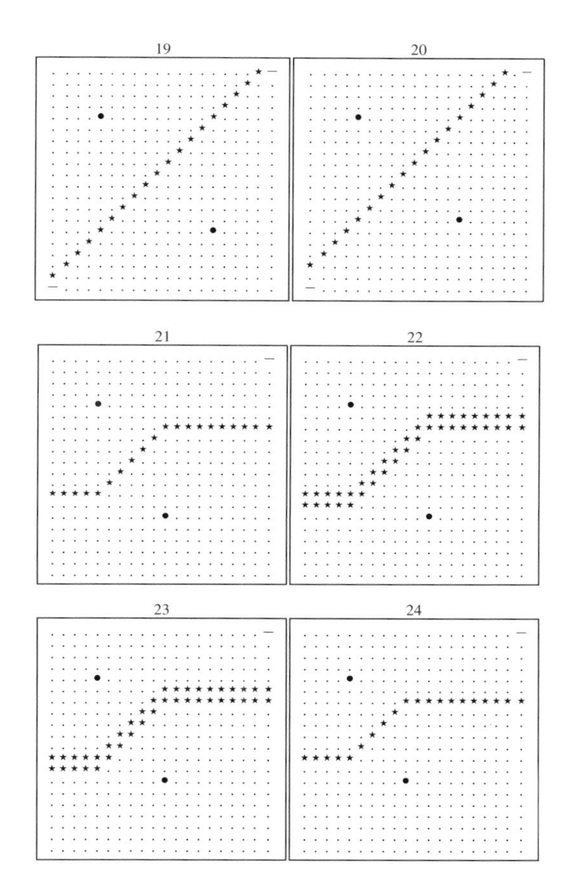

Figure 2.16. The bisectors detected in cellular automaton A_n with cell neighbourhood u_1.

In fact, this is a corollary of the analog statement on the $O(n)$-algorithm. We only have to discuss here the interaction of waves which originate from the elements of the same subset **S**.

Let us consider the example of the computation of the GVD of five subsets in figure 2.19. Every subset is marked with a unique reagent. We have five given subsets, therefore, we must use five different reagents: α, β, γ, ϵ and δ. The precipitate state $\star$ is produced at the sites where two or more different reagents meet each other (figure 2.19).

Recall that a set **P** of the labelled nodes of a lattice is called a disconnected set if there is no path between some elements of **P** along the labelled nodes. Having a

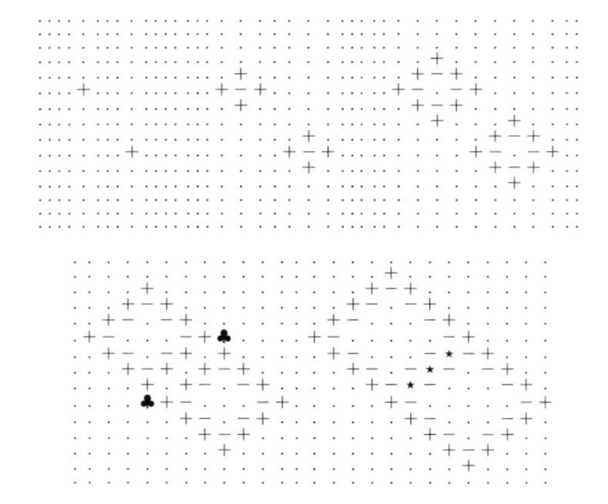

Figure 2.17. An example demonstrating a situation when a bisector cancels its growth. The cells with 'critical' configurations of cell neighbourhoods are marked by ♣. The consecutive snapshots are taken in the evolution of a two-dimensional cellular automaton.

disconnected subset marked with three reagents (figure 2.20) we apply the $O(n)$-algorithm and construct the GVD (figure 2.21).

2.1.8 Inversion of Voronoi diagram

This section aims to show possible restrictions to reaction–diffusion computing rather than to demonstrate its power. Here we discuss quite a peculiar problem related to the Voronoi diagram.

> *Given Voronoi diagram* Vor(**S**) *of some unknown planar set* **S**, *we wish to reconstruct elements of the set* **S**.

Let us start with the planar case.

> *A planar set* **S** *can be extracted from a given Voronoi diagram* Vor(**S**) *iff there is at least one closed Voronoi cell in* Vor(**S**).

In non-degenerate cases any vertex of Vor(**S**) is defined by three points from **S**.

If all cells of Vor(**S**) are open, then for each vertex of Vor(**S**) we can construct at least two different planar sets $\mathbf{S}_1$ and $\mathbf{S}_2$ that determine the same diagram. We assume that a given Vor(**S**) has at least one closed Voronoi cell. To compute **S** it is enough to compute at least one point $p \in \mathbf{S}$ for a closed cell $V(p)$.

Let the closed Voronoi cell $V(p)$ have $m+1$ edges: $e_0, \ldots, e_m$ (figure 2.22). For $i \in \{0, \ldots, m\}$ $\phi_i(p)$ is called the reflection of point p through edge e_i. Edge

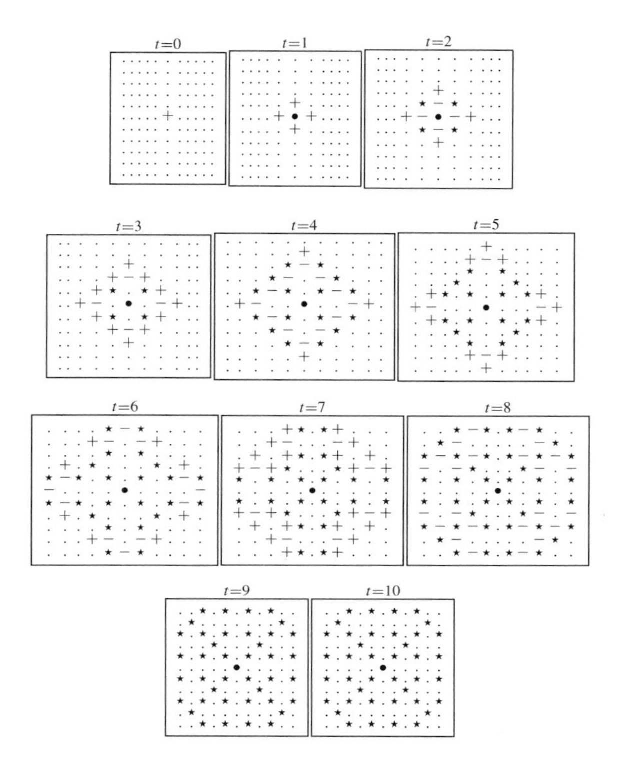

Figure 2.18. A wave generated from a single source repeatedly interacts with itself and a precipitate is produced.

e_i is a bisector of points $\phi_i(p)$ and p; the points are assumed to belong to the approximated set **S**. Assume that we have computed a point p for cell $V(p)$. After that, we construct all reflections of p:

$$\phi(p) = (\phi_0(p), \ldots, \phi_m(p)) = (p_0, \ldots, p_m)$$

all reflections of these reflections

$$\begin{aligned}\phi^2(p) &= \phi(p_0, \ldots, p_m) = (\phi(p_0), \ldots, \phi(p_0)) \\ &= ((p_0^0, \ldots, p_m^0), \ldots, (p_0^m, \ldots, p_m^m))\end{aligned}$$

and so on. At step t of the reflected p we get $\phi^t(p)$. The reflection front *diffuses* on the point configuration. The number of edges of Vor(**S**) is finite, therefore there exists a step τ such that $\phi^\tau(p) = \mathbf{S}$. The value τ depends on Vor(**S**) and the position of the computed point p relative to other elements of **S**. Certainly, $1 \leq \tau \leq n$, and $\tau = 1$ when all Voronoi cells are adjacent to $V(p)$.

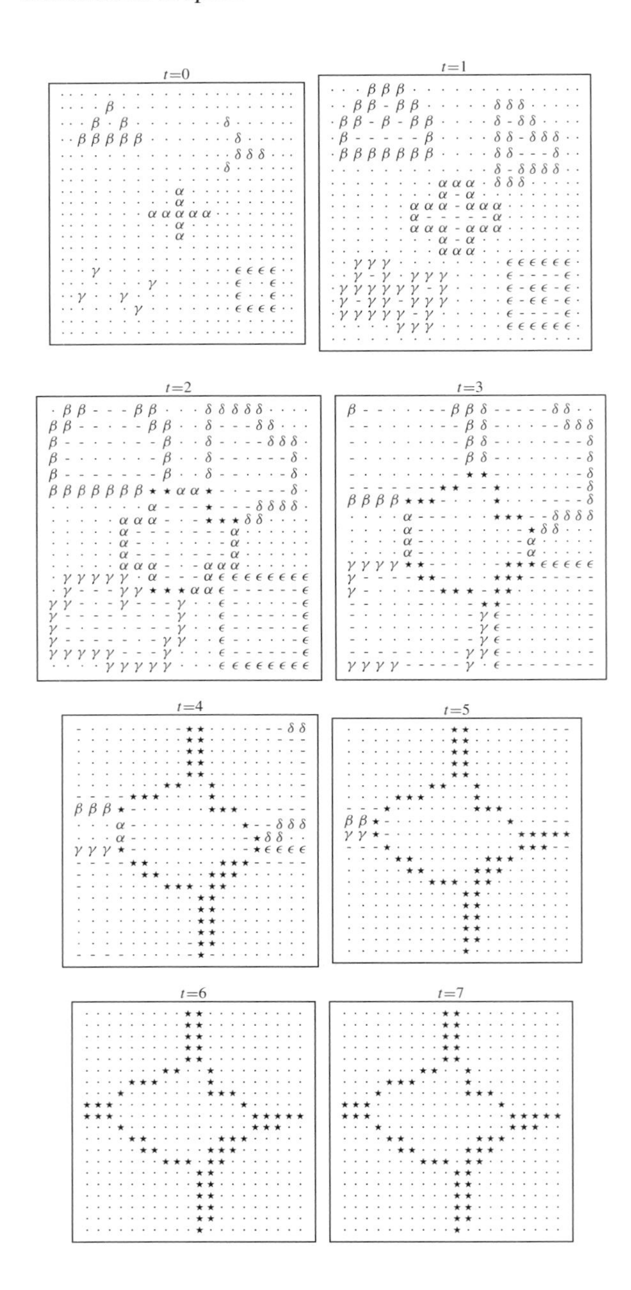

Figure 2.19. Construction of a discrete generalized Voronoi diagram of five subsets in a two-dimensional cellular automaton with seven cell states and an eight-cell neighbourhood.

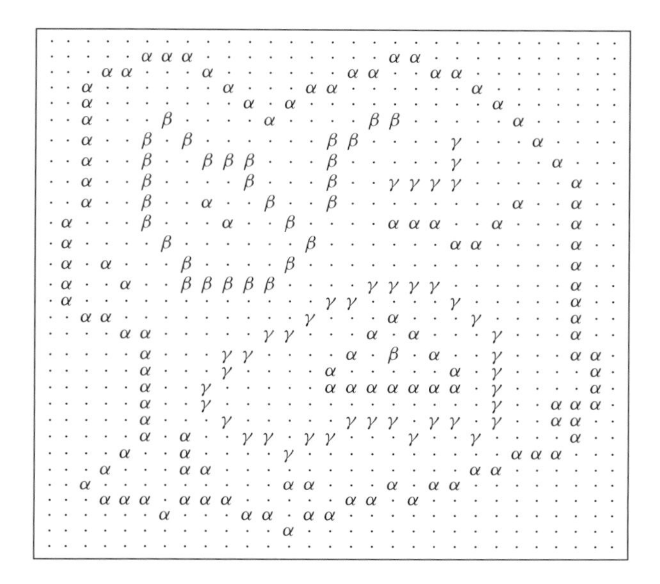

Figure 2.20. Initial stage of computing a generalized Voronoi diagram with five different disconnected subsets of lattice nodes. Five different reagents are employed. In the picture we see the initial configuration of the cellular-automaton model of a reaction–diffusion medium. Cells corresponding to given nodes of the subsets take different reagent states.

Let us look at how the algorithm can be implemented in a rectangular mesh parallel processor with cellular-automaton architecture, i.e. a grid of elementary processors with local connections. Every elementary processor contains the coordinates of the points of the set **P** which is approximated. Each processor has a description of, at most, one Voronoi cell $V(p)$. In this case it approximates the point p. If some processor already has approximated the point p with sufficient precision then it sends the results of its computation to its neighbours, to start the repeated reflection of p. Every processor containing $V(p)$ performs the following operations in the loop of approximating point p (note that p^t means the result of approximation at step t):

- it chooses point p^{t+1} nearest to p^t such that, for each $i \in \{0, \dots, m\}$, the distance between projections of $\phi_i(p^{t+1})$ on l_i and $\phi_{i+1}(p^{t+1})$ on l_i (figure 2.23) is less than the same distance at step t;
- it reads the data computed by its neighbours from its input ports;
- it reflects points approximated by the neighbours into $V(p)$; and
- it stops the computation if the currently computed point satisfies the specified conditions.

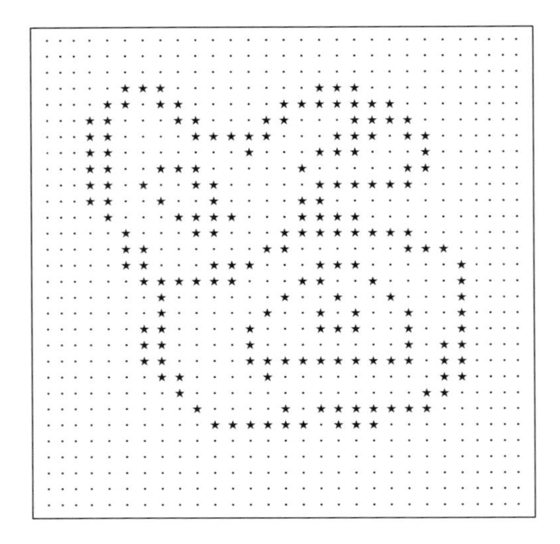

Figure 2.21. Generalized Voronoi diagram with five subsets of lattice nodes, as marked in figure 2.20, computed in a cellular-automaton model of a reaction–diffusion medium with at least five reagents.

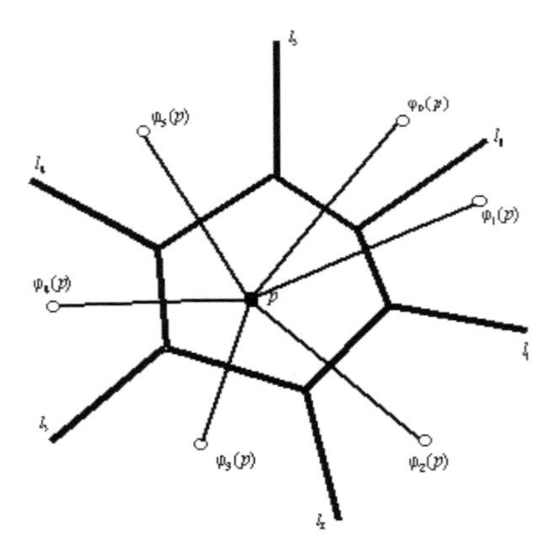

Figure 2.22. A closed Voronoi cell.

Assuming that $\pi_i(\phi_j(p^t))$ is a projection of $\phi_j(p^t)$ onto edge l_i, ϵ and δ are small reals, g is a positive integer, and $q_0, \dots, q_m$ is the set of current approximations of set $\mathbf{S}$ of points computed in neighbouring processors we have the following cartoon algorithm:

`Input:` $\mathbf{V} = (e_0, \dots, e_m, l_0, \dots, l_m)$
`Output:` $p \in \mathbf{R}^2$

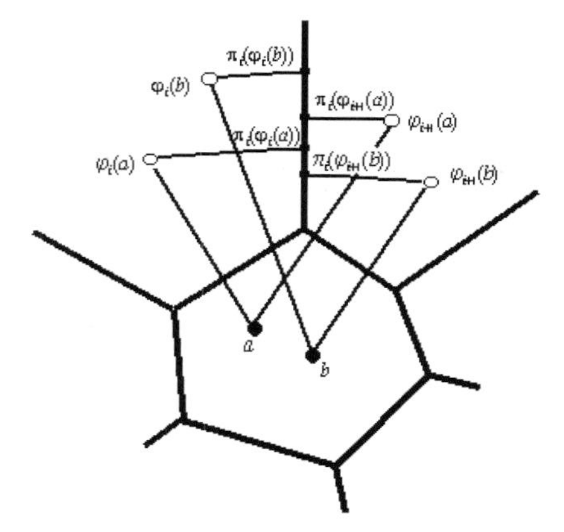

Figure 2.23. Projections.

(i) choose arbitrary point p' inside **V**
(ii) $p^0 \leftarrow p'$
(iii) while $\exists i \in \{0, \ldots, m\}$: $d(\pi_i(\phi_i(p^t)), \pi_i(\phi_{i+1}(p^t))) \geq \epsilon$ do step (iv)
(iv) generate set $\mathbf{P}_a = \{p_1, \ldots, p_g\}$ such that $\forall j \in \{1, \ldots, g\}$: $d(p^t, p_j) = \delta$; $\forall i, j \in \{1, \ldots, g\}$: $d(p_i, p_j)$ is as great as possible; Read points $q_0, \ldots, q_m$ from input ports ; $\mathbf{P}_a = \mathbf{P}_a \cup \{q_0, \ldots, q_m\}$; $d(\pi_i(\phi_i(q)), \pi_i(\phi_{i+1}(q))) = \min_{a \in \mathbf{P}_a} d(\pi_i(\phi_i(a)), \pi_i(\phi_{i+1}(a)))$; Write q to output ports.

So we say that the planar Voronoi diagram DVD(**S**) can be inverted to the approximated set **S**. The planar Voronoi diagram is such a case. In a lattice the situation becomes more complicated.

There is no cellular-automata model, with an eight-cell neighbourhood, of a reaction–diffusion medium that will invert a discrete Voronoi diagram.

It is assumed that waves are generated at the bisector sites, that is at the edges of the Voronoi cells, spread inside the Voronoi cells, interact with each other and form a precipitate at the supposed sites of **P**. Unfortunately, the waves originating from different edges of the Voronoi cell meet each other at sites lying equidistantly from the edges. At the same time we know that, in the construction of the Voronoi edges, the positions of edges of Vor(p) are determined by the distances between p and its closest neighbours.

In fact, a reaction–diffusion processor calculates a generalized Voronoi diagram of a given Voronoi diagram when it tries to invert the Voronoi diagram. Therefore, a skeleton of each Voronoi cell is constructed as a result of reaction–diffusion inversion.

2.2 Skeleton

> *Imagine an ideally homogeneous prairie, dry and ready to burn except for a very wet area A, with the total absence of disturbing effects, such as those of winds and slopes. Assume now that the grass at the edge of A is set afire, all at once, and observe how the fire develops. Since the shape of A is the only factor that we have not excluded, it and it alone will influence the spread of the fire.* [119]

There are particular points, where the circular spread of the fire must be checked: the points where the fire quenches itself. The presence of such points not only indicates the non-convexity of the shape A but the points themselves also represent essential features of the shape A [119].

Thus we can define a skeleton of a planar contour as

a set of centres of bitangent circles which lie entirely inside the contour.

Moreover we can prove that

a closed set is uniquely determined by its convex hull and its skeleton [119]

which is a corollary to Valentine's theorem, stating that two closed sets have the same skeleton if and only if they have the same convex deficiency [618].

The notion of a skeleton was actively explored by Blum in his work related to biological vision (see, e.g., [89]). The outstanding feature of a skeleton is that it concentrates almost all topological information about a given two-dimensional shape [89, 119, 412, 378]. A survey of the algorithms related to skeletonization can be found in [378]. Pleiads of quite efficient techniques for skeletonization of a discrete shapes are given in [413, 169, 617, 548, 54].

Skeletonization is one of the first biologically inspired algorithms and it is also implemented in homogeneous neural networks [419, 369]. This direction is still under exploration. Recent results include the design of a massively parallel neuromorphic processor with self-organized elementary processors [337]. Quite impressive algorithms for skeletonization on parallel processors are presented in [101, 575, 556] as well.

In this section we consider the suitability of a reaction–diffusion paradigm for the problem of skeleton construction.

The data shape is projected onto the lattice of a two-dimensional cellular automaton. The shape is projected in such a manner that the cells corresponding to sites with given edge shapes take states other than the rest state. Let it be an excitation state. The excitation waves should spread inwards through the lattice. The waves react with each other and form segments of a desired skeleton on the interaction sites. The segments are represented by the absorbing states of the cells. These states are not changed during automaton evolution. The cellular-automata configuration becomes stationary. It represents a solution of the problem. A cell

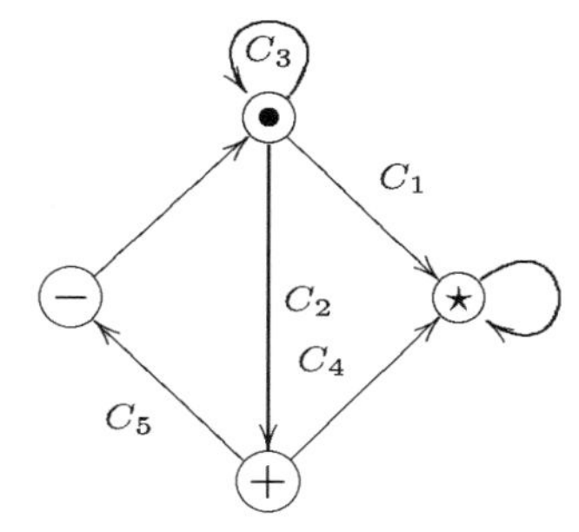

Figure 2.24. A cell state transition diagram of a cellular automaton which approximates a skeleton.

state transition diagram of a cellular automaton that can approximate a skeleton is shown in figure 2.24.

To make the waves spread in towards the given contour we supply cells with a refractory state. Thus our automaton will have four cell states: a rest state (•), an excited state (+), a refractory state (−) and a skeleton state, or a precipitate state, which represents the segments of skeleton (⋆). The transitions $- \to \bullet$ and $\star \to \star$ are unconditional, i.e. the refractory cell takes the rest state independently of the states of its neighbours and once having the skeleton state ⋆ a cell will be in this state throughout the global evolution of the automaton. The excited state can be changed into the refractory state or the skeleton state. The combination of excited and refractory neighbours determines the outcomes of cell state transitions.

Let every cell differentiate between two neighbourhoods: u_1 (a four-cell cruciform neighbourhood) and u_2 (an eight-cell rectangular neighbourhood) as previously defined. We assume that $x \notin u(x)$. We also define several additional functions: $\chi(x^t, \alpha)$=1 if $x^t = \alpha$, otherwise it equals 0, $\alpha \in \{+, -\}$; $\sigma_1^t(x) = \sum_{y \in u_4(x)} \chi(y^t, +)$; $\sigma_2^t(x) = \sum_{y \in u_8(x)} \chi(y^t, +)$; $\sigma_3^t(x) = \sum_{y \in u_4(x)} \chi(y^t, -)$; and $\sigma_4^t(x) = \sum_{y \in u_8(x)} \chi(y^t, -)$.

The conditions for the cell state transitions indicated in figure 2.24 can be written down as follows:

C$_1$: $((\sigma_1^t(x) = 2) \wedge (\sigma_2^t(x) \in \{5, 6\})) \vee (\sigma_1^t(x) \in \{3, 4\}) \vee ((\sigma_1^t(x) = 2) \wedge (\sigma_2^t(x) = 3) \wedge (\sigma_4^t(x) = 1))$.
C$_2$: C$_1$ does not take place and $\sigma_1^t(x) > 0$.
C$_3$: neither C$_1$ nor C$_2$ takes place.
C$_4$: $((\sigma_1^t(x) = 3) \wedge (\sigma_3^t(x) = 1)) \vee ((\sigma_1^t(x) = 2) \wedge (\sigma_3^t(x) = 2))$.
C$_5$: C$_4$ does not take place.

The global evolution of a cellular automaton which approximates a skeleton is shown in figure 2.25. The time complexity of the algorithms is determined by the diameter of the given shape. The objective of the algorithm is to examine all possible combinations of the numbers of excited and refractory states in the

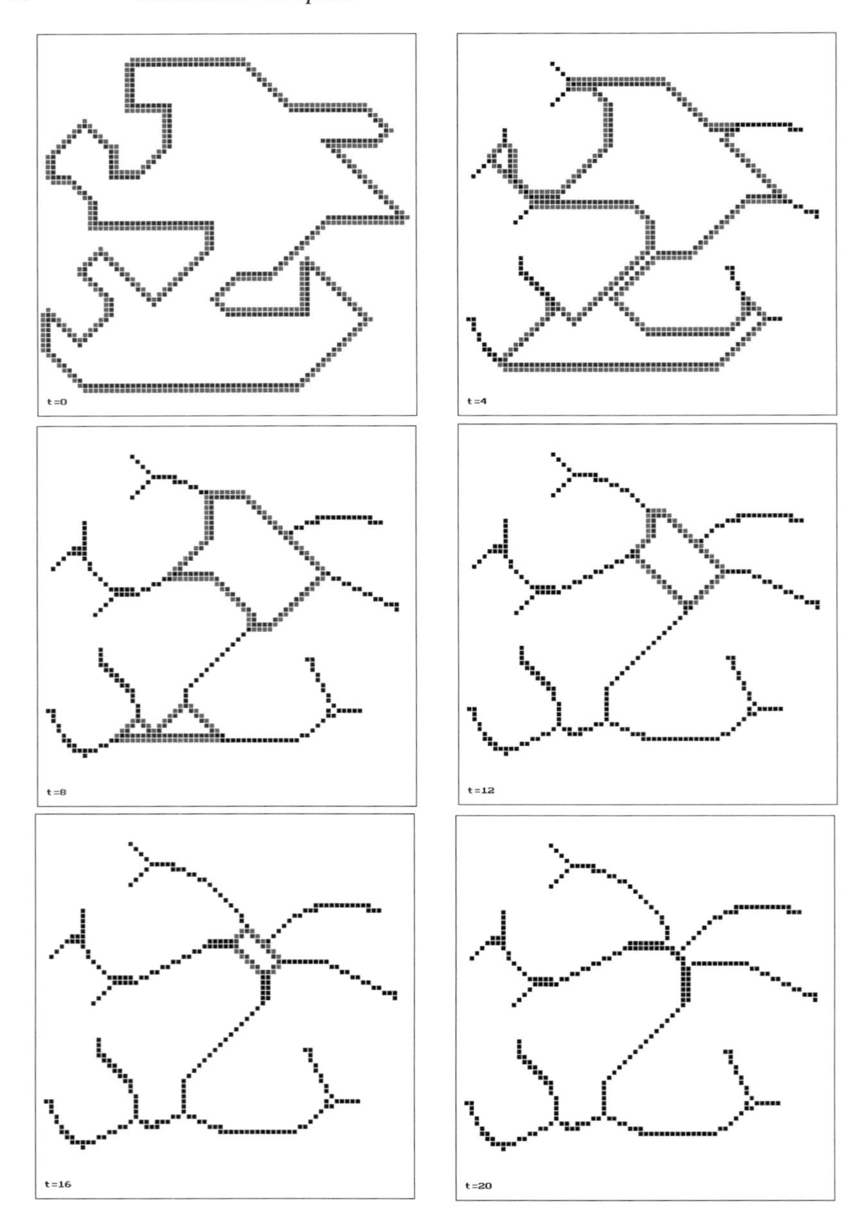

Figure 2.25. Approximation of a skeleton of a planar shape in a two-dimensional cellular-automata model of an excitable medium. The shape is represented as a connected subset of excited cells ($t = 0$). Excitation waves accompanying the post-fronts of the refractory states spread inwards and produce segments of the skeleton as a result of the interaction. Concentrations of precipitate represent the computed skeleton ($t = 20$).

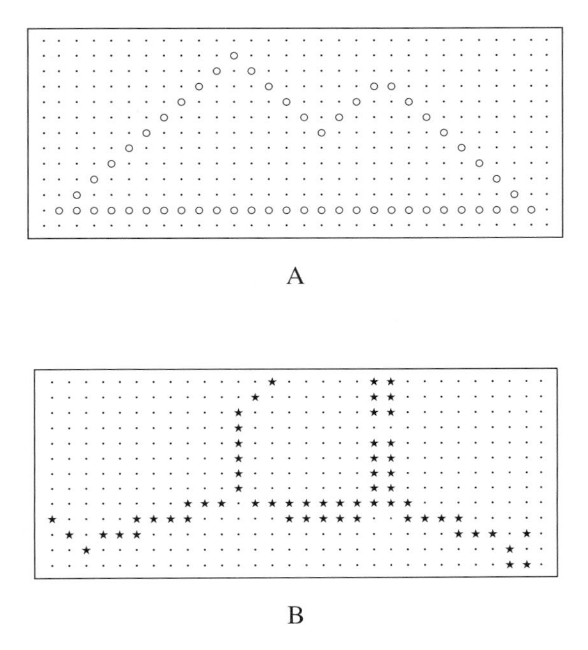

Figure 2.26. Lattice representation of a planar shape (A) and its skeleton computed in a two-dimensional cellular automaton (B).

realization of a cell and to pick out those combinations that are logically essential to skeleton formation. A more detailed example of how a discrete skeleton is produced from a discrete shape is shown in figure 2.26(A) and (B).

2.3 Chemical processors for tesselation and skeletonization

Now we know almost everything about the construction of Voronoi diagrams and skeletons in cellular-automata models of reaction–diffusion and excitable media, what do we know about real wet-ware processors?

2.3.1 Grass fire in Belousov–Zhabotinsky reaction: Rambidi's approach

Blum's ideas may be partially implemented in a Belousov–Zhabotinsky medium for the recognition of a planar shape. The computing medium is assumed to be in trigger mode. In this mode every micro-volume of the reaction liquid switches from one stable state to another in such a manner that the switching state wavefront spreads [514, 515, 516, 517, 519, 520, 523, 524, 525, 526].

Recipe 1. A Rambidi–Maximichev processor for segmentation or reparation of an image.

 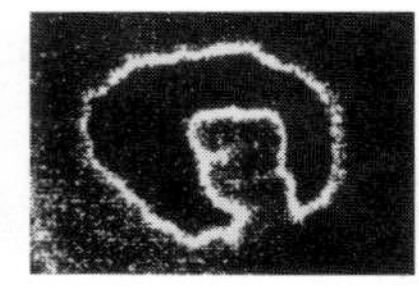 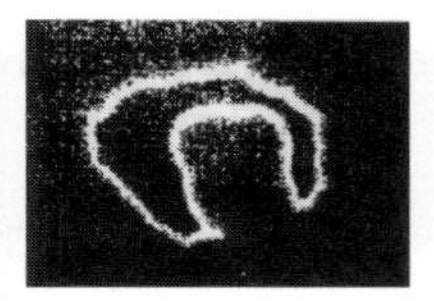 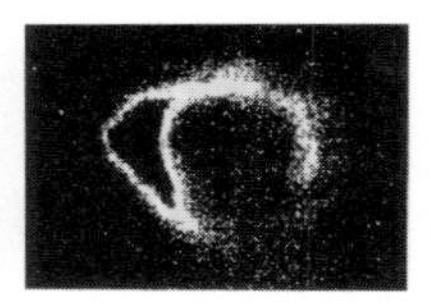

Figure 2.27. Approximation of the skeleton of a planar solid shape in Belouzov–Zhabotinsky medium [521]. Published with kind permission of Nicholas Rambidi.

- Prepare the following solution:
  - H_2SO_4 ... 0.3 M
  - $KBrO_3$ ... 0.2 M
  - malonic acid ... 0.04 M
  - bromomalonic acid ... 0.06 M
  - ferroin ... 0.0025 M
  - Pour the solution into Petri dish to form a 0.5–1.5 mm layer.
- Excite sites in the medium corresponding to a given image contour using a silver probe.
- Record the spacetime dynamics of the reaction medium.

The Rambidi–Maximichev chemical processor is a thin, 0.5–1.5 mm, non-stirred reagent layer, placed in a siliconized reaction vessel. Two different states of the micro-volume can be observed by their colours: red and blue. When the contour of a planar shape is excited in the thin layer of the solution circular waves are generated. The waves run from every point of the contour. Two wavefronts are formed, one spreads in towards the contour, the other runs outwards. Recording the activities of the medium after excitation by various shapes we find that several useful procedures can be realized if the Belousov–Zhabotinsky system is in trigger mode. The first procedure deals with segmentation of an image. Reparation of the image is the second. The third is the detection of a skeleton.

The segmentation procedure is implemented via shrinking: a switching-state wave spreads from the initial excitation sites, a contour, to inner parts of the image. The area of the figure gradually decreases at all sites on the perimeter.

Reparation of defects, e.g. tears in the contours, is done by wave merging. However, we do not know how a Voronoi diagram can be approximated in this situation. Moreover, no stationary structure is formed as the result of computation in such processors. The system is also very sensitive to the many physical characteristics of the experiment.

The skeleton of a planar shape may be detected when wavefronts travelling towards the contour meet each other. Wave-meeting sites may represent the skeleton of the initial shape, see figure 2.27.

2.3.2 Chemical processors for the approximation of Voronoi diagrams and skeletons

We can overcome the disadvantages of the Belousov–Zhabotinsky reaction if we find a set of reactions where a stable precipitate is formed. Here we offer two versions of chemical processors for the approximation of a planar Voronoi diagram and the skeleton of a planar shape. These real processors were intentionally designed from exact cellular-automata computation models [17, 600].

Recipe 2. Adamatzky–Tolmachev processor for approximating a skeleton.

- Prepare a 0.5 mm film of 1% agar gel.
- Mix the gel with $K_4[Fe(CN)_6]\cdot 3H_2O$ (concentration 2.5 mg ml^{-1}).
- Prepare a solution of $FeCl_3\cdot 6H_2O$ (concentration 300 mg ml^{-1}).
- Cut out the contour of the image for electrophoresis in filter paper and saturate it with $FeCl_3\cdot 6H_2O$.
- Put the contour onto the agar film.
- Wait until the whole field inside the contour changes colour.
- Record the distribution of the coloured parts of the reaction space.

An agar gel was chosen as the planar substrate. To make the substrate active we mixed the gel with the reagent $K_4[Fe(CN)_6]\cdot 3H_2O$. At the next step we selected a reagent which would diffuse on the substrate and participate in the formation of skeleton segments. This is the reagent $FeCl_3\cdot 6H_2O$. It forms a light blue precipitate in the reaction with $K_4[Fe(CN)_6]\cdot 3H_2O$:

$$K_4[Fe(CN)_6] + FeCl_3 \rightarrow KFe[Fe(CN)_6] + 3KCl$$
$$3KFe[Fe(CN)_6] + FeCl_3 \rightarrow 3KCl + Fe_4[Fe(CN)_6]_3 \downarrow$$

The filter paper that represents a planar shape is saturated with $FeCl_3\cdot 6H_2O$. When the prepared paper is put on the agar film the reagent $FeCl_3\cdot 6H_2O$ starts to diffuse into the gel.

What happens next? The precipitate $Fe_4[Fe(CN)_6]_3$ is formed. It is coloured blue. Under specified reagent concentrations an exhaustion front is formed which runs ahead of the diffusion front of every wave. Therefore at a site where two or more diffusion waves meet the concentration of $Fe_4[Fe(CN)_6]_3$ becomes too small to be involved in a reaction and such sites in the film remain uncoloured. Thus the skeleton is represented by the uncoloured loci in the film, whereas the other zones of the reaction space are coloured blue.

Two macro-steps in the skeletonization in a chemical processor are represented in figures 2.28 and 2.29. Figure 2.30 displays the skeleton enhanced by standard image-processing techniques.

Recipe 3. Adamatzky–Tolmachev processor for approximating a planar Voronoi diagram.

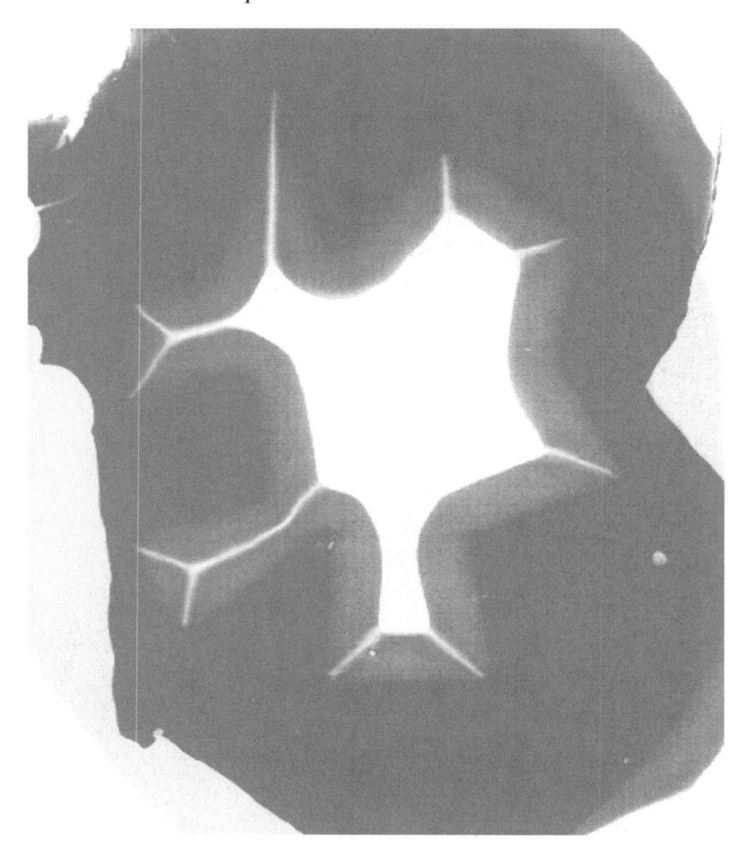

Figure 2.28. Initial phase of the approximation of a skeleton of a planar shape, dark zones. The precipitate sites are shown in grey. The segments of the skeleton which are growing are slightly enhanced.

- Prepare 1.5% agar gel.
- Prepare a solution of palladium chloride (concentration 8.6 mg ml^{-1}).
- Mix the palladium chloride solution with the agar gel.
- Prepare a solution of potassium iodide saturated at room temperature.
- Make a thin film of the agar gel.
- Drop several points of potassium iodide on the gel film.
- Wait until the whole field inside the contour changes colour.
- Take a snapshot of the reaction medium.

A thin layer of agar gel is chosen as the planar substrate on which the reagents diffuse. We prepare a 1.5% agar gel mixed with a solution of palladium chloride (8.5 mg ml^{-1}). A 1–1.5 mm film of the gel is fabricated and placed on

Figure 2.29. The final state of a chemical processor which computes a skeleton. The initial planar shape is shown in a dark colour. The precipitate is grey. Segments and curves of the skeleton are white.

a clear acetate film. Thus we have made an active substrate. Sites corresponding to the planar points of **P** which must be separated by the Voronoi bisectors of Vor(**P**) are supplied with drops of potassium iodide solution saturated at room temperature. The potassium iodide diffuses from the given points. Eventually it covers the whole reaction space. At this stage we can simply scan the film and process it, if necessary, using standard software.

What happens after we have applied drops of potassium chloride to sites of a given planar set **P**? Potassium chloride spreads on the film. It colours the space blue because the precipitate PdI_2 is produced:

$$PdCl_2 + 2KI = PdI_2 \downarrow +2KCl.$$

Figure 2.30. Skeleton from figure 2.29 enhanced by standard procedures.

In this procedure the whole body of every Voronoi cell is coloured blue but the bisectors between every two points (which are the geographical neighbours of each other) remain uncoloured. The cause of this is the exhaustion of $PdCl_2$ in the gel substrate. The exhaustion wave runs in front of the KI diffusion wave. Thus with specially chosen reagent concentrations, the concentration of $PdCl_2$ at the sites where two or more waves of KI meet each other is insufficient to be involved in the reaction.

A real-life picture of a chemical processor, taken when the computation was finished, is shown in figure 2.31. In this monochrome representation the distribution of black and white zones represents the edges of the Voronoi diagram. The extracted diagram is shown in figure 2.32.

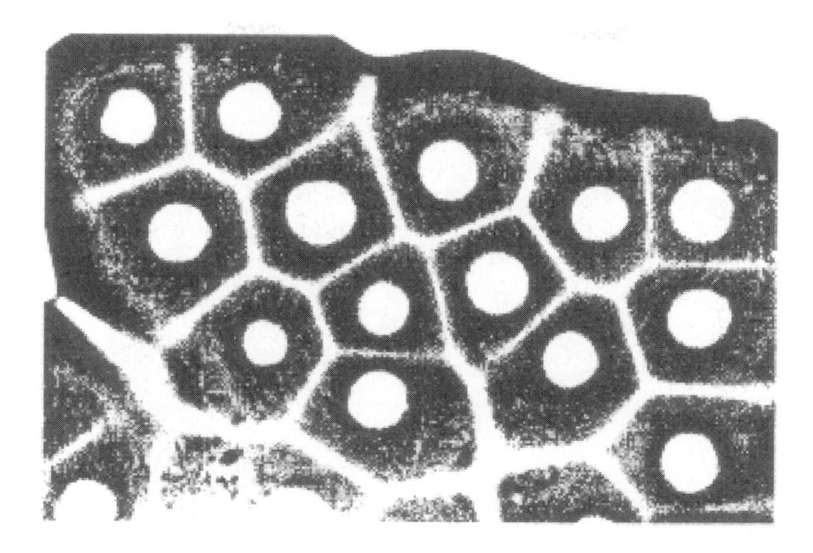

Figure 2.31. An image of the chemical processor which computed a Voronoi diagram. The white discs represent points of a given planar set, the light segments represent the edges of Voronoi cells, the black zones are sites with the precipitate.

The real rate of computation in the chemical processor is proportional to the maximal distance between the geographically neighbouring points of the given planar set and depends on the diffusion rate of the potassium iodide solution.

Practically, we need to satisfy fine stoichiometric constraints to obtain appropriate bisectors, because the tips of the bisectors may be diluted when the concentration of $PdCl_2$ in the agar gel is sufficiently high. This bisector fuzziness prompts us to consider what happens in reaction–diffusion media at the micro-scale level. Again we stick to automata models and their social insect counterparts.

2.4 Voronoi diagram in a lattice swarm

In this section we use the concepts of collectives of automata, reaction–diffusion algorithms and walking or jumping automata on graphs to design an algorithm for the nondeterministic construction of Voronoi diagrams on a lattice.

2.4.1 One experiment with *Tapinoma karavaevi*

This experiment was undertaken by Zakharov [665] in Turkmenia in 1972. It aimed to determine the feeding zone boundaries between several ant families.

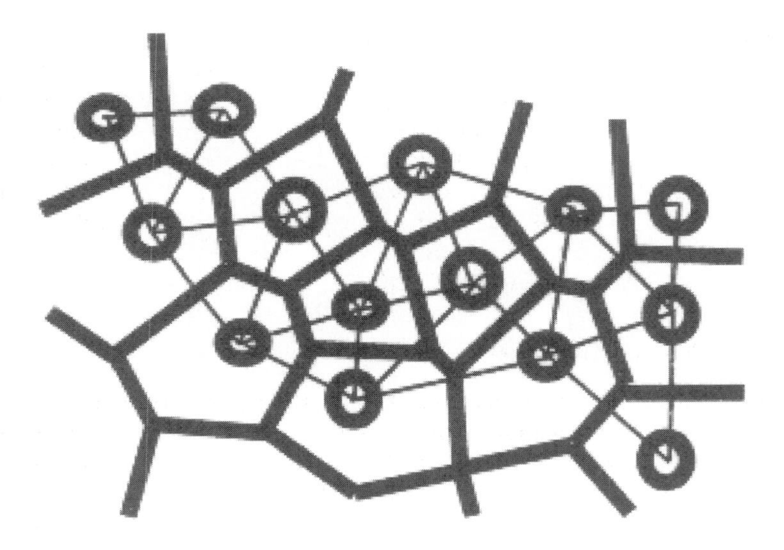

Figure 2.32. Extracted geometrical structure of a Voronoi diagram computed in a chemical processor: the circles are the points of a given set, the thick line segments are the edges of the Voronoi cells. Thin lines connect data points as in Delaune triangulation.

Figure 2.33. Initial setup of a field experiment to determine the boundary between the feeding regions of two nests of *Tapinoma karavaevy*. The two different nest groups are shown by black discs. The artificial food sources are indexed by black squares. Modified from [665].

A quite lucky disposition of two families of *Tapinoma karavaevy* was chosen (figure 2.33). Two ant families are settled on the bank of an irrigation channel. The feeding territory of the families are bounded by water on one side and by a steep bank on another side. The nests are not connected with any trails. To mark an exact boundary between the feeding zones food sources (sugar and insects) are arranged in the chain shown in figure 2.33. The ants start to crowd around the food sources. However, when ants from different families meet near food source number 3 they run away. After some time food sites 1, 2 on one side and 4, 5 on the other side disappear. However, food in the source number 3 remains untouched (figure 2.33). We can speculate that food source 3 represents the bisector of the Voronoi diagram of these two groups of ant nests.

Ideally, this technique can be expanded to construct a two-dimensional Voronoi diagram. We represent planar points by ant nests and cover the whole territory between the nests with food; in field experiments semolina can be successfully used. Allowing some time for the ants to explore the space and to form foraging patterns, we may expect that in a couple of days all grains of food, except those at the boundaries between the foraging zones of the different nests, will be eaten. The remaining grains of food represent the bisectors of a two-dimensional Voronoi diagram. This idea is explored in the next few sections.

2.4.2 Probabilistic diffusion

Let us continue with a simple example. This is an approximation of the bisector of two lattice nodes in a probabilistic cellular automaton which simulates a reaction–diffusion medium. As in the $O(n)$-algorithm each of two given nodes is supplied with a unique reagent. When wavefronts from the two different reagents collide a precipitate is formed at the collision site. The only difference from the automaton which realizes the $O(n)$-algorithm is that in this model the probability of diffusion along an edge between any two neighbouring nodes is 0.1, i.e.

$$\Pr(x^{t+1}) = f(u(x)^t)|x^t) = 0.1$$

where the cell state transition function is defined as in the $O(n)$-algorithm. Several snapshots of the evolution of the approximation of the bisector are shown in figure 2.34. It takes several times longer for the medium to build a bisector than in the deterministic case.

2.4.3 Collectives of automata

Now we can represent diffusion by the spread of collectives of finite mobile automata on a lattice. Three models are discussed here. Each of the models represents some type of a discrete diffusion:

- **$\mathbf{M_1}$ model.** Automata are generated at the sites of a given set $\mathbf{P} \subset \mathbf{L}$, spread randomly on the lattice and react with each other. The result of the reaction of two or more automata originating from different generators (different points of **P**) is a molecule of a precipitate, a particle or a pebble. The number of particles, or pebbles, at a given site of the lattice reflects that site's degree of membership of the set of bisecting nodes. After being involved in the reaction the automata immediately return to their origination sites, are resurrected and continue their progress. The model requires unbounded resources.
- **$\mathbf{M_2}$ model.** Automata are generated at all sites of **P**. Automata can carry pebbles, and each automaton starts its life with only one pebble. Until they

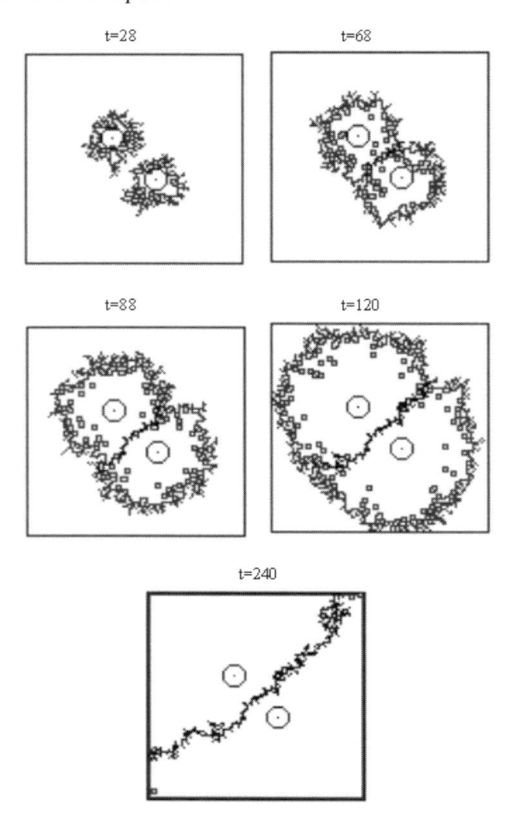

Figure 2.34. Computation of a bisector in a stochastic reaction–diffusion medium. The probability of diffusion along an edge of the lattice is 0.1.

encounter automata from other origination sites (of different colours) each automaton collects all the pebbles it encounters on its way. When automata from different generators meet at a site on the lattice, they drop their pebbles and return to their sources immediately.

- $\mathbf{M_3}$ **model.** This is a conservative model. A fixed number of pebbles are scattered around the lattice, and automata are generated at the sites of **P**. When an automaton finds a pebble it picks it up and carries it until it meets automata from other sources. Each automaton can carry only one pebble at a time.

In all three models automata from different sites of origination have different colours. This is similar to $O(n)$-algorithm.

Let us define a collective of finite automata with probabilistic movement as a tuple

$$M = \langle \mathbf{L}, \mathbf{A}, \mathbf{P}, \mathbf{N}, [0, 1], u, \mu, \beta, \rho, \gamma, \pi \rangle$$

the components of which include a two-dimensional finite lattice **L**, a set of finite automata **A**, a set **P** of labelled nodes of **L**, natural numbers **N**, a real unit interval [0, 1] and six functions:

- neighbourhood function u: $\mathbf{L} \to \mathbf{L}^k$
- colour function γ: $\mathbf{A} \to \mathbf{P}$
- reaction function μ: $2^{\mathbf{P}} \times \mathbf{N} \to \mathbf{N}$
- collection function β: $2^{\mathbf{P}} \times \mathbf{N} \to \mathbf{N}$
- motion function ρ: $2^{\mathbf{P}} \times \mathbf{L}^k \times [0, 1] \to \mathbf{L}$
- probabilistic selection of directions function π: $2^{\mathbf{P}} \times \mathbf{N} \to \mathbf{N}$.

Cells of the lattice **L** take their states from **N** and change their states according to the rule

$$q_x^{t+1} = \mu(\cup_{y \in u(x)} \{\gamma(a) : o_a^t = y\}, q_x^t)$$

where $q_x^t \in \mathbf{N}$, $x, y \in \mathbf{L}$, $t \in \mathbf{N}$, $a \in \mathbf{A}$ and $u(x) \in \mathbf{L}^k$.

Every automaton from **A** has a state, s_a^t, and an output state, o_a^t, which are updated by the following rules:

$$s_a^{t+1} = \beta\Bigg(\sum_{y \in u(o_a^t)} \{\gamma(a' : a' \in \mathbf{A} \text{ and } o_{a'}^t = y\}, s_a^t \Bigg)$$

$$o_a^{t+1} = \rho\Bigg(\sum_{y \in u(o_a^t)} \{\gamma(a' : a' \in \mathbf{A} \text{ and } o_{a'}^t = y\}, \pi(u(o_a^t)|o_a^t) \Bigg)$$

where for any cell x of the lattice **L** and any cell y from cell x's neighbourhood $u(x)$ we have $\pi(y|x) \geq 0$ and $\sum_{y \in u(x)} \pi(y|x) = 1$. At every discrete time step t, the state of M is characterized by the number of pebbles q_x^t at every node x of lattice **L**, the number s_a^t of pebbles carried by every automaton a, and the positions o_a^t of every automaton a on the lattice **L**. An automaton at a given node chooses the next node to move to by selecting from the set of adjacent nodes according to the probability distribution π.

In all models the output state of an automaton from **A** is determined by the rule:

$$o_a^{t+1} = \begin{cases} y \in u(x) \text{ with probability } \pi(y|o_a^t) & \text{if } | \cup_{y \in u(o_a^t)} \{\gamma(a' : a' \in \mathbf{A} \\ & \qquad \text{and } o_{a'}^t = y\}| = 1 \\ p \in \mathbf{P} \text{ with probability } |\mathbf{P}|^{-1} & \text{otherwise} \end{cases}$$

where $o_a^0 = p$ with probability $|\mathbf{P}|^{-1}$. Thus, all automata move according to the same rule regardless of any interaction with each other or any collection of pebbles.

The differences between the models are as follows.

- **M_1 model.** The state of automaton $a \in \mathbf{A}$ remains constant: at any time step t we have $s_a^{t+1} = s_a^t = s$. The state of cell $x \in \mathbf{L}$ increases by one independently of the number of automata of different colours which have their outputs equal to $y \in u(x)$:

$$q_x^{t+1} = \begin{cases} q_x^t + 1 & |\cup_{y \in u(x)} \{\gamma(a) : o_a^t = y\}| = 1 \\ q_x^t & \text{otherwise.} \end{cases}$$

- **M_2 model.** This is a super-model of M_1. The state of any cell $x \in \mathbf{L}$ is increased by the sum of the states of all automata of different colours with outputs in the neighbourhood of x, and is decreased by the sum of the states of the automata in the neighbourhood if all of them have the same colour:

$$q_x^{t+1} = \begin{cases} q_x^t + \sum_{a \in \mathbf{A}: o_a^t \in u(x)} s_a^t & |\cup_{y \in u(x)} \{\gamma(a) : o_a^t = y\}| > 1 \\ q_x^t - \sum_{a \in \mathbf{A}: o_a^t \in u(x)} s_a^t & q_x^t > 0 \text{ and } \forall a \in \mathbf{A} : o_a^t = x \text{ and} \\ & \quad |\cup_{y \in u(x)} \{\gamma(a) : o_a^t = y\}| = 1 \\ q_x^t & \text{otherwise.} \end{cases}$$

At the same time the state of an automaton is changed according to the rule:

$$s_a^{t+1} = \begin{cases} s_a^t + 1 & |\cup_{y \in u(x)} \{\gamma(a) : o_a^t = y\}| = 1 \text{ and } q_x^t > 0 \\ 1 & |\cup_{y \in u(x)} \{\gamma(a) : o_a^t = y\}| > 1 \\ s_a^t & \text{otherwise.} \end{cases}$$

- **M_3 model.** The pebbles are initially placed on the lattice. The overall number of pebbles δ is finite. Therefore, we have $\forall x \in \mathbf{L} : q_x^0 = 1$ with probability δ^{-1}. All other features of the model are the same as in the M_2 model.

In the M_1 model pebbles are formed only as the result of a collision of two automata of different colours. There are initially no pebbles on the lattice. In the M_2 model pebbles are left on the lattice by the automata. There is also a constant supply of pebbles from the nests, i.e. when starting its journey, every automaton carries a pebble. As in the first model the lattice is empty at the start of the computation. So, in models M_1 and M_2 for any $x \in \mathbf{L}$: $q_x^t = 0$. As the lattice $\mathbf{L}$ is finite we assume a quasi-unbounded space: the automata set their outputs on the sites of $\mathbf{P}$ when they reach the boundaries of the lattice: $o_a^{t+1} = \gamma(a)$ if $o_a^t \notin \mathbf{L}$.

In numerical experiments we have analysed several types of distribution of the probability of the orientation of the velocity vector at the next time step. At step t let the automaton at site x_{ij}, where i and j are the coordinates of the sites in the integer coordinate system starting at the upper left-hand corner of $\mathbf{L}$, have

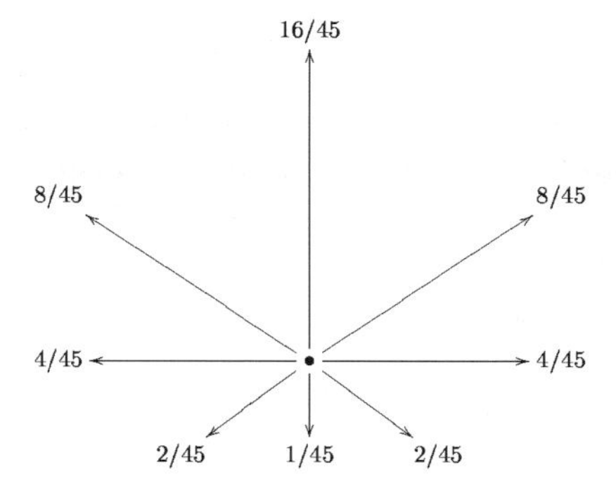

Figure 2.35. Distribution π_ν.

velocity vector $\nu^t = (\overrightarrow{x_{ij}, x_{i'j'}})$, $i' = i + \Delta i$, $j' = j + \Delta j$, $\Delta i, \Delta j \in \{-1, 0, 1\}$. We define a deviation template of the form

$$\Delta = \begin{pmatrix} (-1,-1) & (0,-1) & (1,-1) \\ (-1,0) & (0,0) & (1,0) \\ (-1,-1) & (0,1) & (1,1) \end{pmatrix}.$$

In the first six experiments the velocity vector is computed as a deviation from the vector $\nu^t = (\overrightarrow{x_{ij}, x_{ij-1}})$ independently of the current velocity vector. The following probability distributions are used in the experiments:

$$\pi_4 = \begin{pmatrix} 0 & \frac{1}{4} & 0 \\ \frac{1}{4} & 0 & \frac{1}{4} \\ 0 & \frac{1}{4} & 0 \end{pmatrix} \quad \pi_8 = \begin{pmatrix} \frac{1}{8} & \frac{1}{8} & \frac{1}{8} \\ \frac{1}{8} & 0 & \frac{1}{8} \\ \frac{1}{8} & \frac{1}{8} & \frac{1}{8} \end{pmatrix} \quad \pi_9 = \begin{pmatrix} \frac{1}{9} & \frac{1}{9} & \frac{1}{9} \\ \frac{1}{9} & \frac{1}{9} & \frac{1}{9} \\ \frac{1}{9} & \frac{1}{9} & \frac{1}{9} \end{pmatrix}$$

$$\pi_3 = \begin{pmatrix} 0 & \frac{1}{3} & 0 \\ 0 & 0 & 0 \\ \frac{1}{3} & 0 & \frac{1}{3} \end{pmatrix} \quad \pi_{3_a} = \begin{pmatrix} \frac{1}{3} & \frac{1}{3} & \frac{1}{3} \\ 0 & 0 & 0 \\ 0 & 0 & 0 \end{pmatrix} \quad \pi_2 = \begin{pmatrix} \frac{1}{2} & \frac{1}{2} & 0 \\ 0 & 0 & 0 \\ 0 & 0 & 0 \end{pmatrix}.$$

All distributions except π_9 define non-recurrent random probabilistic walks, and only π_8 and π_9 are spatially invariant. In another set of experiments, every automaton (when it starts its journey on the lattice) is supplied at random with one of the eight vectors $\nu = (\overrightarrow{x_{ij}, x_{i'j'}})$, $i' = i+\Delta i$, $j' = j+\Delta j$; $\Delta i, \Delta j \in \{-1, 0, 1\}$. This initial vector ν remains unchanged during the experiment. The automaton computes its next vector at time $t + 1$ using the probability distribution

$$\pi_\nu = \begin{pmatrix} \frac{8}{45} & \frac{16}{45} & \frac{8}{45} \\ \frac{4}{45} & 0 & \frac{4}{44} \\ \frac{2}{45} & \frac{1}{45} & \frac{2}{45} \end{pmatrix}.$$

In the last experiment we assume that the vector ν is the current vector of the automaton. The distribution π_ν is shown in figure 2.35. It describes the motion of an automaton in which the probability of deviating from the current direction of movement by some amount is inversely proportional to the degree of deviation. In the first distribution, π'_ν, the deviation is calculated from the current velocity vector. In the second case, π''_ν, the deviation is calculated from the initially selected direction of movement.

2.4.4 Analysis of algorithms

It will be convenient to introduce the notion of fuzzy cellular automata at this stage. Let us imagine that at every node of a lattice there is an automaton x that changes its state depending on the states of its neighbours lying in $u(x)$. Such an automaton can be used for the on-line transformation of a Voronoi diagram as follows. If the number of pebbles on the lattice is unbounded we can say that the automaton x takes its states from $\mathbf{Z}_+$ and the state x^t of the automaton at time t represents the fuzzy degree of membership of node x to the Voronoi diagram. Elementary automata are arranged on the lattice and are locally connected with one another. Such a lattice of automata belongs to the family of fuzzy cellular automata (see e.g. [11, 234, 235]). Most techniques developed for dealing with fuzzy cellular-automata and fuzzy cellular-neural networks (see e.g. [656, 657]) can therefore be applied. If the number of pebbles is bounded, the range of possible states of the automaton x can be normalized down to the unit interval, and the state x^t of the automaton at time t therefore represents the probability of the membership of node x to the Voronoi diagram.

First of all we note that in models M_1 and M_2 a fuzzy approximation of DVD(**P**) is computed, whereas M_3 model computes DVD(**P**) probabilistically. In all these models it is simple for us to assess the extent to which a given node belongs to DVD(**P**) by considering the numbers of pebbles at every node—that is, by operating at a global level. However, we require instead that a local assessment is used—that every cell of **L** should, in effect, decide for itself whether or not it belongs to DVD(**P**). In model M_3 the total number of pebbles is finite, and each cell of the lattice can know beforehand. Any cell can thus estimate its probability of being a bisecting node at any step of the computation. In M_1 and M_2 local normalization will not be correct, but nevertheless every cell knows the fuzzy degree of its membership to DVD(**P**). Moreover, the surface of the membership function can be partitioned into a series of α-slices of the fuzzy DVD(**P**):

$$\mathrm{DVD}(\mathbf{P})^t_a = \{x \in \mathbf{L} : q^t_x \geq \alpha\}.$$

Let us consider the Laplacian equation:

$$\frac{\partial^2 w}{\partial x^2} + \frac{\partial^2 w}{\partial y^2} = -g(x, y)$$

in a field $w|_\Gamma = \Psi(x, y)$ with square boundaries Γ. On substituting partial derivatives with second differences we obtain a system of linear algebraic equations:

$$w_{ij} = \tfrac{1}{4}(w_{i-1j} + w_{ij-1} + w_{i+1j} + w_{ij+1} + l^2 g_{ij}) \tag{2.1}$$

where $1 \le i, j \le m-1$, w_{ij} is an approximate value of w_{iljl} and l is the step size of the lattice; if the site (il, jl) is on the boundary then $w_{ij} = \Psi(il, jl)$, $g_{ij} = g(il, jl)$. The equation has the form of a mathematical expectation (when $g_{ij} = 0$):

$$w_{ij} = \tfrac{1}{4}(w_{i-1j} + w_{ij-1} + w_{i+1j} + w_{ij+1}).$$

However, it can also be interpreted as a cell state transition rule for a cellular automaton with a four-cell cruciform neighbourhood:

$$w_{ij}^{t+1} = \tfrac{1}{4}(w_{i-1j}^{t} + w_{ij-1}^{t} + w_{i+1j}^{t} + w_{ij+1}^{t})$$

where w_{ij}^{t+1} is the state of a cell at time $t+1$ and w_{i-1j}^{t}, w_{ij-1}^{t}, w_{i+1j}^{t} and w_{ij+1}^{t} are the states of its neighbours at time step t. Moreover w_{ij}^{t+1} may represent either the concentration of reagents or the number of pebbles at the site (i, j) of the lattice $\mathbf{L}$ at time step $t+1$. w_{ij} is an average value of the random variable ξ_{ij} which is obtained in the following way:

(i) starting at node (il, jl), set the initial value of a counter to $\frac{l^2 g_{ij}}{4}$;
(ii) move to one of the four neighbouring nodes at random with equal probability, adding the corresponding value $\frac{l^2 g_{ij}}{4}$ to the counter;
(iii) repeat (ii) two until the current position crosses the boundary.

When movement has ceased, the value of the counter gives us the sample value of ξ_{ij}. Obviously, $M\xi_{ij} = u_{ij}$ and $D\xi_{ij} < \infty$. Taking into account $|u_{ij} - u_{il,jl}| = O(l^2)$, indicating that the rate of convergence is $O(l^2)$, it should take $\sqrt{\epsilon}$ operations to reach the error ϵ of this random walk solution.

By placing generators of automata at the points of $\mathbf{P}$, we obtain a system of wave equations, the solution of which will be represented by a system of expanding discs. Assuming a uniform velocity for the waves, we can show that the first contacts between discs originating from geographically neighbouring points of $\mathbf{P}$ will take place at the intersections of the bisectors of these points and the straight lines connecting the points (figure 2.36).

By allowing the continuing expansion of the discs, we will detect all the points belonging to the boundaries of the Voronoi cells. Some Voronoi cells can be open; this requires infinite boundaries. Therefore we can assume that the time complexity of the algorithm will be bounded by $O(d(\mathbf{P})/D)$, where $d(\mathbf{P})$ is the diameter of the set $\mathbf{P}$. This factor, combined with our subsequent numerical results, show that algorithms based on the model M_1 are convergent.

The time complexity of the reaction–diffusion deterministic computation of the Voronoi diagram on a lattice is bounded by the size of the largest discrete Voronoi cell when the boundaries are periodic, and by the linear size of the lattice

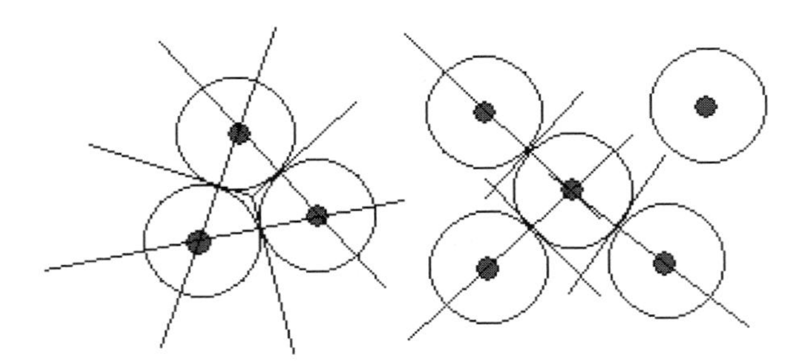

Figure 2.36. At time t discs originating from points $a, b \in \mathbf{P}$ contact one another; $d(a, b) = 2tD$.

in the case of absorbing boundaries. The time complexity is therefore $O(l^{1/2})$ when the lattice has l cells. In models M_1, M_2 and M_3 the time complexity will be determined by the internal complexity of an automaton, which is proportional to the average time spent on a single random walk.

When speaking about the internal time complexity of the models, we should note that the expected time for executing a single random walk step is independent of the current node (in the case of a conventional random walk) at which the automaton is located, up to a multiplicative constant factor. That is, the time spent on selecting the next orientation of the velocity vector exceeds the time spent by the automaton on all other operations. If the vertices of the graph are ordered into a list, and the automaton has random access to this list, then it compares two $O(\log l)$ numbers in each time unit. The automaton therefore performs its computation in logarithmic space and $O(\log l)$ time. If we assume that the automata are really located at the nodes of the lattice, then the time complexity bound will be $O(\log k)$, where k is the size of the node's neighbourhood. In the case of a probabilistic walk, the situation with 'random access' becomes more complicated, as the automaton may use a list of random numbers or may generate them internally. The general method for simulating a discrete random value ξ is based on

$$\mathrm{Prob}\left(\sum_{z=0}^{g-1} p_z \leq \alpha < \sum_{z=0}^{g}\right) = p_g \qquad p_g = \mathrm{Prob}(\xi = x_g),\ g = 0, 1, \ldots$$

which has complexity $S = 1 + \sum_{z=0}^{\infty} z p_z$.

For an integer variable we have $S = 1 + M\xi$; for example, S is proportional to $1+bp$, $1+\lambda$, $1/p$, $1+b_1 h/b$ for a binomial distribution with parameters (b, p), a Poisson distribution with parameter λ, a geometric distribution with parameter p and a hyper-geometric distribution with parameters b_1, h, b, respectively. It is possible to increase the efficiency by changing the order of sampling. Thus we can obtain

$$S \approx \sqrt{\frac{2\lambda}{\pi} + 1.5}$$

for a Poisson distribution and

$$S \approx \sqrt{\frac{2bp(1-p)}{\pi} + 1.5}$$

for a binomial distribution.

Assuming now that every automaton changes its position and state in one unit of discrete time we can discuss the overall time complexity of the models. From this point on, we will assume that the size of the lattice is $l = n^2$. The time complexity of the graph algorithms can be expressed in terms of the space complexity of the transition functions of the automata and the size of the lattice. Because of the similarity of our models to the so-called jumping automata on graphs, it may be useful to indicate here that jumping graph automata with limited supplies of pebbles can determine the threadability (i.e. the so-called (st)-connectivity) of an l-node graph in $O(\log^2 l)$ space. At the same time the overall storage (i.e. the sum of the number of pebbles and states of the automaton) has the lower bound $\Omega(\frac{\log l^2}{\log\log l})$ [151]. The most general bound on the cover time of an arbitrary graph is $O(le)$, where l is the number of nodes and e is the number of edges [44]. In the lattice case this is $O(kl^2)$, where k is the size of the neighbourhood, i.e. the degree of a node. The bounds $O(l^2)$ and $O(l \log l)$ are for the expected cover times of complete and linear graphs respectively. It has also been shown that any connected graph has the expected cover time $O(l^2 \frac{\log l}{(1-\lambda_2)})$, where λ_2 is the second eigenvalue of the transition probability matrix of the Markov process corresponding to the random walk [105]. We also know that the expected cover time is $O(l^2 d'(1/d)')$, where d' and $(1/d)'$ are the average degree and the average of the inverse degree [219, 220]. As we shall see later, none of these bounds correctly predicts the cover time on small lattices.

Why consider small lattices only? Because in real-life situations we can reduce the lattice size to the diameter of the set **P** of given nodes. The time complexity will therefore be determined by the size of the largest Voronoi cell. However, by deleting the empty columns and rows of the lattice, we are able to compress the Voronoi diagram preserving the evenness and oddness of the distances between the neighbouring cells.

What about covering small lattices up to 100×100 nodes? For small lattices of $l \leq 100 \times 100$ nodes, the average cover time has upper bound $l^{3/2}m^{-1}$. This gives us the convergence rate for the computation of a Voronoi diagram on a lattice of l nodes by a collective of m finite automata.

In figure 2.37 we see comparative graphs of the cover time using a probabilistic walk with all possible probabilities of the orientations of the velocity vectors. From the numerical results we can suggest the following hierarchy of the

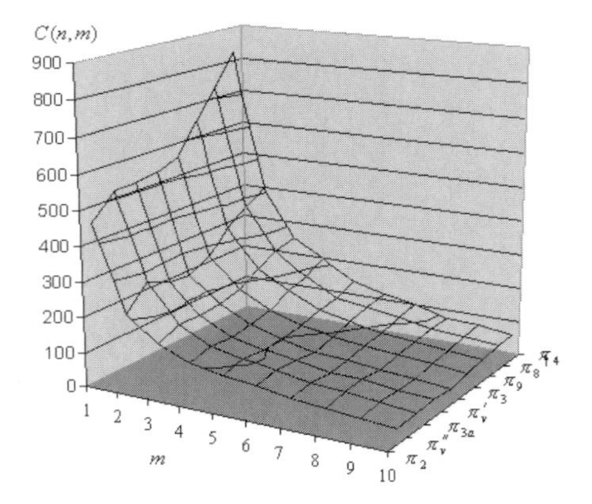

Figure 2.37. Average cover time of a lattice of 100×100 nodes for model M_1 of m finite automata, $m = 1, \ldots, 10$; the velocity vectors of the automata are changed by the distributions defined earlier.

cover times of a two-dimensional discrete torus by a probabilistic walk:

$$\pi_4 \lhd \pi_8 \lhd \pi_9 \lhd \pi_3 \lhd \pi'_\nu \lhd \pi_{3_a} \lhd \pi''_\nu \lhd \pi_2.$$

From the hierarchy we are able to propose that, for small lattices:

(i) a space-invariant walk takes longer to cover the lattice (this is especially clear when we consider the π_3 and π_{3_a} distributions);
(ii) a recurrent walk takes longer to cover the lattice than a non-recurrent one;
(iii) increasing the number of possible orientations of the velocity vectors does not reduce the cover time.

The positions of π'_ν and π''_ν at the lower levels of the hierarchy reflect the superiority of asymmetric walks. But, at the same time, the avoidance of strong asymmetry in the transition from π'_ν to π''_ν surprisingly does not increase but in fact reduces the experimental cover time. The higher convergence rate of asymmetric walks is due to the fact that the random walk with drift has $O(E(C))$ expected cover time, where $E(C)$ is the expected cover time of the drift-free walk when both walks are on a discrete torus.

2.4.5 Shapes of Voronoi diagram

We should note here that despite the strong analytical and numerical results obtained on the shape of random walks, namely the shape and shape variance factors, which are computed from the principal components of the shape tensor and gyration, the problem itself is still far from any final solution. Another fact relating to shape is that the recurrence of a random walk on a two-dimensional

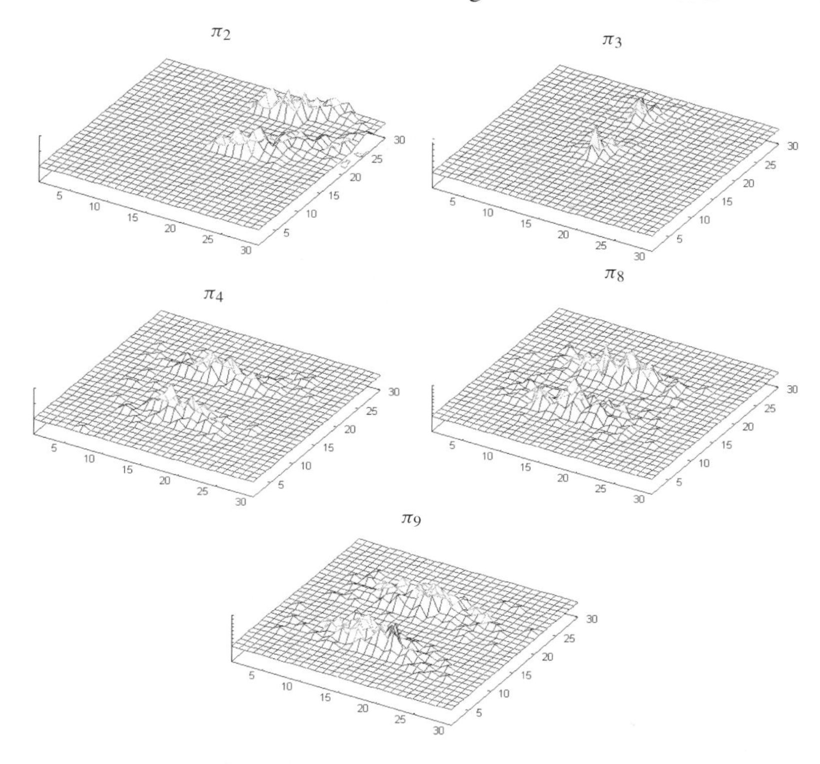

Figure 2.38. Concentrations of pebbles representing the fuzzy Voronoi diagrams of three nodes $(\frac{j}{2}, \frac{i}{4})$, $(\frac{j}{2}, \frac{i}{2})$ and $(\frac{j}{2}, \frac{3}{4}i)$, on a lattice of 30×30 nodes, $m = 200$, $t = 500$. The Voronoi diagrams were computed using the M_1 model for the probability distributions π_2, π_3, π_4, π_8 and π_9.

lattice means that the two-dimensional mean displacement vector is zero. The numerically computed shapes of the Voronoi edges are shown in figure 2.38. They reflect the number of degrees of freedom for every type of motion.

Model M_1 allows us to compute a fuzzy Voronoi diagram and perfectly reconstruct the Voronoi cells (figure 2.39) which look like the cells of a conventional Voronoi diagram. Figure 2.39(B) clearly shows us how the degree of membership of a lattice node to DVD(**P**) depends on the distance of this node from the elements of **P**.

The results are changed somewhat when we use model M_1 with π'_ν and π''_ν distributions (figure 2.40(A) and (B)). For the same dispositions of elements of **P**, the density of pebbles on bisecting nodes is much higher. Remarkably, the M_1 model with the π''_ν distribution (which is perhaps the closest to the characteristics of real-life movement) gives us the best convergence rate among the models with

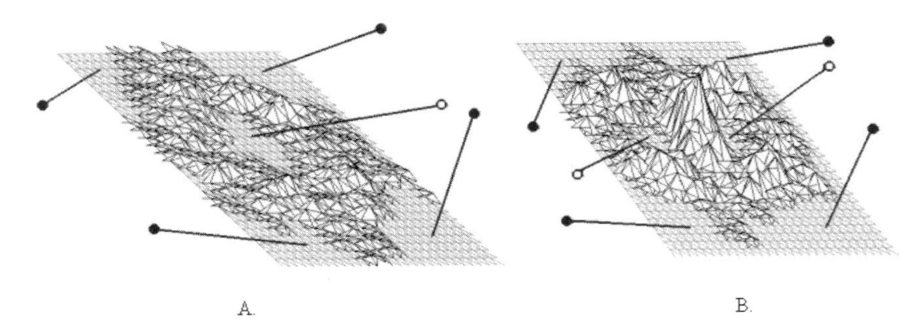

Figure 2.39. The fuzzy Voronoi diagrams of five (A) and six (B) nodes of the set **P** computed using the M_1 model with π_8. The nodes of **P** are marked by pins. Nodes having closed Voronoi cells are indicated by pins with hollow heads. Other parameters are as follows: $l = 30 \times 30$, $m = 200$, $t = 500$.

Figure 2.40. The fuzzy Voronoi diagram of the five nodes of set **P** (arranged as in figure 2.39) computed using model M_1: (A) π'_ν and (B) π''_ν; $l = 30 \times 30$, $m = 200$, $t = 500$.

spatially invariant rules of motion and very convenient degrees of membership (figure 2.40(B)).

Model M_2 requires unlimited resources. The pebbles are redistributed from point-sources around the lattice, and a high concentration of pebbles is expected to occur on the bisecting nodes. From the results of numerical simulations (figure 2.41) we see that when automata are scattered randomly and uniformly around the lattice, fuzzy bisectors are built. We did not find any obvious signs indicating that these fuzzy bisectors will disappear with time; this problem requires further study. Another interesting question concerns the case where the sites of **P** are allowed to move around the lattice. Due to the dynamic characteristics of the algorithm, we predict that if the velocity of the elements of **P** is sufficiently small (relatively to the velocities of the automata) the old bisectors will be dissolved, and pebbles from the old bisectors will be utilized to build new bisectors.

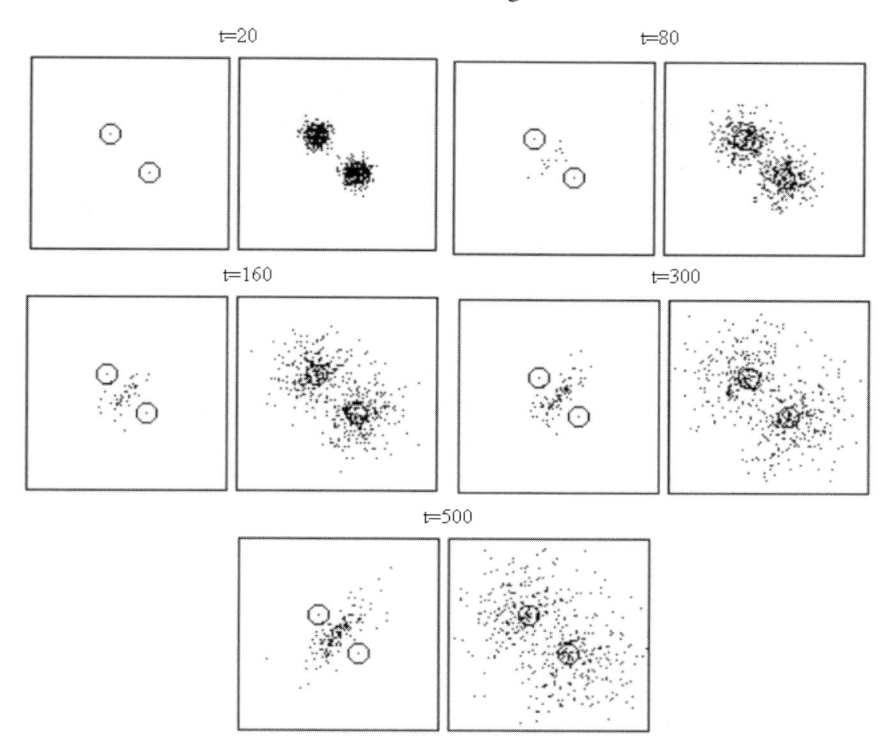

Figure 2.41. Computation of a bisector in the model M_2 of a lattice swarm for $l = 100 \times 100$ and $m = 100$. For each time slice ($t = 20, 80, 160, 300, 500$) two diagrams are recorded. For each time slice the diagram on the left shows the pebbles as black pixels, whereas the diagram on the right shows the spatial distribution of the automata of M_2. The two given sites of **P**, which are separated by the bisector, are encircled.

The two examples in figure 2.42 show us how a bisector is constructed in the M_3 model. Here we assume elastic boundaries. When an automaton collides with the boundaries (which might represent the walls of a room) it does not drop its pebbles. Every automaton spends more time near the boundaries than in open space, since some neighbouring nodes are occupied by the boundaries, and therefore the number of possible nodes to move to is reduced. After colliding with the boundary, an automaton chooses the next node to move to at random according to the probability distributions π_8. The number of possible orientations of the velocity vectors is reduced to $\frac{5}{8}$ when the automaton hits a wall and $\frac{3}{8}$ when it is in a corner of the lattice. When the number of automata is limited, the automata have a higher probability of meeting one another near the boundaries than in open space (excluding the region of a bisector). This is reflected in the high density of pebbles shown in figure 2.42. On restricting the carrying capabilities

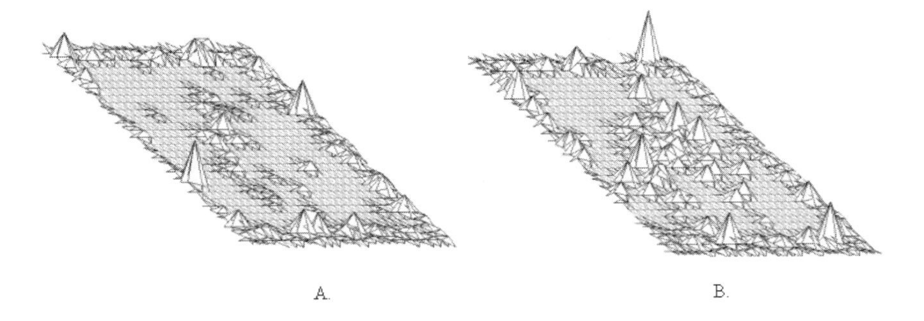

Figure 2.42. The distribution of pebbles represents the Voronoi diagram of two nodes $(\frac{n}{3}, \frac{n}{3})$ and $(\frac{2n}{3}, \frac{2n}{3})$ on a lattice of $n \times n = 30 \times 30$ nodes with impassible boundaries. It is computed using model M_3 (with distribution π_8) in standard form (A) and with a slight modification (B). In (B), an automaton picks up a pebble only if a single pebble lies at the current site of the lattice.

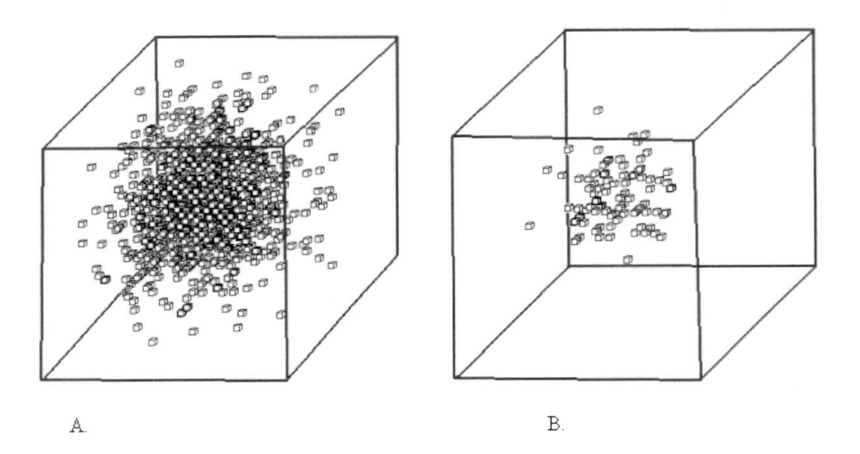

Figure 2.43. An incomplete bisector in a three-dimensional lattice with cell size $30 \times 30 \times 30$. The two sites of the set lie on the left- and right-hand faces of the cube. Sources placed on these sites generate automata with probability 0.9 at each step of development, but the overall number in the system is limited to 500: (A) the cluster of pebbles represents an incomplete bisector after 500 time steps; (B) after filtration using model M_4, all sites containing only one pebble have been cleared.

of the automata we obtain better results because the density of the distribution of the pebbles changes more smoothly. Of course, the boundary phenomena can be completely eliminated if we deform the lattice into a torus.

The transition to a three-dimensional world does not change the main rules

Figure 2.44. Digitized map of the British Isles (A) and a simplified configuration of the major cities used in computer experiments (B).

of cellular-automaton evolution. We expand the square neighbourhood to a cubic one, and adjust the probability distributions of the next steps. An example of the approximation of the bisector of two points in a three-dimensional lattice is shown in figure 2.43(A). When we assume an active underlying automaton, model M_4, which implements erosion

$$x^{t+1} = \begin{cases} 0 & \sum_{y \in u(x)} y^t = 0 \\ x^t & \text{otherwise} \end{cases}$$

we obtain filtered bisector nodes, see figure 2.43(B).

2.5 Rational geography

Let us consider one curious example which shows how discrete 'intelligent' diffusion, sometime called ant-based, algorithms are applied to a real-life problem. Here we tackle the problem of the rational administrative subdivision of the British Isles (figure 2.44(A)). We have chosen around 50 major cities (figure 2.44(B)). The problem is to assign a certain unique region to every city in such manner that all parts of the region are closer to that city than to other cities of the country. If the regions are assigned in an optimal fashion then the administrative management of the territory surrounding the city is assumed to be rational.

We make the subdivision using two techniques: reaction–diffusion approximation of a discrete Voronoi diagram ($O(n)$-algorithm) and a mobile automata, or ant-based, approximation (M_1-model with a π_8 distribution for the walk).

To start a reaction–diffusion subdivision we drop some amount of a unique reagent into every city. The reagents start to diffuse on the lattice representation of the map and form a precipitate as a result of the interaction (figure 2.45). Distribution of the precipitate in the problem space represents the boundaries between administrative regions (figure 2.46).

The problem can be solved in a similar way with mobile automata. We assume that automata generators, like ant nests, are situated at the cities. Each automaton can recognize automata from its native site and automata from other sites. Automata start moving at random from their sites and they carry pebbles. When two automata from different sites meet one another they drop their pebbles and return to their original sites. Thus, a distribution of pebbles is formed (figure 2.47). The pebble density represents the boundaries of the administrative regions of the cities (figure 2.48).

2.6 Convex hull

A planar convex hull CH(**S**) of a planar set **S** is a smaller region containing **S**, namely it is a subset of **S** that satisfies the following conditions:

> *for every point $u \in$ CH(**S**) there is at least one such point of $v \in$ CH(**S**) that all points of **S** are on the same side of line $\overline{uv}$ or belong to this line.*

Three possible groups of convex-hull computation approaches are usually explored.

The first one is a parallelization of serial algorithms, mostly divide-and-conquer algorithms [172, 171, 216] and the detection of extremal sites of a data set in a parallel (see e.g. $O(\log n)$ time algorithm in [394]). Several algorithms for finding the proximity points, which may be used in the standard approximation of a planar convex hull, are designed in [294, 332]. Parallel processors and algorithms for computation of the convex hull are discussed in [395, 354, 483, 175, 49]. Those who prefer multiprocessor systems, or coarsely grained parallel processors, may find the paper [183] quite useful.

An up-to-date review on parallel algorithms for convex hull computations mainly intended for multiprocessor architectures can be found in [128], randomized approaches are also discussed there. Approximation of a planar convex hull in parallel is discussed in [122, 129], where computation involving $O(n)$ processors is implemented in $O(\log n)$ time.

The second group deals with the iterative solution of the convex hull problem in many-dimensional cellular arrays [312].

The third group includes self-organizing multilayer unsupervised and supervised learning neural networks [167, 389].

One of the first intuitive algorithms for the computation of a *discrete* convex hull is offered in [295, 296].

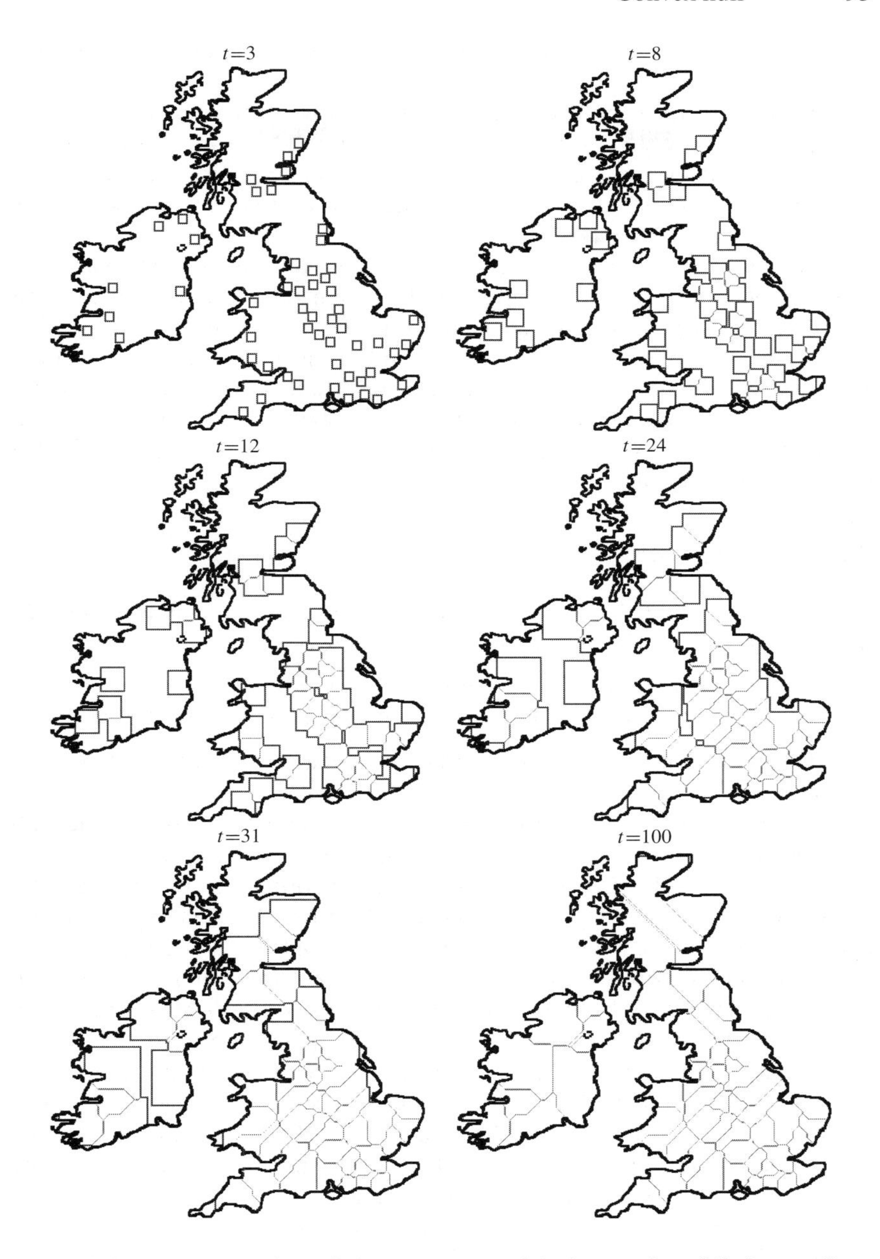

Figure 2.45. Evolution of a cellular-automata model of a reaction–diffusion medium. This reaction–diffusion processor calculates the Voronoi diagram of 51 cities by the $O(n)$-algorithm. The boundaries of the British Isles are in thick black, diffusing waves are black and the precipitate, which represents bisectors, is grey.

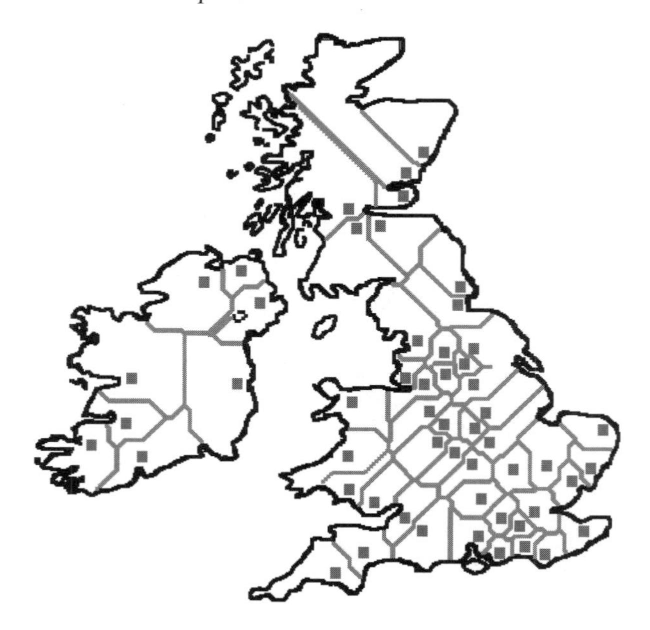

Figure 2.46. Subdivision of the British Isles computed in the cellular-automata model of a reaction–diffusion medium.

The algorithms for cellular-automata-based computation of discrete convex hulls, discussed in this section, were originally presented in [12] and [9].

Our ideas on the inductive computation of discrete convex hulls have much in common with the computation of convex hulls in neuromorphic processors [168], the approximation of a convex hull [652] and the randomized techniques for convex-hull computation [174]. Three-dimensional convex hulls have been left out of this book. However, we suppose that the possible extension into three dimensions will not be complicated. Some traditional techniques for the computation of three-dimensional convex hulls can be found in [171].

How can the convex hull be calculated in a cellular-automata processor? Let the configuration of **S** be mapped onto the processor. The elementary processors, or cells, communicate locally. Every processor containing the coordinates of some points from **S** can share its knowledge with its closest neighbours.

The first straightforward approach is based on the following condition:

> *point* $\nu \in \mathbf{S}$ *is not an element of* $\mathrm{CH}(\mathbf{S})$ *if* ν *belongs to the interior of the triangle* $\triangle abc$*, the vertices of which lie in* **S**, *and point* ν *is not a vertex of* $\triangle abc$.

This means that every processor which has the coordinates of the point ν from **S** checks whether point ν belongs to the triangle $\triangle abc$ for some $a, b, c \in \mathbf{S}$. The processor is informed about the coordinates of a, b and c by its neighbours.

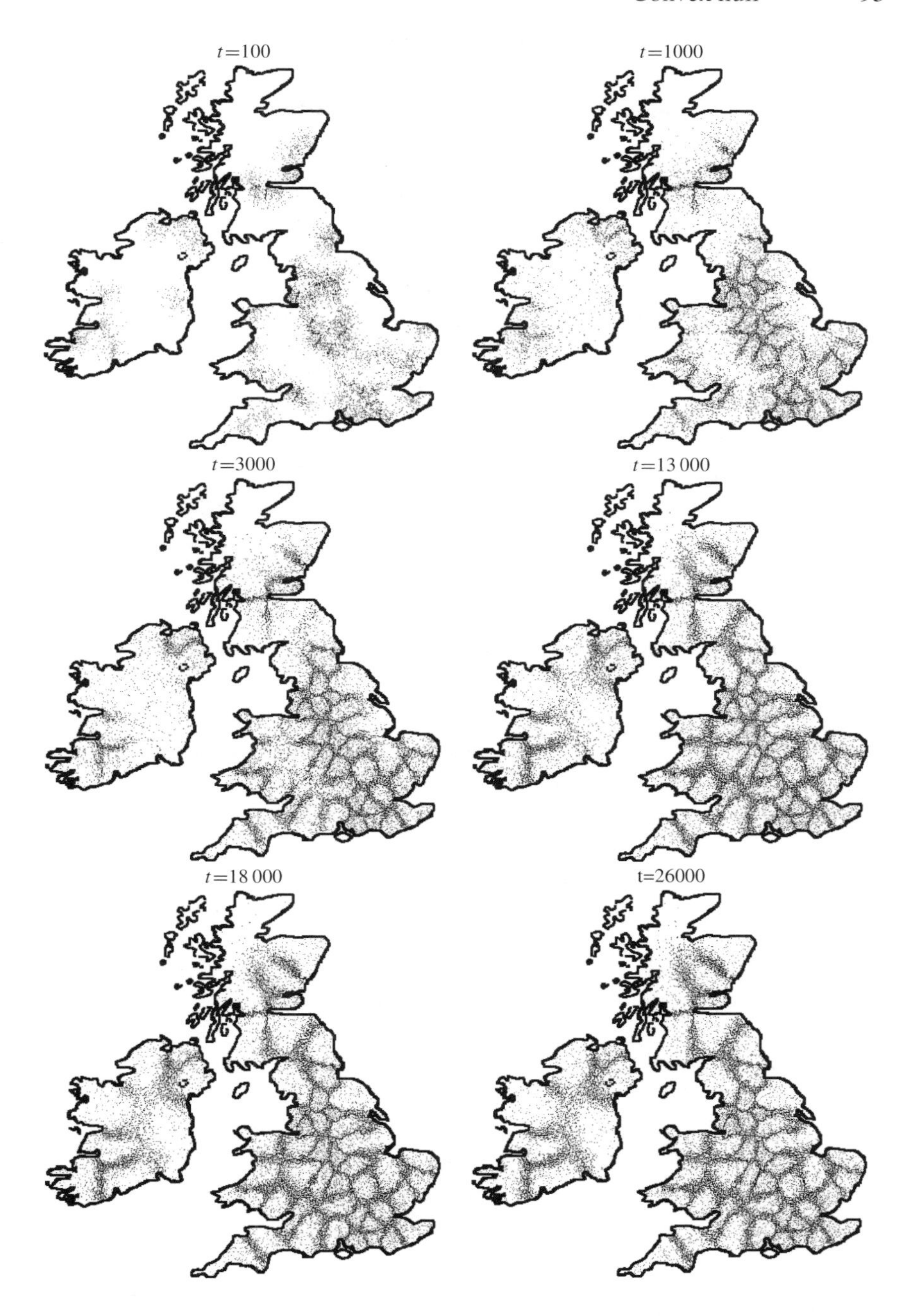

Figure 2.47. Approximation of a Voronoi diagram of British cities in a finite mobile automata model of a discrete diffusion.

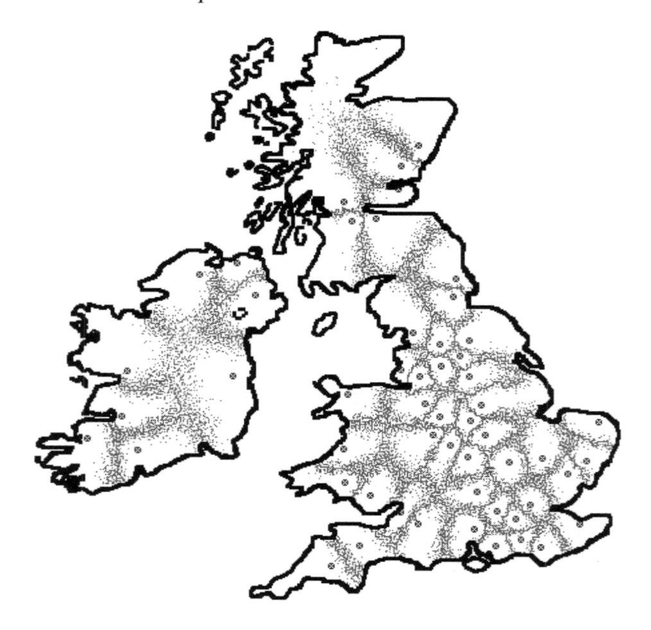

Figure 2.48. Subdivision of the British Isles computed in a finite mobile automata model of a discrete diffusion.

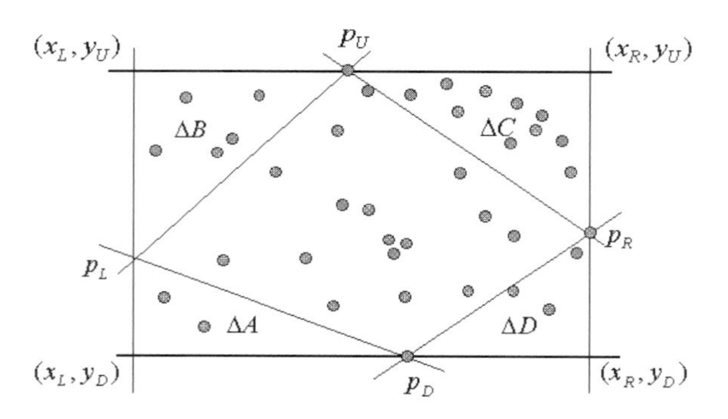

Figure 2.49.

There is, however, another approach, where set **S** is wrapped from eight points simultaneously. Let us define the so-called external subsets of **S** as

$$\mathbf{U} = \{p \in \mathbf{S} | y_p \geq y_q \forall q \in \mathbf{S}\} \qquad \mathbf{D} = \{p \in \mathbf{S} | y_p \leq y_q \forall q \in \mathbf{S}\}$$
$$\mathbf{R} = \{p \in \mathbf{S} | x_p \geq x_q \forall q \in \mathbf{S}\} \qquad \mathbf{L} = \{p \in \mathbf{S} | x_p \leq x_q \forall q \in \mathbf{S}\}.$$

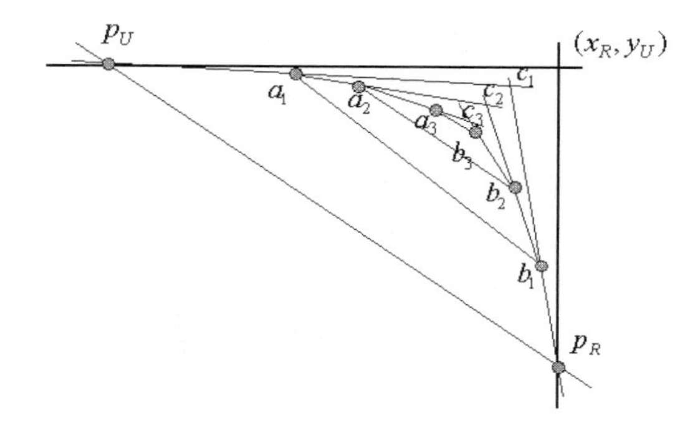

Figure 2.50.

Points of **S** are incident to the same line only if **U** = **D** or **R** = **L**. Let the external subsets be singletons

$$\mathbf{L} = \{p_L\} \qquad \mathbf{R} = \{p_R\} \qquad \mathbf{U} = \{p_U\} \qquad \mathbf{D} = \{p_D\}.$$

Let us detect the intersections of horizontal and vertical lines incident to the elements of external subsets (figure 2.49):

$$(x_L, y_U), (x_R, y_U), (y_R, y_D), (x_L, y_D)$$

and consider triangles $\triangle A$, $\triangle B$, $\triangle C$ and $\triangle D$ (figure 2.49). Another element of **S** must be searched in the interior of these triangles. Consider the computations inside the triangle $\triangle C$; the triangle contains some non-empty set of the points of **S** (figure 2.50). We assume that we have found two points, a_1 and b_1, inside $\triangle C$:

$$a_1 : x_{a_1} \min_{q \in \mathbf{S} \cup int \triangle C} \{x_q | x_q > x_U\} \qquad \text{and} \qquad y_{a_1} = \min_{q \in \mathbf{S} \cup int \triangle C} \{y_q | y_q < y_U\}$$

$$b_1 : x_{b_1} \min_{q \in \mathbf{S} \cup int \triangle C} \{x_q | x_q > x_R\} \qquad \text{and} \qquad y_{b_1} = \min_{q \in \mathbf{S} \cup int \triangle C} \{y_q | y_q < y_R\}.$$

Let us compute the intersection of the lines $(p_U a)$ and $(p_R b)$. Let it be c_1. Then we analyse triangle $\triangle C_1 = \triangle a_1 b_1 c_1$, then $\triangle C_2 = \triangle a_2 b_2 c_2$, $\triangle C_3 = \triangle a_3 b_3 c_3$ *etc* until the current triangle is empty. Starting with $\triangle A$, $\triangle B$, $\triangle C$ and $\triangle D$ we generate four monotonic sequences of nested triangles:

$$\triangle A \supset \triangle A_1 \supset \cdots \supset \triangle A_{m_1} \qquad \triangle B \supset \triangle B_1 \supset \cdots \supset \triangle B_{m_2}$$
$$\triangle C \supset \triangle C_1 \supset \cdots \supset \triangle C_{m_3} \qquad \triangle D \supset \triangle D_1 \supset \cdots \supset \triangle D_{m_4}.$$

The set **S** is finite, therefore the sequences are finite and they are generated until the final triangle has no triangle inside it. This is how things stand with the planar convex hull. Now we return to the lattice world.

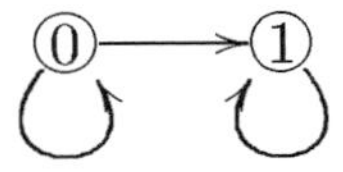

Figure 2.51. A cell state transition diagram of the cellular automaton with binary cell states. The automaton approximates the discrete convex hull.

Discrete 45-*half-planes of the integer lattice* **L** *are sublattices of* **L** *that are bounded by the lines which intersect the coordinate axes at an angle that is an integer multiple of* 45°.

These half-planes can be defined as the sets of lattice nodes x_{ij} that satisfy the inequality $a \cdot i + b \cdot j \leq c$ for $|a|, |b| \leq 1$ and integer c.

A 45-*convex hull is the intersection of all discrete* 45-*half-planes.*

We will designate the 45-convex hull by CH 45(**S**).

Given a labelled subset of integer lattice, $\mathbf{S} \subset \mathbf{L}$, we wish to find the minimal cellular automaton that evolves from a configuration representing **S** and finishes its evolution in a configuration representing CH 45(**S**).

The desired automaton must have at least two states, say $\circ$, the rest state, and $\bullet$, the non-rest state. The distribution of non-rest states should represent **S** at the beginning of evolution and CH 45(**S**) **S** at the end of the evolution. It would be better if the configuration representing CH 45(**S**) be a fixed point in the evolution, i.e. the process of computation is eventually stopped.

Given $\mathbf{S} \subset \mathbf{L}$, find the cellular-automata function f which, for some $\tau < \infty$, satisfies the following conditions:

- $\forall x \in \mathbf{L}$ takes place $x^0 = \bullet \Leftrightarrow x \in \mathbf{S}$.
- $\forall x \in \mathbf{L}$ takes place $x^\tau = \bullet \Leftrightarrow x \in \mathrm{CH}\,45(\mathbf{S})$.

In the two next sections we consider the cellular automata that approximate a discrete convex hull to some degree of accuracy. Both automata have binary cell states and the cell state transition diagram shown in figure 2.51. The state 1 is an absorbing state in the evolution of the cell. The transition $0 \to 0$ happens only if the condition for the transition $0 \to 1$ is not satisfied. Therefore we will mainly concentrate on the condition for the transition $0 \to 1$.

2.6.1 Topology-based algorithm

One of the first cellular-automata-based algorithms for computation of a discrete convex hull was offered in [295]. The cell state transition function is defined on the elements used to structure the iterative computation of the 45-convex hull of a four-connected binary image. Every cell has eight neighbours, takes two states,

```
1 1 1   1 1 0   1 0 0   0 0 0   0 0 0   0 0 1   0 1 1   1 1 1
1 0 0   1 0 0   1 0 0   1 0 0   0 0 1   0 0 1   0 0 1   0 0 1
0 0 0   1 0 0   1 1 0   1 1 1   1 1 1   0 1 1   0 0 1   0 0 0
```

Figure 2.52.

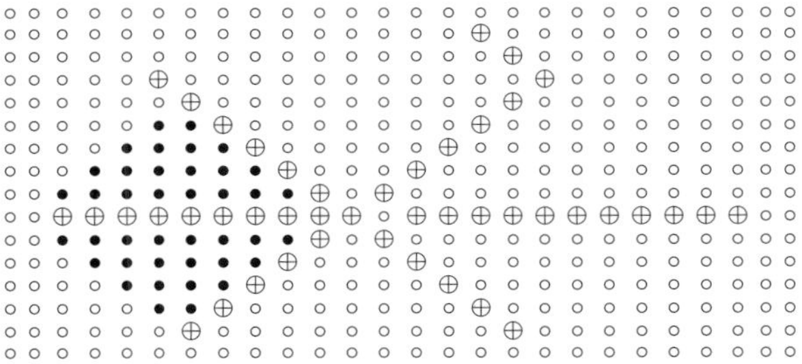

Figure 2.53. An example of a convex hull set which is not a constructible topology based algorithm. Symbols $\circ$ and $\bullet$ represent the usual states of the cell whereas $\oplus$ represents the nodes of a given set **S**.

say 0 and 1, and updates its states by the following rule:

$$x^{t+1} = \begin{cases} 1 & u(x)^t \in \mathbf{U} \vee x^t = 1 \\ x^t & \text{otherwise} \end{cases}$$

where the elements of **U**, the cell neighbourhood states, are represented in figure 2.52.

In figure 2.52 we can see that these structuring elements are designed in such a manner that the state 1 simply fills the lacunae inside the corners of the binary image. This results in a convex hull pattern.

Sometimes the automaton fails to complete the construction of the convex hull when given set **S** has some specific structure, as, e.g., in figure 2.53.

2.6.2 Two-state algorithm

Let the nodes of a given set **S** be marked by 1s and the other cells of the cellular automaton lattice **L** by 0s. Imagine that we draw a picture on ruled paper: the cell filled with black colour (black pixel) is assumed to have 1s and the blank cell 0s. In such a manner the configuration of **S** is mapped onto the initial configuration of a cellular automaton. In the evolution of the automaton all cells corresponding to a node of CH 45(**S**) have to pass into 1s within a finite time interval. As usual we choose an eight-cell neighbourhood. The minimal neighbourhood determines the unit radius of interaction in the metric L_∞. At the same time the nodes of the set **S** may be spread over the lattice and the set **S** may have quite a large diameter.

```
1 0 0   1 0 0   0 0 0   0 0 1   0 0 1   0 0 0   1 1 1
1 0 0   1 0 0   1 0 0   0 0 1   0 0 1   0 0 1   0 0 0
1 0 0   0 0 0   1 0 0   0 0 1   0 0 0   0 0 1   0 0 0

1 1 0   0 0 0   0 0 0   0 0 0   0 1 1   0 0 0   1 1 0
0 0 0   0 0 0   0 0 0   0 0 0   0 0 1   1 0 0   1 0 0
0 0 0   1 1 1   1 1 0   0 1 1   0 0 0   1 1 0   0 0 0

0 0 0   1 0 0   0 1 0   0 0 1   0 0 0   0 0 0   0 0 0
0 0 1   0 0 0   0 0 0   0 0 0   0 0 1   0 0 0   0 0 0
0 1 1   0 0 0   0 0 0   0 0 0   0 0 0   0 0 1   0 1 0

        0 0 0   0 0 0   1 0 0   0 0 0
        0 0 0   1 0 0   0 0 0   0 0 0
        1 0 0   0 0 0   0 0 0   0 0 0
```

Figure 2.54. The states of a neighbourhood for which a central cell (of the neighbourhood) in the state 0 does not take the state 1 at the next step of the evolution.

How do we switch all these rest cells inside the convex hull into the non-rest state? Possibly, by using waves of 1-states. The 1-state waves are generated in one or more cells inside the convex hull, spread in all directions and halt their growth in the boundaries of CH 45(**S**).

> *Let* **L** *be an integer lattice of $n \times n$ nodes and* **S** *be a connected subset of the nodes of* **L**, $|\mathbf{S}| > 3$, *such that there exists at least one node of* **S** *at least four neighbours of which are members of* **S**. *Then the cellular automaton with $n \times n$ cells computes nodes of* CH 45(**S**) *in $O(n)$ time.*

The main principle of the local evolution states: 'if you are a 0-state cell and have your neighbourhood state as in figure 2.54—do not change your state onto 1'. Why is this so? Because these combinations of 1-states near to the 0-state central cell are typical for the boundaries of the convex hull, i.e. 0-state cells have a neighbourhood such as the one in figure 2.54 are close to the chains of 1-state cells 'intersecting' the positive axis at angles 45°, 90° etc. Thus, every cell $x \in \mathbf{L}$ switches from 0-state to 1-state if and only if the arithmetical sum over the states of its neighbours ($u(x)$ includes x) exceeds three:

$$\forall x \in \mathbf{L} : x^0 = 1 \qquad \text{if } x \in \mathbf{S}$$

$$x^{t+1} = \begin{cases} 1 & \text{if} \left(\sum_{y \in u(x)} y^t > 3 \right) \\ x^t & \text{otherwise.} \end{cases}$$

After generation, the 1-state waves run towards the distant elements of **S** and stop when the boundaries of the convex hull have been shaped. The 1-state

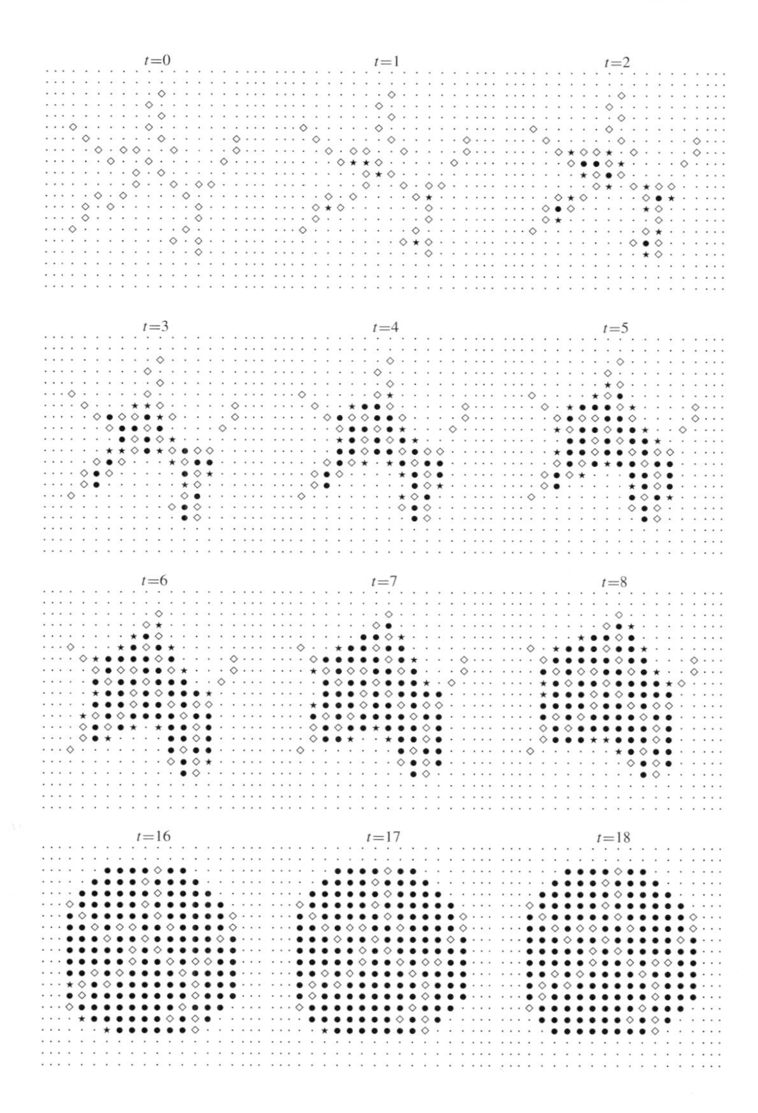

Figure 2.55. Evolution of a cellular automaton that constructs CH 45(**S**). Cells in 0-state are labelled by dots, the cells in 1-state are marked by $\diamond$, $\bullet$ and $\star$ symbols. A cell that has just changed its state from 0 to 1 is marked by $\star$ symbol, cells in the 1-state that represent an element of **S** are marked by $\diamond$, all other 1-state cells are $\bullet$s.

waves reach the extreme elements of **S** in a time proportional to diameter $d(\mathbf{S})$ of the set **S**. Therefore we have an $O(n)$ upper boundary on the complexity of the algorithm. An example of the construction of a convex hull in such a cellular

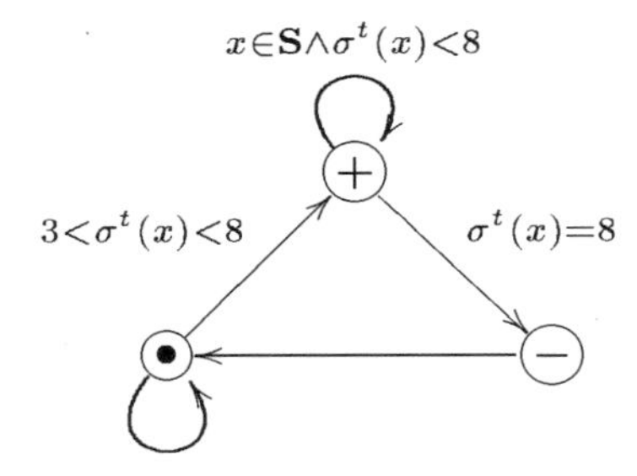

Figure 2.56. A cell state transition diagram of the excitable lattice which approximates a discrete convex hull. External inputs are not considered in the diagram. Arcs corresponding to the transitions which happen 'otherwise' or unconditionally are not labelled.

automaton is shown in figure 2.55. At time $t = 18$ the automaton reaches the fixed point of its evolution—an entirely stationary configuration that represents the discrete convex hull CH 45(**S**).

2.6.3 Excitable computation

In excitable lattices, without a precipitate state, cells do not usually keep their state unconditionally. Therefore we must excite the cells corresponding to the elements of a given set **S** externally. So the medium should have external inputs.

Let M be a homogeneous excitable lattice with external inputs. The cells of M take three states: the rest state, •; the excited state, +; and the refractory state −; and have the standard eight-cell neighbourhood. Any rest cell of the medium becomes excited when at least four but no more than seven neighbours are excited. The cell remains in the excited state if the sum of excited neighbours is less than eight. The cell switches to the refractory state otherwise. The cell recovers from the refractory state to the rest state unconditionally. Cells corresponding to the elements of **S** are excited externally at every step of the evolution. Thus, every cell of M changes its state by the following rule:

$$x^{t+1} = \begin{cases} + & ((x^t = \bullet) \wedge (3 < \sigma^t(x) < 8)) \vee ((x^t = +) \wedge (\sigma^t(x) < 8)) \\ & \quad \vee (x \in \mathbf{S}) \\ - & (x^t = +) \wedge (\sigma^t(x) = 8) \\ \bullet & \text{otherwise} \end{cases}$$

where

$$\sigma^t(x) = \sum_{y \in u(x)} \chi(y^t, +)$$

and $\chi(y^t, +) = 1$ if $y^t = +$ and $\chi(y^t, +) = 0$, otherwise.

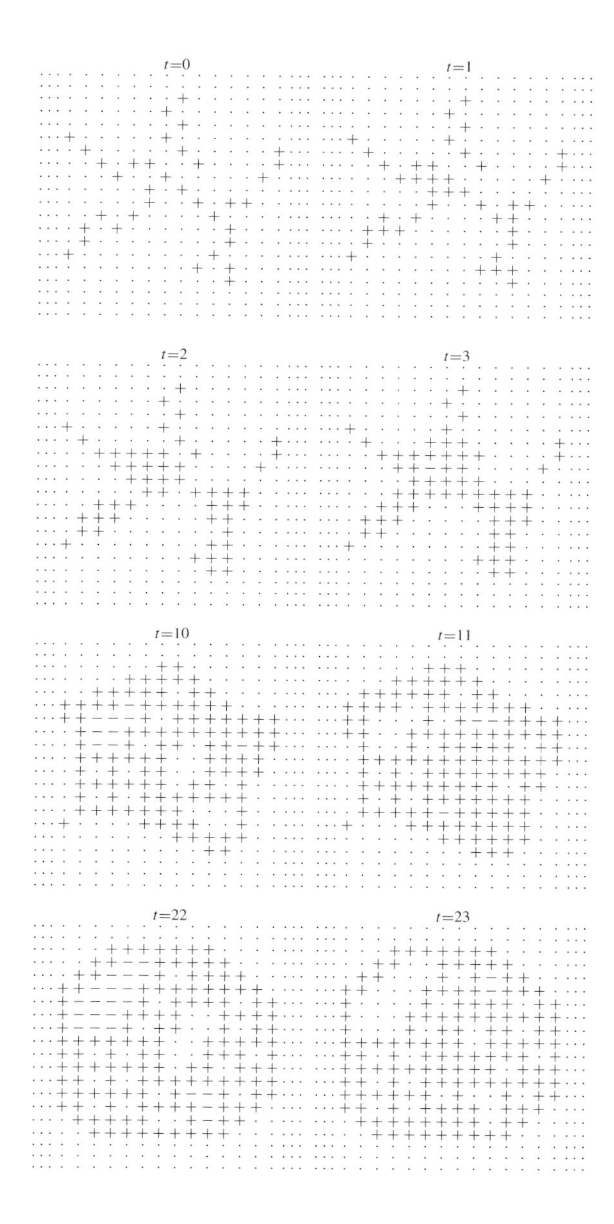

Figure 2.57. Evolution of a two-dimensional excitable lattice which approximates the discrete convex hull of a discrete connected set. Excited cells are shown by +, refractory by −; other states are ·.

This model is an analogue of a homogeneous neural network where there are two possibilities for a neuron to be locally excited (figure 2.56):

- externally, if the neuron has the same coordinates as one of the elements of **S**;
- locally internally, i.e. by its neighbours, if the number of excited neighbours belongs to the interval $[4, \ldots, 7]$.

The neuron passes to the refractory state if all of its eight neighbours are excited.

At the beginning of the computation only the cells representing the nodes of **S** are excited; all other cells are at rest. The cells representing **S** are the generators of the excitation waves. The excitation waves spread on the lattice, interact with each other and create new waves. However, they never leave the boundary of the convex hull. Because of the specifics of the transitions between the excited and refractory states the model M never falls into the stationary configuration but exhibits periodic activity instead. So, the computation will never be finished in a classical sense. However, the problem is decidable in the sense of inductive Turing machines. We say the cell x represents the element of $\mathrm{CH}\,45(\mathbf{S})$ if x becomes excited at least once during the evolution. An example of the evolution is shown in figure 2.57.

> *Let* **L** *be an integer lattice of $n \times n$ nodes and* **S** *be such a connected subset of the nodes of* **L** *that for at least one node $y \in$* **S** *at least four and at most seven closest neighbours are elements of* **S**. *Then there exists a homogeneous excitable medium of $n \times n$ cells (each one of which has eight closest neighbours, external inputs and takes three states) that compute* $\mathrm{CH}\,45(\mathbf{S})$ *in $O(n)$ time.*

Chapter 3

Computation on and with graphs

Diffusion, random walks, travelling waves and electrical flows have much in common. They are employed in this chapter to solve various problems of graph construction and optimization on graphs, particularly the shortest-path problem. Generally, the spread of and competition between disturbances in nonlinear media for a skeleton of all algorithms discussed here, although particular realizations vary from automata models of excitable media to morphogenetic computing by neurites to collective problem-solving in ant families. The text is spiced up by a game application, a practical problem of load balancing in communication networks and recipes for real chemical processors.

3.1 Shortest paths and trees

Given a path with weighted edges, we wish to find a path from one vertex to another through the minimal number of edges with minimal sum of their weights.

The shortest-path problem, which has developed substantially since Dijkstra's pioneering work [187], dominates this chapter.

Dealing with graphs we assume a given graph to be oriented (there are two different edges from vertex x to vertex y and from y to x), and therefore it is useful to define the *source* from which we may start our path and the *destination* towards which we run and where we stop. The shortest-path problem has three well-known versions.

- A single-source shortest path (S^3P). We wish to compute the shortest path from a given vertex of a graph to all others; there is a single source and many destinations.
- All-pairs shortest path (APSP). We wish to find the shortest paths between all pairs of graph vertices.
- A single-source single-destination shortest path (S^3DSP). We wish to compute the shortest path between two given vertices of the graph.

The S^3P algorithm in Dijkstra's version runs in $O(n)$ time when every edge of a given graph has a non-negative weight. Effective S^3P algorithms exist for the sparse graphs which work in $O(\min\{n^{1+1/k}+m, m \log n\})$ [334], $O(m \log \log D)$ [343] and $O(n \log n + m)$ [244] upper times, where the input graph has n vertices and m edges, and D is the maximal length of the shortest path between two vertices; k is a non-negative integer. Using Floyd's algorithm [236] it is possible to solve an APSP problem in $O(n^2 \log n)$ and $O(n^2)$ upper times (see [464] and [486], respectively); the last bound is obtained when a shortest path is computed in a mesh, the edges of which have constant weights.

One of the most straightforward methods to compute a shortest path in parallel is to do it by solving matrix equations (with elements from the so-called path algebra) in systolic arrays (see e.g. [317]). Thus the computation of an APSP on $O(n)$ processors takes $O(\log n)$ time [513]. It has been demonstrated in [242] that the shortest-path tree can be found in a distributed network of n vertices using $O(n^2)$ messages; the boundary is reduced to $O(n^{5/3})$ when the given network is planar [242]. Dijkstra's serial algorithm for S^3P is parallellized in [623]: the algorithm is implemented in a network of finite automata; a network of $O(n)$ processing elements finds a solution of the S^3P problem for the class of weighted graphs of bounded degree and size n in $O(n)$ time. Early results on graph problems solved on processor arrays can be found in [55].

To demonstrate the potential usefulness of a reaction–diffusion algorithm for optimization on graphs, we have chosen the problem of computing a minimal spanning tree, which belongs to a family of proximity graphs [598, 597].

A minimal spanning tree of a planar point set is defined as an acyclic undirected graph, formed from straight-line edges connecting the points of the set, for which the sum of the lengths of the edges is minimal.

A Steiner tree is a minimal spanning tree where the edges can intersect anywhere in the plane except the points of a given set. The basic idea of using distributed competitive algorithms for the computation of planar minimum spanning trees was introduced in [4], where the spanning tree is constructed using a model inspired by the growth of neuron terminals. This is quite similar to the technique using artificial amoeba or the vibrating potential field composed of a potential function and a wavefunction, elaborated in [659] and applied successfully to the design of a simulated flexible robot avoiding obstacles [661] and the solution of the n-salesmen–m-cities problem [655].

Here we construct a minimum spanning tree, a relative neighbourhood graph and other proximity graphs. Let us define these structures.

The minimum spanning tree of the planar set $\mathbf{V} \in \mathbf{R}^2$, $|\mathbf{V}| = n$, is the connection of elements of $\mathbf{V}$ by straight lines such that

- *there is a path of connected lines between every pair in the set*
- *line segments connecting points do not intersect anywhere in the plane except at points of $\mathbf{V}$*

- *the graph is acyclic*
- *the sum of lengths (or weights) of all edges is minimal.*

The edges of the minimum spanning tree of a given planar set form a set MST(**V**). If we omit the last condition in the previous definition we get the definition of a simple spanning tree. In some situations, where the minimality of the constructed graph is not obvious, we talk about spanning trees not minimum spanning trees.

> *The relative neighbourhood graph of a planar set* **V** *is the graph* RNG(**V**) *such that any two points* p *and* q *from* **V** *are connected by an edge of the graph,* $(pq) \in$ RNG(**V**), *if and only if the intersection of spheres centred at points* p *and* q *with radii equal to the distance between* p *and* q *does not contain any points of* **V**. *If* $B(x, r)$ *is an open sphere with its centre at* x *and radius* r*, then* $(pq) \in \mathbf{V}$ *iff* $B(p, d(p, q)) \cap B(q, d(p, q)) \cap \mathbf{V} = \emptyset$. [330]

The Steiner tree of a planar set **V** is defined in the same way as the minimum spanning tree except that the line segments connecting points of **V** may intersect anywhere in the plane except at points of **V**.

Thousands of papers about spanning trees and other proximity graphs have been published. We briefly mention a tiny fraction of them here.

The original algorithms for computing minimum spanning trees are described in [370, 509, 187, 271]. Distributed algorithms for finding the edges of the minimum spanning tree run in $O(n \log n)$ time on $O(n)$ processors, see e.g. [251]. Ahuja and Zhu [40] have presented a distributed technique, based on the echo algorithm, for the computation of the minimum spanning tree in a general undirected graph. It runs in $O(2d \log n)$ time, where d is the diameter of the graph (maximum of all the distances between vertices), with $O(2m+2(n-1)\log(n/2))$ messages (information exchange) between the processors in the worst case, and in $O(2d)$ time with $O(2m)$ messages in the best case, where m is the number of edges of the data graph. Based on Tarjan's on-line approach, Huang [316] proposed a systolic algorithm with complexity $O(m + n)$, where the edges of the graph are fed sequentially into the array in an arbitrary order in each systolic cycle, and the edges of the spanning tree are taken from the array at one per cycle.

A detailed description of the algorithm for the computation of a relative neighbourhood graph can be found in [331]. In 1980 Toussaint [596] proved that

$$\mathrm{MST}(\mathbf{V}) \subset \mathrm{RNG}(\mathbf{V}) \subset \mathrm{DT}(\mathbf{V})$$

where DT stays for Delauné triangulation.

The most typical approach to the construction of a minimal spanning tree is first to compute the relative neighbourhood graph and then to delete some edges of the graph to transform it into a minimal spanning tree [331, 587]. Some ideas for the massively parallel computation of a relative neighbourhood graph (and also of other special graphs like the Gabriel graph and the influence graph) can be found in [6]. Exact and approximate algorithms for Steiner tree computation can also be found in [448, 138].

3.2 Random walk, electricity and tree approximation

The connection between random walks and electrical flows has a long history; see [196] for a detailed overview. The principal results may be roughly stated as follows.

> *Trees are approximated by random walks. Random walks are similar to electrical flows.*

Let $\mathbf{G} = \langle \mathbf{V}, \mathbf{E} \rangle$ be a finite graph with vertices from $\mathbf{V}$ and edges from $\mathbf{E}$. If we start a random walk at x_0 and absorb it at $\mathbf{Z} \subset \mathbf{V}$, and S_{xy} is the number of transitions from x to y then

$$\mathcal{E}[S_{xy} - S_{yx}] = I_{xy}$$

where I is the current when a potential is applied between x_0 and $\mathbf{Z}$ [404]. Moreover, if $\mathbf{G}$ is a connected graph and V is a voltage function derived from the unit current flow from x_0 to ∞ then for any x there is a path of vertices from x_0 to x along which the function V is monotonic [404].

A probabilistic interpretation of an electrical current is clear [404]:

- particles enter the graph at x_0
- they do Brownian motion
- they spend more time on the edges with smaller conductances
- they are removed when they hit any vertex from the set $\mathbf{Z}$ of destination vertices
- the net flow of particles along an edge is the current on that edge.

Let weights be equivalent to conductances (or resistances), and the batteries be connected to $\mathbf{S}$ and $\mathbf{Z}$, $\mathbf{S}, \mathbf{Z} \subset \mathbf{V}$, in such a manner that the voltages (potentials) are zero at $\mathbf{S}$ and one at $\mathbf{Z}$. Then the current flows along the edges. Ohm's law states that, for any edge $(x, y) \in \mathbf{E}$, the following relation between current and resistances holds:

$$I_{xy} = \frac{V_x - V_y}{R_{xy}}$$

where I_{xy} is the flow along (x, y) and R_{xy} is the resistance of the edge. By Kirchhoff's node law we have

$$V_x = \frac{1}{\deg(x)} \sum_{y:(x,y)\in\mathbf{E}} C_{xy} V_y$$

where $\deg(x)$ is the degree of the vertex x; that is, the voltage function represents a solution of the Dirichlet problem [404].

Moreover, Kirchhoff's cycle law also holds. If path $x_1, x_2, \ldots, x_n, x_1$ is a cycle then

$$\sum_{i=1}^{n} I_{x_i x_{i+1}} R_{x_i x_{i+1}} = 0.$$

In words we can say that if a voltage is established between vertices x_0 and x_1 such that it is zero at x_0 and one at x_1 then the voltage at vertex x equals the probability that a random walk visits x_1 before it visits x_0 when it starts at x.

Let us look now at a well-known algorithm for the generation of random trees as represented in [404]. The algorithm is similar to a probabilistic technique known almost a century ago—the Bienamé–Galton–Watson construction of random trees. Given a set of probabilities we begin with one individual node and let it reproduce itself, making connections with neighbouring nodes, according to the given probabilities; we then let each such child reproduce itself in the same way etc. Lyons and Peres [404] discuss how to choose one of the random trees from the numerous trees of the given graph by using some of the properties of Markov chains. In the case of a directed graph, a spanning tree is a subgraph that includes any vertex of **G** and has a root vertex x_0 such that every other vertex is a tail of exactly one edge in the tree. To choose the tree we pick some vertex x_0, and another vertex y, and draw the path from y to x_0 (it exists by irreducibility); then pick another vertex z not contained in the path from y to x_0 and draw the path from z to x_0 and so on. We continue until all vertices have been used. The path can be chosen probabilistically according to the weights of the edges. If all weights are the same the approximated tree will be chosen with uniform probability. If the weights are different and if at every step we choose an edge with the locally minimal weight, the minimum tree will be approximated.

A basic result noted by Aldous [43] and Broder [106] states that

> *a random walk on a finite connected graph constructs the uniform random spanning tree.*

This is because a random walk on **G** is seen as the discrete-time Markov chain with transition probability matrix P calculated as follows: $P_{xy} = \deg(x)^{-1}$ if $(x, y) \in \mathbf{E}$, otherwise it is zero.

These ideas provide the basis for the algorithms for the approximation of spanning trees during neuron morphogenesis described in [4], as described in section 3.5.

3.3 Field computing on graphs

The main idea of field computing on graphs and network lies in applying a voltage to a graph (the edges and nodes are assumed to have a certain resistance) and measuring the resistance or capacities of the networks. This technique has been used, at least implicitly, from the beginning of the 20th century or even earlier

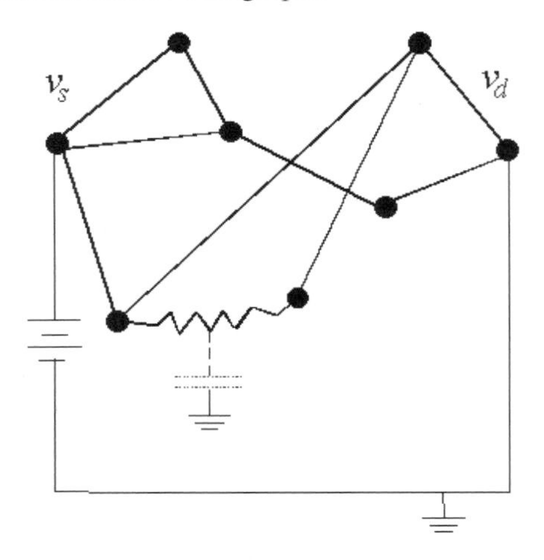

Figure 3.1. An electrical machine that computes the shortest path. From [624].

but the earliest publication which emphasizes the algorithmic part is the paper by Vergis *et al* [624]. This paper discusses the complexity of an analog computation and considers, in this context, the first toy example of an analog computer that solves the graph problem with polynomial resources.

To solve the well-known (s, t)-connectivity problem we construct an electrical model of the given graph (figure 3.1) that works as follows:

- replace the edges with wires and connect them at the nodes;
- apply a voltage between the nodes v_s and v_d;
- measure the current;
- assuming that the resistance is proportional only to the length of a wire or edge we encounter a near null current, that reflects a very high resistance, if there is no path between v_s and v_d.

Vergis *et al* [624] proposed that (if the lengths of wires increase linearly with the number of graph nodes) the total capacity of the voltage source and total resistance have the upper bound $O(|\mathbf{E}^2|)$; this leads to the total size and power consumption $O(|\mathbf{E}^4|)$; that is, the analog computer operates polynomial resources [624].

An electrical computation of a shortest path has been implemented practically in silicon by Marshall and Tarassenko [431] to solve the problem of path planning in robot navigation. In their prototype a navigation space is represented by a rectangular network of locally connected resistors where the obstacles are the nodes with high value resistors and the space free of obstacles is represented in the nodes with minimum values of the resistors. The destination-node is grounded and a positive voltage is applied to a source-node. The path is detected from the direction of maximum current flow at a current node of the grid. Experiments using a VLSI chip representing the resistive grid proved the viability of the algorithm. Based on the experimentally observed settling time (0.35 μs) of the silicon resistive grid, Marshall and Tarassenko [431] assume a 10^4 increase in speed in a resistive grid chip of 400 nodes comparing to software implementations of robot navigation algorithms.

A systematic analysis of field-routing in telecommunication networks can be found in [132]. Based on techniques for robot navigation planning using fields (see e.g. [108]), Chong [132] studies the applicability of a resistive network for optimal routing in a packet-switched network. In general, the resistive network scheme is similar to that described in [624] except that additional pins and the corresponding wires are added to each node of the network. The obvious advantages of field-routing is that the resistive network propagates traffic information to all nodes of the network. The ability to avoid high communication cost is the main point in the efficiency evaluation of field-routing. However, encoding the analog and digital data in the routing process may add some extra costs. The paper [132] suggests that the possibility of implementing field-routing in digital networks is quite real: each node periodically sends messages/packets in all directions to update information about the field source at this node. As soon as electricity is related to the random walk (the random walk itself is a key process in diffusion), we can easily apply most of the previous results to reaction–diffusion computing.

Before investigating automata models for the reaction–diffusion solution of the shortest-path problem we would like to mention several relevant results on the analog solution of the shortest-path problem:

- the solution in a network, isomorphic to a data graph, of variable turn-on-voltage diodes [112];
- digital path planning circuits, which encode distances by hardware time delays [466];
- routing in a lattice of continuous state processors [362];
- the shortest path on a resistive network [340];
- determination of the shortest path in a two-dimensional array of a cellular neural network, i.e. resistively coupled Chua circuits, using the propagation of excitation waves [496];
- path planning in a terrain using the potential field approach [504];
- cellular-automata algorithms of shortest-path computation based on the diffusion of abstract substances from a target site [555, 560, 621].

Algorithms implemented in recurrent neural networks, where the weights of a given graph are represented via the spacetime distribution of time delays [440], and in networks of spiking neurons [549], should also be highlighted in this context.

A liquid metal robot, designed by Ishida *et al* [320], is an impressive example of a real-life implementation of the field-routing approach. The liquid mobile robot, which has no structure at all and exhibits easily changeable shapes, is based on the paradigm of artificial amoeba, developed by Yokoi and Kakazu

Figure 3.2. Two snapshots of a liquid metal robot. Published with kind permission of Hiroshi Yokoi.

[660]. Each robot is a mercury drop. The robot's environment is an array of electrodes, connected to an external controller. By varying the potential pattern of the electrode array we can guide the mercury drop around the experimental arena, cause the drop to split into several daughter drops, the fusion of drops and other types of possible robot behaviour (figure 3.2). This design is consistent with the behaviour of real amoeba, e.g. *Amoeba proteus*, the behaviour of which shows signs of a positive galvanotaxis [361].

3.4 Computation of shortest paths in cellular automata

Let us supply a cellular automaton with a 'pointer' and a finite vector. Actually, both elements may be expressed via the enrichment of a set of cell states but this might lead to unnecessary complication. Therefore we update the vector and pointer states separately.

We consider a cellular automaton on a d-dimensional integer lattice $\mathbf{L}$ of n cells (nodes). Every cell x of $\mathbf{L}$ is connected to the closest cells (which form a neighbourhood $u(x)$ of the cell x), takes states from a finite non-empty set $\mathbf{Q}$ and has a 'pointer' p_x, which takes its states from a finite non-empty set $\mathbf{Y}$; the cell also has a vector w_x. The vector w_x represents the weights of the input edges of a graph vertex which are mapped to x. The sets $\mathbf{Q}$ and $\mathbf{Y}$ have the following structures:

$$\mathbf{Q} = \{+, \#, \circ, 0, 1, 2, \dots, \nu\} \qquad \text{and} \qquad \mathbf{Y} = \{1, 2, \dots, k, \lambda\}.$$

Every entry w_{xy} of the vector w_x, $y \in u(x)$, takes one of the following states: $\infty, 0, \dots, \nu$. We write $w_{xy} = \infty$ when cells x and y are not connected to one another by an edge oriented from y toward x.

Let $G = \langle \mathbf{V}, \mathbf{E} \rangle$ be an oriented graph of n vertices arranged on the d-dimensional lattice. Every vertex $v \in \mathbf{V}$ is connected to at most k neighbouring vertices $v_1, v_2, \dots, v_k$ by input edges $(v_1 v), (v_2 v), \dots, (v_k v) \in \mathbf{E}$. These edges have weights $w(v_1 v), w(v_2 v), \dots, w(v_k v) \in \{0, 1, \dots, \nu, \infty\}$. If there is no edge $(v'v'')$ connecting the vertices v' and v'' we write $w(v'v'') = \infty$. To be successfully mapped onto cellular-automaton lattice **L** the graph G must satisfy the rule: $k \ll n$, i.e. the degree k of any vertex is much lower than the number n of vertices in the graph. Let $p = (v_0, \ldots, v_m)$ be a path from the vertex v_0 to the vertex v_m in the graph G and $l(p)$ be the length of the path p. Then $l(p) = \min\{v_0, \ldots, v_m\}$.

Let us map the graph G onto the cellular automaton in a way that preserves the topology, i.e. if the vertex v is mapped onto cell x then all the vertices $v_1, \ldots, v_k$ connected to v by the edges of **E** are mapped to the cells $y_1, \ldots, y_k$; the cells $y_1, \ldots, y_k$ form the neighbourhood $u(x)$ of the cell x. The weights of the input edges of the vertex v are collected in the vector

$$w_x = (w_{xy_1}, \ldots, w_{xy_k}) = (w_{x1}, \ldots, w_{xk})$$

where $w_{xi} = w(v_i, v)$, and v_i corresponds to the cell $y_i \in u(x)$.

Let cells x_s and x_d be the source and destination cells in the cellular-automaton representation of the data graph, corresponding to the source and destination vertices of the required path on the graph G. At first, we discuss the single-source single-destination shortest-path (S^3DSP) problem. After that we show that this technique can easily be applied to the problems of single-source shortest-path (S^3P) and all-pairs shortest-path (APSP).

At the beginning of the computation of S^3DSP we assume that $x_s^0 = +$. The computation must stop when the cell x_d takes the state # or when every cell of the lattice **L** takes either the state $\circ$ or the state #; the second constraint may be used to stop the computation process when there is no path between the source and destination cells.

The cellular automata considered in this section simulate some type of excitable media. Therefore we call $+$ an excitation state (cell x is excited when it takes the state $+$) and $-$ a refractory state; cell x switches to state when it takes this state $-$. Following the paradigm we allow excitation waves, represented by states $+$, to run towards the source cell x_s. Ideally, the excitation wave travels along the lattice until it reaches the destination cell x_d. After that the computation is assumed to be finished. When the travelling excitation pattern reaches cell x the pointer p_x changes its initial state to one of the states from $\{1, 2, \ldots, k\}$ in order to point to the neighbour of cell x that excited this cell, i.e. the neighbour y which transferred excitation state $+$ to the cell x. At the end of the computation the required path can be easily extracted from the pointer states, namely, by back-tracking over pointers from cell x_d to cell x_s.

How does cell x update its states? At time step t let cell x be in a state $x^t = \circ$ (this is a rest-like state) and some of its neighbours be in the state $+$. The cell x finds neighbour y such that

$$y^t = + \text{ and } w_{xy} = \min\{w_{xy'} : y' \in u(x) \text{ and } y'^t = + \text{ and } w_{xy'} \neq \infty\}.$$

Then cell x takes the state w_{xy}. Starting in state w_{xy}, which is a positive integer, cell x decreases its state at every step of discrete time, $x^{t+1} = x^t - 1$, until it takes state 0, or encounters neighbour y such that

$$y^t = + \text{ and } w_{xy} = \min\{w_{xy'} : y' \in u(x) \text{ and } y'^t = + \text{ and } w^t_{xy'} < x^t\}.$$

The cell transitions from state 0 to state + and from state + to state # happen unconditionally, i.e. without any influence from the cell neighbourhood.

At the beginning of the computation the pointers of all cells take the state λ. The pointer of a cell changes only when some of the cell neighbours are excited. The pointer state becomes constant and can no longer be modified after the cell takes state 0.

We say the computation of S^{3}P is completed at the time step t if $\forall x \in \mathbf{L}$: $x^t \in \{\#, \circ\}$, i.e. if no cell is excited.

Now we are ready to discuss cell state transition rules. Every cell x of the cellular automaton updates its state x^t to the next state x^{t+1} by the following rule:

$$x^{t+1} = \begin{cases} \# & \text{if } x^t \in \{\#, +\} \\ + & \text{if } x^t = 0 \\ \circ & \text{if } x^t = \circ \text{ and } \forall y \in u(x) : y^t \neq + \\ w_{xy} & \text{if } (x^t = \circ \text{ or } x^t > 0) \text{ and } \exists y \in u(x) : y^t = + \text{ and } w_{xy} \neq \infty \\ & \text{and } w_{xy} < x^t \text{ and } w_{xy} = \min\{w_{xy'} : y' \in u(x) \text{ and } y'^t = +\} \\ x^t - 1 & \text{if } x^t > 0 \text{ and } \forall y \in u(x) : y^t \neq + \text{ or } w_{xy} < x^t. \end{cases}$$

At the same time the pointer p_x of the cell x changes its state by the following rule:

$$p^{t+1} = \begin{cases} y & \text{if } \exists y \in u(x) : y^t = + \\ & \text{and } w_{xy} = \min\{w_{xy'} : y' \in u(x) \text{ and } y'^t = +\} \\ & \text{and } (0 < w_{xy} < x^t \text{ or } x^t = \circ) \\ p^t_x & \text{otherwise.} \end{cases}$$

Recall that $x^0_s = +$ and $\forall x \in \mathbf{L}, x \neq x_s$: $x^0 = \circ$ and $p^0_x = \lambda$. Moreover, we know that state # is an absorbing state and that the cell state transitions $+ \to \#$, $\# \to \#$ and $0 \to +$ are unconditional.

Let us consider an example. We wish to solve the S^{3}DSP problem on a two-dimensional grid, the edges of which are randomly assigned weights from the set $\{0, \infty\}$, i.e. either an edge exists or it does not. The graph is shown in figure 3.3; we wish to find the shortest path from the upper left node of the graph to the bottom right node. To calculate the shortest path we design a two-dimensional cellular automaton, every cell of which is indexed by two positive integers i and j, $1 \leq i, j \leq n$, and every cell x_{ij} has a cruciform neighbourhood as shown below:

$$u(x_{ij}) = (x_{i-1j}, x_{i+1j}, x_{ij-1}, x_{ij+1}).$$

In this particular case every cell x takes states from the set

$$\mathbf{Q} = \{\circ, +, \#\}$$

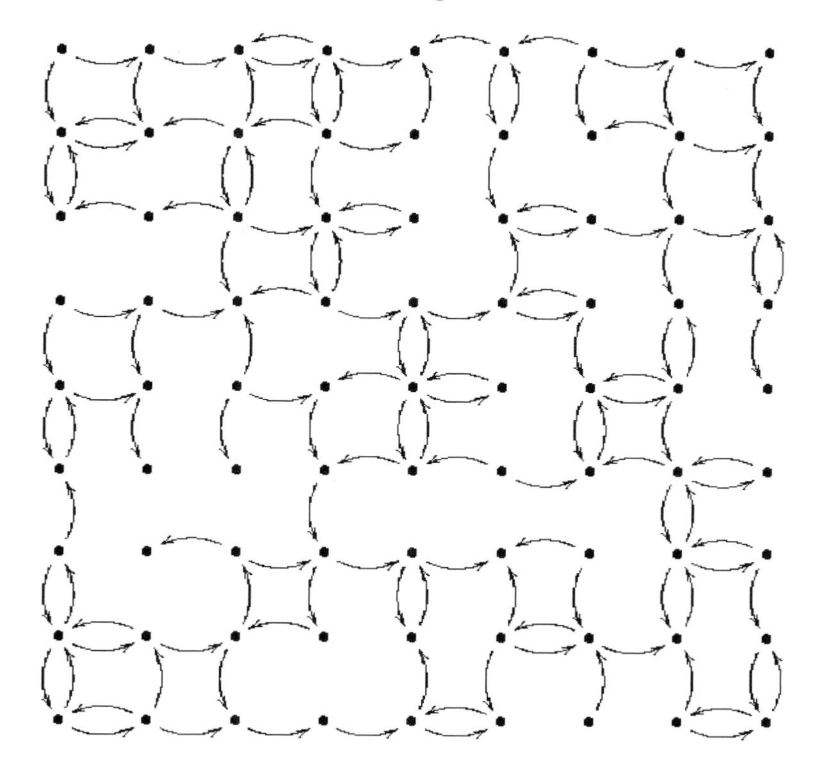

Figure 3.3. A planar digraph G of 81 nodes: the arrows indicate the orientation of the edges; the nodes are shown by •s. The required path must start at the left upper corner and finish at the right bottom corner of the graph G.

its pointer p_x takes states from the set

$$\mathbf{Y} = \{\uparrow, \leftarrow, \downarrow, \rightarrow, \lambda\}$$

and its vector has four elements.

The spacetime evolution of the cellular automaton is shown in figure 3.4. At the beginning of the evolution we excite the upper left cell of the lattice (figure 3.4, $t = 0$). The excitation spreads across existing edges of the lattice (figure 3.4) and the pointers of the cells are therefore modified. At the 27th step of the evolution the excitation reaches the destination cell, x_d, of the lattice and we stop the computation.

The path between the cells x_s and x_d can be easily extracted from the stationary configuration of pointers shown in figure 3.4, $t = 27$. The resultant path is shown in figure 3.5.

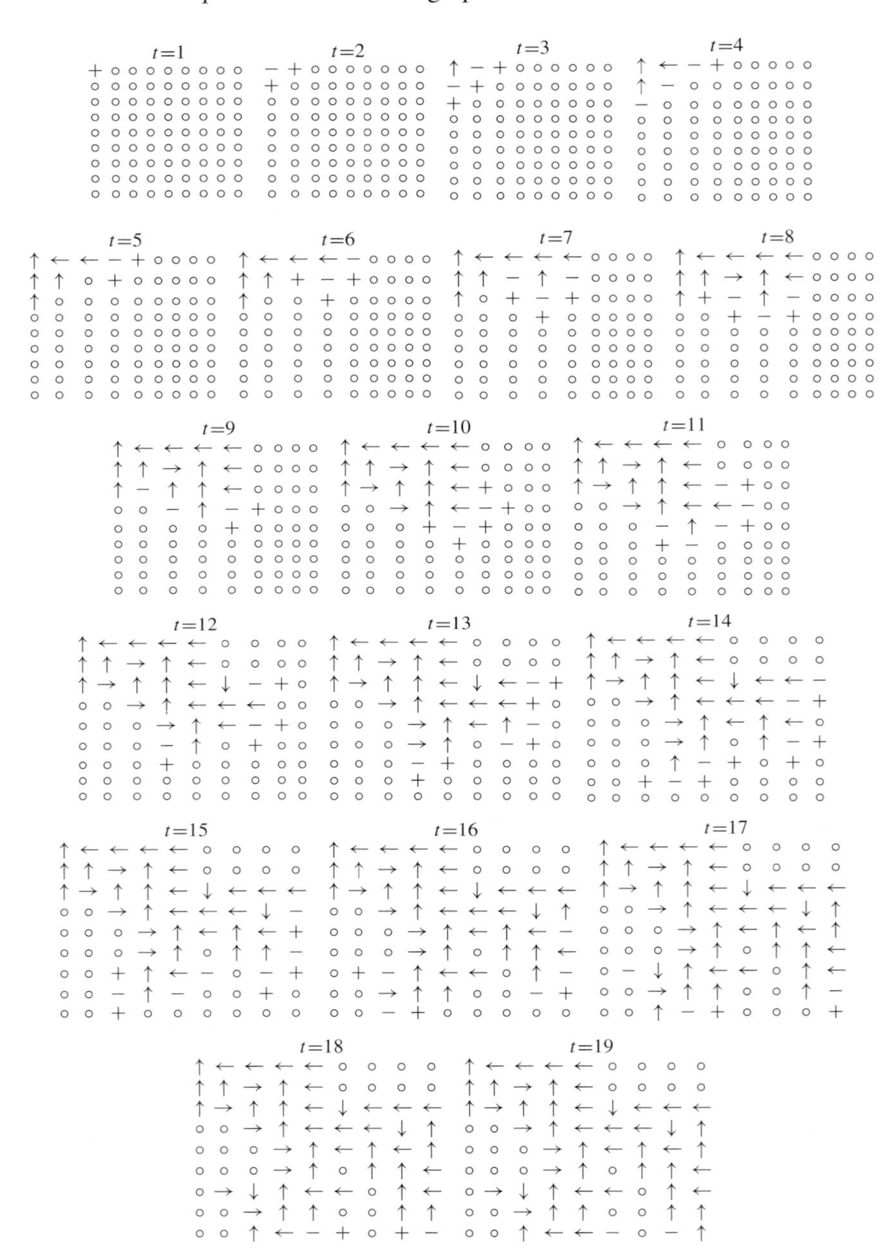

Figure 3.4. Evolution of a two-dimensional cellular automaton of 9 × 9 cells, which computes the S^{3}DSP on the graph G, shown in figure 3.3. The cell states +, ∘ and − are shown explicitly everywhere except when a cell takes the state #. In this case the state of the cell pointer is indicated instead of the cell's state.

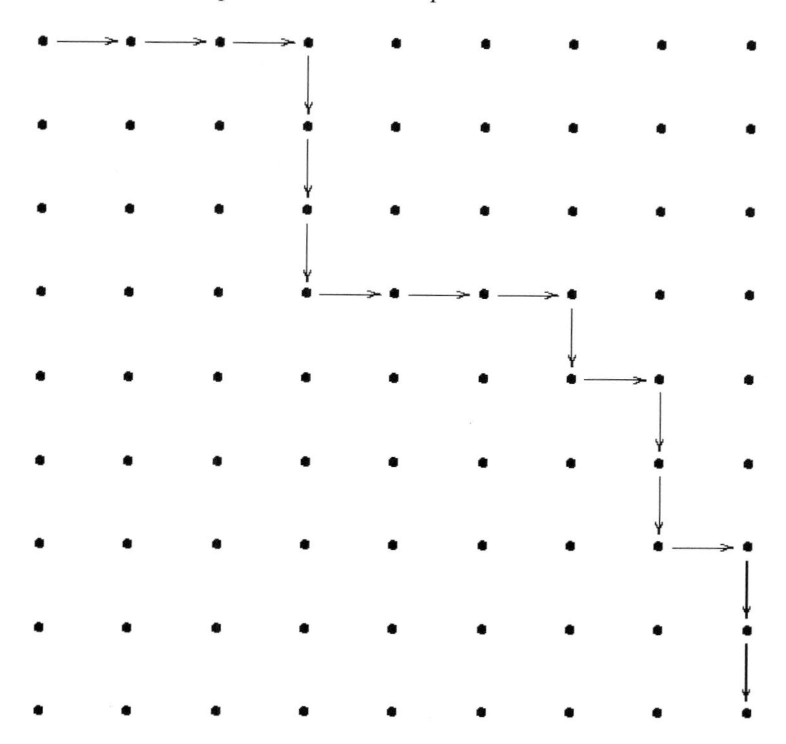

Figure 3.5. The shortest path from left upper node to right bottom node computed on the graph G (figure 3.3) in a two-dimensional cellular-automaton model of an excitable medium.

> *Let G be a two-dimensional integer grid of n nodes. Some edges of G are cut out. Then there is a two-dimensional cellular automaton of n cells (each of which has a four-cell neighbourhood and nine states) that finds the S^3DSP on G in $O(n)$ upper time. The automaton finds the single-source (all destinations) shortest paths in $O(n)$ upper time and extract paths in $O(n^2)$ steps.*

Under the conditions of the proposition, the graph edges have binary weights: either 0 (there is an edge) or ∞ (there is no edge). Therefore the cell state $\mathbf{Q} = \{+, \circ, \#\}$ is sufficient.

Every vertex of a rectangular lattice is connected with no more than four immediate neighbours (northern, western, southern and eastern). Thus, $\mathbf{Y} = \{\uparrow, \leftarrow, \downarrow, \rightarrow\}$. The state of the cell x can be thought of as a tuple (x^t, p_x^t) of a proper state x^t and that of the pointer p_x^t. If $x^t = \circ$ then, necessarily, $p_x^t = \lambda$. However, p_x^t may equal any element from $\mathbf{Y}$ when $x^t \in \{+, \#\}$. Thus, a cell of the automaton has at most nine states. The longest path consists of $n - 1$ nodes, therefore it can be computed in $O(n)$ time. A simple application of the technique to all vertices of G gives us the $O(n^2)$ bound for the APSP problem.

> *Let G be a two-dimensional rectangular grid of n nodes. Let the edges of G take weights from the set $\{0, 1, \ldots, \nu, \infty\}$. Then the S^3DSP on G can be found in $O(\nu n)$ time in a two-dimensional cellular automaton of n cells, each of which has four neighbours and takes $6+5n$ states.*

That is we construct a cellular automaton such that it supports auto-wave dynamics. An excitation wave is generated in a source cell. The wave spreads across the edges of the graph. The wave reaches the destination cell travelling along the shortest path in a minimal time.

As soon as time is involved we assume that there is a delay in switching cell x from state $\circ$ to state $+$. This delay is proportional to the weight $w_x y$ of the edge (yx) along which the excitation has been transmitted; namely, if cell x was excited by cell y then cell x will take state $+$ 'physically', i.e. state $+$ becomes visible to all x's neighbours, after a delay proportional to w_{xy}. After that the state of the cell will decrease every step of the evolution, $w_{xy} \leftarrow w_{xy} - 1$, until it becomes zero or there is a neighbour y'' of cell x such that

$$y''^t = + \text{ and } w_{xy''} = \min\{w_{xy'} : y' \in u(x) \text{ and } y'^t = +\} \text{ and } w^t_{xy''} < x^t.$$

In this last case cell x modifies the state of the pointer to $w_{xy''}$. For example, assume that cell x has a neighbourhood

$$u(x) = (y_\mathrm{N}, y_\mathrm{W}, y_\mathrm{S}, y_\mathrm{E}).$$

At time step t the neighbourhood takes the state

$$u(x)^t = (+, +, +, +)$$

and $x^t = +$. Moreover, we have

$$w_x = (w_{x\mathrm{N}}, w_{x\mathrm{E}}, w_{x\mathrm{S}}, w_{x\mathrm{W}}) = (4, 9, 3, 1).$$

Cell x finds a neighbour y, which has state $+$ and is connected to x by an edge with the minimal weight w_{xy}. In this case, cell x chooses y_S and $x^{t+1} = 3$, $x^{t+2} = 2$, $x^{t+3} = 1$, $x^{t+4} = 0$, $x^{t+5} = +$, $x^{t+6} = \#$, $p^t_x = \downarrow$, $p^{t+1}_x = \downarrow, \ldots$. By definition, the weights are bounded by an integer ν. The longest path consists of n vertices and every edge has a delay ν of excitation transition. This gives us $O(n\nu)$ time for the cellular-automaton solution of the S^3P problem.

Let us consider another example. Now the planar graph has edges weighted by elements of the set $\{0, 1, 2, 3, 4, \infty\}$ (figure 3.6). We map the graph onto a cellular-automaton model of an excitable processor and let the automaton evolve.

As we see in figure 3.7 the excitation starts at the source cell and travels from one lattice node to another with a velocity proportional to the weights of the edges connecting neighbouring nodes.

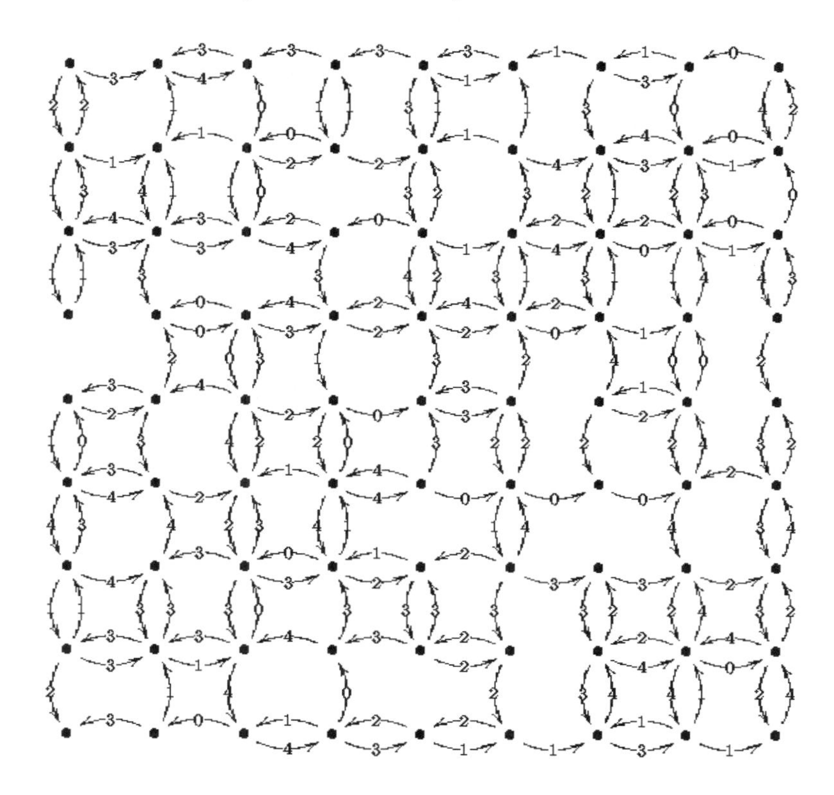

Figure 3.6. A planar weighted digraph G of 81 nodes. The required path must start at the left upper corner and finish at the right bottom corner of the graph G.

A minimum-weight shortest path, computed in the evolution of a two-dimensional cellular automaton, is shown in figure 3.8.

Let G be an oriented graph of n vertices arranged on a discrete lattice. Every vertex has at most degree k, where k is a positive integer. Then there is a cellular automaton of n cells, each of which has k neighbours and takes $O(k\nu)$ states. The automaton finds the S^3DSP on G in $O(n(\nu + k))$ upper time. An S^3P can be extracted from the final configuration of the automaton in $O(n^2)$ time.

In this section we have demonstrated that there is a mathematical model of a reaction–diffusion or excitable medium which solves the problem of shortest-path computation in a reasonable time. The model—a cellular automaton—is capable of selecting nodes that form the shortest path between a pair of nodes. The source node, or the starting point of the path, generates an excitation wave. The wave

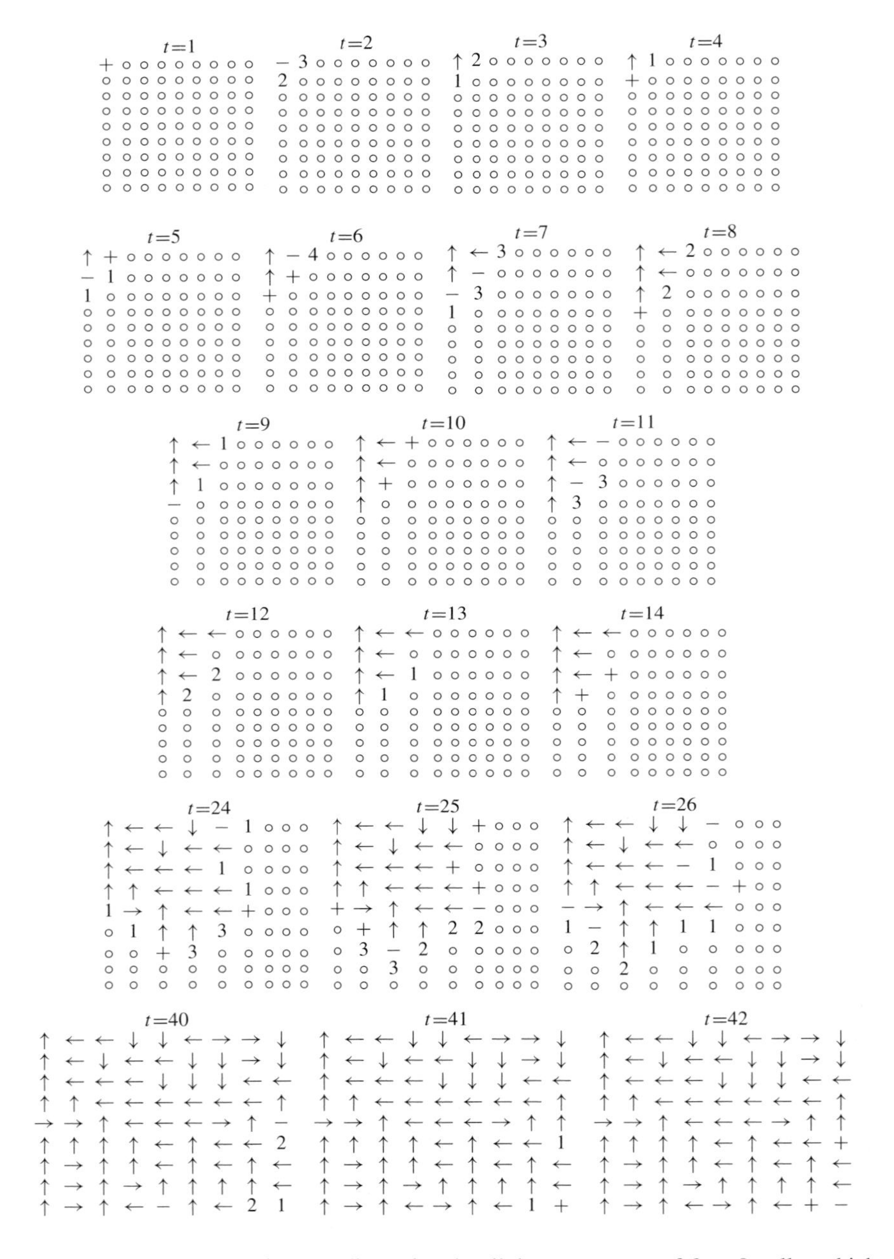

Figure 3.7. Evolution of a two-dimensional cellular automaton of 9×9 cells, which computes the S^3DSP on the weighted graph G, shown in figure 3.6. The states of the cells are shown explicitly everywhere except when a cell takes the state #. In this case the state of a cell pointer is indicated instead of the cell's state.

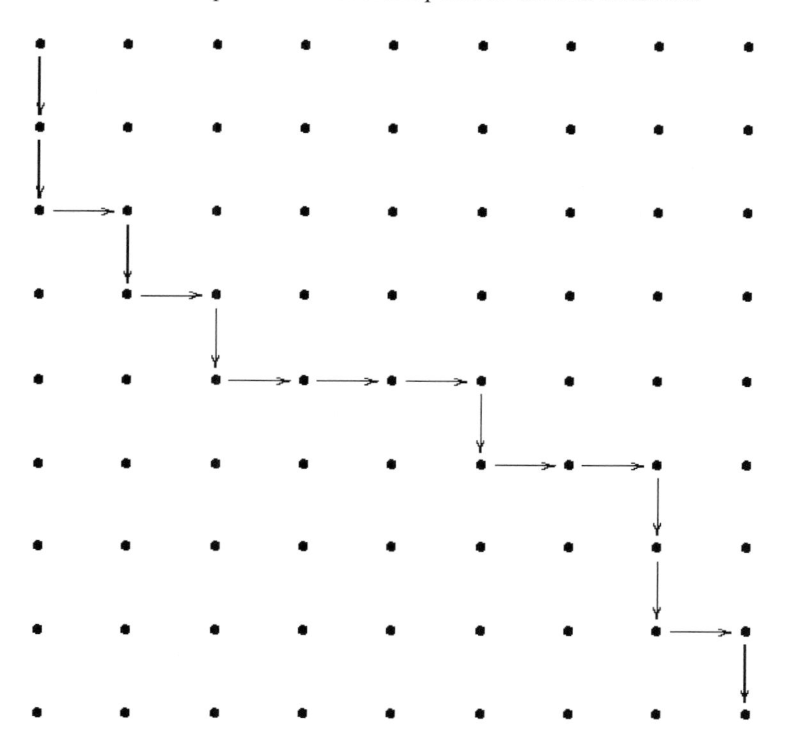

Figure 3.8. The shortest path from the left upper node to the right bottom node computed on the weighted graph G (figure 3.6) in a two-dimensional cellular-automaton model of an excitable medium with delayed excitation.

spreads through the graph. The velocity with which the excitation is conducted from one node to another is proportional to the weights of the edges. Every cell of the automaton is supplied with a pointer, or a rotating arrow, which (after the cell has been excited) points to the neighbour which excited it.

The S^3DP is quite simple. After the excitation dies, the cellular automaton takes a global stationary state. We run backward following the pointers and mark the cells we pass. This might be done in $O(n^2)$ time.

How complex is the cellular automaton that finds the shortest path on a mesh with weighted edges? The size of the cell neighbourhood equals the degree of a mesh node. Every cell of the automaton takes the number of states equal to the product of the degree of a mesh node and the maximum weight of an edge. Thus in a worst case we have an $O(n(\nu + k))$ time bound, where k is the degree of a node, ν is the maximum integer weight and n is the number of nodes.

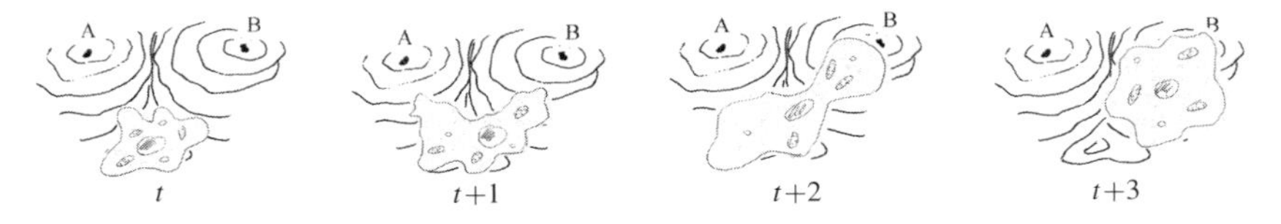

Figure 3.9. A simple decision undertaken by an amoeba. The concentration of the substrate is higher at site B than at site A.

3.5 Computation of trees by neurites

In the previous sections we discussed how to approximate shortest paths by reaction–diffusion and excitation waves spreading in a data space. What happens if we enclose the solution of a reaction–diffusion medium in a membrane sack and represent data points by sources of other reagents distributed on a plane, on which the enclosed medium is placed? The reagents diffusing from the data points form a gradient field, which may guide the membrane systems from one point to another in the same way as an electrical field does. Will this system be capable of solving shortest-path and spanning-tree problems? The chemotactic behaviour of amoeboids gives a clue to a positive answer.

Pseudopodial motion is typical for many cellular structures: *amoeba* are the most famous examples. An amoeba moves due to the virtual separation of its cytoplasm into a sol and a gel. When a pseudopodium is formed, some parts of the amoeba's cytoplasm are transformed into a sol, then the endoplasm starts to flow to this area. The membrane expands and the pseudopodium is extended forward. When the sol-like cytoplasm reaches the end of the pseudopodium, it is transformed into gel. This recurrent conversion of sol into gel and back again allows the organism to move purposefully.

The behaviour of the amoeba is usually associated with chemotaxis. When an amoeba, say for example *Physarum polycephalum*, moves, it usually has one linear pseudopodium, which expands during organism motion. If some attracting chemical is presented in a substratum, then the linear pseudopodium is split into several lateral pseudopodia, which explore the space around the main linear tip. One of the lateral pseudopodia, usually positioned at a site on the substrate with a relatively high concentration of attracting reagents, is stabilized and becomes responsible for the amoeba's linear motion. Eventually the amoeba finds the site with the highest concentration of attracting reagents, as shown in figure 3.9.

A cellular slime mould *Dictyostelium discoideum* offers another classical example of morphogenesis and aggregation induced by travelling waves in reaction–diffusion systems. The *Dictyostelium* shows positive chemotaxis, which is coupled to the dynamics of cAMP waves, see e.g. [193, 194, 563, 471, 580]. At a certain step in the mould's development a wave generator is formed. The amoeboids move toward this pacemaker and eventually aggregate around it (see figure 3.10) [563, 471, 420].

Figure 3.10. Waves of *Dictyostelium discoideum* cells [563]. Published with kind permission of Cornelis Weijer.

These particulars of amoeboid behaviour have already been incorporated into distributed algorithms for shortest-path computation. One successful demonstration of such a computation with *Physarum polycephalum* was presented in [476]. In these experiments a leading pseudopodium is selected not on the basis of the position of the pseudopodia relative to sites with substrate concentration but on the distance of the pseudopodia tips from the amoeba's body 'centre', i.e. exactly as in our early models on spanning tree computation by a growing neuron [4].

Unfortunately, amoeboids do not leave permanent traces, which could represent the edges of a constructed graph, neither do they simultaneously explore all possible parts of the data space at the same time. To approximate, for example, a spanning tree we need to employ the number of amoeboids proportional to the number of given nodes that must be covered by the tree. These options are available in neuron morphogenesis. When an axonal or dendritic tree develops in a neuron, then the growing tips move, in general, like amoeboids, the traces of which are represented by axons and dendrites.

The morphogenetic algorithm discussed in this section is based on the formation of a neurite tree in the development of a single neuron [4]. The algorithm directly employs the notion of diffusion, spreading neuron processes and reaction, or a competition, between the different processes for marked sites of a space.

Let us consider how a spanning tree can be constructed in the growth of a dendritic tree of a single maturing neuron. During its development every neuron undergoes three basic transformations: medulloblast to neuroblast to neuron. A medulloblast can divide itself into several new cells, a neuroblast has neurofibrilles, a large nucleus and a small volume of cytoplasm while a mature neuron has axonal and dendritic trees and exhibits electrical activity. The growth of axonal and dendritic trees is the principal feature of the neuroblast-to-neuron transformation. These trees grow by means of 'growth cones'. Every growth cone is a bulb on the top of a neuronal process filled with organelles and capable of amoeboid motion. The growth cone supports the relationship with the neural soma via the transport of macromolecules and microtubules determined activity. However, such communications between a growth cone and a soma are sufficiently slow that one can propose that growth cones are almost independent of soma from the information point of view. Such an abstraction would be appropriate in this particular context.

Axon guidance by chemicals released by target tissues was first suggested in [567] and then extensively developed in myriads of publications, either experimentally or theoretically (see e.g. [57, 445, 353, 270, 268, 113, 445, 269, 299]). Most experimental results deal with two-dimensional examples of axonal growth on planar substrates *in vitro* (see e.g. [57]) or the growth of axons of retinal ganglion cells in the tectum [269]. The following assumptions, mostly supported experimentally, may form the basis of computation with growth cones:

- a growth cone as a whole employs spatial-sensing mechanisms rather than temporal ones [269];
- filopodia are the basic subjects of gradient sensing [269]; each filopodium of the growth cone uses a temporal gradient [269];
- the behaviour of each filopodium has a high degree of randomness, which facilitates space searches;
- a neurite's branching may be gradient dependent: a growth cone sprouts new filopodia in the direction of maximal concentration of chemoattractants [269];
- axons may *compete* for neurotrophins, which are released by targets [485];
- chemorepellents may be involved in axon guidance as well as chemoattractants.

A beautiful example of several growth cones travelling in a two-dimensional space is shown in figure 3.11.

A neuronal growth cone is an elementary mobile processor of a morphogenetic computing device.

We place a neuroblast somewhere on the plane amongst drops of chemical attractants, the positions of which represent the points of a given planar set. What happens next?

The neurite tree starts to grow. Usually several growth cones sprout from the medulloblast body. The growth cones crawl out of their origination sites. Every growth cone is an analogue of an 'intelligent' particle, the wandering of which is partially controlled by gradients of implicit reagents diffusing from the marked sites of the space. The system is not conservative and each growth cone can divide itself into several new growth cones.

Every growth cone has a wide range of motion and branching modes. Moreover, each growth cone itself decides which way to elongate and when to branch out and generate a pool of new growth cones—second-order cones.

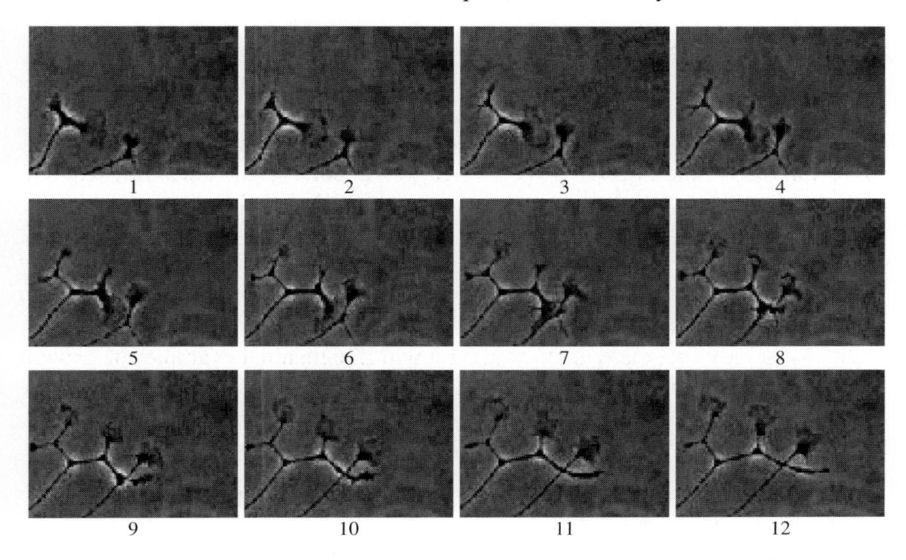

Figure 3.11. Several snapshots of growth cones spinning-out axons during migration. Published with kind permission of Louise Cramer [156, 157].

Constructing an algorithm fully employs biological findings on axons guiding the developing neural tissues [269], particularly those related to the spatial strategy used by real growth cones and the increase in growth-cone sensitivity with concentration of attracting reagents [445].

We identify projections of drops of chemical attractants with the elements of a given planar set **V**, which must be spanned by a tree. We assign the following variables to a growth cone C: E is the 'energy' of the growth cone (it is proportional to the distance from the cone to the neuron soma) and x is the current position of the cone on the plane; the constant γ is an upper limit for the degree of branching, ρ is the maximal distance at which the growth cone can recognize afferent terminals (elements of **V**), θ is the threshold of energy decrease, δ is a small real.

At every step of its development the growth cone C can implement one of the following procedures:

- crawl to the next position;
- generate no more than γ daughter growth cones;
- detect a point of **V** and occupy it (crawl to this point);
- resolve conflicts, if any, with other growth cones that try to occupy the same position as growth cone C.

The energy E of the growth cone decreases during evolution: $E^0 = E_0$ and $E^{t+1} = E^t - 1$, where E_0 is some appropriate initial value. The energy simply represents the lengths of the branches and distances from current positions of

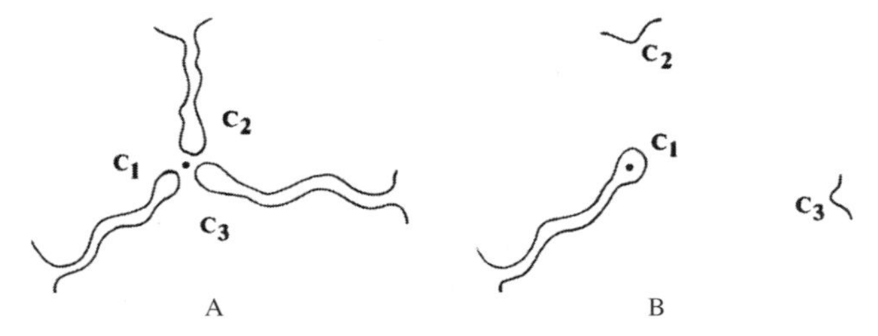

Figure 3.12. Competition of growth cones for data points marked by chemical attractant.

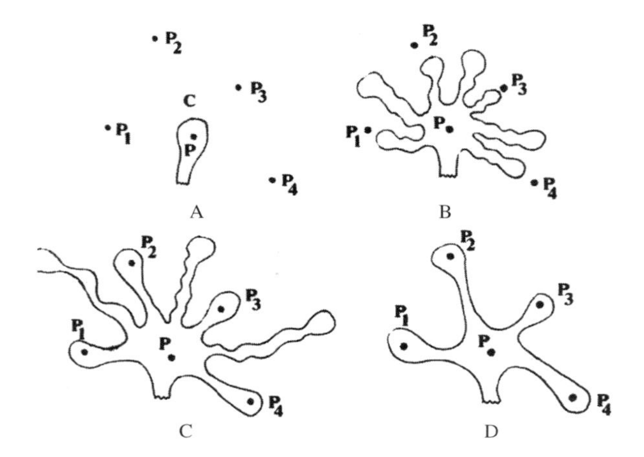

Figure 3.13. Development of branches of a growth cone.

growth cones to the neuron soma, or root of the growing tree, along the branches of the tree. Energy is used as the main criterion in the competition of several growth cones for the same site on the plane. For example, three cones C_1, C_2 and C_3 are in the neighbourhood of the point $p \in \mathbf{V}$ (figure 3.12(A)). All three cones pretend to take over the site p. We assume that the cones have energies E_1, E_2 and E_3: $E_1 > E_2 > E_3$. Each of the cones recognizes a leader of the group from its energy. Cones with smaller energies lose the competition, pull back to their parent branches and the leading cone occupies the site p (figure 3.12(B)). The energy E_i of the growth cone C_i is an abstract entity in our model. Ideally, it can be linked somehow to the concentration of repellents, released by each growth cone.

So we can see that growth cones may be destroyed as a result of competition with other cones. Neither does a growth cone survive long without finding an incoming terminal from another neuron—it dies when its energy drops below a certain threshold. Let us look at some typical situations (figure 3.13). The growth cone C occupied the point p (figure 3.13(A)) and generated several second-order cones. These new cones search for the projections of afferent terminals, which represent points of **V** (figure 3.13(B)). Some of the second-order cones have found the points and crawled to their final positions (figure 3.13(C)). Hence some other cones have found nothing. They continue to search until their energies became less than some specified threshold: $|E_i - \theta| \leq \delta$. As they run out of energy the cones pull back into the parent cone C (figure 3.13(D)).

In the design of the algorithm we use the following assumption, which is not entirely true from a biological point of view but simplifies the solution of the problem.

> *Any growth cone terminates its actions when all its daughter cones run out of energy and pull back. A neuron stops growing its dendritic tree when all growth cones stop moving and branching.*

We now provide a sketch of the algorithm describing the behaviour of the growth cone C. It is implied that $\delta = \delta_0$, $\rho = \rho_0$, $\theta = \theta_0$ are constant, $x \in \mathbf{R}^2$, B is a Boolean variable, $\mathbf{C}$ is the set of energies of those cones which intend to occupy the position chosen by the cone C, $D(x, \rho)$ is a planar disc with radius ρ centred at the point x.

- Procedure Growth Cone C
  - $E = E_0$; $B =$ `False`; $\mathbf{C} = \{\emptyset\}$;
  - Repeat
    - $E = E - 1$;
    - Repeat
      - $x =$ `next` (x)
    - Until (x is free of branches and cones);
    - Collect $\mathbf{C}$;
    - If $D(x, \rho) \cap \mathbf{S} \neq \emptyset$ and ($\mathbf{C} = \{\emptyset\}$ or $\max\{E' \in \mathbf{C}\} < E$) then $B =$ `True`;
  - Until B or $|E - \theta| \leq \delta$;
- If B then
  - Occupy site x and stop moving;
  - Branch out (generate not more than γ new growth cones);
- else pull back to parent cone;

When every growth cone implements this code a dendritic tree is formed such that it has branching points only at points of the given set **V**. It is also acyclic and has minimal (amongst all possible) lengths for its branches. This dendritic tree represents the minimum spanning tree of the given planar set.

If we want to implement the algorithm in a conventional parallel processor, e.g. with cellular-automata architecture, we need to assume that the dendritic tree grows on a rectangular two-dimensional lattice. In this case the discrete motion of the growth cone is simulated in a cellular-automaton processor in such a manner that if a cone is at node (i, j) a processor P_{ij} is activated; that is, the cone does not crawl continuously but jumps from one node of the lattice to another. The processor P_{ij} searches for the next node to transfer the growth cone. The following constants and variables are kept in processor P_{ij}'s local memory: ρ, θ, γ and δ are already defined; D is the length of the path from the initial point (soma or root of the tree) to the current point of the set $\mathbf{V}$ (this is an analogue of the energy E); $\mathbf{L}$ is the list of coordinates (addresses) of the processors, which are the neighbours of P_{ij} and contains points of $\mathbf{S}$ which are the targets for branching (i.e. points where daughter cones will be sent to); $\mathbf{LP}$ is the list of the coordinates of points stored in the processors, the addresses of which are in the list $\mathbf{L}$; L_{mst} is a list of points which are geographical neighbours of points stored in the processor P_{ij} and which belong to the spanning tree; P_{mst} is the list of the addresses of processors, which contain elements of L_{mst}; A is a Boolean variable, A is true if there is a growth cone in the node (i, j), and it is false otherwise.

Consider a brief example. Every elementary processor has exactly eight neighbours:

$$u(P_{ij}) = (P_{i-1j-1}, P_{ij-1}, P_{i+1j-1}, P_{i-1j}, P_{i+1j}, P_{i-1j+1}, P_{ij+1}, P_{i+1j+1}).$$

Let processor P_{ij} store the coordinates of point $p_1 \in \mathbf{V}$ and processors P_{i-1j-1}, P_{i+1j-1} and P_{ij+1} store coordinates of the points p_2, p_3 and $p_4 \in \mathbf{V}$, respectively. Other processors from the neighbourhood $u(P_{ij})$ of the processor P_{ij} have nothing. Then $\mathbf{L}_{ij} = \{(i, j-1), (i+1, j), (i, j+1)\}$ and $\mathbf{LP}_{ij} = \{p_2, p_3, p_4\}$. Assume that the processor P_{ij} connects pair of points (p_1, p_2) and (p_3, p_4) by an edge of the minimal spanning tree. So, we have $L_{\text{mst}\,ij} = \{p_2, p_4\}$ and $P_{\text{mst}\,ij} = \{(i, j-1), (i, j+1)\}$.

If processor P_{ij} knows the coordinates of the point p it acts as follows. If the Boolean variable A is true (i.e. there is a growth cone at the node (i, j)) then P_{ij} scans its neighbourhood $u(P_{ij}) = (P_{i-r\,j-r}, \ldots, P_{i+r\,j+r})$ clockwise, increases the neighbourhood radius on a unit, $r = 1, 2, \ldots$, and implements the following operations:

- updates the set $\mathbf{L}_{ij}$ with the addresses of neighbouring processors, which contain the coordinates of such points of $\mathbf{V}$ lying at distance at most $\min\{d(p, q) + \epsilon | q \in \mathbf{LP}\}$;
- adds the points contained in the processors of the list $\mathbf{L}_{ij}$ to $\mathbf{LP}_{ij}$;
- writes its own address and coordinates of p in the lists L_{mst} and P_{mst} of the processors, the addresses of which are stored in $\mathbf{L}_{ij}$;
- activates the processors by assigning the truth value to their Boolean flag variables.

Computation of the spanning tree is finished globally in the massively parallel processor when each of the elementary processors has its Boolean variable A in the value `False`, i.e. neither of the elementary processors has an active growth cone.

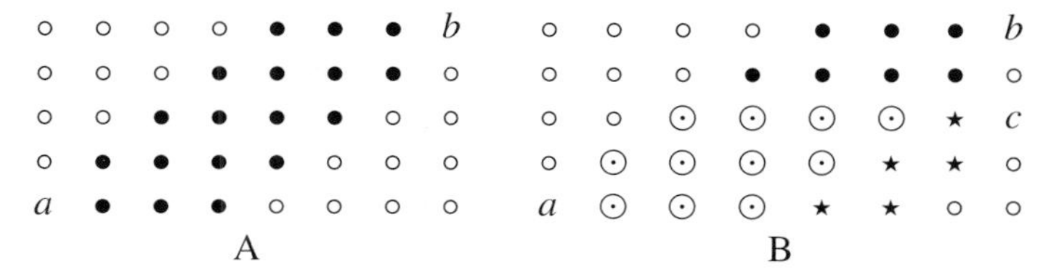

Figure 3.14. Shortest paths on a lattice: (A) shortest-path-related set of nodes (marked by •) between the nodes a and b, calculated in metric L_∞; (B) overlap between shortest paths.

3.6 Spanning trees on random lattices

A cellular-automata approximation of a spanning tree on a lattice is related to the idea of the visibility graph, which is a critical component of shortest-path problems in the plane. Recall that when we transfer conventional representations of planar graphs to a discrete lattice we usually use the metrics L_∞ or L_1 instead of the Euclidean metric. This transition generates a lot of artifacts that can distort the resultant picture. Each edge of a planar graph is the shortest of all the possible curves connecting the two nodes. In the planar case there is exactly one minimal curve. This is the straight line joining the nodes. The line forms the edge. In a lattice the edge is a set of nodes. The concept corresponding to the planar graph definition of an edge connecting two given nodes a and b is defined on a lattice $\mathbf{L}$ as the set of all possible paths from the node a to the node b which contains a minimal number of nodes. In discrete metrics there may be more than one minimal path between two nodes of the lattice. Thus, if the distance is measured in the metric L_∞ (i.e. every node of the lattice has an eight-node neighbourhood) a unique single path through the nodes connecting a to b exists only when the nodes a and b lie in the same column or the same row of the lattice, or when the distance (in nodes) between a and b is equal along the x- and y-axes.

Otherwise, there will be several shortest paths as in the example shown in figure 3.14(A), where the nodes of the lattice covered by one or more shortest paths are marked by •. In other words, the lattice version of the line connecting a and b is a set of nodes lying inside the 45-convex hull, i.e. the intersection of discrete half-planes intersecting the axes at angles divisible by 45°.

Sometimes, as demonstrated in figure 3.14(B), the lines of nodes connecting sets of nodes may overlap one another. Thus, in figure 3.14(B), the nodes exclusively covered by the shortest paths connecting a to b are marked by •; nodes exclusively covered by the shortest paths connecting a and c are marked by $\star$; the nodes covered by at least two paths, $a - b$ and $a - c$, are marked by $\odot$.

A discrete approximation of a planar graph on a lattice $\mathbf{L}$ is a tuple $G = (V, E)$, where $V \subset L$ is a set of graph nodes and $E \subset L$ is a set of edges that for every $a, b \in V$, $\overline{ab} \in E$ is the unique path from $s(a, b)$, where $s(a, b)$ is the set of nodes constituting the shortest path from a to b. In a discrete approximation of the spanning tree we should have $\overline{ab} \cap \overline{cd} = \emptyset$, when $\overline{ab}, \overline{cd} \in E$ and all four nodes are different. If the intersection is not empty we have some discrete analogue of a Steiner tree.

The Steiner tree of n given nodes of the subset $\mathbf{V}$ *of a two-dimensional integer lattice of m nodes can be approximated by a two-dimensional cellular automaton with* $O(n)$ *cell states in* $O(m^{1/2})$ *time.*

The computation algorithm is based on the diffusion of reagents from the sites of $\mathbf{V}$. Every cell of the automaton is capable of recognizing from which site of the set $\mathbf{V}$ each type of reagent comes. We, therefore, require $O(n)$ cell states to represent n reagents. In the worst case, when $\mathbf{V}$ is binary and two of its elements lie on opposite edges of the lattice, it takes $O(m^{1/2})$ steps of discrete time for a diffusion wave to reach one site from another.

To construct a spanning tree we need to modify the algorithm for shortest-path computation, discussed in the previous section; that is, we construct a cellular automaton in which every cell has an eight-cell neighbourhood. At every discrete-time step t a cell x has a two-component state $x^t = \langle s_x^t, p_x^t \rangle$, where $s_x^t \in \{\circ, -\} \cup \mathbf{C}$, $\circ$ is the rest state, $-$ is the refractory state, $\mathbf{C} = \{C_1, \ldots, C_n\}$ is the set of concentrations of n reagents $1, \ldots, n$. The p_x^t is the state of a pointer, $p_x^t \in \mathbf{P}$. The set

$$\mathbf{P} = \{\lambda, \uparrow, \leftarrow, \downarrow, \rightarrow, \nearrow, \searrow, \swarrow, \nwarrow\}$$

is a set of pointer states. The state of the pointer may be blank, λ, or may point to the neighbour of its cell that excited this cell. For example, the state $\nwarrow$ means that the pointer is oriented north-west and λ means that the pointer is oriented nowhere.

A cell updates its state according to the following rule:

$$s_x^{t+1} = \begin{cases} \eta(u(x)^t)) & \text{if } \epsilon(u(x)^t) > 0 \\ - & \text{if } s_x^t \in \{-\} \cup \mathbf{C} \\ \circ & \text{otherwise} \end{cases}$$

where

$$\epsilon(u(x)^t) = \sum_{y \in u(x)} \chi(y^t, \mathbf{C})$$

reflects the number of reagents in nodes neighbouring x. Moreover,

$$\chi(y^t, \mathbf{C}) = \begin{cases} 1 & \text{if } y^t \in \{1, \ldots, n\} \text{ and } C_{y^t} > 0 \\ 0 & \text{otherwise} \end{cases}$$

and

$$\eta(u(x)^t) = \begin{cases} C_{v(x)} & \text{if } x \in \mathbf{V} \\ \max_{y \in u(x)}\{y^t : C_{y^t} > 0\} - 1 & \text{otherwise} \end{cases}$$

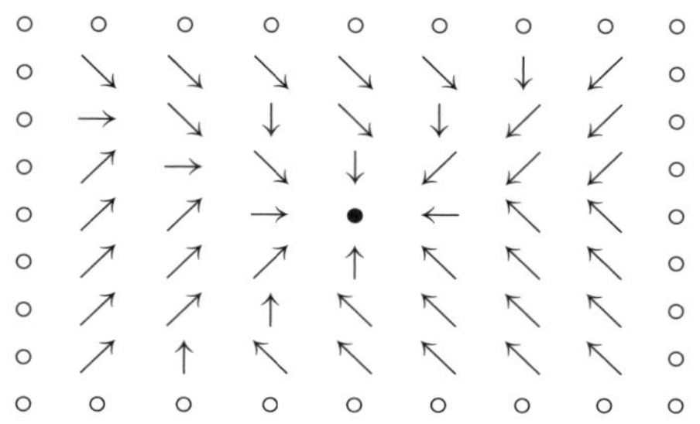

○ ○ ○ ○ ○ ○ ○ ○ ○
○ ↘ ↘ ↘ ↘ ↘ ↓ ↙ ○
○ → ↘ ↓ ↘ ↓ ↙ ↙ ○
○ ↗ → ↘ ↓ ↙ ↙ ↙ ○
○ ↗ ↗ → • ← ↖ ↖ ○
○ ↗ ↗ ↗ ↑ ↖ ↖ ↖ ○
○ ↗ ↗ ↑ ↖ ↖ ↖ ↖ ○
○ ↗ ↑ ↖ ↖ ↖ ↖ ↖ ○
○ ○ ○ ○ ○ ○ ○ ○ ○

Figure 3.15. An example of pointer orientations at the fourth step of the diffusion of a reagent from the site marked by •.

and $\nu : \mathbf{L} \to \{1, \ldots, n\}$. Cell x changes the state of its pointer by the following rule:

$$p_x^{t+1} = \begin{cases} y & \text{if } \eta(u(x)^t) + 1 \\ p_x^t & \text{otherwise.} \end{cases}$$

Every node, or cell, of the given set $\mathbf{V}$ receives a drop of its own unique reagent. All other nodes of a lattice are the subjects of diffusion and interaction with the reagents. As soon as the different reagent waves reach node x the node accepts such a reagent that its concentration is the maximum over all reagents in the nodes neighbouring x. At every step of the diffusion the reagent concentration decreases with the distance from the source of this reagent. Being activated by the reagent cell x directs its pointer toward the cell which transmitted this reagent to it. After this the cell changes its state to the refractory state – and no longer changes its state. So, as soon as diffusion waves have passed through a cell the cell fixes the orientation of its pointer forever.

To start the automaton, we excite a single (start or source) cell on the lattice with some reagent. A reagent wave spreads over the lattice and excites other cells, and the concentration of the reagent decreases as it diffuses away. After a cell has been excited, the pointer in this cell indicates the excited neighbour with the maximal concentration of the exciting reagent. Note that in contrast to conventional models of excitable media, the cells in our model do not change their refractory states to the rest state but remain in the refractory states forever. Therefore, when all cells of the lattice take the refractory state, the computation is finished. As soon as the computation terminates each cell of the given set $\mathbf{V}$ is connected by a single oriented path of pointers to another single cell of $\mathbf{V}$ from which the reagent which excited the first cell originated.

Let us show that the edge connecting two given nodes a and b of $\mathbf{V}$ is the shortest path between these nodes. Let the diffusion of the reagents start at node a. The reagents spread in all directions with a constant velocity and, therefore,

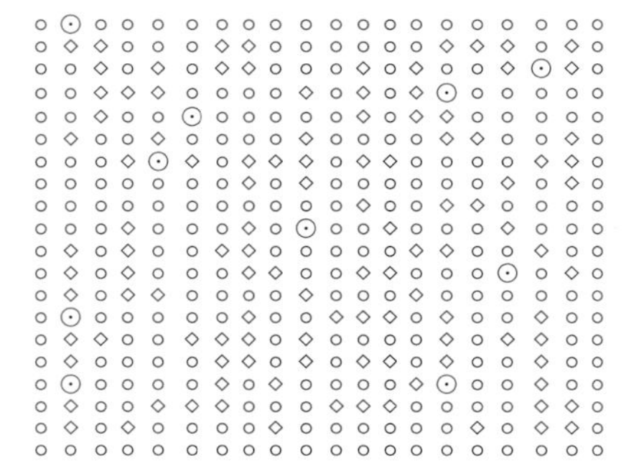

Figure 3.16. Distribution of data points, ⊙, and obstacles (removed or inaccessible nodes), ⋄, on a two-dimensional lattice.

the diffusion wave reaches node b at time $t = |a - b|_{L_\infty}$. Our cellular-automaton model is deterministic, therefore we have to allow a cell to scan its neighbours' states serially. Let us do so clockwise, starting from the north-west neighbour. For a point source of diffusion, the orientation of the pointers at time step 4 is shown in figure 3.15.

In other words, after t steps of temporal evolution, every point b lying at a distance of not more than t sites from the point a is connected by an oriented path of cells $p(a, b)$, identified by the orientation of their pointers, to the cell a. The size of the set $p(a, b)$ is minimal because $p(a, b) \subset s(a, b)$ and every cell calculates the orientation of its pointer via a sequential analysis of its neighbourhood. There are no cycles because the diffusive wave spreads unidirectionally with constant velocity.

Let there be three elements of the given set $\mathbf{V} = \{a, b, c\}$, $|a - b|_{L_\infty} = t$, $|a - c|_{L_\infty} = t/2$ and $|b - c|_{L_\infty} = t/2$. The node a is a root of the tree. A wave of the reagent a reaches node c at time $t/2$ and excites cell c, which in its turn becomes excited and starts to spread the reagent c. The initial concentration C_0 is the same for all reagents. In the general case, the waves of reagents a and c will be in the neighbourhood of the cell b during the same time t. However, neighbours of b with pointer paths leading to a will have concentration $C_0 - t/2$ of the reagent c. Therefore, cell b will be excited by the neighbour with reagent c and it will became connected to the cell c via the path of pointers. This is the outcome of wave competition.

We call a lattice random if some of its nodes are made inaccessible, or are deleted, at random. An example of a random lattice with its data points is shown in figure 3.16.

A spanning tree of n given nodes of a two-dimensional random lattice of m nodes can be approximated by a two-dimensional cellular automaton with $O(n)$ cell states in $O(m)$ time.

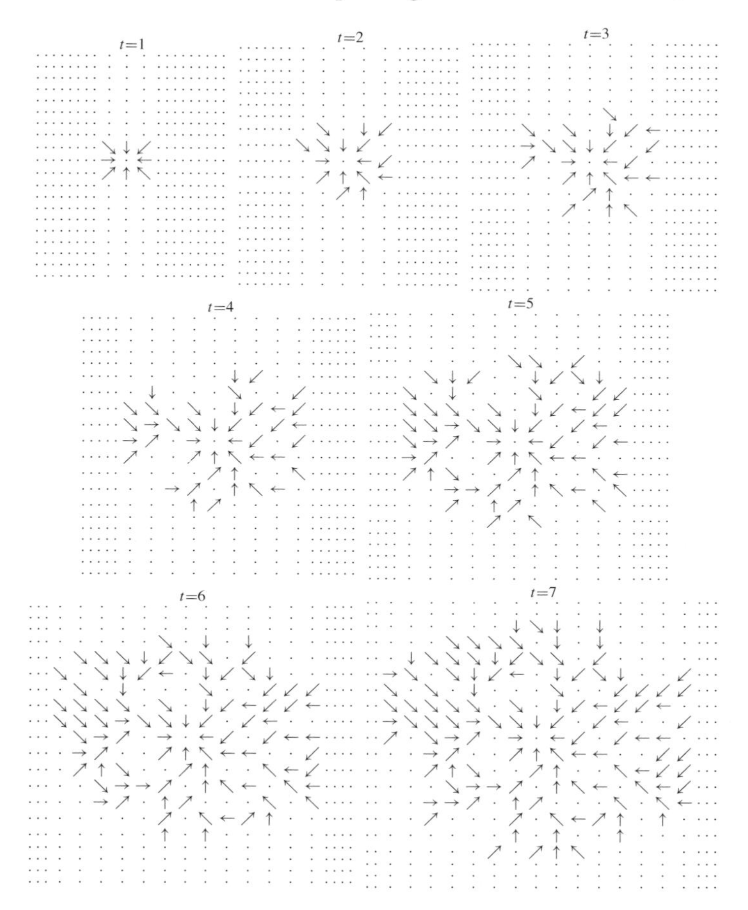

Figure 3.17. Evolution of a cellular-automaton model of a two-dimensional reaction–diffusion processor, which approximates a spanning tree on a lattice.

Removed nodes may be simulated in a cellular automaton with the help of an additional, 'obstacle', state $\diamond$. A cell of the automaton takes the state $\diamond$ if the position of the cell corresponds to the position of the removed node of the data graph. Any cell which starts the evolution in state $\diamond$ remains in this state throughout the evolution.

The evolution of a two-dimensional cellular automaton (with an eight-cell neighbourhood), which approximates a spanning tree on a random lattice (figure 3.16), is shown in figure 3.17. In this figure we can observe how the diffusion wave starts at the root of the tree. The spot of reagents, reflected in the growing spot of oriented pointers, grows in size and is distorted by the absence of some randomly deleted nodes. The computation is assumed to be finished when

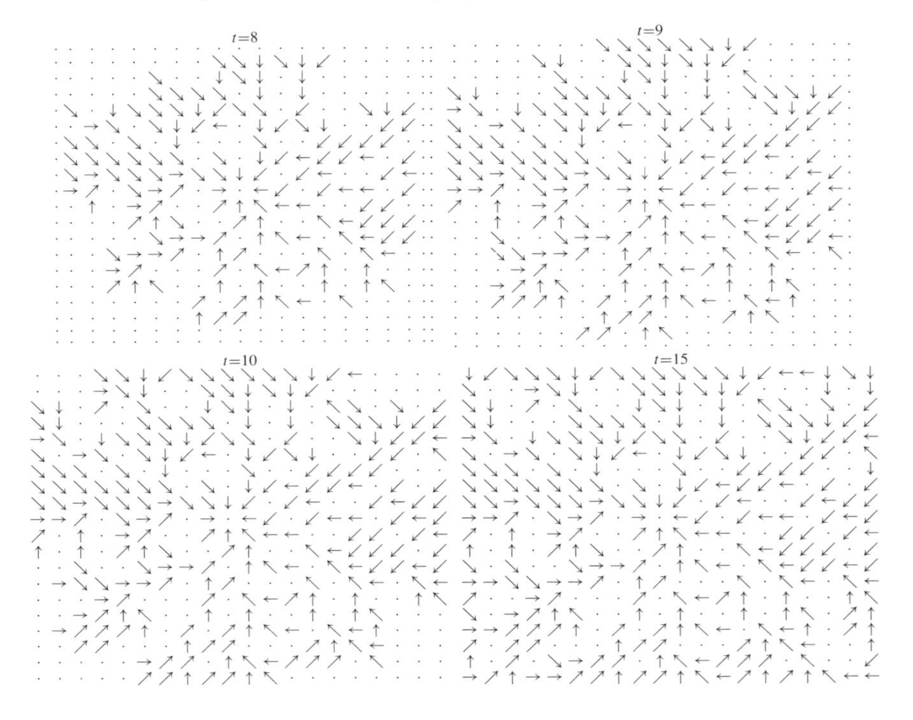

Figure 3.17. (Continued)

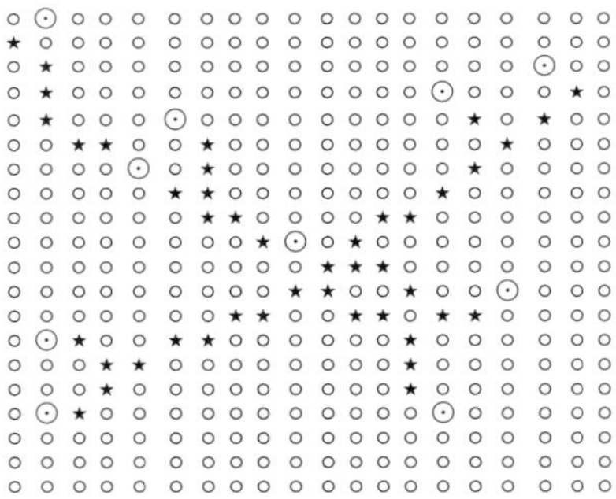

Figure 3.18. A spanning tree constructed in a cellular-automaton model of a two-dimensional reaction–diffusion processor. The nodes of the tree are shown by ⊙, the edges are formed by nodes marked by ⋆.

the diffusion front reaches the edge sites of the lattice (figure 3.17, $t = 15$). After this we can extract the desired spanning tree, as shown in figure 3.18, from the configuration of the pointers.

A discrete spanning tree of a set **V** *can be computed in a two-dimensional cellular automaton with m cells and four cell states in* $O(m^{1/2})$ *time if the elements of the set* **V** *are arranged on vertices of the convex shape in a square lattice of m nodes.*

In this situation we can use a cellular-automaton algorithm for constructing skeletons of planar shapes, as in the previous chapter. Here the edges of the convex shape, for which the skeleton is required, generate excitation waves with refractory states behind them. The waves travel inside the shape and form so-called precipitate states, which are the absorbing cell states. The precipitate states represent the nodes of a spanning tree.

3.7 Construction of graphs by diffusive families of ants

Previously we have studied families of reaction–diffusion algorithms for computating shortest paths and spanning trees. These problems form the core of a graph optimization field. Now we will try to enhance these results in two directions. First, we make the molecule of a diffusing reagent more intelligent—as intelligent as an ant can be. Second, we expand the domain of addressable problems to a wide range of proximity graphs. Therefore, we represent waves of diffusing reagents or excitation by the foraging patterns of ants. The interaction of these 'ant waves' may lead to either the formation of new nests, the elimination of old ones or the modification of trail-based graphs, connecting ants' nests.

The key features of 'ant-based' methods centre around minimum entities, rather than mobile finite automata, called ants which can move through some spatial domain. The domain is typically a continuous or discrete Euclidean space, or a graph of nodes embedded in a Euclidean space.

Each ant may carry with it a set of attributes and variables which may be affected by what it encounters as it moves over the graph or in a space. Typical attributes are source, destination and type. Typical variables are age, distance travelled by ant and route taken. The ants move according to rules which may take into account any of the following:

- the intrinsic inhomogeneities of a space,
- collisions and interactions with other ants,
- the local qualities of the domain space derived from some autonomous process, which may be affected by alterations made by ants and
- the constraints which may be required for a particular problem.

The determination of movement is typically stochastic rather than deterministic with the likelihood of the choice of a particular direction being a function of the ants' internal state, and local characteristics, which may include fixed factors and factors due to or influenced by ant activity. The motion may also be affected by a fixed or variable amount of noise. An ant may be brought into being at a time and location determined by external factors, by local factors or by another ant subject to some other contingency.

Ants may be terminated under certain conditions. For example, ants may be terminated when they reach their designated destination site. Termination may be accompanied by changes to the domain, for example updating segments of the ants' route, or launching new ants which may inherit information from the terminating ant. An ant may alter the local qualities of the domain in ways, at times and by amounts which may include the following:

- the quantity modified may be a site of a space;
- the time of modification may be immediate or after termination of the ant; and
- the amount of modification may be fixed or may vary with some local or global attribute

The effects of any such modification may persist indefinitely until changed or may change autonomously with time.

The solution to the problem may be the effect on some autonomous process of the activities of the ants or may be a static configuration derived from the modified domain, perhaps by probing with a mobile agent making deterministic movement decisions. This degree of complexity must clearly support a number of distinct phenomena affecting the results. Ant-based methods are usually used as alternatives to established methods for simple problems or for such problems where conventional algorithms are unsatisfactory. The types of difficult problems for which ant algorithms seem to be particularly well suited are those combinatorial optimization problems, which can be expressed in a suitable graphical form. All have the underlying characteristic that ants associated with each part of the problem or solution both affect and are affected by one another, principally by using and modifying route segment information on the graph.

Behaviourally, ants are usually considered to be rather simpler and less competent individually than solitary insects. They are dominated by specific sensory inputs, and often behave as if implementing simple `if-then` rules and functions, but with a strong admixture of random behaviour. Although short- and long-term learning is often used, some real ants appear to become attached to particular tasks. In any assumptions, ants are essentially automata. However, any implementation of interacting high-level agents which uses large numbers of space- or graph-based mobile agents which can interact directly or indirectly via environmental modification will inevitably show some lower level mass behaviours. A nice recent example is described in [298] which deals with pedestrian crowds (high-level agents behaving like simple automata).

An idea of the competition between ants' nests for food sources is illustrated in figure 3.19. There are two episodes in the territorial wars of American honeypot ants as shown in [313]. The first example (figure 3.19(A)) shows how workers from the distant Nest 2 gain food from the food site but the workers from local Nest 1 are engaged in a tournament close to the opposite nest. In the second

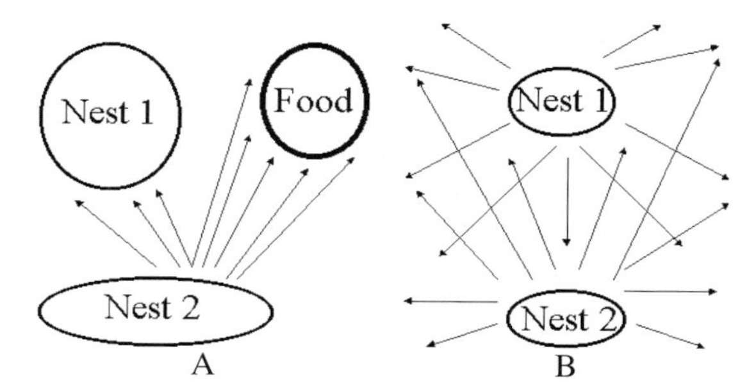

Figure 3.19. Tournament for the food sources: (A) shading of foraging ants; (B) overlapping foraging areas (adapted from [313]).

tournament (figure 3.19(B)) overlapping foraging zones lead to the extirpation of one of the colonies. If we increase the number of colonies and initially connect all nests with food sources by the edges of some virtual graph we can obtain a tournament-based algorithm for the elimination of graph edges, where the edges are weighted by the distances between nests and food sources. These ideas are biologically realized in many ant colonies. However, rather than starting from scratch with a model arbitrarily abstracted from some ant species, it may be more useful and instructive to take as a model a type of ant behaviour, which already appears to operate in a way similar to reaction–diffusion computation, especially to the algorithms on proximity graphs. Such a 'reaction–diffusion friendly' type of ant behaviour is realized in 'diffusive' colonies of ants.

> *A diffusive colony, usually a colony-clan, is a colony distributed on small halls or nests with a flux of inhabitants; that is, it is dispersed into the diffusive colonies or groups of small temporary nests with high exchange by workers among them.* [666]

For example, ants of the species *Cryptocerus texanus* live in the abandoned holes of beetle pupas and impregnated females disperse themselves around the micro-nests [667].

Let $\mathbf{V} \subset \mathbf{R}^2$ be a set of nests of the diffusive family, and $\epsilon : \mathbf{V} \times \mathbf{V} \rightarrow \mathbf{N}$ be an exchange function such that for $p, q \in \mathbf{V}$ $\epsilon(p, q,)^t$ is the number of workers arriving at the nest q from p at time t. The directed graph $\langle \mathbf{V}, \mathbf{E}, \epsilon \rangle$ in which the edges are weighted by ϵ is the exchange graph for the diffusive ant family $\mathbf{V}$.

3.7.1 Graph dynamics in real ants

Let us consider two examples of modification of the worker exchange graph. They are based on observations of Australian ants *Rhytidoponera terros* and *Rhytidoponera metallica* published in [666]. The ants of these species are atypical. There is typically no queen with a monopoly of fertility. Instead,

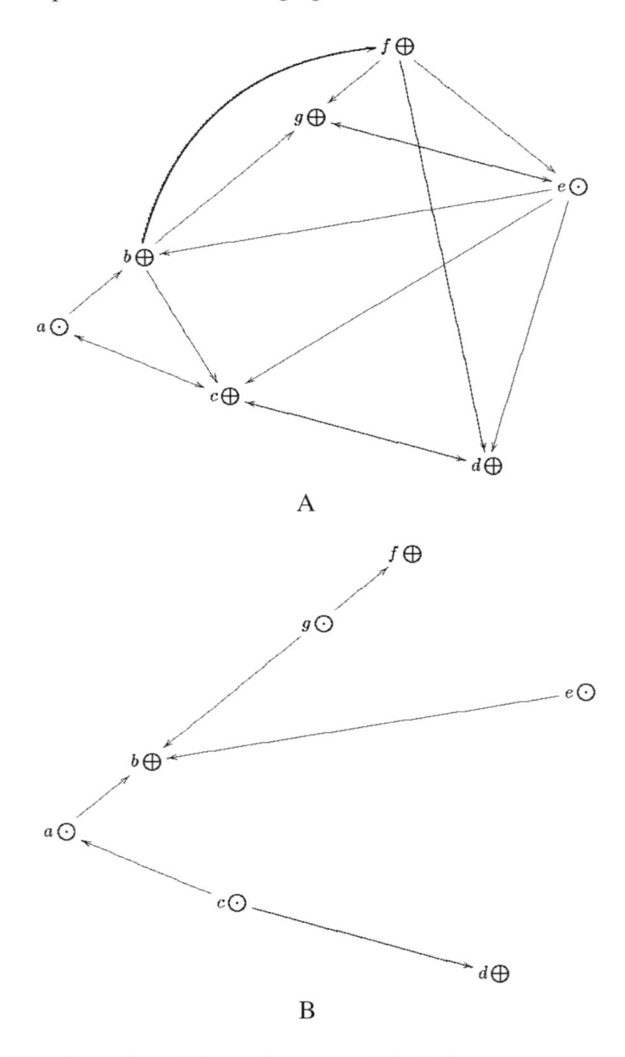

Figure 3.20. Transformation of the exchange graph in *Iridomyrmex purpureus*: (A) first observation; (B) observation of exchange trails made three days later.

reproduction is via fertile workers (all workers are female) which are fertilized by males who sit permanently inside the nests [291]. It is assumed that a high percentage of reproductive individuals with low fertility is a prerequisite for the weak organization of the community as a whole. The colony of *Rhytidoponera* is a group of nests interconnected in the entire system by regular transport by the workers. On average, such a colony consists of 1800 ants with a relative intensity of exchange between nests at the level of about 1.19 ants/min [666].

Therefore, *Rhytidoponera* is a very useful object for investigating re-configuration phenomena in ant societies. It is shown that half of the ants will be redistributed within 18–20 hr or 2–3 light days when every ant participates only once in the transfer. Nests may have both positive and negative balance in the exchange. The sign of the balance may be changed with probability 0.49.

> *The exchange graph of the diffusive colony of* Rhytidoponera *is a system with dynamical equilibrium.*

The graph is highly sensitive to changes in the environment. Its edges change depending on the oscillation of the intensity of the transfers between particular nests, the generation of new nests and food sources. In figure 3.20 we demonstrate how an exchange graph with cycles is transformed into a spanning tree within three days when the intensities decrease in the diffusive family of *Iridomyrmex purpureus*. The graphs are the exchange graphs of the ant family, the nodes are the nests and the edges are the exchange paths. Here we present results modified from [666]. The orientations of the edges have not been taken into account here. In figure 3.20 the symbol $\oplus$ labels nests with a positive exchange balance while $\odot$ marks nests with negative balance. The family under study consists of seven nests. At first observation, the exchange graph (figure 3.20(A)) between the nests contains cycles and the intensities of exchange by workers along the edges are as follows: edges ab, bg, fg, eg and cd have the exchange rate 0.11–0.15 ants/min; edge be has more than 0.15 ants/min rate; edges ac and fe have the rate 0.06–0.10 ants/min; other edges have the exchange rate 0.01–0.05 ants/min. The graph is transformed into an acyclic graph in about three light days.

> *The food sources are the targets where new nests of* Rhytidoponera *are formed. This leads to redistribution of the family and reconfiguration of the exchange graph.*

We can use this assumption from the paradigm of ant diffusive families in the computation of some proximity graphs, namely the relative neighbourhood graph and the minimum spanning tree, on the Euclidean plane. To do this we set food sources on the points of a given set **V**, choose a starting point, for example, the point nearest to the geometric centre, and set one of the nests on the start point. Due to the centrifugal tendency of diffusive families the colony expands, new nests are built close to the food sources and the intensity of the exchange connections is modified. At the first macro-step the colony occupies food sources which lie near the primary nest. It forms second-order nests near food sources. Then third-order nests lying close to the second-order ones are activated etc.

The second example presents our interpretation and understanding of Zakharov's data on the recurrence of the primary federation (a type of stable super family structure) of *Iridomyrmex purpureus* [666]. To investigate the re-computation of the exchange graph after a disturbance, Zakharov [666] removed foragers from the nests and later returned them to the territory spanned by the colony. After some time, the returned foragers again moved between nests according to the previously established exchange graph. It is noted in [666] that with time the volume of exchanges between nests tended to decrease, and the graph of active exchanges tended to be transformed from a highly interconnected graph to one in which exchanges along certain edges were not seen. This led us to investigate the strategy of artificially reducing the volume of traffic on the exchange graph to induce the formation of a spanning tree.

Let an ant colony consist of five nests and C be the matrix of connection intensities between the nests. Initially, the matrix looks as follows (the values were recorded in field experiments in [666]):

$$C^0 = \begin{pmatrix} 0 & 0.8 & 0 & 0 & 0.05 \\ 0 & 0 & 0.063 & 0.05 & 0.058 \\ 0 & 0 & 0 & 0.47 & 1.11 \\ 0 & 0 & 0 & 0 & 2.36 \\ 0 & 0 & 0 & 0 & 0 \end{pmatrix}.$$

We change this matrix in the following iteration. For each pair (i, j) of the nests i and j, $1 \leq i, j \leq 5$, $C_{ij} \leftarrow C_{ij} - \min\{C_{z'z''} : 1 \leq z', z'' \leq 5\}$ and take $C_{ij} = 0$ if $C_{ij} \leq 0$. In the result we have

$$C^1 = \begin{pmatrix} 0 & 0.75 & 0 & 0 & 0 \\ 0 & 0 & 0.58 & 0 & 0.008 \\ 0 & 0 & 0 & 0.42 & 1.06 \\ 0 & 0 & 0 & 0 & 2.31 \\ 0 & 0 & 0 & 0 & 0 \end{pmatrix}$$

$$\vdots$$

$$C^3 = \begin{pmatrix} 0 & 0.28 & 0 & 0 & 0 \\ 0 & 0 & 0.16 & 0 & 0 \\ 0 & 0 & 0 & 0.34 & 0 \\ 0 & 0 & 0 & 0 & 1.6 \\ 0 & 0 & 0 & 0 & 0 \end{pmatrix}.$$

So, in three iterations we obtain the weight matrix of a spanning tree where zero weight means the absence of an edge or connection. In fact, we should mention that even trivial ant foraging patterns have distinctive tree-like structures (see e.g. [272]) for routes from a single nest to food sources.

3.7.2 Ant-approximation of planar proximity graphs

The works by ant biologists which we took as a basis in the formation of the paradigm of diffusive ant-based computing were purely empirical. The exchange patterns between the nests were reported without undertaking any investigation of the mechanisms underlying the formation and preservation of the patterns. In order to adapt the empirical observations to the generation of structures related to proximity graphs, we choose to investigate the consequences of making the following assumption:

> *The intensity of exchanges between nests of a diffusive family is inversely proportional to the distance between them.*

The ant model we now use is not a biological model. It is instead a formal abstraction of a mobile agent with certain simple functions known to be found in certain ant species. We assume that

- an ant can change from one behaviour to another on receipt of a certain signal or signal combination,
- an ant can work directly away from the nest,
- an ant can work directly back to the nest from any position which it might have reached,
- an ant responds primarily to proximal signals, from other ants or from the environment,
- all ants walk at the same speed and
- ants communicate with each other directly, via physical interactions, and indirectly, via environment, e.g. pheromone trails.

We study two basic techniques for constructing a minimum spanning tree. In the first method we construct a relative neighbourhood graph and then derive the minimum spanning tree from the graph. In the second method the minimum spanning tree grows directly, without constructing any intermediary graphs. Both methods are inspired and implemented by ant-based diffusion algorithms.

> *A relative neighbourhood graph of the set* $\mathbf{V}$ *of* n *planar points can be constructed by a diffusive ant family in* $O(1)$ *activations of the nests and in* $O(D(\mathbf{V}))$ *time, where* $D(\mathbf{V})$ *is the diameter of the set* $\mathbf{V}$*. The minimum spanning tree is constructed in* $O(n)$ *activations of the nests and in* $O(D(\mathbf{V}))$ *time.*

Let the ants be at rest, distributed over the nests which are at the points of a given set $\mathbf{V}$. To compute the relative neighbourhood graph, all the nests of a family are activated simultaneously by some signal which causes the workers to walk away from the nests. We assume that the directions in which the workers leave a given nest is effectively distributed uniformly. The result of the receipt of the signal at a given nest will therefore be an expanding circular wave of workers spreading away from the nest. It will be convenient to explain what happens next in the simplified context of a system with only two nests, a and b, forming a two-element set $\mathbf{V}$. First, the wavefronts will meet. If we assume that the ants continue walking in their original directions, the wavefronts will pass through each other and continue to expand. Next, the wave from a will reach b, and *vice versa*, each wave having travelled a distance $d(a,b)$ which is the diameter of $\mathbf{V}$. We

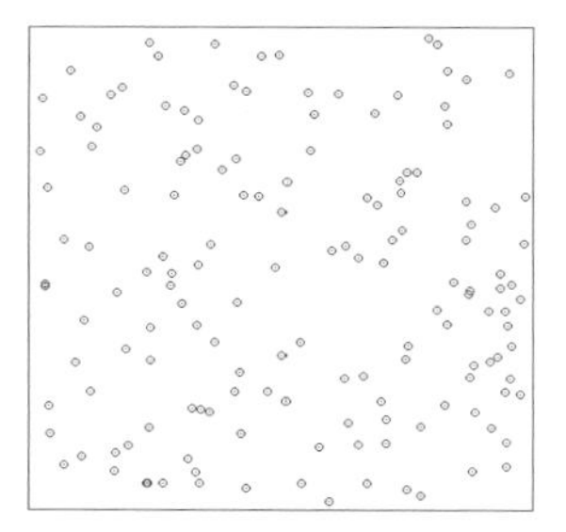

Figure 3.21. Distribution of planar points used in the construction of a relative neighbourhood graph and minimum spanning trees in diffusive families of ants.

assume that an ant reaching a nest immediately reverses its direction and starts walking back towards its own nest, laying a trail as it does so, and that an ant which sees its neighbour reversing also immediately reverses unless at the same time it encounters an approaching ant wavefront. Therefore, the ants in the arcs of the wavefronts, which have passed through each other will reverse the directions of their motion, returning to their original nests, and a trail will be laid in each direction between a and b. When all such trails have been laid, the network of trails forms the relative neighbourhood graph. This may be summarized formally as follows.

Decentralized computing of the relative neighbourhood graph

- `Input`: Given family $\mathbf{V}$ of n nests, on a plane.
- `Procedure`: Diffusion takes place and the exchange graph is computed.
- `Output`: The relative neighbourhood graph $\mathrm{RNG}(\mathbf{V})$ is represented by the set of exchange trails.

- $\mathrm{RNG}(\mathbf{V}) = \emptyset$
- For any nest $v_i \in \mathbf{V}$ `do`
  - `For any nest` $v_j \in u(v_i)$ `do`
    - `If` $B(v_j, d(v_i, v_j)) \cap B(v_i, d(v_i, v_j)) \cap \mathbf{V} = \emptyset$ `then` $\mathrm{RNG}(\mathbf{V}) = \mathrm{RNG}(\mathbf{V}) \cup \{v_j\}$

An example of the growth of the relative neighbourhood graph during the evolution of an exchange graph between 150 nests (distribution of the nest is shown in figure 3.21) of a diffusive ant colony is shown in figure 3.22. In this experiment the diameter of the planar set is 400 units, and the waves of workers spread with unit velocity. All nests, or micro-nests, are activated at the same time and exchange trails are formed mutually between relatively close nests.

Once the relative neighbourhood graph, which is also the exchange graph of the diffusive family, is constructed it is necessary to prune it so that cycles are eliminated and the sum of the remaining edges is minimal.

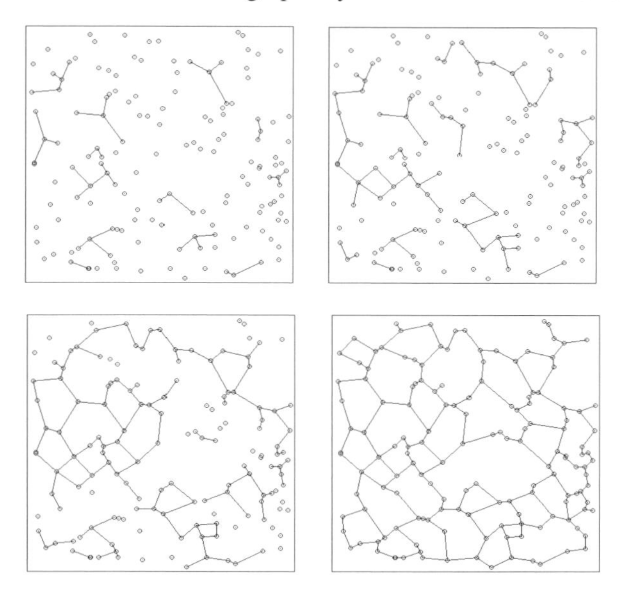

Figure 3.22. Several snapshots of a relative neighbourhood graph constructed by a diffusive ant family.

This technique for constructing the relative neighbourhood graph intends ants to form trails of minimal length to connect the nests of the family in the whole unit. This is consistent with field-based experiments [666]. The second technique employs another biological paradigm.

Initially, the ants are based at the same single nest representing one, arbitrarily chosen, point of the given planar set **V**. All other points of the set **V** are assumed to be represented by food sources. Therefore, in contrast with the previous instalment, we originally have only one nest but $n-1$ food sources. The ants start exploring the space searching for the food sources and, when they find one, they set up another nest at the place of this source of food and continue to explorate the space. Such experiments have never been done with real ants but nevertheless the idea does not contradict biological intuition.

In the following algorithm it is assumed that $\alpha(v) \in \{\texttt{Rest}, \texttt{Active}\}$ characterizes the activity of the nest v and $\mathbf{S}_i$ is a set of vertices connected to v_i.

Decentralized relative neighbourhood graph and rooted minimum spanning tree

- `Input:` The graph RNG(**V**) of exchanges between nests which represent elements of **V**
- `Output:` The graph of exchanges representing the MST(**V**)

- Choose one element of $\mathbf{V}$ and make it `Active`
- Repeat
  - For any nest $v_i \in \mathbf{V}$: $\alpha(v_i) = \texttt{Active}$ do
    - * For any nest $v_j \in \mathbf{V}$: $(v_i v_j) \in \mathrm{RNG}(\mathbf{V})$ do $\mathbf{S}_i = \mathbf{S}_i + \{v_j\}$
    - * For any pair $v_i, v_j \in \mathbf{V}$: $(v_i \neq v_j)$ and $\alpha(v_i)$ and $\alpha(v_j)$ do
    - * If $\mathbf{S}_i \cap \mathbf{S}_j \neq \emptyset$ do if $d(v_i, v_z) \leq d(v_j, v_z)$ then $\mathbf{S}_j = \mathbf{S}_j - \{v_z\}$ else $\mathbf{S}_i = \mathbf{S}_i - \{v_z\}$
  - For any $v_i \in \mathbf{V}$: $\alpha(v_i) = \texttt{Active}$ do
    - * $\alpha(v_i) = \texttt{Rest}$
    - * For any $v_z \in \mathbf{S}_i$ do $\alpha(v_z) = \texttt{Active}$
- Until $\exists v \in \mathbf{V}$: $\alpha(v) = \texttt{Active}$.

The computation process resembles the spread of excitation on graphs. Competition between nests starts at one arbitrarily chosen nest, the root nest, and then spreads from nest to nest by the exchange paths. The elimination of edges as a result of nest competition leads to the elimination of cycles. The elimination of the longest edges leads to the minimality of the constructed tree.

Now we can consider an algorithm for a straightforward construction of a minimum spanning tree without employing any other auxiliary graphs.

> *A minimum spanning tree of n planar points of the set* $\mathbf{V}$ *can be approximated by the development of a diffusive ant family in* $O(n)$ *reallocations of nests, or occupations of food sources, in* $O(d(\mathbf{V}))$ space.

Take the set $\mathbf{V}$ of labelled planar points and place a nest of a diffusive family close to one of the points, and food sources at each of the others. Following the centrifugal development tendency of diffusive families, the ants will build new nests near the food sources. Nests will communicate with each other by exchange graphs, as previously discussed.

Rooted minimum spanning tree

- `Input:` The primary nest v and the food sources $\mathbf{V} - \{v\}$
- `Procedure:` The development of the distribution of a diffusive ant family
- `Output:` The minimum spanning tree $\mathrm{MST}(\mathbf{V})$

Let $\mathbf{S}(v)$ be a set of nests activated by nest v and $\alpha(v) = \texttt{Active}$ initially.

- Repeat
  - For any $v \in \mathbf{V}$: if $\alpha(v) = \texttt{Active}$ do
    - * $\mathbf{S} = \emptyset$
    - * Find $v' \in \mathbf{V}$ such that $d(v, v') = \min\{d(v, v'') : v'' \in \mathbf{V}$ and $B(v, d(v, v'')) \cap B(v'', d(v, v'')) \cap V = \emptyset\}$
    - * $\mathbf{S}(v) = \mathbf{S}(v) \cup \{v'\}$

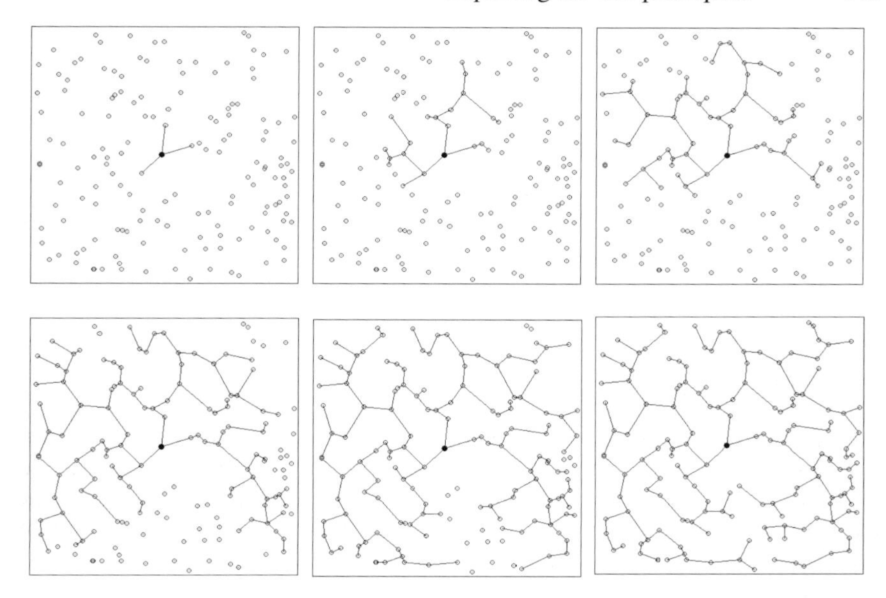

Figure 3.23. Several snapshots of the construction process of a minimum spanning tree by a diffusive ant family.

- For any $v \in \mathbf{V}$: if $\alpha(v) = \texttt{Active}$ do
  - $*$ $\alpha(v) = \texttt{Rest}$
  - $*$ For any $v' \in \mathbf{S}(v)$ do $\alpha(v') = \texttt{Active}$

- Until $\exists v \in \mathbf{V}$: $\alpha(v) = \texttt{Active}$, i.e. there are some food sources not occupied by the nests

The graph, computed in this way, will be an approximation of the minimum spanning tree. An example of the growth of the spanning tree by a diffusive ant family is shown in figure 3.23.

At the end of this section we would like to produce some curious pictures. Those made along the lines of a rational geography paradigm look quite impressive. They are the relative neighbourhood graph (figure 3.24(A)), Gabriel graph (figure 3.24(B)) and minimum spanning tree (figure 3.24(C)) computed by diffusive ant families. To compute the graphs we place either nests or food sources at the main cities of the British Isles and let the ants run, expand their family diffusively and form the exchange graphs.

3.8 Exploring the complex space

There are surprisingly few results that make any connections between cellular automata and games. The main reason for this is that most spatial games

Figure 3.24. Relative neighbourhood graph (A), Gabriel graph (B) and minimum spanning tree (C) of British cities constructed by an algorithm derived from diffusive ant families.

are strictly serial and became extremely complicated if parallelism is involved. Published results are concentrated around the chip-firing game that is based on the physical phenomena related to sand-pile models [588, 51, 85]. Here we consider one game that is not simply played by cellular automata but also benefits from a reaction–diffusion paradigm [18].

The game of Minesweeper is familiar to everyone. The rules are extremely simple: mines are distributed on the cells of a finite two-dimensional lattice. Uncovering the site of the lattice we explode ourselves (if the site contains a mine) or discover the number of mines around the empty site. Initially all mines are hidden. We wish to mark all of them. In this section we design a two-dimensional deterministic cellular automaton which marks all mines populated on the lattice in a time proportional to lattice size. Every cell of the automaton has a 25-cell neighbourhood and takes at most 27 states.

Let us look at the standard kit of the very simple games supplied with any distribution of Microsoft Windows. There we find the Minesweeper game with its clear rules and hidden parallel nature. Starting with the *Help* section we read

> *You are presented with a mine field, and your objective is to locate all the mines as quickly as possible. To do this, you uncover the squares that do contain mines. The trick is determining which squares are which.* [452]

We say the Minesweeper problem is solved deterministically if all the moves that can be done without guessing are done, i.e. by playing device halts. We demonstrate the following.

> *The Minesweeper problem is solved deterministically in a two-dimensional cellular automaton in a time proportional to the lattice size.* [18]

Given mines distributed inside a bounded region and empty sites filled with ciphers indicating how many neighbouring mines exist around a lattice cell, we wish to construct a two-dimensional cellular automaton that, starting with the few uncovered sites, marks all mines populated on the $n \times n$-node lattice in $O(n)$ time. So, to prove the proposition we construct a cellular automaton, every cell of which has a 25-cell neighbourhood and 27 states, and demonstrate that a cellular automaton usually uncovers the majority of the lattice nodes without exploding itself, subject to the absence of impassable barriers.

Before starting to design the cell state transition rules we should remember the following tip.

> *Knowing when to mark a square as a mine is the key to winning the game.* [452]

When we ask whether a cell contains a mine, we should analyse not only the closest neighbours of this cell but also the second-order neighbours. Failing to do this leads to a global explosion (figure 3.25). So, the neighbourhood of cell x will consist of 25 cells, i.e. it will be the complete neighbourhood of radius 2:

$$u(x_{ij}) = (x_{i-2j-2}, \ldots, x_{ij}, \ldots, x_{i+2j+2})$$

$i, j = 1, \ldots, n$.

The state of a cell should indicate whether the cell contains a mine and, if not, how many mines are around the cell. Therefore, the cell state set is a direct product

$$\mathbf{Q} = \{\bullet, \star, \circ\} \times \{0, 1, \ldots, 8\}$$

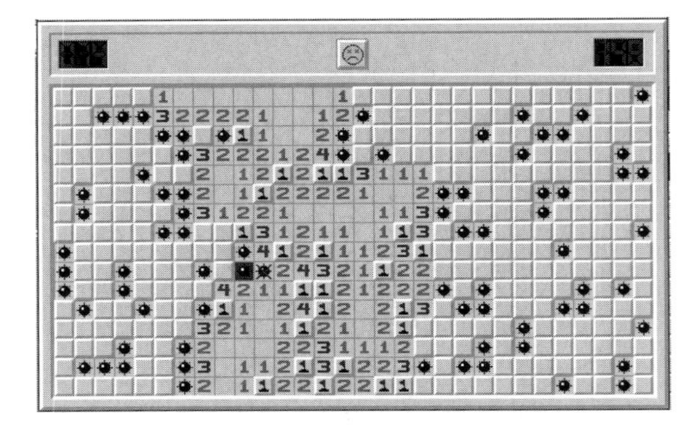

Figure 3.25. Unsuccessful finish of a Minesweeper game: the wrong cell has been uncovered.

where

- $x^t = \star$ means *it is known at time t that the cell x contains a mine*;
- $x^t = \circ$ means *at time t site x is open*;
- $x^t = \bullet$ means *site x is closed at time t*;
- ciphers $0, 1, \ldots, 8$ indicate the number of mines in the first-order neighbours: $(x_{i-1j-1}, \ldots, x_{i+1j+1})$.

A duple $\langle x^t, \nu(x)\rangle$, where $x^t \in \{\bullet, \circ, \star\}$ and $\nu(x) \in \{0, 1, \ldots, 8\}$ completely determines the state of cell x at a moment t. Due to the rules of the game the neighbours of x are allowed to know $\nu(x)$ only when $x^t = \circ$.

When we play the game, we decide whether a site is mined or not, basing our decision on the correlations between the ciphers drawn in the neighbouring open sites and numbers of mines around them. The deterministic player uncovers the site if he/she finds that there is at least one neighbouring open site, a cipher which corresponds to the number of already marked mines, or there is a neighbouring open site with zero cipher. If a player does not mean to uncover a site he/she may mark it by clicking on the mine.

There is a straightforward way to detect a mine:

> *the site is definitely mined if it is a single covered neighbour of such an uncovered site, on which the cipher is more than the number of marked mines around.* [452]

The second entry of the duple $\langle x^t, \nu(x)\rangle$ is predefined and does not change in the game. A cell once uncovered cannot be covered again. This means that the $\circ$ state is an absorbing state in a cell state transition diagram. The state $\star$ is also absorbing. Therefore we determine the conditions for two transitions: $\bullet \to \circ$ and $\bullet \to \star$, and assume that the transition $\bullet \to \bullet$ happens otherwise. The following hints may be useful.

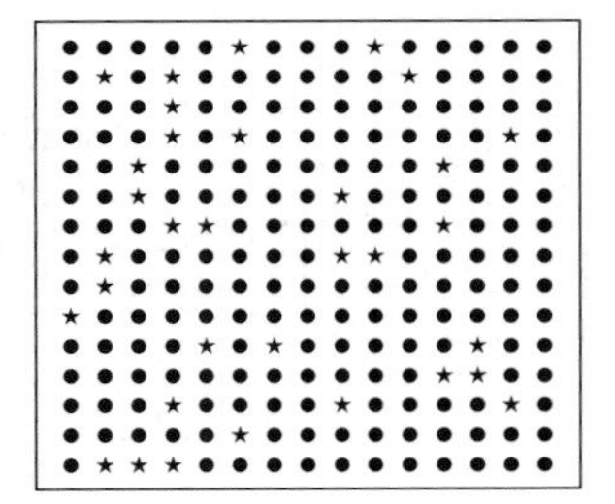

Figure 3.26. Distribution of the mines is unknown for the cellular automaton before the game. The mines are shown by $\star$s.

If an uncovered square already has the correct number of adjacent mines marked, clear around it. [452]

If an uncovered square is labelled 1, *and there is only one covered square touching it, that covered square must be a mine.* [452]

We employ the first hint when determining the transition $\bullet \to \circ$. The second hint plays its role in implementation of the transition $\bullet \to \star$.

We offer the following rule for cell state transitions:

$$x^{t+1} = \begin{cases} \circ & x^t = \bullet \text{ and } \exists y \in u_1(x) : y^t = \circ \text{ and } (\nu(y) = 0 \\ & \text{or } \nu(y) = \sum_{z \in u_1(y)} \chi(z^t, \star) \\ \star & x^t = \bullet \text{ and } \exists y \in u_1(x) : (y^t = \circ \text{ and } \sum_{z \in u_1(y)} \chi(z^t, \bullet) = 1 \\ & \text{and } |\nu(y) - \sum_{z \in u_1(y)} \chi(z^t, \star)| = 1) \\ \bullet & \text{otherwise} \end{cases}$$

where $u_1(x_{ij}) = (x_{i-1j-1}, \ldots, x_{i+1j+1}) \backslash x_{ij}$, $u_1(x_{ij}) \subset u(x_{ij})$, and $\chi(z^t, a) = 1$ if $z^t = a$, $a \in \mathbf{Q}$, and $\chi(z^t, a) = 0$ otherwise.

An entirely covered lattice will never be uncovered.

A cell x changes its state $\bullet$ to the state $\circ$ if it has at least one neighbour in the state $\circ$. Therefore, for every cell x of the lattice we have $x^t \neq \circ \to x^{t+1} \neq \circ$. At the first step of the game we help the cellular automaton and deliberately uncover several lattice sites without mines for it because only zero-cipher sites from the given ones can be the sources, or generators, of the uncovering waves.

Let us consider a two-dimensional lattice with mines randomly distributed in its cells (figure 3.26).

Initially we uncover two sites, which do not contain mines, for the automaton (figure 3.27, $t = 0$). These sites generate waves of cell uncovering (figure 3.27,

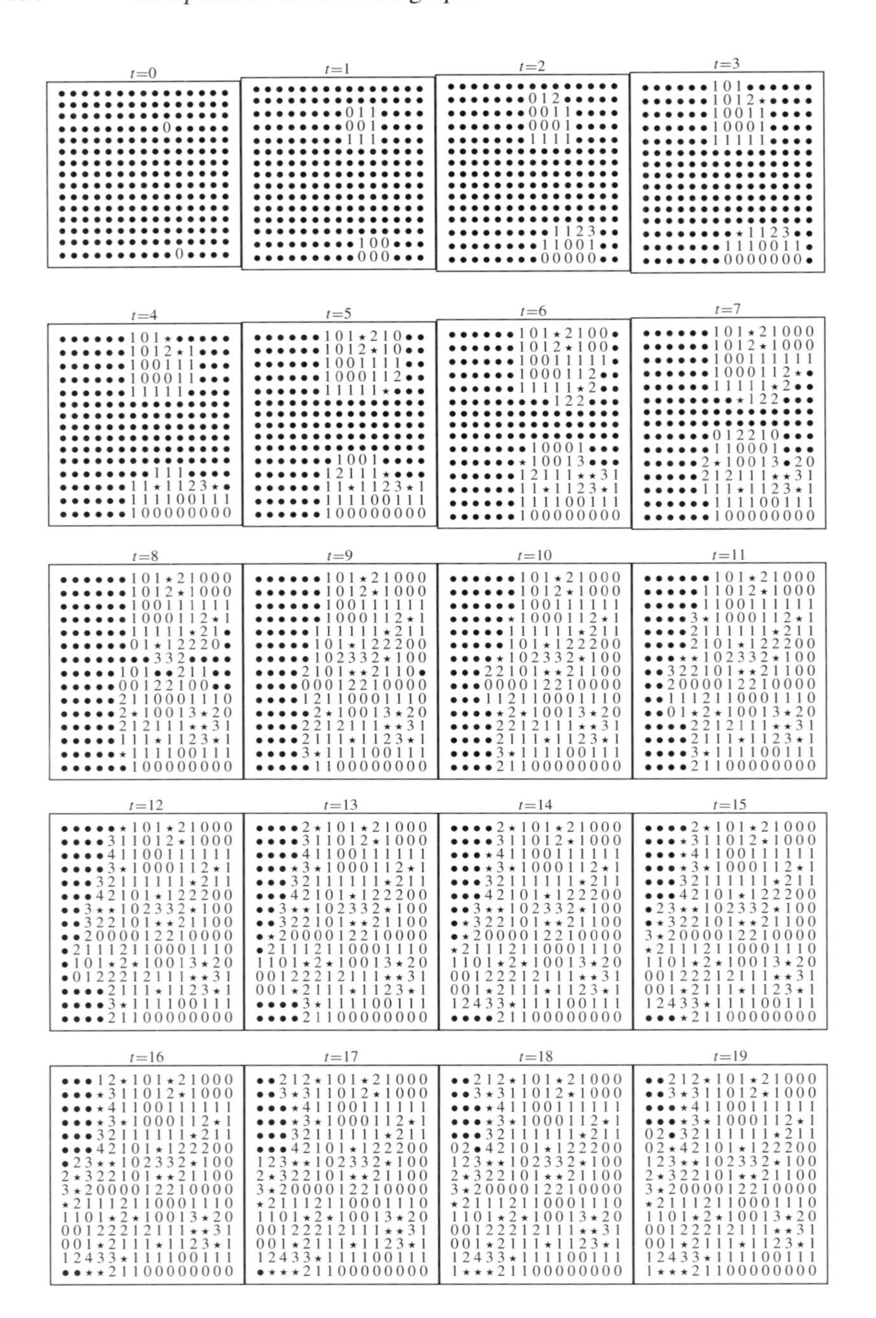

Figure 3.27. Evolution of a two-dimensional cellular automaton which plays Minesweeper.

t=20

```
••212⋆101⋆21000
••3⋆311012⋆1000
•••⋆41100111111
01•⋆3⋆1000112⋆1
02⋆32111111⋆211
02⋆42101⋆122200
123⋆⋆102332⋆100
2⋆322101⋆⋆21100
3⋆2000012210000
⋆21112110001110
1101⋆2⋆10013⋆20
00122212111⋆⋆31
001⋆2111⋆1123⋆1
12433⋆111100111
1⋆⋆⋆21100000000
```

t=21

```
••212⋆101⋆21000
••3⋆311012⋆1000
114⋆41100111111
013⋆3⋆1000112⋆1
02⋆32111111⋆211
02⋆42101⋆122200
123⋆⋆102332⋆100
2⋆322101⋆⋆21100
3⋆2000012210000
⋆21112110001110
1101⋆2⋆10013⋆20
00122212111⋆⋆31
001⋆2111⋆1123⋆1
12433⋆111100111
1⋆⋆⋆21100000000
```

t=22

```
••212⋆101⋆21000
•⋆3⋆311012⋆1000
114⋆41100111111
013⋆3⋆1000112⋆1
02⋆32111111⋆211
02⋆42101⋆122200
123⋆⋆102332⋆100
2⋆322101⋆⋆21100
3⋆2000012210000
⋆21112110001110
1101⋆2⋆10013⋆20
00122212111⋆⋆31
001⋆2111⋆1123⋆1
12433⋆111100111
1⋆⋆⋆21100000000
```

t=23

```
•1212⋆101⋆21000
1⋆3⋆311012⋆1000
114⋆41100111111
013⋆3⋆1000112⋆1
02⋆32111111⋆211
02⋆42101⋆122200
123⋆⋆102332⋆100
2⋆322101⋆⋆21100
3⋆2000012210000
⋆21112110001110
1101⋆2⋆10013⋆20
00122212111⋆⋆31
001⋆2111⋆1123⋆1
12433⋆111100111
1⋆⋆⋆21100000000
```

t=24

```
11212⋆101⋆21000
1⋆3⋆311012⋆1000
114⋆41100111111
013⋆3⋆1000112⋆1
02⋆32111111⋆211
02⋆42101⋆122200
123⋆⋆102332⋆100
2⋆322101⋆⋆21100
3⋆2000012210000
⋆21112110001110
1101⋆2⋆10013⋆20
00122212111⋆⋆31
001⋆2111⋆1123⋆1
12433⋆111100111
1⋆⋆⋆21100000000
```

t=25

```
11212⋆101⋆21000
1⋆3⋆311012⋆1000
114⋆41100111111
013⋆3⋆1000112⋆1
02⋆32111111⋆211
02⋆42101⋆122200
123⋆⋆102332⋆100
2⋆322101⋆⋆21100
3⋆2000012210000
⋆21112110001110
1101⋆2⋆10013⋆20
00122212111⋆⋆31
001⋆2111⋆1123⋆1
12433⋆111100111
1⋆⋆⋆21100000000
```

Figure 3.27. (Continued)

$t = 3$). They travel on the lattice. The waves of cell uncovering change shape with time (figure 3.27, $t = 3, 4, \ldots$) due to the random distribution of the mines around the mine field. The whole wavefront becomes fractured like a coast line. Ideally, the waves of uncovering or marking the mines span the whole lattice. A game is assumed to be finished when the cellular automaton reaches the fixed point of its global evolution, $c^{t+1} = c^t$, i.e. its configuration becomes stationary (figure 3.27, $t = 27$).

Starting in an arbitrary configuration the cellular automaton, playing Minesweeper, finishes its evolution in a stationary configuration.

This is because two of the three cell states are absorbant, and once a cell changes its state • to any other state it never changes state again.

Our cellular automaton is deterministic. Unlike a human player, it does not guess. This leads to the situation when there are regions which cannot be opened. It is particularly clear in the case of non-periodic boundaries. The impassable regions have such combinations of closed, opened and marked sites on their boundaries that make it impossible for waves to move through it (figure 3.28). Computational experiments show there is little chance for a cellular automaton to evolve in such a configuration where the whole field is open and has an island of uncovered sites. Rather, particularly when the ratio of mines is large enough, the game is finished when tiny islands, originating from the initially uncovered single sites, of open and marked sites appear in the ocean of covered sites.

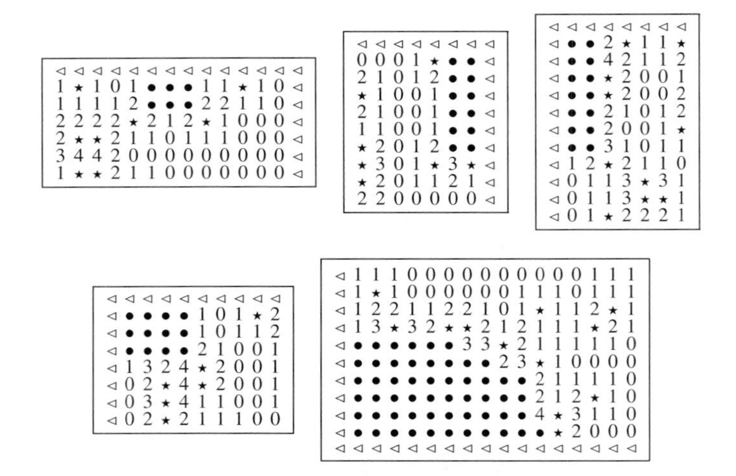

Figure 3.28. Islands with the impassable barriers. The symbol ◃ indicates boundaries of the lattices. The sub-configurations are cut out from the larger configurations of a game field.

A computation wave spreads with unit velocity; therefore, having one site uncovered at $t = 0$ we can expect that in the worst case the computation wave reaches the boundaries of the game field in $O(n)$ steps.

The cellular automaton starting with three uncovered unmined sites loses at least one of two games when the site of a lattice is mined with probability at least 0.15.

This is the result of numerical experiments. We found that the critical probability of mine distribution is 0.10–0.16. In about half of the games around half of the sites remain uncovered when the playing field is a torus, and every lattice site is mined with the probability 0.10–0.14.

One of the reasons is that the cell-uncovering waves spread only from the sites indexed with 0, i.e. with no mines around. If, for example, all three open sites have one or more cipher, then there is no uncovering wave at all, and such a configuration for the cellular automaton remains stationary. On the other hand, the cellular automaton cannot uncover or mark a site in the situation of local uncertainty. Therefore, increasing the ratio of mines, we increase the probability that the initially uncovered site with a zero cipher becomes encircled with impassable barriers.

3.9 Load balancing in communication networks

The idea of reaction–diffusion computing with a swarm of simple agents, or intelligent particles, can be developed further in the direction of massively parallel computing in agent collectives, i.e. we propose to employ enormously large numbers of elementary entities. They are able to diffuse, flow or randomly walk through a space or over a graph or network, possibly being affected by some local conditions of their environment. In the limit this situation reduces to electrical or chemical diffusion and their automata models [624, 431, 132, 604, 577, 140]. A brilliant example is field computing, where a graph is expressed as a network of resistances, and the current flow in response to applied voltages is interpreted as a solution to some difficult problem.

Scarcely more complex are the reaction–diffusion systems of real or simulated chemicals which diffuse away from nodes and form immobile precipitates when they encounter different chemicals from other nodes; these systems can construct various useful spatial structures such as Voronoi partitions and skeletons (e.g. [17, 600]). This class of models was probably first developed by Tsetlin [604], Stefanyuk [576] (for a historical review see [577]) and Rabin (see [140]). In particular, Tsetlin's book [604] identifies many of the features associated with biologically inspired automata: randomness, self-organization and spatial distribution. The absence of direct interactions between agents was also implied. These principles were applied to the same types of problems that seem attractive now—to the control of telecommunications networks [115] and groups of radio stations [576]. Rabin also introduced the idea of jumping and walking pebble automata that could solve problems on graphs and lattices by interacting with the consequences of their previous actions using what is now called stigmergy.

The ant-based computation paradigm was developed in mathematical biology and biology-inspired robotics [76] independently of primary physical and automata models, and has only recently moved toward the mainstream of distributed computation models (see e.g. [139]). The basic principles underlying the possible use of ant algorithms for the control of telecommunication networks were set out in [53]: mobility of the agents, interactions mediated by the environment with no direct contact between the agents, decay of messages with time, stochastic components of agent behaviour, and large numbers of agents. These principles can be found in more recent works on network control [553, 552, 96] and distributed combinatorial optimization [139, 191, 252]. Another significant area of application of ant-like algorithms and computation with mobile agents deals with the problems of computational geometry, particularly the approximation of proximity graphs [22, 20].

In this section we explore the potential of some new methods for controlling packet-switched networks with dynamically changing topology and variable traffic patterns. The routing tables controlling packet-switching are updated by mobile agents, or intelligent particles, without any central control and with no *a priori* knowledge of the topology of the network.

3.9.1 Diffusion-like algorithms

The *echo* algorithm by Ahuja and Zhu [40] is one of the first diffusion-based techniques for the construction of the minimum weight spanning tree. During the execution of the algorithm messages are sent by the *root* down the tree (down-wave). When the messages arrive at the leaves they are turned into echoes and run back up the tree (up-wave) [40]. In the down-wave the root sends information to all the nodes, and it receives information collected from all the nodes in the up-wave phase. The down- and up-waves are repeated until the minimum spanning tree is constructed. We discuss now how a similar technique can be realized in reaction–diffusion systems.

Let **G** be a lattice where every node is connected by undirected edges with its four closest neighbours. The state of every node x at time step t is characterized by the tuple $\langle c_x^t, p_x^t \rangle$ where c_x^t is the concentration of a reagent and p_x^t is the orientation of a pointer. The concentration variables take their values from the real interval [0, 1] and p_x^t points either to the neighbour y, $p_x^t = y$, $(x, y) \in \mathbf{E}$, that is the closest to the root x_s of the spanning tree that is under construction, or nowhere, $p_x^t = \cdot$. Initially all nodes except x_s have no reagent and their pointers are not oriented, i.e. for any $x \in \mathbf{V}$ we have $c_x^{t=0} = 0$ and $p_x^{t=0} = \cdot$, whereas $c_{x_s}^{t=0} = 1.0$. The diffusion equations can be written as follows:

$$\dot{c_x} = \delta\left(\sum_{y:(x,y)\in\mathbf{E}} c_y - c_x \right)$$

where $\delta = 0.25$. Space is therefore discrete but time is continuous.

The pointers change their states by the rule:

$$p_x^{t+1} = \begin{cases} y & (x, y) \in \mathbf{E} \text{ and } c_y^t = \max_{z:(x,z)\in\mathbf{E}}\{c_z^t\} \text{ if } p_x^t = \cdot \\ p_x^t & \text{otherwise.} \end{cases}$$

An example of a concentration profile and a pointer field for the case of static uniform weights of edges is shown in figure 3.29; for $x_s^{t=0} = 1.0$, **G** is a grid of 31×31 nodes, $x_s = (15, 15)$, and $\delta t = 0.1$; the results are shown at the 30th step of the simulation.

In this version the distance minimal tree is approximated. However, we can also take loading, or spare capacity as the inverse of loading, at the nodes in consideration. Let σ_x be the spare capacity at node x. Then the diffusion equation takes the form

$$\dot{c}_x = \delta\left(\sum_{y:(x,y)\in\mathbf{E}} c_y - c_x \right) - \sigma_x^{-1}$$

assuming $\sigma_x > 0$ and $c_x := 0$ if $c_x \leq 0$. After such a modification the maximum spare capacity of the spanning tree rooted at x_s is approximated. An example of a numerical simulation is given in figure 3.30 ($c_s^t = 500.0$ every time step and σ_x is a random variable distributed uniformly in the interval [1, 10.0]).

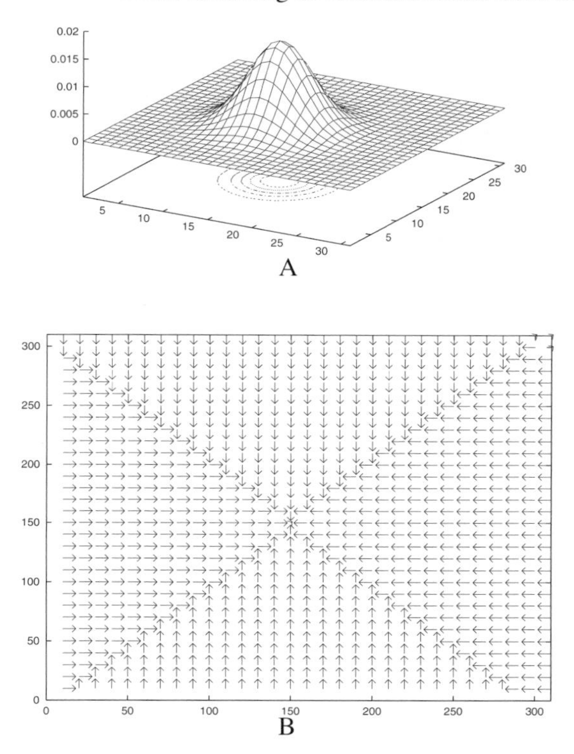

Figure 3.29. Concentration profile (A) and orientations of pointers (B) at the 30th step of a reaction–diffusion tree constructor.

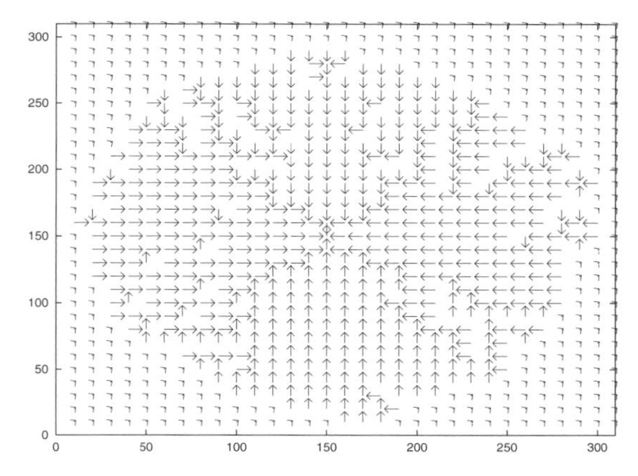

Figure 3.30. An example of a random distribution of load capacities in a rectangular communication network.

As we see from the pointer equation the state of the pointer is changed only once; that is, the technique is not adaptive. To make it adaptive we have to reset the pointer states before every round of the diffusive approximation of minimum weight/capacity trees or spread the orientation of a pointer amongst all neighbours of the node, making it fuzzy or probabilistic.

We can now discuss routing tables rather than simply deterministic pointers. The routing table based at node x of a grid $\mathbf{G}$ has four entries (one entry per neighbour node because we are approximating a tree with a *given root*) per root x_s. Here we still consider a single root x_s, that is we are interested in only sending packets to node x_s, therefore we deal with a real valued vector of order 4.

Let us now use three reagents, α, β and γ. The α reagent diffuses from the root node in the same manner as it did in the former model. Imagine now that every edge (x, y) consists of two tubes (x, y) and (y, x) filled with reagent β, β is a catalyst:

$$\alpha + \beta \rightarrow \gamma + \beta$$

the γ reagent is produced as a result of the reaction between α and β. The concentration e_{xy} of γ at tube (y, x) of edge (x, y) reflects the amount of the reagent α that has passed through the tube:

$$\dot{e}_{xy} = f(c_y - c_x).$$

A numerical implementation of function f can vary from model to model. If we need to know to which node a given node x is connected at time step t of system's evolution we can compute the state of the pointer as

$$p_x = y : e_{xy} = \max_{z:(x,z)\in\mathbf{E}} e_{xz}.$$

If we are anxious about the growing values of e_{xy} we can make them 'probabilistic' via normalization:

$$e_{xy} := \frac{e_{xy}}{\sum_z e_{xz}}.$$

Now we can establish the state of the pointer at every real step of the evolution time. Let us check again how good this technique is.

Using spatial analogies we can think of overloaded nodes as obstacles placed on the grid $\mathbf{G}$. Several different versions of obstacles are represented on a grid (figure 3.31, overloaded nodes are diamonds) by groups of nodes with $\sigma_x = 2$; the rest of the nodes have $\sigma_x = 100$.

A drop of reagent is placed at node $x_s = (15, 15)$: $c^0_{x_s} = 100$. The reagent diffuses in such a manner that the diffusion wave velocity is inversely proportional to the loading of the nodes (see figure 3.32). After a certain amount of diffusion time the distribution of the pointer orientations (figure 3.33) finds the optimal tree rooted at x_s. Moreover looking at the dynamics of the diffusion we see that the

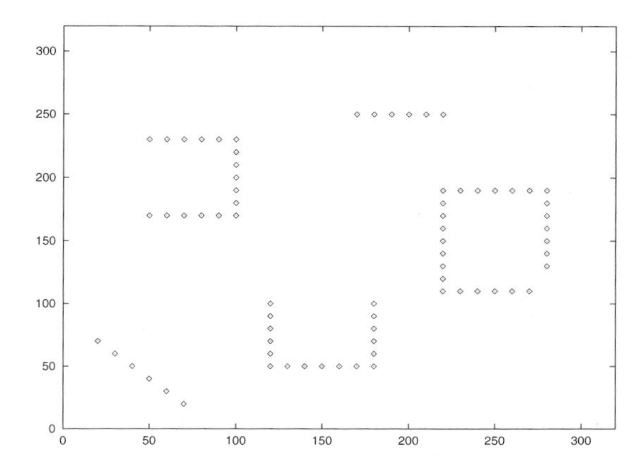

Figure 3.31. Overloaded nodes on the communication mesh are shown by $\diamond$s.

reagent flux through the obstacle nodes is virtually blocked and almost all the diffusion goes through the 'open' parts of the grid.

Therefore we know that a spanning tree is approximated using diffusion, that this is the minimum weight tree when the nodes or edges are weighted, and that it can be dynamically recomputed at any time in the system's evolution. The system reacts smoothly to changes in loading or spare node capacity.

Let us introduce a second entity that moves along the branches of the tree towards the root. This may be the diffusion of another reagent or the movement of packets. We know from the introductory sections that we can hardly see the difference between diffusion and a random walk, or diffusion along gradients and packet flows; however, as soon as we deal with communication networks, we have to consider packets of information that are discrete entities.

3.9.2 Ants on the net

Here we discuss a modified version of the Schoonderwoerd–Holland–Bruten–Rothkrantz (SHBR) algorithm, initially designed for circuit-switched networks [552, 553], on a packet-switched network.

The original version of the SHBR algorithm is tested on the British Telecom synchronous digital hierarchy network (figure 3.34). For a comparison with previous results we also ran our algorithms on this network.

In this section we aim to find the answers to the following questions:

- Does 'discrete' reaction–diffusion perform well when the numbers of diffusing components are limited?
- What are the advantages of auto-catalytic self-localization of minimum load routes, using the pheromone paradigm?
- Do probabilistic agents perform better than deterministic ones?

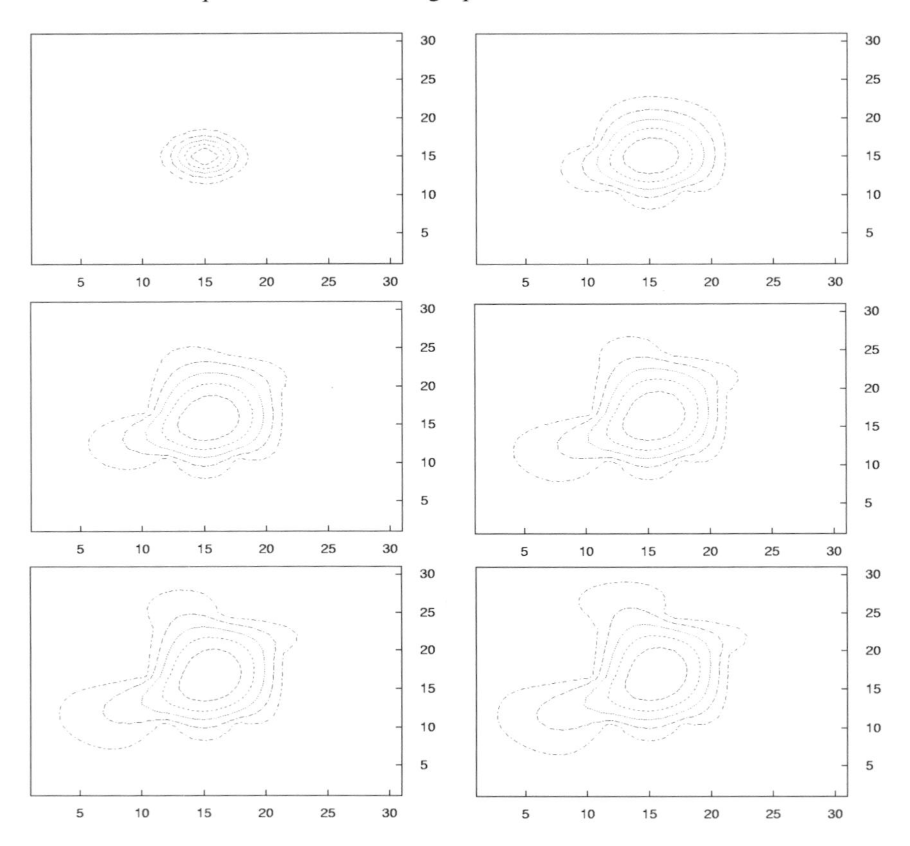

Figure 3.32. Diffusion of a reagent on a mesh with overloaded nodes. The concentration profiles are shown for several steps of a medium's evolution.

The basic model considered here, the S1-model, involves two types of messages: data packets, emitted with some defined generation rate, and control packets, or ants, which are generated at the rate of one ant per node per time step. Every ant can be represented by the tuple (x_s, x_d, x, a), where x_s and x_d are the source node and destination node of the ant, x is the current position of the ant in the network, and a is the so-called *age* of the ant. Every node x of the network has a routing table R with entries that look like R_{xyx_d}. The entries of the routing table are updated by the ants:

$$R_{xyx_s} \leftarrow R_{xyx_s} + f(a^{-1})$$

so that some quantity proportional to the age of an ant is added to the appropriate entry. The updating of routing tables by ants is an analogue of pheromone-laying on ant trails.

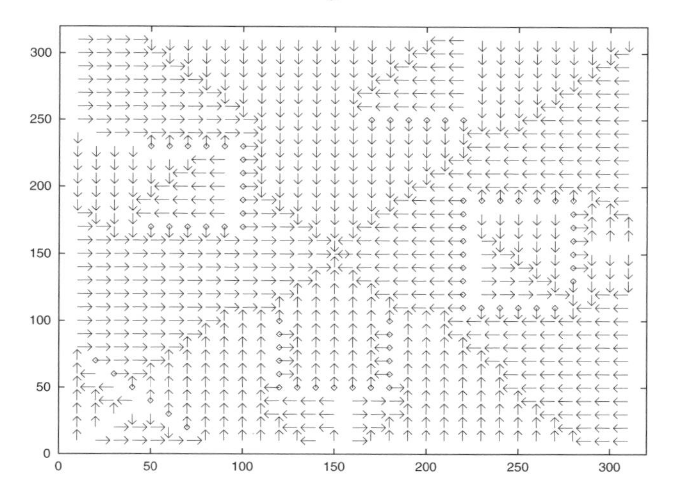

Figure 3.33. Overloaded nodes (◇s) and orientation of pointers in the reaction–diffusion model of load balancing.

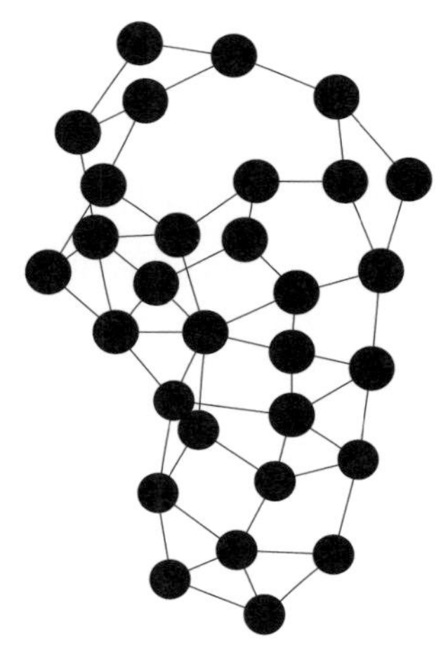

Figure 3.34. Topology of synchronous digital hierarchy network of British Telecom used in the simulations.

At every step of the simulation the routing tables are normalized to make them look like probabilities: for every x_d

$$\sum_{y:(x,y)\in\mathbf{E}} R_{xyx_d} = 1.$$

Ants may also age when they encounter queues at nodes: $a \leftarrow a + g(l)$, where l is the length of a queue. Travelling to destination x_d, and being at node x

at the current step of simulation time a packet is routed to the node y neighbouring x where the entry R_{xyx_d} is maximal amongst all other entries R_{xzx_d}, $(x, z) \in \mathbf{E}$.

To test our ideas we investigated three modifications of the SHBR algorithm:

- *S2-model.* The ants move as in the S1-model but they do not have a load-dependent age. Age is simply increased by distance measured in nodes: when an ant leaves its current node x its age increases by one unit, $a_x \leftarrow a_x + 1$.
- *S3-model.* The ants perform as in the S1-model except that they choose the next node at random. Being in node x at time step t, an ant goes to node y with probability k^{-1}, where k is the degree of x.
- *S4-model.* The ants follow their routing tables deterministically; that is, the ant (x, x_s, x_d) chooses to go to node y corresponding to the maximal entry of the routing table, i.e. y is such that

$$R_{xyx_d} = \max_{z:(x,z)\in\mathbf{E}} R_{xzx_d}.$$

In all except this, the ants behave as in the S1-model.

Here we present some results on the simulation of ant-based control on the British Telecom topology network (figure 3.34). The simulation parameters were chosen to conform in general with the data-packet and ant-packet generation parameters in the Schoonderwoerd *et al* model [552, 553]. In our toy model

- every node of the network has a capacity of 40 packets equally split amongst the queues of the outgoing links;
- the capacities of the incoming links are not taken into consideration;
- the packets are launched from randomly uniformly chosen nodes towards destination also chosen at random;
- at every time step in the simulation no more than 1.2 K packets can be on-line (including packets waiting in queues and newly launched packets);
- every node generates one ant per time step and the ant is assigned a destination node chosen at random,
- we are now also able to change the call patterns for a *hot session*.

The simulation is run for 10 K units of time, and the results are averaged for every 100th step of the simulation. In the [4000, 4500] time interval a hot session is set up on the network. In the simulation hot sessions were organized as follows. Every time step node y is chosen to be the destination for every packet generated at this time step with a probability 0.3; any other nodes of $\mathbf{V} - \{y\}$ may be the destinations of packets with probability 0.7. The number of packets dropped at each time step is chosen to be a measure of the efficiency of the load balancing scheme.

As we see from figure 3.35 the effectiveness of these four techniques varies significantly.

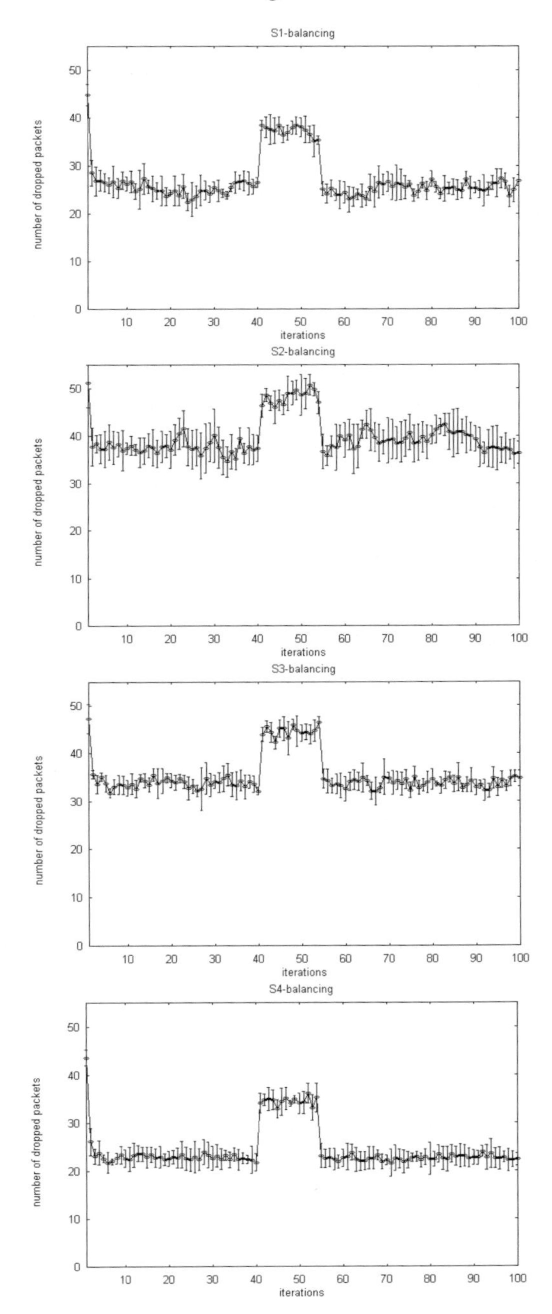

Figure 3.35. Mean number of dropped packets in the S1-, S2-, S3- and S4-versions of the ant algorithm. Vertical lines represent standard deviations.

Using the mean number of dropped packets as a parameter we can order the models in the following hierarchy: S4 (24.4 packs/step), S1 (26.8 packs/step), S3 (35.3 packs/step) and S1 (39.8 packs/step).

The huge gaps between S4 and S1, on one hand and S2 and S3, on the other, prove that simple mass transfer may not necessarily work when the number of control agents is small. In words, we could not solve a large-scale problem in a reaction–diffusion processor which contains only a few molecules of a reagent. That is, there appears to be a trade-off between the number of agents and the information passed by a single agent. The agents which are aging proportionally to the loading of the queues outperform, by a factor of 1.6 in our example, the agents whose ages are simply incremented.

When we consider the so-called *shock response* (the increase in number of dropped packets during the hot session compared to that during normal operation of the network) we find that the mass flow modifications (S2 and S3 models) are quite tolerant: 8.2 packs/step and 9.2 packs/step. The deterministic ants are slightly better than the probabilistic ants (10.1 packs/step and 10.8 packs/step, respectively). Surprisingly, the networks controlled by the probabilistic ants (S1-model) exhibit higher amplitude (and certainly aperiodic) oscillations than the systems managed by the deterministic ants (S4-model).

Finally, what about adaptation to the changing environment? If we compare the behaviour of the systems during the hot session we notice that the systems with probabilistic ants start adapting (i.e. the number of dropped packets gradually decreases) immediately after the start of the hot session (figure 3.35, S1). This does not happen with deterministic ants (figure 3.35, S4), which appear to have little or no adaptive abilities.

3.9.3 How good is ant diffusion at balancing communication networks?

Despite numerous results and expressive demonstrations we are still quite far from a solid theory describing the connections between mass flow techniques and computations with mobile agents. Thus we need to identify several points that appear to be important for further consideration.

3.9.3.1 Diffusion

As we saw in previous sections the discrete diffusion algorithms are similar to the technique used in the experiments by Subramanian *et al* [585]. In their experiments all ants are uniform; they move at random on the network, and they die when their fixed lifetime expires.

3.9.3.2 Balancing multiprocessor networks

Another form of load balancing occurs in multiprocessor networks, and involves finding a way to move tasks among processors to make the number of tasks approximately uniformly distributed amongst the processors. The tasks should usually walk (i.e. move between the adjacent nodes) but not jump. The *diffusion schedule* states that the tasks have to be moved along the edges depending on the number-of-tasks gradient at the edge [162, 185, 186, 540]. The method is claimed to be asynchronous, fault tolerant and topology stable [185]. The simplest scheme looks like a typical model of diffusion on graph. Writing ψ^t for the transposed vector of the whole workload on the nodes of $\mathbf{G}$ at time step t we have $\psi^{t+1} = \Delta\psi^t$ (we can write this in the more familiar form $\psi^{t+1} = \psi^t + \Delta\psi^t$, assuming $\delta_{ii} = 0$), where Δ is the doubly stochastic and symmetric matrix with entries from [0, 1]. The number of tasks moved between nodes i and j along the edge $(i, j) \in \mathbf{V}$ equals $\delta_{ij}(\psi_i^t - \psi_j^t)$. The scheme converges if the second largest eigenvalue of the diffusive matrix Δ is less than one in absolute value. We can conjecture that the convergence rate is inversely proportional to the diameter of the graph $\mathbf{G}$; thus, e.g., it is inversely proportional to the number of nodes in the case of a linear graph [653, 654] and should obviously be twice less when a linear graph or a mesh is formed into a torus.

At the same time in the ant-based model, the ant at one node walks to an arbitrary uniformly chosen localization node, and never stays at the same node for the next time step (except in a queue, i.e. at the exit). How does this relate to diffusive schemes? In [185] it was demonstrated in numerical experiments that for different graph families with standard topologies the entries of the diffusion matrix are the same, and that they are inversely proportional to the degrees of the nodes. There are no general principles for determining the number of tasks left on a node at every iteration, so, the *leave no ants* principle seems to be a good assumption. It has also been proved [653] that the convergence-rate optimal diffusion coefficients have the form $\frac{1}{2d}$ when the network is a d-dimensional cube.

3.9.3.3 Reinforcement learning

The so-called distributed reinforcement learning scheme for load balancing [397] utilizes an *estimated delivery time* which the nodes exchange locally. A node sends a message to its neighbour with an estimated lowest delivery time, and asks its neighbour to send back an updated estimate. Its neighbour does the same when sending the message to its neighbour, and so on. When at a distance p nodes from the target node x_t the node x_s will finish the phase of updating after p steps. We can easily reformulate this in terms of ants: ants travelling from x_t to x_s are messages that deliver information about the estimate and inform the data messages about it by leaving pheromone in the routing tables. In terms of reaction–diffusion, the estimated delivery time along a path is proportional to the gradient along this path and this is reflected in the concentration of the precipitate at the edges.

3.9.3.4 Complexity

Here we write m for $|\mathbf{E}|$ and n for $|\mathbf{V}|$. The communication and time complexity of the construction of spanning tree is quite obvious. We need $O(m)$ messages to traverse m edges, and it takes $O(n)$ time steps because $\mathbf{G}$ can have diameter n. One of the first algorithms for the distributed construction of spanning trees was published in [251]; in the execution of the algorithm the tree fragment formed forests of trees and the minimum outgoing edges between the fragments and combinations of fragments via minimum weight edges are carried out repeatedly. It requires $O(2m + 5n \log n)$ messages and $O(5n \log n)$ time steps. Gallager *et al's* [251] algorithm was modified in [131].

There the network starts its evolution when every node knows the weights of its adjacent edges and finishes the computation with the spanning tree. The nodes exchange different sorts of messages. The message complexity has upper bound $O(m + m \log n)$ and time complexity $O(n \log n + 3n)$; in the synchronous version every message contains at most one edge weight or one node identity and $O(\log n + \log \log n + 9)$ bits. The message complexity in the echo algorithm [40] is $O(2m - 2(n - 1) \log(n/2))$; the time complexity is $O(2d \log n)$, where d is a diameter of $\mathbf{G}$.

The *excitation* algorithm [16] for the construction of the spanning tree needs $O(m)$ messages and $O(n)$ times steps in the worst case. Actually $\sum_{t=1}^{T} S_t = n$, where T is the time complexity of the algorithm and S_t is the number of messages generated at the tth time step. When averaging the time complexity we can use the boundary $O(\log n)$ [590] in a very wide range of probabilistic assumptions because the computation time is bounded by the length of the longest path among the shortest paths.

For the case of the *diffusion schedule*, namely for first-order schemes, it has been shown that the optimal scheme, i.e. the optimal diffusion matrix, can be determined by a time polynomial in the number of nodes and logarithmic in the optimality parameter [185].

In several papers published recently we can find *extensive* modifications of the original ant algorithms where the capabilities of the single ant are increased enormously. Interpreting this in reaction–diffusion terms we can say that a molecule becomes even more intelligent. For example, ants can travel back towards their source (see, e.g., [638]) after they have visited the destination. These algorithms obviously have little difference if any of the centralized or slightly decentralized algorithms for the computation of shortest paths are used for the routing optimization of communication and computation networks (see e.g. [40]). In this case the ants are the messages that deliver information about the local states of the network to the central controller, which in their turn recalculate the matrices of the shortest paths.

3.10 Chemical processors that compute shortest paths

A now classical work by Steinbock, Tóth and Showalter [582] presents the first ever experimental evidence of the feasibility of chemical processors applied to the shortest-path problem. Their basic proposition states that

> *the optimal path is determined from the velocity field, which is generated by specifying the direction of wave propagation at each point in the reactor space.* [582]

Recipe 4. The Steinbock–Tóth–Showalter processor for computation of an optimal path in the labyrinth [582].

- Prepare a solution of $NaBrO_3$ (0.2 M), H_2SO_4 (0.4 M), malonic acid (0.17 M), NaBr (0.1 M), ferroin (1.0 mM).
- Cut a corridor configuration of the labyrinth out of a thin membrane (Gelman Metrial), 140 μm thick with pores of 0.4 μm.
- Soak the membrane with the prepared solution.
- Touch the membrane with a silver wire at the source point x.
- Record the propagation waves in monochromatic light ($\lambda = 500$ nm), i.e. make a series of snapshots (the number of snapshots determines the accuracy of the approximation).
- Encode the snapshots into an elapsed time configuration, where every point z of the labyrinth gets the time the excitation wave takes to reach point z from source point x.
- The minimum path length from any point of the labyrinth to the source point x equals the product of the elapsed time, assigned to each point of the labyrinth, and a constant wave velocity.

Steinbock *et al* [582] use a similar procedure for path backtracking as that implemented in our cellular-automata models [16]: for every point $p(t)$ we search for a point indexed by $t - \tau$, the point $p(t - \tau)$, where τ is the time increment between snapshots of wavefronts.

Another implementation of a chemical processor for shortest-path problems is offered in [38]. There Agladze *et al* demonstrate how the fastest path between two points, x and y, in an arbitrary velocity field can be found in a planar Belousov–Zhabotinsky reactor.

In the experimental setup by Steinbock *et al* velocity fields were constructed by an external computing device. Agladze *et al* take a step forward and put more work on the chemical medium. Their approach is based on the construction of a motion chart, a set of isochrones, which covers the problem space by lines, the points of which are at the same distance from the two data points.

> *To create the motion chart we start an excitation wave at point x and record the configuration of wavefronts every τth step of the evolution until the wave reaches the point y.* [38]

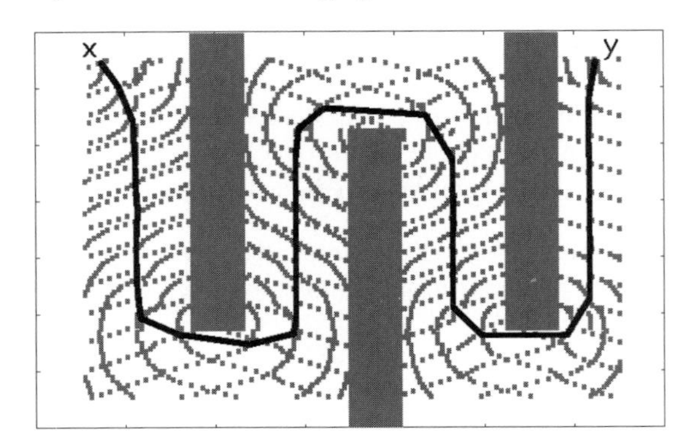

Figure 3.36. Computation of the shortest path between points x and y in a simple labyrinth in a cellular-automaton model with trigger waves. The intersections of isochrones are connected by line segments to highlight the path.

After that we can start the excitation wave at the point y and build a series of wavefront snapshots. They represent the set of isochrones generated by the point y. Having two sets of isochrones we are able to find the shortest path connecting x to y.

The set of sites where isochrones generated by the point x cross isochrones generated by the point y represents the shortest path between the points x and y.

An example of the computation of the shortest path in a simple labyrinth is shown in figure 3.36.

Recipe 5. Agladze–Magome–Aliev–Yamaguchi–Yoshikawa ($AMAY^2$)-processor for computation of the shortest path in a medium with obstacles [38].

- Prepare a solution of $NaBrO_3$ (0.3 M), H_2SO_4 (0.15 M), $CH_2(COOH)_2$ (0.1 M), $CHBr(COOH)_2$ (0.1 M), and $Fe(phen)_3$ (0.005 M).
- Pour 5 ml of the solution into a Petri dish, approximately 9 cm.
- Create obstacles by adding drops of KCl or by strong illumination.
- Initiate a wave at point x by a silver wire.
- Illuminate the reactor from below. Record the wave propagation on a video device coupled with a computer. A set of snapshots of the wave travelling from point x to point y must be recorded.
- Initiate a wave at point y with a silver wire.
- Record a set of snapshots of the wave propagating from point y to point x.
- Connect the intersection of the opposite waves (recorded at the same time intervals) with line segments. This will be the shortest path between the sites x and y.

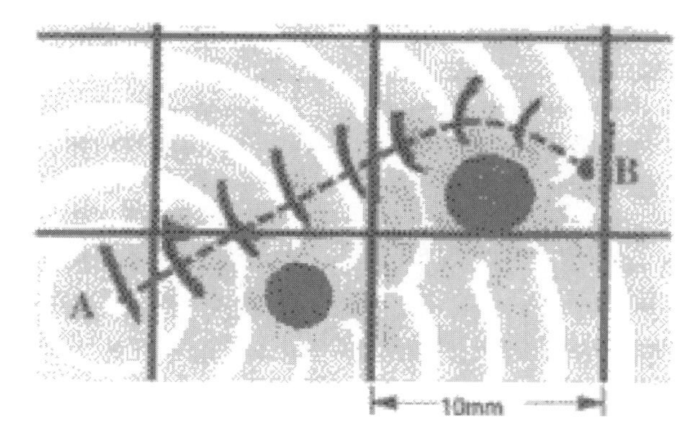

Figure 3.37. An optimal obstacle-avoiding path between sites A and B detected in experiments with an AMAY^2-processor [38]. Published with kind permission of Rubin Aliev.

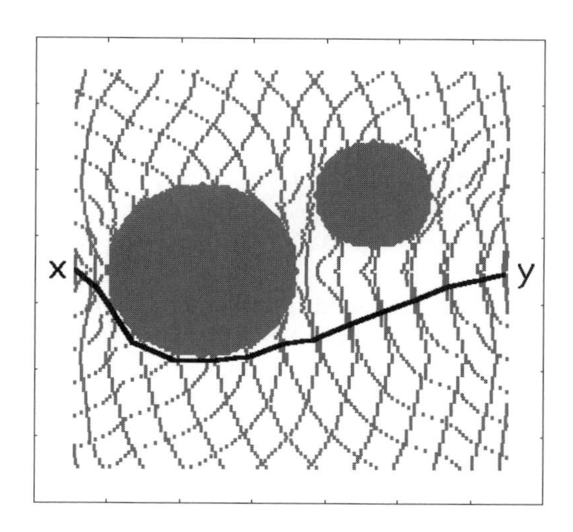

Figure 3.38. Superposition of wavefronts travelling from point x to point y and from point y to point x in a cellular-automaton model of a chemical medium with two obstacles (shown by full discs). The approximated shortest path is shown by the black curve.

Snapshots of the laboratory experiment and a cellular-automaton simulation are shown in figures 3.37 and 3.38.

Some estimations made in [38] may be useful. It takes about 15 min to detect a path between two points at a distance of 20 mm. Moreover, if the size of an elementary micro-volume of a Belousov–Zhabotinsky reactor is about 0.1 mm then the standard Petri dish ($\approx$ 9 cm) is approximately 10^5 micro-volumes; that is, the chemical massively parallel processor has around 10^5 elementary processors.

As soon as a trigger wave propagates with a speed 0.05 mm s^{-1}, the elementary processor updates its state once every 2 s.

The processors designed by Steinbock *et al* and Agladze *et al* employ the propagation of the trigger waves. Trigger waves are controllable but slow. The reported speed of a trigger wave in a thin layer is about 5×10^{-2} mm s^{-1} [527]. Rambidi and Yakovenchuck [527] tried to employ *phase waves*. The phase waves propagate independently of diffusion along a phase gradient in an oscillatory medium [527]. They move faster than trigger waves but are more difficult to control. The Rambidi–Yakovenchuk chemical processor for computation of a path from the root of a tree to one of the leaves is based on some experimental evidence that controllable phase waves could be initiated by heterogeneous reactor illumination. In words, the main idea of the algorithm is to control the phase processes by distribution of the light radiation and to exploit the connectivity of the tree to eliminate those routes which do not lead to the output, the leaf of the tree.

Now we present a brief outline of the computation of a path in a tree in a Belousov–Zhabotinsky medium.

Recipe 6. A Rambidi–Yakovenchuk processor for computating a path on planar tree [527].

- Prepare a solution of $KBrO_3$ (0.3 M), H_2SO_4 (0.5 M) and malonic acid (0.2 M).
- Immobilize the light-sensitive catalyst $Ru(bipy)_3Cl_2$ (0.001 M) on thin solid film.
- Put the film with the catalyst at the bottom of the reactor and pour the solution to make a 0.5–1.5 mm non-stirred reagent layer.
- Project the data labyrinth onto the surface of the planar reactor to make a light-intensity gradient from the input to the output parts of the tree.
- Record images of the reactor at regular intervals.
- Using the image-processing software eliminate the components which are not connected to the output.

The light-intensity gradient in the reactor is chosen in such a manner to allow the phase wave to run from the source site of the tree (i.e. the root) to the destination site, i.e. the leaf. It is claimed [527] that the velocity of the phase wave depends on the slope of the light-intensity gradient. So, the light-induced phase wave runs toward the leaves of the tree.

An example of the evolution of the Rambidi–Yakovenchuk processor, which computes a very simple path from a tree root to a tree leaf, is shown in figure 3.39. This works as follows. An image of a tree-like labyrinth is projected onto a reactor. The distribution of the light intensity on the surface corresponds to the labyrinth structure. The image is projected during the whole computation period. Initially, we excite a wave at one of the labyrinth sites, the source point, then we record images of the medium, and store snapshots of waves travelling through

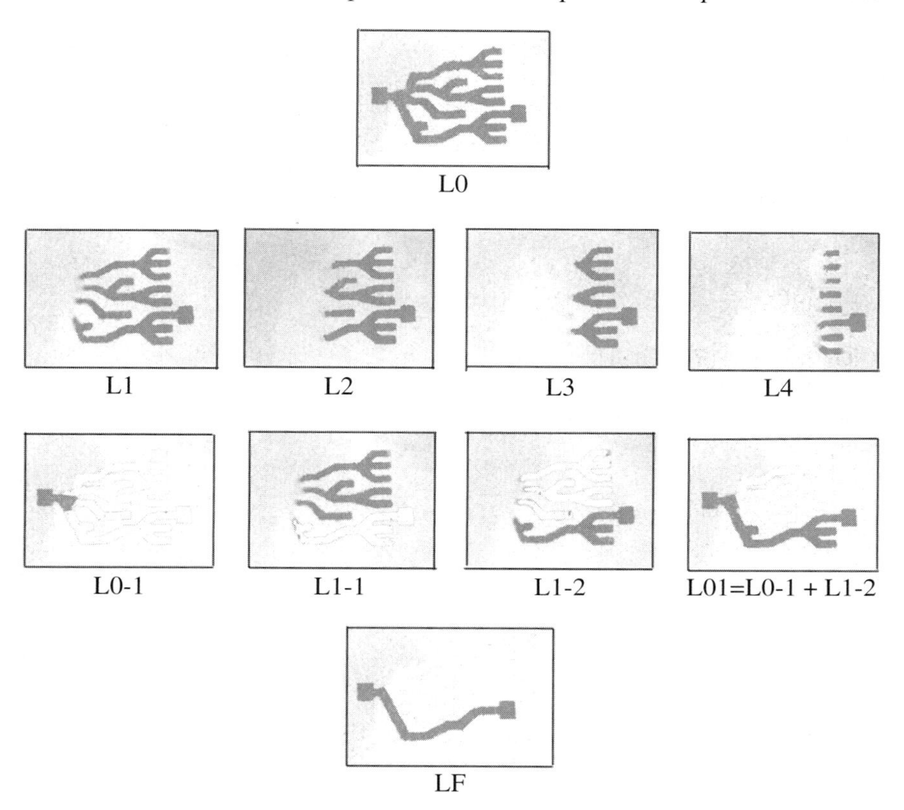

Figure 3.39. Computation of the shortest path on a tree in a Belousov–Zhabotinsky medium [527]: (L0), original image of a labyrinth projected onto the medium; (L1–L4), snapshots of medium evolution showing the wave spreading into the labyrinth; (L0-1), a path covered by the wave in its first step; (L1-1), extracted parts of the labyrinth; (L1-2), deletion of routes not connected to the target site (final leaf); (LF), shortest path in the labyrinth. Published with kind permission of Nicholas Rambidi.

the labyrinth. The computation is assumed to be finished when the wave reaches the destination point. Then we numerically process the images and determine the desired shortest path between source and destination points.

How is the path extracted from serial images of the travelling wave? The method for extracting the results chosen in [527] is a tricky one. When a wave spreads through a branching point of the labyrinth the labyrinth image separates into several fragments. One of the fragments is connected to the source point, the others are not. The connectivity is determined from the snapshots of the medium's evolution using numerical techniques [527].

There are two advantages of the processor:

- The computing medium is homogeneous, no data are stored in an allied external structure but they are represented by light intensity.
- The feasibility of light input may allow us to operate data structures which dynamically change.

They open new horizons in the design of re-usable chemical processors. However, the method by which the resultant path is extracted still has the potential for further improvements.

Chapter 4

Computational universality of nonlinear media

Computational universality implies the ability to compute any logical function.

In the common sense the universal computer is *one that can be programmed to perform any digital computation* [79]. To validate the computational universality of some abstract or real physical, chemical or biological system we should construct a functionally complete tuple of Boolean functions, e.g. $\{\overline{x}, x \wedge y\}$ or $\{\overline{x}, x \vee y\}$, realizable in the spacetime dynamic of the system.

Throughout this chapter we assume that the computing device is universal if it calculates any Boolean function (function of the algebra of logic) presented in the form $f : \{0, 1\}^k \rightarrow \{0, 1\}$. From the basics of Boolean algebra we know that every Boolean function can be represented in the disjunctive normal form:

$$f(x_1, \ldots, x_k) = \bigvee_{(\sigma_1,\ldots,\sigma_k) f(\sigma_1,\ldots,\sigma_k)=1} x_1^{\sigma_1} \wedge \ldots x_k^{\sigma_k}.$$

$f(x_1, \ldots, x_k) \equiv 0$ is represented by $f(x_1, \ldots, x_k) \equiv x_1 \wedge \overline{x_1}$. Therefore every Boolean function is represented via negation, conjunction or disjunction, that is, e.g., the systems $\{\overline{x}, x_1 \wedge x_2, x_1 \vee x_2\}$ and $\{\overline{x}, x_1 \wedge x_2\}$ are functionally complete.

Given a real or imaginary system, to prove its computational universality we have to represent

- quanta of information (e.g. `True` and `False` values of the Boolean variables);
- routes of information transmission;
- logic gates where quanta of information are processed in the states of the given system.

Several basic requirements for physical models or realizations of computational devices are discussed by Islam *et al* [321]. Not all requirements are equally necessary, however each of them should at least be considered when an unconventional universal computer is designed [321].

- The device should be cascadable. An output of one unit serves as the input of another unit.
- The device must have fun-out. The output of one gate switches at least two similar gates.
- The device should be computationally complete.
- The ports of the device should be orthogonal either in space or time. In an optical device this can be done via polarization or tuning of wavelengths.
- The inputs should be isolated from the outputs. The reflection of a signal from one of the units should not disturb operations of another unit.
- Logic-level restoration is desirable. A wide range of inputs may be accepted as `Truth` and `False` values of logical variables, however outputs must be standard. This may prevent accumulation of noise at each stage of the computation.

The standard method of stationary computation, or computation with stationary architectures, embeds the entire architecture of a Boolean circuit into a given system in such a manner that wires, gates etc are represented in the stationary states of the system, namely in the absorbing states (the states that will never change) of the elementary units, and do not change their position with time [65, 279, 281, 436, 533]. We consider some remarkable examples of conventional computing in reaction–diffusion media. The examples include recent findings by Blittersdorf *et al* [88], Tóth and Showalter [602] and Steinbock *et al* [583], where the computation is implemented via mass transfer and chemical kinetics or chemical waves travelling along the tubes and interacting with each other in the tube junctions. In the experiments the reaction space is inhomogeneous. The wires are predetermined and there is little difference between such chemical models and logic operations realized in living axons or dendrites.

> *A dynamical computation assumes that autonomous signals travel in the space and perform computation by colliding with other travelling signals. There are no specially determined wires, however we can use mirrors to specify the trajectories of the signal.* [243]

The billiard ball model [243] and the glider computation in the *Game of Life* [81] historically form the core of contemporary theory of dynamical computation, particularly in discrete models of natural systems.

Fredkin and Toffoli [243] suggested a pioneer proof to a proposition stating that there is no need to keep the computer architecture stationary and unchanging: everything can move, collide and compute by colliding. The discovery of gliders and glider guns in the *Game of Life* led to the possibility of simulating universal logic gates via collisions between gliders [81] or other self-localized movable patterns [124, 125, 198].

In this chapter we answer the following questions:

> *Is it possible to make the two- and three-dimensional reaction–diffusion or excitable processors dynamically computationally universal? If yes, then what natural materials can be used to build a dynamical universal processor?*

We start with a discrete version of an active nonlinear medium—excitable lattices. We demonstrate that excitable lattices exhibit a wide range of movable self-localized excitation patterns (particle-like waves) and that the universal logic gates are realized at the sites of the collisions between these mobile patterns. Keeping in mind that our models of excitation are far from the disappointing reality we nevertheless decided to play here all possible scenarios of the collisions. That is we investigate almost all possible binary and ternary collisions of the particle-like waves in two- and three-dimensional excitable lattices and compile a catalogue of the interaction gates that are implemented as a result of the collisions. Most of the gates prove to be logically universal and so our simple models of lattice excitation are universal computers. The complexity of the models is analysed and a comparison with existing cellular-automata models of universal computation is provided. At the end of the chapter we disclose a great variety of real-life models of dynamical universal computers, e.g. solitons and breathers, and demonstrate that

> *most of our results obtained in a cellular-automaton model of an excitable medium have real world counterparts.*

4.1 Artificial and natural universality

Generally speaking, a universal computer is a device that implements computable recursive functions and composition rules, i.e. null functions, successor functions, projection operations, function composition, primitive recursion. Fisher [230] and Mazoyer [436, 533] proved that even a one-dimensional cellular automaton makes arithmetic computations and simulates recursive functions. So, such an automaton is universal. However, the signals that carry information and the proper cell states are separated in these models. This means that the static architecture of the computer is actually embedded in configurations of cellular automaton.

Durand-Lose [199] demonstrates that partitioning cellular automata simulates reversible cellular automata and *vice versa*, i.e. it is shown that there is a reversible cellular automaton that simulates any reversible (partitioning) cellular automaton and there is an irreversible cellular automaton that simulates any (partitioning) cellular automaton. Remarkably, these two universal cellular automata have 16 and 81 cell states, respectively [199].

The entirely new field of particle models of computation was developed by Steiglitz *et al* [492, 578, 568, 326, 569]. Their *particle machines* are one-dimensional but truly *dynamical* computing devices: the particles propagate and collide and implement computation as a result of collisions. In particle models quanta of information are encoded in the vector states of the travelling particles.

An abstract machine is called universal, i.e. simulation universal, if it simulates the universal Turing machine. Thus, e.g., the cellular automaton is universal if it simulates the Turing machine and any other cellular automaton (see e.g. [478]).

Neural and reaction–diffusion networks (in their purely abstract form) have proved to be universal. The simulation universality of the neuron-like network is demonstrated by Siegelman and Sontag [562]: every neuron of the network has a linear combination of the inputs and saturated linear threshold function. Assuming that the universality of the neural networks is proved, Goles and Matamala [281] show that the three-state automata network is universal because it simulates any neural network. The sand-pile model, which is discussed later, is a more realistic species of the reaction–diffusion models. A proof of universality of the one-dimensional reaction–diffusion media can be accepted in general via the proof of universality of the sand-pile model [279].

Continuous and hybrid systems constitute another class of the universal computers. One of the first substantiated claims on the universality of partial differential equations was raised by Omohundro [484]: he proved that any two-dimensional cellular automaton with an $(8 + 1)$-cell neighbourhood is simulated by a system of ten coupled partial differential equations. The cellular automaton in its turn simulates the Turing machine. Therefore, that system of differential equations is universal.

As a corollary to the simulation of discrete systems by continuous and hybrid systems (which are obviously a superclasses of neural networks) Branicky [103] showed that the Turing machine is simulated by ordinary differential equations in three-dimensional real-valued space. Elaborating the Koiran–Cosnard–Garzon idea [358] on the simulation of the universal Turing machine by a two-dimensional piecewise linear iterated function, Orponen and Matamala [487] confirmed that one- and two-dimensional networks of coupled oscillators simulate the Turing machine and therefore arrays of coupled oscillators are universal.

A variety of physical models are considered to be universal computers, for example, a hard sphere gas, a lattice gas, systems of partial differential equations, and single particles moving in a room with plane and parabolic mirrors [79].

A chemical kinetic-based approach is presented in [303], where logic gates are designed by combining binary McCulloch–Pitts-like neurons. The states of the neurons are represented by concentrations of chemical reagents. However, no spatial phenomena are involved and all reactions are assumed to be undertaken in stirred reactors.

Some results on computational universality are obtained by studying the computational complexity of physical phenomena; that is, if we want to prove that a problem is P-complete (i.e. may be solved in a serial computer in a time polynomial in the size of the problem) we demonstrate that the problem is reducible to a circuit value problem. The circuit value problem can be stated as follows.

> *Given binary-valued inputs of a Boolean circuit, evaluate the output of the circuit.*

That is when we show that the problem is P-complete we demonstrate that the algorithm for the problem can be used to simulate an arbitrary Boolean circuit, i.e. to be computationally universal. Machta and Greenlaw [405, 406] prove that the natural decision problem assigned to diffusion-limited aggregation is P-complete by simulating a universal computing device in the model of diffusion-limited aggregation. They show that the trajectories of particle walks can be programmed in such a manner to imitate NOR gates; the topology of the wires are also arranged via particle motion [406].

4.2 Architecture-based universality

The concept of a *stationary computation*, when the architecture of a computing device is initially predetermined, static and expressed in special states of a computational space, is widely accepted. The concept is much more widespread than dynamical computing because it relies on existing conventional electronic circuits, mechanical analogues and biological prototypes, particularly neural networks. That is why we still do not have laboratory prototypes of many-dimensional dynamical (i.e. collision-based) computers but are able to enjoy a couple of working examples of chemical computers with a stationary architecture. In the next three sections we discuss three examples of architecture-based computers.

- The first one—sand-pile models—is rather theoretical. However, it demonstrates what happens in a modelled physical system.
- The second example deals with mass transfer computation. We can think here about kinetic-based logic gates.
- The third example is related to very attractive ideas on the wave-based construction of elementary logic gates.

4.3 Sand-piles

The sand-pile model [279, 467] is amongst the most curious constructs of universal computations. It originates from a physical sand-pile model [58, 59], combinatorial games with discs and balls [51] and chip-firing games [280, 85]. Both logic gates and registers are simulated in the model via configurations of sand grains on the lattice.

A classical sand-pile model [58, 59] looks as follows. Given a lattice where every node has a non-negative number of sand grains. The sand grains can topple from a node to its neighbouring nodes. Assume a node x has x^t grains at time step t and the node x is connected to k nodes then k grains in the node x will go to k neighbouring nodes at the time step $t+1$ only if $x^t \geq k$. If this happens, each of k neighbours receives a grain and the number of grains in the node x is reduced by k. This process is called toppling. All sites of the lattice (or graph) topple their grains in parallel. If the concentration of sand grains around the lattice is disproportional at the beginning of the experiment an avalanche of sand grains continues until every node of the lattice has less grains than its neighbours. As soon as all lattice nodes update their states—to topple or not topple the grains—in parallel the sand-piles may be simulated in cellular automata (see e.g. [181]) or in automata networks [267].

It is quite reasonable to simulate sand-piles on graphs in automata networks, where every node x_i updates its states depending on the states of its closest neighbours x_{i-1} and x_{i+1}:

$$x_i^{t+1} = x_i^t - \gamma(x_i^t, k) + k^{-1} \sum_{j \in u(x_i)} \gamma(x_j, k)$$

where $\gamma(x_i^t, k) = k$ if $x_i^t \geq k$ and it equals zero otherwise. The neighbourhood $u(x_i)$ of the node x_i includes all those nodes that are connected by the edge of the graph to the node x_i or have a physical nearest contact in the case of real systems.

Goles and Margenstern [266] were probably the first to demonstrate that the general cellular-automata model of a sand-pile can perform universal computation. They proved that it is possible to implement register machine using automata models of sand-piles. Moore and Nilson [467] generalized Goles–Margenstern's results to an arbitrary multidimensional case. However, even in multidimensional constructions of sand-pile machines we need to mark the wires explicitly. This means that the computation is essentially one-dimensional and it is unreasonable to discuss multidimensional architectures.

Let us look at a detailed description of the computation process. A signal in a one-dimensional linear graph, each node of which has exactly one grain, is represented by the pattern 02, if it moves east, or the pattern 20, if it moves west. The spacetime configuration of the grains in the one-dimensional lattice, where the signal 02 moves, is shown in figure 4.1.

To build AND and OR gates we connect one node of our linear graph to the top node of another linear graph. In the result almost all nodes of the construct have two neighbours but only the junction node has three neighbours. The architecture of the AND and OR gates is shown in figure 4.2. They have a similar graph topology but different initial distributions of grains on the graphs. In addition, the right-hand shoulder of the AND gate is intentionally made longer than the left-hand one because we need to keep an appropriate 'phase' difference between the signals to implement the conjunction operation in the gate.

When both input variables have the logical value False, $x = 0$ and $y = 0$, both gates have stationary configurations, i.e. nothing is changed. This is because

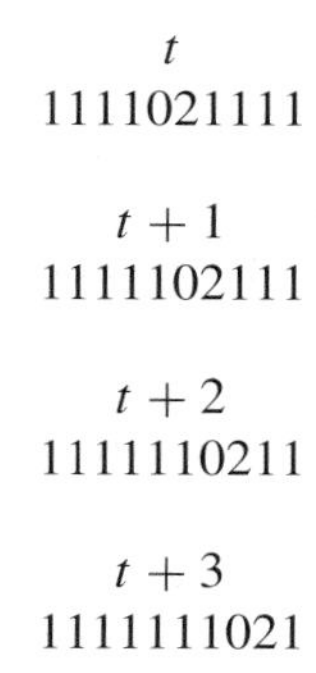

Figure 4.1. Movement of the signal along a linear sand-pile graph. The numbers represent the amount of grains in the nodes.

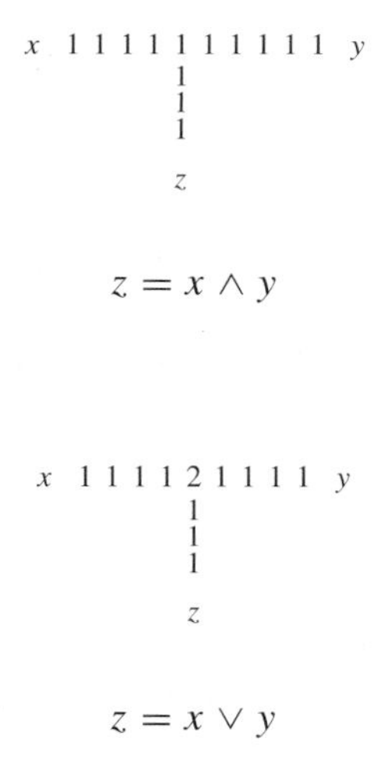

Figure 4.2. Architecture of logic gates, AND gate and OR gate, realized in the sand-pile model. The numbers represent the amount of grains in the nodes.

every node of the sand-pile graphs, representing the gate, contains less grains than its neighbours. In figure 4.2 we see that in the configuration of the resting gate AND (the 'resting gate' means that neither the 'wire' nor the 'junctions' transmit a signal) every node has exactly two or three (if it is a junction node) neighbours and a grain of sand. The only node which forms an intersection of the wires x, y and z has three neighbours. In the configuration of the resting OR gate (figure 4.2) every node except the junction node has one grain and the junction node has two grains; in all cases the number of grains in a node does not exceed the number of node's neighbours.

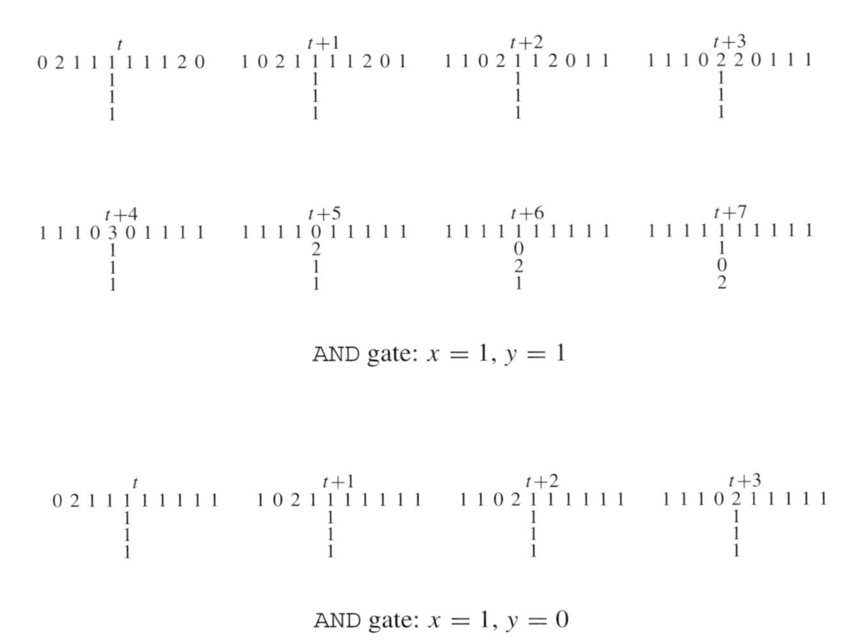

Figure 4.3. Evolution of the AND gate for two sets of input variables. The numbers represent the amount of grains in the node. x corresponds to a left-hand wire (shoulder), y is a right-hand one.

Let us investigate the behaviour of the AND gate when both input variables bear the value `True`, $x = y = 1$ (figure 4.3). We start two small avalanches of the form 02 and 20 on both shoulders of the gate (figure 4.3, t). The grains topple and two toppling waves move toward the junction of the x and y wires (figure 4.3, $t + 1$ and $t + 2$). The lengths of the gate shoulders are different. In our example shoulder x has four nodes and shoulder y has five nodes. Therefore signal 02 arrives at the junction node one step earlier than signal 20 (figure 4.3, $t + 3$). At time step $t + 3$ the junction node has two grains. However, they cannot topple because the junction node has three neighbours. At step $t + 4$ (figure 4.3, $t + 4$) signal 20 arrives (along the wire y) at the junction node. Thus at step $t + 4$ the junction node has three grains of sand. It is enough to topple! All nodes connected to the junction node get a grain each. The signal 02 (we see it as $\begin{smallmatrix}0\\2\end{smallmatrix}$) starts to travel along the wire z. Eventually, we have $z = 1$ at the time step $t + 7$.

If $x = 1$ and $y = 0$ in the AND gate (figure 4.3) signal 02 reaches the junction node and stays there forever; that is, the signal never reaches the wire z. Thus we have $z = 0$ when $x = 1$ and $y = 0$. The gate's behaviour with other combinations of input variables can be analysed by analogy.

The evolution of the OR gate is even more interesting. When both input

```
        t                  t+1                  t+2                  t+3
0 2 1 1 2 1 1 2 0   1 0 2 1 2 1 2 0 1   1 1 0 2 2 2 0 1 1   1 1 1 0 4 0 1 1 1
        1                   1                   1                   1
        1                   1                   1                   1
        1                   1                   1                   1

        t+4                  t+5                  t+6
1 1 1 1 1 1 1 1 1   1 1 1 1 2 1 1 1 1   1 1 1 1 2 1 1 1 1
        2                    0                    1
        1                    2                    0
        1                    1                    2
```

OR gate: $x = 1$, $y = 1$

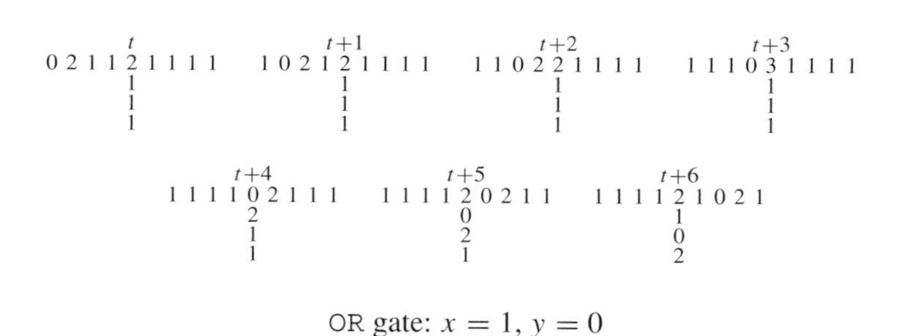

Figure 4.4. Evolution of the OR gate for two sets of input variables. The numbers represent the amount of grains in the node. x corresponds to a left-hand wire (shoulder), y is a right-hand one.

variables have a `True` value: $x = y = 1$, the signals travel along the shoulders and reach the junction node in the same time (figure 4.4). Therefore at step $t + 3$ the junction node has four grains (figure 4.4, $t + 3$). This sand-pile at the junction node topples at the next time step (figure 4.4, $t + 4$). The signal is generated in an output wire and travels down the wire (figure 4.4, $t + 4$, $t + 5$ and $t + 6$). Thus, we have $z = 1$.

When $x = 1$ and $y = 0$ the signal 02 only travels along the wire x. When the signal arrives at a junction node at time step $t + 3$, its arrival increases the number of grains at the node up to three (figure 4.4). At time step $t + 4$ we observe quite an interesting configuration of the grains in the junction node and three of its neighbours (figure 4.4, OR gate, $t + 4$): the signals travel along the output wire and along the previously resting y wire. As a result of the evolution we obtain $z = 1$ and, as a byproduct, $y = 1$. This means that the OR gate can also be used for multiplication of the signals.

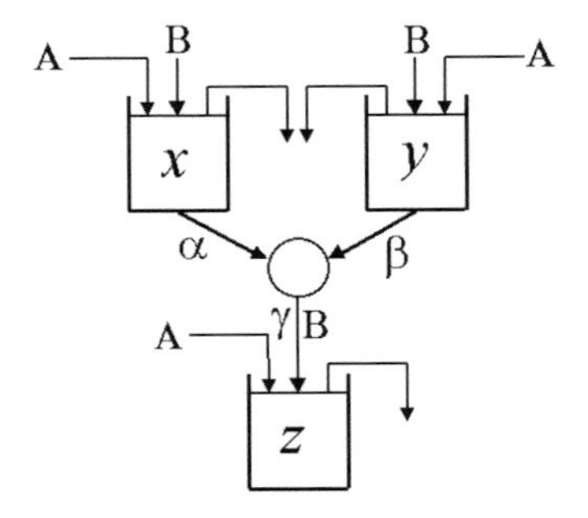

Figure 4.5. Mass-transfer-based chemical device for computation of basic logic gates. See detailed description in the text. Modified from [88].

As soon as sand-pile graph models can simulate any logical gate by their spacetime configurations they are computationally universal.

While finishing this section we should mention a typical example of a conventional architecture-based computing structure, which is implemented in a non-standard medium. This is the design of logical circuits in configurations of the lattice in the Minesweeper game, suggested in [346, 347]. Look again at section 3.8 to recall the cell state transition rules of a cellular automaton which plays Minesweeper. In the design, offered in [346, 348], a wire configuration is built on a two-dimensional lattice of three chains of cells in non-zero states. A chain of a periodic configuration of two uncovered and unmarked cells and one cell in state 1 is sandwiched between two chains of cells in state 1. The orientation of the wire is determined before evolution. As soon as the first cell is uncovered the process of uncovering travels along the wire configuration deterministically. A mine represents the value `True` and an absence of a mine the value `False`. Therefore, a travelling signal emerges. For detailed constructions of all logic gates on a Minesweeper lattice see [348].

4.4 Mass-transfer-based gates

Neutralization is employed in [88] to build logic gates in a network of three chemical reactors. Two of the reactors represent input variables and the third reactor is the output. The binary states of two input reactors and one output reactor are represented by levels of acidity in reagent solutions: low pH levels correspond to the zero (`False`) state and the high levels of pH are identified as state 1 (`True`).

A computing device is assembled from three continuous-flow stirred reactors, x, y and z (figure 4.5). The reactors x and y are connected to the reactor z via a syringe pump. All three reactors are supplied by a reagent solution of HCl containing the indicator phenolphthalein (the feeding tubes are marked by

x	y	z
a	a	a
a	b	a
b	a	b
b	b	b

Figure 4.6. Correspondence between the values of variables x, y and z and pH levels in the reactors when the gate $z = x \wedge y$ is calculated in real experiments [88]. The pH level a represents the `False` value and level b represents the `True` value, $a = 2$, $b = 12$.

A in figure 4.5) and by reagent NaOH (the feeding tubes are marked by B in figure 4.5). The solution A is delivered with a constant rate. In the reactors x and y the flow rate of the reagent B is adjusted to represent the input variables. In the experiments [88] the red (alkaline) state of the reagent solution is identified as one and the colourless (acidic) state is identified as zero.

The switching process is seen as a change of colour of the phenolphthalein indicator. Blittersdorf *et al* [88] calculate the flow rate γ of NaOH from the pump to the output reactor z via the coupling strengths (measured in s^{-1}) α and β between the reactors x and z, and between the reactors y and z, respectively (figure 4.5):

$$\gamma = \alpha p_x + \beta p_y + b.$$

In the equation the terms p_x and p_y represent pH values in the reactors x and y, respectively, and b is a bias (measured in s^{-1}).

> *The coupling coefficients α and β and the bias b determine which type of logical gate is realized in the reactor network.*

Thus, for example, $\alpha = 2.3 \times 10^{-6}$, $\beta = 4.3 \times 10^{-6}$ and $b = 1.0 \times 10^{-6}$ when the reactor network implements AND gate xy (see other parameters in [88]).

> *To perform the calculation we set up the pH levels in the reactors x and y, calculate the flow rate γ, and measure the pH in the output reactor z.* [88]

The pH values registered in real experiments [88] on implementation of the AND gate are shown in figure 4.6.

4.5 Wave gates

A *state-switching threshold*, already explored in the sand-pile logic gates, is employed in the design of *chemical wave logic gates* by Showalter *et al*

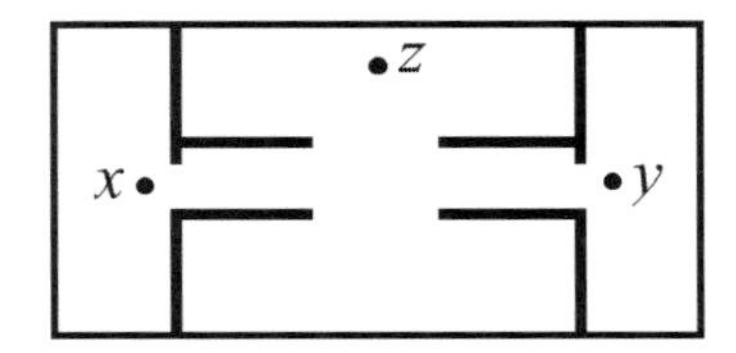

Figure 4.7. Design of AND and OR gates [602]. The size of every channel input is reduced to prevent backpropagation of the excitation.

[583, 602]. The concept of wave gates is tested on the Barkley model of an excitable media:

$$\frac{\partial u}{\partial t} = \nabla^2 u + \epsilon^{-1} u(1-u)\left[u - \frac{v+b}{a}\right]$$

and

$$\partial v/\partial t = \nabla^2 v + u - v$$

where ϵ, a and b are the parameters of the model.

In the framework of wave gates [583, 602]

> *the states* 1 (`True`) *and* 0 (`False`) *are represented by the presence or absence of a phase wave at a specified location. The computation is based on the fact that a critical nucleation size must be taken into account to generate a successful wave.*

The critical nucleation size is derived from the equation which links the normal wave velocity c, the velocity c_0 of a planar wave, the curvature κ of the wavefront and the diffusive coupling of the autocatalyst u [602]:

$$c = c_0 + D\kappa.$$

As soon as the wavefront velocity tends to zero when the front curvature approaches its critical value, we have the following critical radius of wave curvature: $r_c = D/c_0$ [602]. If the radius of wavefront curvature is less then r_c the excitation wave collapses; otherwise it expands.

A schematic diagram of a logical gate is shown in figure 4.7. The reaction medium is blank whilst the barriers, impenetrable by diffusing reagents, are shown in black. Dots are inputs x and y, and output z. To represent input pattern 01 ($x = 1$ and $y = 0$) we generate an excitation wave at site y and do not generate a wave at x. If an excitation wave appears at site z this means the output variable z takes the value 1 (`True`).

The outlines of AND and OR gates look the same—how then do the gates produce different results? Tóth and Showalter [602] show that the planar wave velocity in the Barkley model can be changed by adjusting the parameter a. The lower the value of a is, the higher the velocity c becomes and the smaller the critical radius r_c. That is, in numerical studies the channel radius is 0.065 (see figure 4.7), the width of the gap between the channels is 0.195 and $a = 0.6$ for the OR gate and $a = 0.55$ for the AND gate. This means that for the set of parameters defining the OR gate the critical radius is sufficiently large to allow a wave, propagating along one of the channels, to start the excitation process in the gap between the channels and to reach the output site z. That is, if we excite the sites x and y separately or together the excitation wavefront reaches the output site z in any of the initial configurations except for the case of no initial excitation.

To modify the OR gate to the AND gate we slightly decrease a. The critical radius increases. It is enough to be sure that a single excitation wave spreading along either channels x or y (but not both) does not give birth to an excitation in the chamber between the channels. That is we have $z = 0$ for $(x, y) = (0, 0), (0, 1), (1, 0)$. Two excitation waves leaving the opposite channels simultaneously initiate an excitation in the gap between the channels. This gives us $z = 1$ for $(x, y) = (1, 1)$.

We refer a reader to the original paper [602] to examine the attractive design of NOT and XNOR gates and to inspect the particulars of the experimental realizations of the gates in a Belousov–Zhabotinsky medium.

The ideas of geometrically constrained excitable media are employed in the architectures of wave logic gates, implemented with a Belousov–Zhabotinsky reaction, described in [583]. There the geometry of wires and junctions is represented by a catalyst, which is simply printed using an ink-jet printer on a polysulphone membrane with agarose gel saturated with $NaBrO_3$, H_2SO_4, malonic acid and NaBr. The logic operations are determined by the geometry of the chambers, which connect the input and output channels, and are expressed via (a)synchronous occurrences of excitation waves on the output channels (figure 4.8).

As expected the critical radius of the excitation wave can be employed in the construction of other logical and 'electronic' devices. Thus, for example, a diode based on the Belousov–Zhabotinsky reaction is designed in [39]; the construction of the diode uses the same principle of the critical value of the gap to control unidirectional diffusion penetration.

4.6 Employing collisions: billiard ball model

The billiard ball model, with its *binary 'signals' travelling on the grid and interacting with one another* [243] is probably the most well-known model of universal physical computation. Created by Fredkin and Toffoli [243] in the context of a conservative logic, the model is based mainly on idealized elastic collisions between the balls and immobile reflectors. Balls of finite diameter move with a constant velocity along the straight trajectories on the planar grid. A bit of information is represented by the presence or absence of the ball at a given time

Figure 4.8. Example of a wave realization of $x \vee y \vee z$ gate. The waves in the output chambers are asynchronous in the case of (`False`, `False`, `False`,) input. If one of the inputs bears a `Truth` value, represented by an excitation wave, the output chambers exhibit synchronous waves, that represent `Truth`. Modified from [583] with kind permission of Oliver Steinbock.

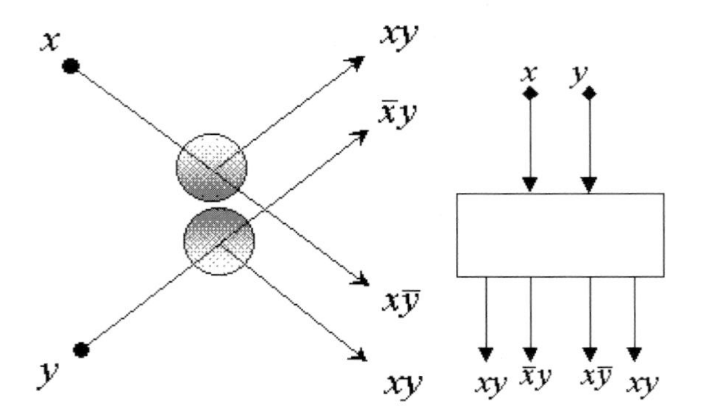

Figure 4.9. Billiard ball interaction gate.

at a given site. The routes of information flow are represented by the trajectories of the balls (crossover is trivially prevented) and routing is implemented by the orientations (initial configuration) of the reflectors.

The elementary interaction logical gate of the billiard ball model that computes $\overline{x}y$, $x\overline{y}$, xy (we write xy for $x \wedge y$) functions is shown in figure 4.9(A) with its ball representation figure 4.9(B). The switch and the so-called Fredkin (conservative) gates are constructed from the elementary gates and mirrors (that reflect the balls). One of the practical applications of the Fredkin gate is found in magnetic-bubble logic and conservative logical circuits [124]. Chau and Wilczek [125] have also designed a Fredkin gate using a sequence of six two-body quantum gates. Some recent results related to the universality of billiard ball model in the sense of the Margolus neighbourhood are presented in [198].

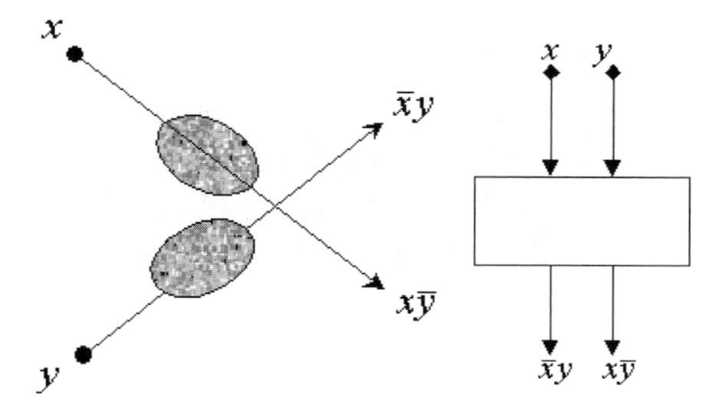

Figure 4.10. Glider interaction gate.

4.7 *Game of Life*

The *Game of Life* is one of the first cellular-automata models to have been proved to be dynamically computation universal [81], so we devote a separate section to it. The *Game of Life* is a two-dimensional automaton with an eight-cell neighbourhood and a binary cell state set. The *Life* cell state transition rule is attractively simple. Every cell in the 1-state does not change its state if it has exactly two or three neighbours in the 1-state. The cell switches from the 0-state to the 1-state when there are exactly two neighbours in the 1-state. The self-localized oscillating translating patterns (gliders and spaceships, and generators of the movable patterns—glider guns) usually emerge in the spacetime evolution of the *Game of Life* automaton that starts from the random initial configuration. The gliders can be considered as analogues of the balls, which represent, transmit and process information. The gliders, however, do not collide elastically. Both gliders disappear as a result of collision and the interaction gate has not four but two outputs as in figure 4.10. Therefore, only the $\overline{x}y$ and $x\overline{y}$ functions can be computed in a single collision of two gliders.

The universality of the *Game of Life* is proved in [81] by the construction of universal logical functions via collisions of glider streams. If the distances between neighbouring gliders in a stream are fixed, then every glider will represent 1, or `Truth`, and the absence of a glider 0, or `False`. If two gliders collide with one another they annihilate. The glider streams beating a constant 1-state from the glider gun can be considered. Therefore, the $\overline{x}$ gate is constructed in the following way. A data stream x of gliders runs across a stream of gliders generated by a glider gun. When a glider from the data stream collides with a glider from the glider gun it is annihilated. If it does not meet a glider from the glider gun it continues to move undisturbed. As a result of the stream interaction, we obtain a stream $\overline{x}$ of gliders and holes that represents the result of the computation. Combining the glider streams from the glider gun with the data streams x and y we produce the streams $x \vee y$ and xy.

4.8 The excitable lattice

The lion's share of this chapter will deal with a very simple but behaviourally rich model—that of an excitable lattice. We show that not only can most localized excitation phenomena be expressed in two- or three-dimensional cellular automaton, each cell of which takes only three states and has only eight neighbours, but the model itself may form a solid basis for the theory of collision-based computing.

Let every cell of a discrete lattice **L** take three states (•, rest; +, excited; and – refractory), and change its states deterministically and in a discrete time depending on the states of its closest neighbours. Every cell x of the lattice **L** has an eight-cell neighbourhood in a two-dimensional lattice and a 26-cell neighbourhood in a three-dimensional lattice, i.e. the neighbourhood of the cell x includes all cells at distance 1 (in the L_∞ metric) from x and does not include the cell x itself.

A rest cell becomes excited if exactly two (in a two-dimensional lattice) or four (in a three-dimensional lattice) of its neighbours are excited and passes from an excited state to a refractory state and from a refractory state to the rest state unconditionally, i.e. independently the states of its neighbours.

A cellular-automata model of the lattice excitation is the tuple

$$\mathbf{M}^+ = \langle \mathbf{L}, \mathbf{Q}, u, f \rangle$$

where

(i) **L** is a two- or three-dimensional array of cells,
(ii) **Q** is a set of cell states, $\mathbf{Q} = \{\bullet, +, -\}$,
(iii) $u : \mathbf{L}^k \to \mathbf{L}$ is a neighbourhood function, which assigns k different cells of **L** to every cell of **L** and
(iv) $f : \mathbf{Q} \times \mathbf{Q}^k \to \mathbf{Q}$ is a function of the cell state transitions.

Every cell x has a neighbourhood $u(x) = (y_1, \ldots, y_k)$ such that

$$\forall y \in u(x) : |x - y|_{L_\infty} = 1.$$

The size of the neighbourhood is $k = 8$ cells in a two-dimensional lattice and $k = 26$ cells in a three-dimensional lattice. The cell x in state x^t at time step t calculates its next state x^{t+1} in accordance with the local transition function:

$$x^{t+1} = f(x^t, u(x)^t).$$

The transitions $+ \to -$ and $- \to \bullet$ are unconditional. Therefore we have to define the only condition for transition $\bullet \to +$, i.e. the excitation condition. This rule has the following form:

$$x^{t+1} = \begin{cases} + & x^t = . \text{ and } |\{y \in u(x) : y^t = +\}| = \theta \\ - & x^t = + \\ \bullet & \text{otherwise} \end{cases} \tag{4.1}$$

where $\theta = 2$ (in two dimensions) and $\theta = 4$ (in three dimensions). The value of θ determines the version name of the model: the 2^+-medium and the 4^+-medium [13, 14, 19], respectively.

4.9 Minimal particle-like waves

Let $\mathbf{L}$ be an infinite lattice and

$$\sigma^t = \{x \in \mathbf{L} | x^t \in \{+, -\}\}$$

be a finite subset of cells in the states $+$ and $-$ such that

$$\forall x \in \sigma^t \exists y \in \sigma^t : x \neq y \text{ and } x \in u(y) \text{ and } y \in u(x) \qquad t \in \mathbf{N}.$$

Then σ^t is called a particle-like wave if it is translated by the parallel application of the function f to the cell neighbourhoods in the lattice $\mathbf{L}$.

Let $\mathbf{E} = \{0, \ldots, k\}$ be the set of all possible sums of excited elements of $u(x)$. The function $f : \mathbf{Q} \times \mathbf{Q}^k \rightarrow \mathbf{Q}$ belongs to the family Ξ of multiple threshold excitations if it is determined by the rule

$$x^{t+1} = \begin{cases} + & x^t = \bullet \text{ and } |\{y \in u(x) : y^t = +\}| \in \Theta \\ - & x^t = + \\ \bullet & \text{otherwise} \end{cases} \tag{4.2}$$

where $\Theta \in 2^{\mathbf{E}}$. The definition implies that the rest cell becomes excited if the sum of its excited neighbours matches one of the elements of Θ. The family Ξ includes all other possible functions of the lattice excitation with unconditional state transitions $- \rightarrow \bullet$ and $+ \rightarrow -$.

Example 1. Let f be the function of the interval excitation defined by the rule

$$x^{t+1} = \begin{cases} + & x^t = \bullet \text{ and } \Gamma_1 \leq |\{y \in u(x) : y^t = +\}| \leq \Gamma_2 \\ - & x^t = + \\ \bullet & \text{otherwise} \end{cases} \tag{4.3}$$

where $\Gamma_1 \leq \Gamma_2$ and $\Gamma_1, \Gamma_2 \in \mathbf{E}$. To prove $f \in \Xi$ we show that $\Theta = \{a \in \mathbf{E} | \Gamma_1 \leq a \leq \Gamma_2\}$. Assuming that $\Gamma_2 = \max \mathbf{E}$ we demonstrate that all functions of the conventional threshold excitations are the elements of Ξ [25].

Recall that the configuration of the cellular automaton is a mapping

$$c : \mathbf{L} \to \mathbf{Q}.$$

Given the sequence

$$c^0 \to \cdots \to c^p$$

of the configurations of the excitable lattice $\mathbf{M}^+$ we state that the function $f \in \Xi$, determined by Θ, is minimal in the family Ξ if for any other $f' \in \Xi$, determined by Θ', we have

(i) $|\Theta| \leq |\Theta'|$
(ii) $\min \Theta \leq \min \Theta'$.

In words, f is minimal if its corresponding set Θ has a minimal number of elements and the minimal element among the elements of all other characteristic sets of the elements of Ξ. In the terms of the interval excitations this means that the cells have the only excitation interval, which is a singleton and the least possible element of $\mathbf{E}$.

> *The cell state transition rule of the excitable lattice* $\mathbf{M}^+$ *is the minimal rule of excitation of two- and three-dimensional excitable lattices with near neighbour interactions; it is minimal amongst the lattices that exhibit particle-like waves (self-localized excitations).*

Let us inspect the configurations of the particle-like waves. In a two-dimensional lattice a minimal translating pattern, the so-called 2^+-particle [13, 14, 19], consists of two excited and two refractory states. The 2^+-particle occupies four lattice nodes. The particle moves along the coordinate axes. The configuration of the excited and refractory states encodes the orientation of the velocity vector as follows: north $\left(\begin{smallmatrix}+&+\\-&-\end{smallmatrix}\right)$, south $\left(\begin{smallmatrix}-&-\\+&+\end{smallmatrix}\right)$, west $\left(\begin{smallmatrix}+&-\\+&-\end{smallmatrix}\right)$, east $\left(\begin{smallmatrix}-&+\\-&+\end{smallmatrix}\right)$. To demonstrate non-stop movement we consider how many excited neighbours there are around the rest cells surrounding the particle. Let us look at the 2^+-particle moving north:

$$\begin{array}{cccc} 1 & 2 & 2 & 1 \\ 1 & + & + & 1 \\ 1 & - & - & 1 \\ 0 & 0 & 0 & 0. \end{array}$$

In this scheme the states of excited and refractory cells are shown explicitly whereas every rest cell (closest to the particle) contains the number of its excited neighbours. We see that only two northern rest neighbours of the excited cells have two excited neighbours and, therefore, they are excited at the next step of discrete time; the currently excited cells become refractory and the pattern is shifted north.

In three-dimensional lattices the minimal particle-like waves, the so-called 4^+-particles, consist of four excited cell states and four refractory cell states. The 4^+-particle occupies eight lattice cells.

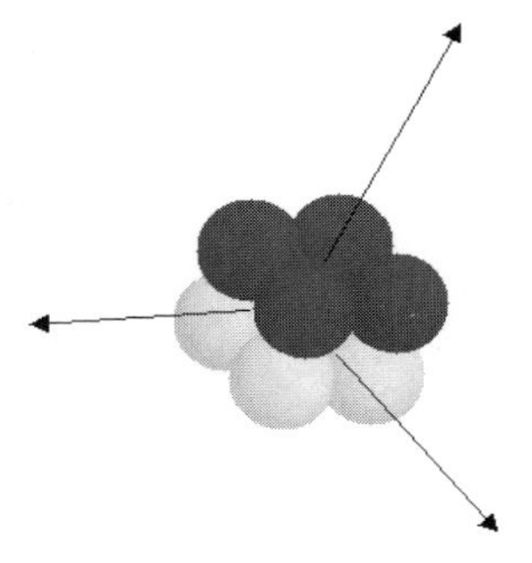

Figure 4.11. 4^+-particle: black spheres represent the excited cells, light grey spheres are the refractory cells; rest cells are not shown.

Every 4^+-particle can be imagined as a duplex of 2^+-particles (moving in parallel planes). The velocity vector of the 4^+-particle is parallel to one of the coordinate axes, perpendicular to the plane in which the block of $+$ states lies and the orientation of the vector is encoded in the configuration of the $+$ and $-$ cells. Thus, e.g., the 4^+-particle, the velocity vector of which is collinear with the y-axis, has the configuration

$$\begin{pmatrix} + & + \\ - & - \end{pmatrix}^{(x,y,z)} \quad \begin{pmatrix} + & + \\ - & - \end{pmatrix}^{(x,y,z+1)}.$$

A more attractive picture of the 4^+-particle is shown in figure 4.11.

The excitation rule of $\mathbf{M}^+$ satisfies the minimality condition (i) because Θ is a singleton. So, we need to prove the condition (ii) of minimality.

In two-dimensional lattices $\theta = 2$, therefore the only candidate to test is $\theta = 1$. Let the rest cell become excited if exactly one neighbour is excited. On infinite $\mathbf{L}$ any finite connected set σ^0, $|\sigma^0| \geq \theta$, of excited cells causes unbounded growth of excitation because the rest cells nearest to the extreme elements of the set have exactly one excited neighbour. Considering $\theta = 1, 2, 3$ in the three-dimensional case we obtain similar results.

More detailed issues on the structure of minimal wave generators can be found in the next chapter dealing with the parametrization of excitation.

The excitation rule of $\mathbf{M}^+$ is minimal only amongst the models with *spatially invariant* cell state transition rules because some of the asymmetric neighbourhoods smaller in size can also give us the translating patterns.

The minimal particles (2^+- and 4^+-particles) move along the rows or columns of the cell array.

Fortunately, there are also patterns that move along the diagonals of the arrays. These are 3^+-particles (in a two-dimensional lattice) and 6^+-particles (in a three-dimensional lattice). They consist of three $+$-states and three $-$-states, and six $+$-states and six $-$-states, respectively. Any 3^+-particle has four possible

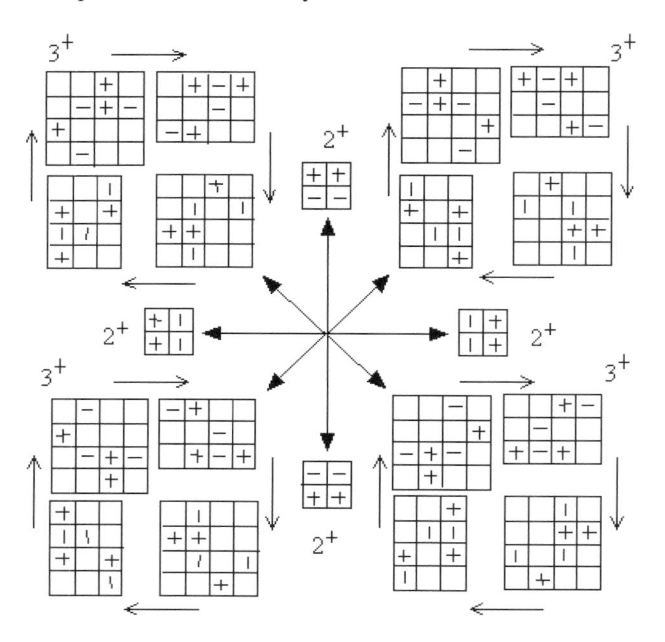

Figure 4.12. Encoding of the directions of movement in the states of minimal particles: two-dimensional lattice, 2^+-medium.

configurations (particle states) that are changed step by step in the loop. The diagonal movement is approximated by a series of ladder shifts (figure 4.12).

The 6^+-particle can be imagined as the duplex of the 3^+-particles. The 6^+-particle moves in the plane parallel to one of the coordinate planes and only two of its three coordinates are changed during the movement.

> *The* 2^+-, 3^+-, 4^+- *and* 6^+*-particles are the minimal indivisible compact movable excitation patterns.*

The mobility was proved when we considered the number of excited neighbours of the rest cells around the currently excited states. If, in the evolution of the model, we remove one of the excited states from any of the particles the pattern disappears or is transformed into the other pattern (not minimal). This was demonstrated by straightforward perturbations of the patterns [13]. The neighbourhoods of the non-rest cells in the particles overlap; therefore these patterns are compact. The 2^+- and 4^+-particles are minimal because the number of excited cells in these patterns equals θ. Any compact excitation pattern that includes less than θ excited cells will be entirely at rest after several steps of the evolution. Using an exhaustive search we prove that the 3^+- and 6^+-particles are the minimal patterns that move along the diagonals of the lattice.

The velocity of 3^+*- and* 6^+*-particles is four times less than the velocity of* 2^+*- and* 4^+*-particles.*

More generally we can say that the patterns moving along the columns or rows of the lattice have unit velocity whereas the velocity of the patterns running along the diagonals is equal to $\frac{1}{4}$. This is because the diagonal movement is approximated by the ladder shifts along columns and rows [13].

How big are the minimal quanta of information in an excitable lattice?

The minimal quantum of information in excitable lattices occupies 2×2 *cell volume in two-dimensional lattices and* $2 \times 2 \times 2$ *cell volume in three-dimensional lattices.*

4.10 Particle guns: the generators of the particle-like waves

Constant **1** is realized by a stream of particle-like waves. The streams are semi-infinite and they are produced by so-called particle guns.

A particle gun is a finite compact pattern that periodically generates one or more types of particles or mobile self-localizations.

There is still no rigorous technique for designing particle guns (however identification algorithms [9] and integer programming [99] could possibly help) and we make the most of our propositions from an exhaustive search of almost all the possible localized patterns we have investigated in numerical experiments [26].

There are no immobile guns in a 2^+*-medium.*

This is because there are no immobile localized patterns in a 2^+-medium at all.

The minimal particle gun in a 2^+*-medium, the so-called* G_2*-gun, moves along one of the coordinate axes with unit velocity and has a size of* 6×9 *cells and the weight of* 26 *non-rest states. The gun generates a* 2^+*-particle every fourth step of the evolution. The generated particles move in the opposite direction to the direction of the gun motion.*

The configuration of a cellular automaton with a G_2-gun moving west and emitting 2^+-particles eastwards is shown in figure 4.13. The weight of the gun is maximal just before it delivers the new particle. Because the period of generation is four time steps the 2^+-particles are at a distance of eight cells from each other in the particle stream. This determines the maximal frequency of a possible computing device built from an excitable lattice.

In certain cases it may be extremely useful to have a gun that generates particles in the direction perpendicular to the velocity vector of the gun. Two

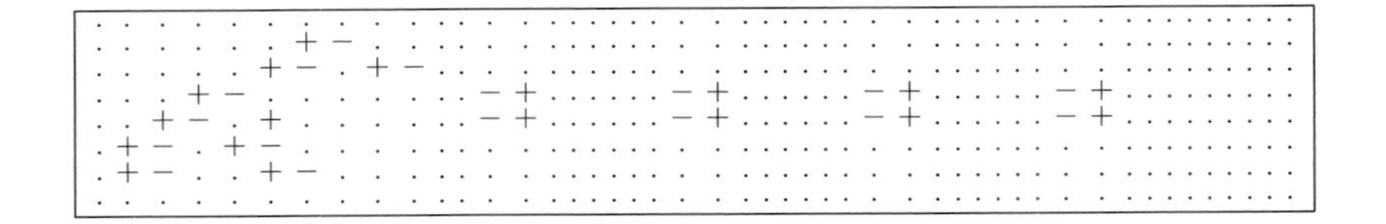

Figure 4.13. Two-dimensional mobile gun G_2.

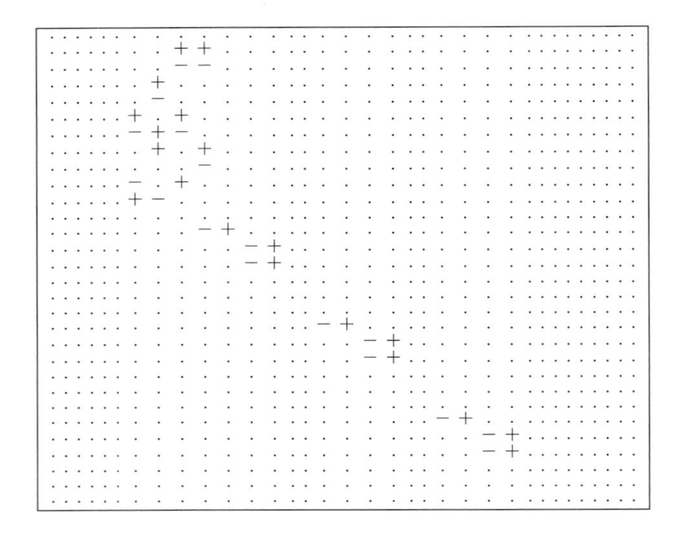

Figure 4.14. Two-dimensional mobile gun G_3.

minimal guns that generate particles in this manner are shown in figures 4.14 and 4.15. The first gun, the G_3-gun, gives birth to $(2 + 1)^+$-particles (figure 4.14). As we will see later the $(2 + 1)^+$-particle is transformed into the 2^+-particle via collision with another 2^+-particle. The G_3-gun has 18 non-rest-state weight, 5×11 cell maximal size, and it generates particles every fourth step of the evolution. The generated particles are grouped into a particle front. The distance between two particles in the front is eight cells along the axes parallel to velocity vector and four cells along another axes.

The third gun, G_2^7-gun, simultaneously generates 2^+-particles and more complex patterns (of seven excited states) every eighth step of the evolution (figure 4.15). The velocity vectors of all generated patterns are perpendicular to the velocity vectors of the gun. The 2^+-particles and 7^+-patterns move in opposite directions.

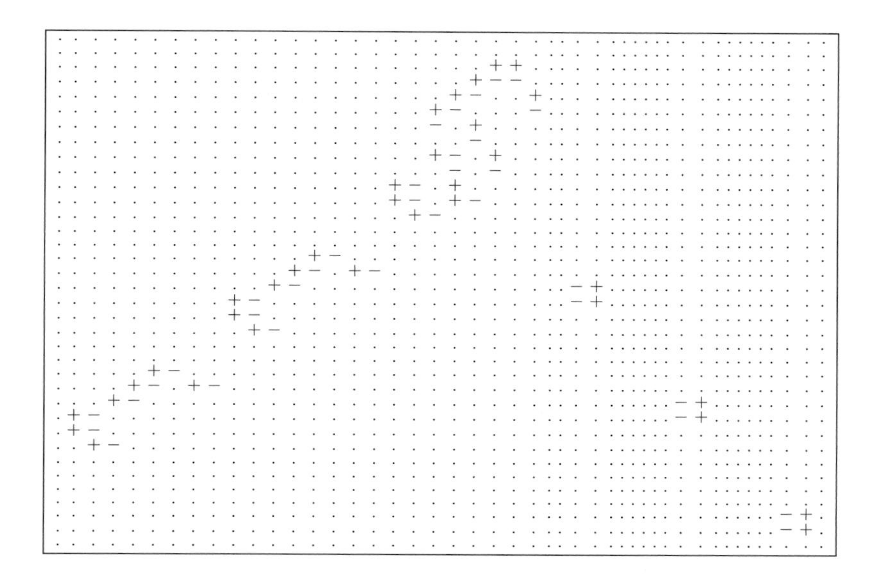

Figure 4.15. Two-dimensional mobile gun G_2^7.

When talking about a minimal gun we should emphasize the fact that the G_3-gun has actually less size and weight than the G_2-gun. Though it does not produce elementary particles directly it does generate pro-particles that need additional collisions to be transformed into elementary particles. Hence, we cannot accept the G_3-gun as a minimal gun.

Is it possible to generate a particle gun in a natural way, that is in the collision of several elementary 2^+*- and* 3^+*-particles?*

There is no certain answer for two-dimensional lattices. In a three-dimensional lattice the stationary gun is generated by the collision of two elementary particle-like waves.

In the 4^+*-medium the minimal mobile gun has the weight of 16 non-rest states, a* $4 \times 4 \times 3$ *cell volume, and generates* 4^+*-particles every second step of the evolution. The generated* 4^+*-particles move in the opposite direction to the velocity vector of the gun.*

The gun is shown in figure 4.16. It moves along one of the coordinate axes. It is very elegant and can be 'built' from 4^+-particle by adding four complementary excited states and placing them at the angles of the quadruple of the refractory states (see figure 4.16, t). The new 4^+-particles are launched on the back of the gun and they move in opposite directions to the gun velocity vector. The gun

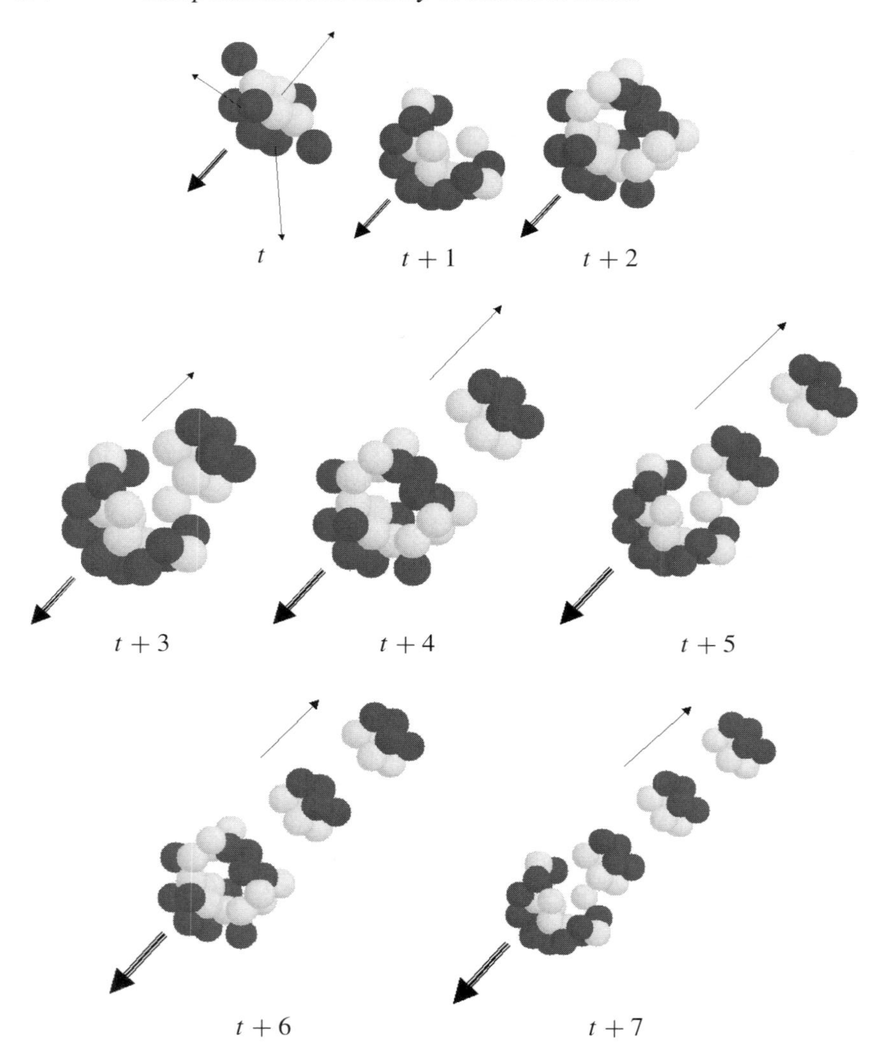

Figure 4.16. Mobile gun in a 4^+-medium (three-dimensional excitable lattice). The coordinate axes are indicated on the first image. The double arrow shows the direction of gun motion. Black spheres represent excited cells, light grey spheres are the refractory cells; the rest cells are not shown.

forms an extremely dense stream of 4^+-particles, where there is only two empty cells between the neighbouring particles.

> *In the 4^+-medium the minimal stationary gun, G_2^4-gun, has a weight of 12 non-rest states, a $4 \times 4 \times 3$ cell volume and generates two 4^+-particles every third step of the evolution.*

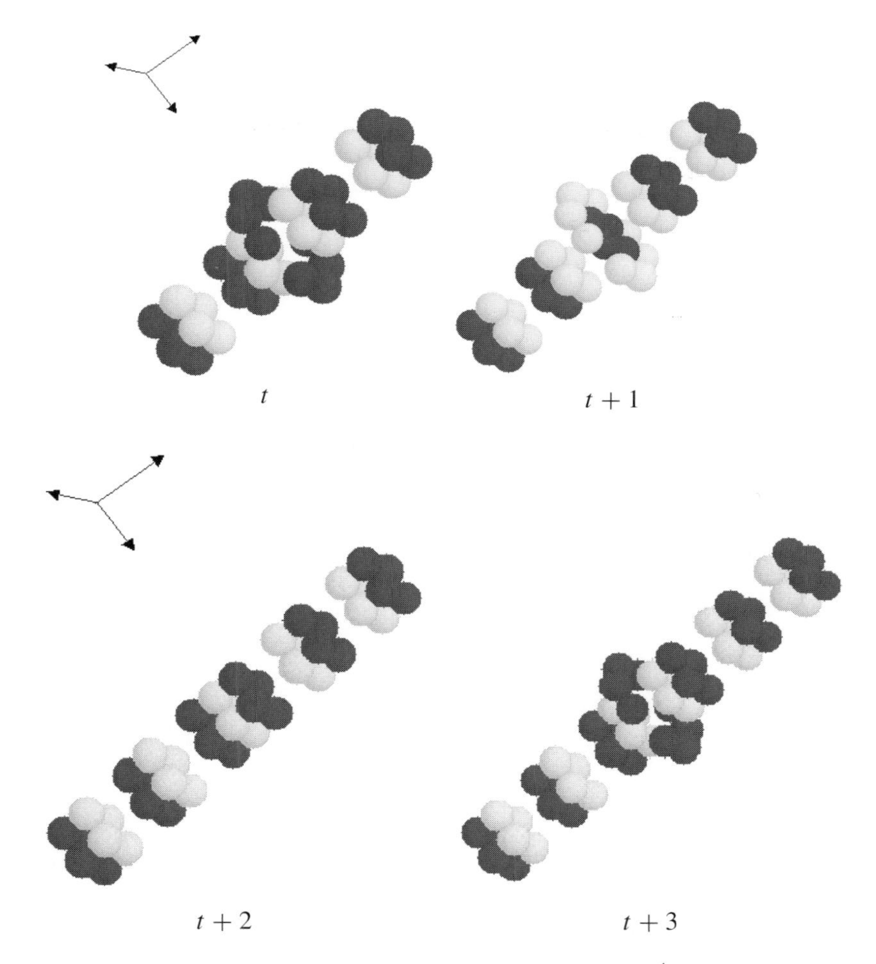

Figure 4.17. Stationary three-dimensional gun G_2^4.

The development of the G_2^4-gun is shown in figure 4.17. When generating the particles the gun changes its state in a cycle of period three. It subsequently takes the following proper states:

(i) an empty square with excited boundaries or eight excited states (figure 4.17, t),
(ii) a plate of four excited states surrounded by eight refractory states (figure 4.17, $t+1$),
(iii) a plate of four refractory states sandwiched between the plates of four excited states (figure 4.17, $t+2$).

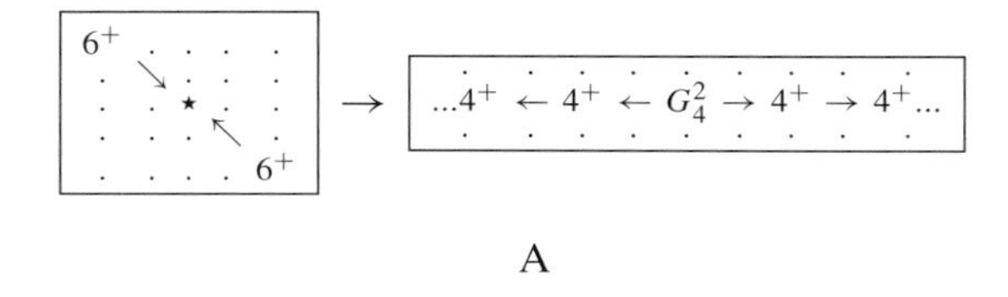

A

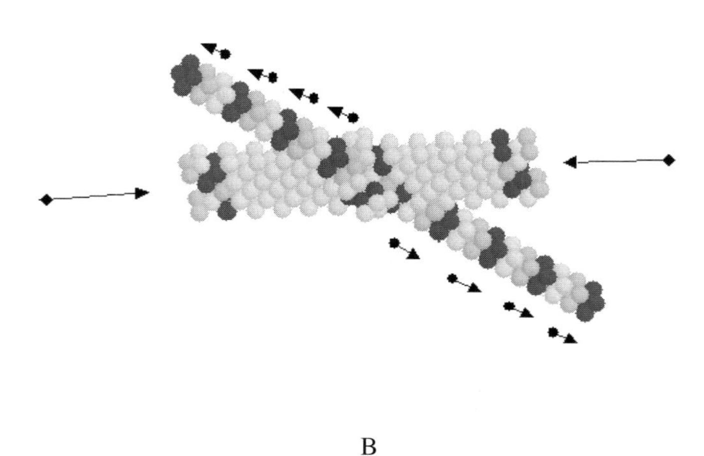

B

Figure 4.18. The scheme (A) and the skeleton (B) of the frontal collision of two 6^+-particles that leads to the formation of the stationary G_2^4-gun, which generates a stream of 4^+-particles.

The three-dimensional stationary gun, G_2^4-gun, is generated as a result of the frontal collision of two 6^+-particles.

The scheme and collision skeleton are shown in figure 4.18. Here, as well as in forthcoming sections, we present the explicit parameters of the elementary particles just a few steps before the collision. Assuming a three-dimensional lattice of $n \times n \times n$ cells entirely at rest we describe every collision by the sets $\mathbb{C}_+$ and $\mathbb{C}_-$ of the coordinates of cells in rest, $+$, and refractory, $-$, states, respectively. The coordinates are represented relative to the $(0, 0, 0)$ coordinate centre placed in the cell with index $(n/2, n/2, n/2)$. The frontal collision of two 6^+-particles with the formation of a stationary gun has the following explicit parameters:

$$\begin{aligned}\mathbb{C}_+ &= \{(12,10,0),(12,11,0),(10,12,0),(12,10,1),(12,11,1),\\ &\quad (10,12,1),(0,0,0),(0,0,1),(0,1,0),(0,1,1),(2,-1,0),(2,-1,1)\}\\ \mathbb{C}_- &= \{(11,11,0),(13,11,0),(11,13,0),(11,11,1),(13,11,1),(11,13,1),\\ &\quad (-1,0,0),(-1,0,1),(1,0,0),(1,0,1),(1,-2,0),(1,-2,1)\}.\end{aligned}$$

It is remarkable that the slight relative space shift in the colliding 6^+-particles changes the orientation of the G_2^4-gun by 90°.

All guns are destroyable.

This can be trivially proved by colliding 2^+- and 4^+-particles with particle guns. The collision is followed by a short-range perturbation (dissipation of energy) after which all excitation patterns disappear. The following important problem remains open.

Are there generators of 3^+- *and* 6^+*-particles in excitable lattices?*

We did not find any.

4.11 Collisions and interaction gates

Self-localized excitations, or particle-like waves, represent the bits of information (or unit impulses) and the logic gates are realized when two or more particles collide with one another. The logic operations are implemented at the collision sites when two or more particle-like waves interact with one another. Therefore the logic gates realized in excitable media are called interaction gates. In general, the result of the collision is determined by the types and phases of colliding particles and the angle of collision. To build the structure of all interaction logic gates in excitable lattices all combinations of particles were collided in an entirely at rest environment. Surprisingly, we found that excitable lattices support a much wider range of interaction gates than any other automata model.

The typical interaction gate has two input wires (by '*wire*' we mean the trajectory of the particle-like wave) and at least two but usually three output wires (output trajectories). Two output wires represent the particle trajectories when they continue their motion undisturbed and the third output wire represents the trajectory of a new particle formed as a result of the collision of two incoming particles.

To describe the interaction gates we use two symbols: $\triangle$ and $\bigtriangledown$. The symbol $\triangle$ indicates that the results of the collision appear simultaneously on different output wires/trajectories. $\bigtriangledown$ shows that the resulting particles appear exclusively on one of the outputs. Thus, e.g., the billiard ball gate (figure 4.9) is represented by

$$\langle x, y\rangle \rightarrow x\overline{y} \bigtriangledown \overline{x}y \bigtriangledown (xy \triangle xy).$$

The glider interaction gate (4.10) of the *Game of Life* is described in the following form:

$$\langle x, y\rangle \rightarrow x\overline{y} \triangledown \overline{x}y.$$

When specifying the product of the interaction gates we also indicate (in the square brackets) which particle-like waves represent the terms of the gates. The specification is monotonic in the case of the billiard ball gate:

$$\langle \text{ball}, \text{ball}\rangle \rightarrow \text{ball} \triangledown \text{ball} \triangledown (\text{ball} \vartriangle \text{ball})$$

and the glider gate

$$\langle \text{glider}, \text{glider}\rangle \rightarrow \text{glider} \triangledown \text{glider}$$

but it becomes very interesting when excitable media are included in the consideration.

The next result deals with a two-dimensional excitable lattice.

The following types of interaction gates are realized in a 2^+*-medium:*

$$\text{g1}(x, y) = \overline{x}y \triangledown x\overline{y}$$
$$[\langle 2^+, 2^+\rangle \rightarrow 2^+ \triangledown 2^+]$$

$$\text{g2}(x, y) = (xy \vartriangle xy) \triangledown \overline{x}y \triangledown y\overline{x}$$
$$[\langle 2^+, 2^+\rangle \rightarrow (2^+ \vartriangle 2^+) \triangledown 2^+ \triangledown 2^+]$$

$$\text{g3}(x, y) = xy \triangledown \overline{x}y \triangledown x\overline{y}$$
$$[\langle 2^+, 2^+\rangle \rightarrow 3^+ \triangledown 2^+ \triangledown 2^+]$$

$$\text{g4}(x, y) = g3(x, y)$$
$$[\langle 2^+, 3^+\rangle \rightarrow 3^+ \triangledown 2^+ \triangledown 3^+]$$

$$\text{g5}(x, y) = xy \triangledown \overline{x}y \triangledown x\overline{y}$$
$$[\langle 2^+, 2^+\rangle \rightarrow (2+1)^+ \triangledown 2^+ \triangledown 2^+]$$

$$\text{g5}a(x, y) = xy \vartriangle xy$$
$$[\langle (2+1)^+, 2^+\rangle \rightarrow 2^+ \vartriangle 3^+]$$

$$\text{g6}(x, y) = xy \triangledown \overline{x}y \triangledown x\overline{y}$$
$$[\langle 3^+, 2^+\rangle \rightarrow 2^+ \triangledown 2^+ \triangledown 3^+]$$

$$\text{g7}(x, y) = y \vartriangle [x\overline{y} \triangledown xy]$$
$$[\langle 3^+, 2^+\rangle \rightarrow 2^+ \vartriangle (3^+ \triangledown 2^+)].$$

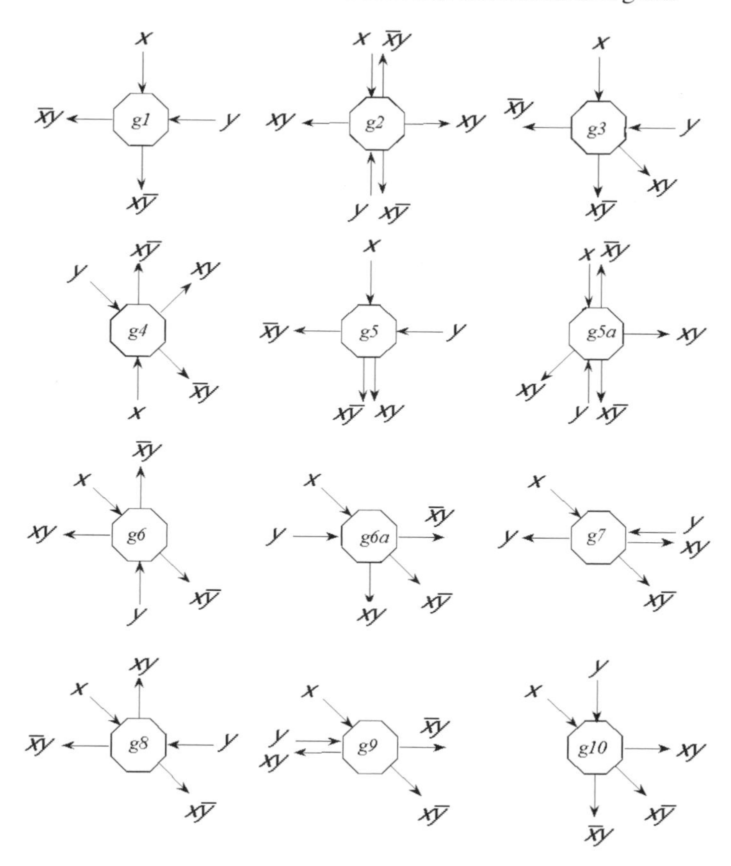

Figure 4.19. The interaction gates realizable in a two-dimensional excitable lattice with a 2^+-medium.

A graphical representation of the gates is shown in figure 4.19. The absolute orientation of the wires reflects the orientation of the velocity vectors of the particles approaching and leaving the collision site. The vertical and horizontal arrows represent the trajectories of 2^+-particles (or 2^+-like particles as, e.g., $(2+1)^+$-pattern), and the diagonal arrows are the trajectories of the 3^+-particles. In all examples of the evolution of an excitable lattice, which realizes one of the logic gates, we assume that both input variables have `Truth` values, i.e. two particles approach a collision site simultaneously.

The first gate, g1 is a typical glider gate, as in the *Game of Life*. When two 2^+-particles collide (figure 4.20, t) they destroy each other (figure 4.20, $t+4$) and disappear. The $\overline{x}y$ is computed as a result of such a collision.

If 2^+-particles collide frontally (figure 4.21, t) we implement xy in the gate

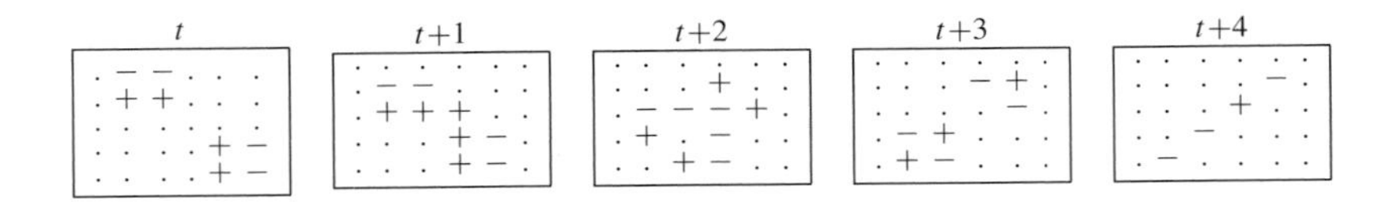

Figure 4.20. Gate g1.

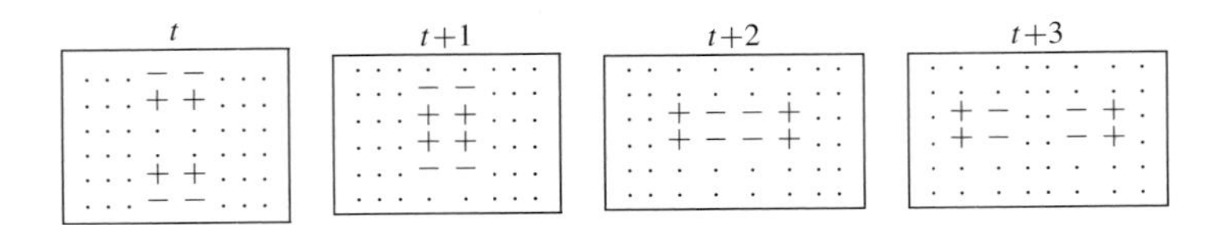

Figure 4.21. Gate g2.

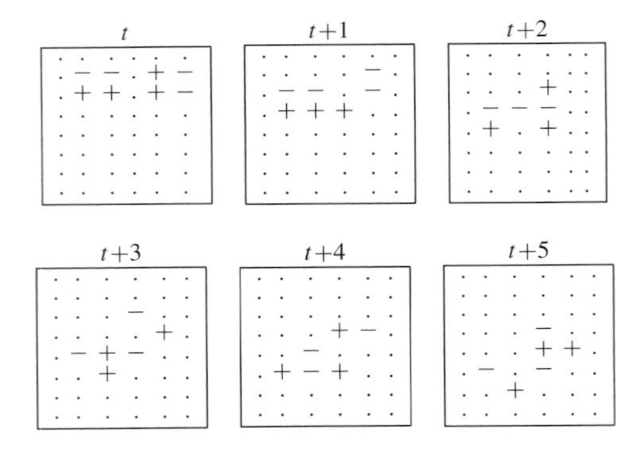

Figure 4.22. Gate g3.

g2. The 2^+-particles collide quasi-elastically and change the direction of their motion as a result of the collision. The first two gates involve only 2^+-particles. All other gates may produce 3^+-particles as the outcome of the collision.

In gate g3 a 3^+-particle is formed when two 2^+-particles are in a head-by-side collision as specified in figure 4.22. The 3^+-particle represents the xy product of the logic operation. When the 2^+-particle moving west crashes into the 2^+-particle moving south (figure 4.22, t) a 3^+-particle is formed and it moves south-east (figure 4.22, $t + 5$).

The g4 gate is a very typical example of excitable interaction gates (figure 4.23). The x variable is represented by the 3^+-particle moving south-east when $x = 1$ and the y variable is represented by the 2^+-particle moving north when $y = 1$. When the particles collide ($t + 1$) the new self-localized

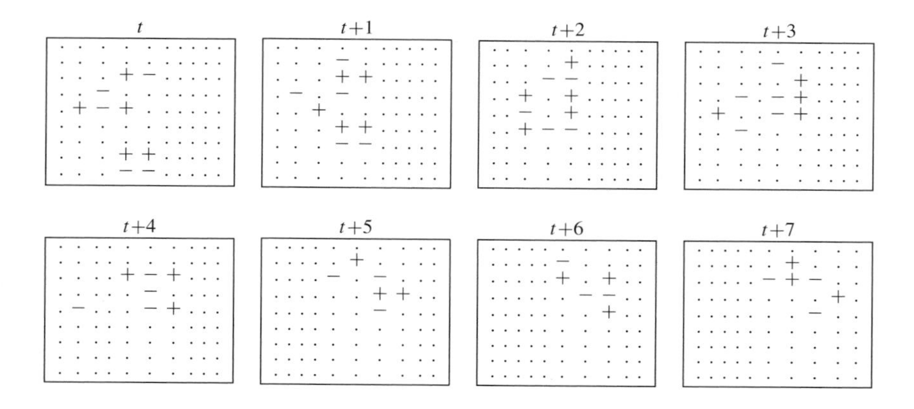

Figure 4.23. Gate g4.

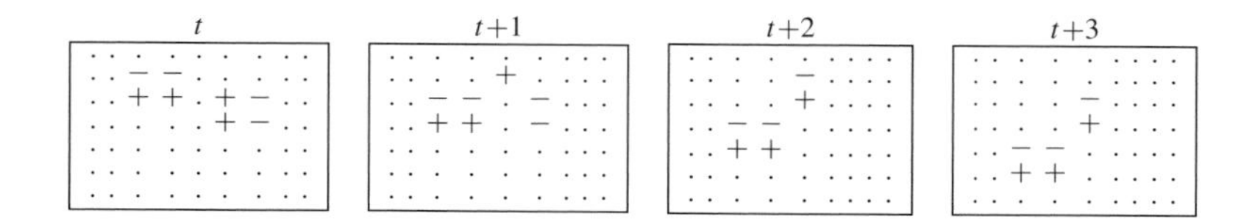

Figure 4.24. Gate g5.

excitation—3^+-particle—is formed ($t+6$). This particle moves north-east. If $y=0$ and $x=1$ then the 2^+-particle simply continues its motion and we have 1 on the $x\overline{y}$ output trajectory. If $x=0$ and $y=1$ we see a 3^+-particle on the $\overline{x}y$ output trajectory.

The g5 gate is one of the most attractive. On first sight it is similar to gate g1 because we get $\overline{x}y$ and $x\overline{y}$ on the same wires as in the g1 gate (figure 4.19). The gate g5 also computes the xy function. However, if $x=1$ and $y=1$ the product xy is represented by the $(2+1)^+$-pattern (figure 4.24). This is the only gate where we allow representation of the variable value by a non-elementary particle-wave. This is because the 2^+- and $(2+1)^+$-patterns can easily be differentiated in the collision from another 2^+-particle (figure 4.25). Let p be an unknown pattern (either a 2^+- or $(2+1)^+$-pattern) that represents the result of the computation in the g5$(1, y)$ gate. Then we compute g5a$(p, 1)$. If p is represented by the $(2+1)^+$-pattern, then the south-west 'wire' of gate g5a gives us a 3^+-particle. If p is represented by a 2^+-particle (i.e. $x\overline{y}$ result of g5 gate) then we have nothing on the south-west wire of the g5a gate.

The next two gates—g6 and g6a—are equivalent to one another in almost all (figures 4.26 and 4.27) orientations except that of the input wires: the x and y

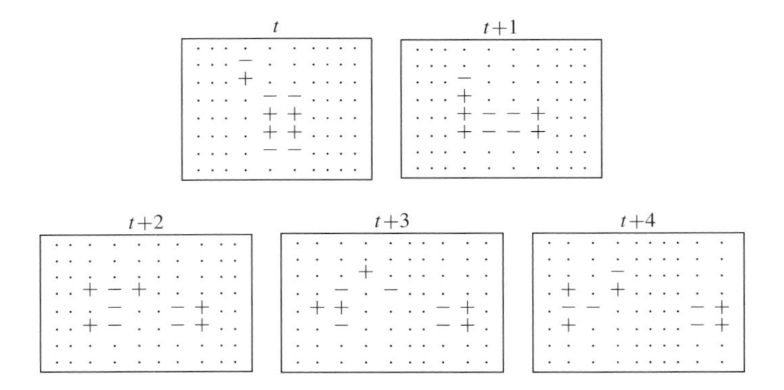

Figure 4.25. Gate g5a.

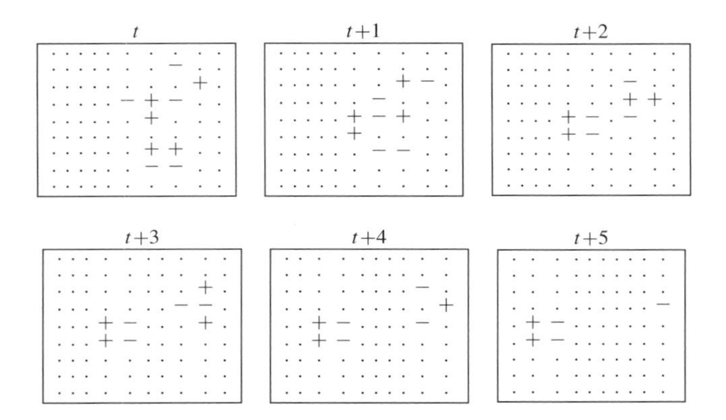

Figure 4.26. Gate g6.

variables are represented by 3^+- and 2^+-particles, the $x\overline{y}$ term is represented by the 3^+-particle and $\overline{x}y$ and xy terms by the 2^+-particles, respectively.

Gate g7 gives us an example of the deflection of one particle by another where the second one does not change its trajectory (figure 4.28). When $x = 1$ and $y = 1$ in gate g7 the 3^+-particle moving south-east collides with the 2^+-particle moving west. As a result of the collision the 2^+-particle continues its journey undisturbed but the 3^+-particle is transformed into a 2^+-particle that runs east.

The three remaining gates (figure 4.19), g8, g9 and g10, are actually equivalent to one of the previous gates except for the relative orientation of the trajectories of incoming particles.

All gates except g2, g5a, g7 and g9 do not alter data.

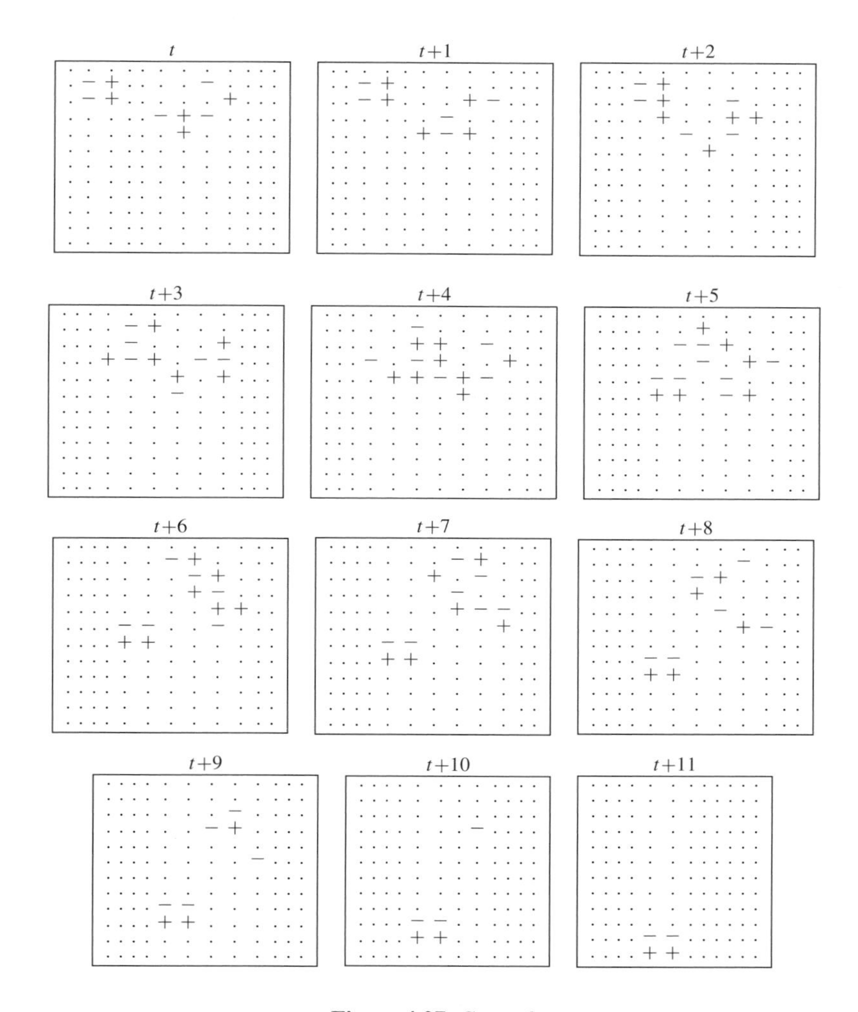

Figure 4.27. Gate g6a.

The data are assumed to be altered as a result of the computation when at least one of the outcoming particles travels along the same trajectory as one of the incoming particles. The data are altered in two different ways. The g2 and g5a gates alter data weakly (figure 4.19) because the particle representing $\overline{x}y$ goes along the trajectory of x only when $x = 1$. The g7 and g9 gates alter data sufficiently because the particle representing xy goes along the trajectory of the particle representing y when $x = 1$ and $y = 1$ (figure 4.19).

The gate g2 conserves excitation.

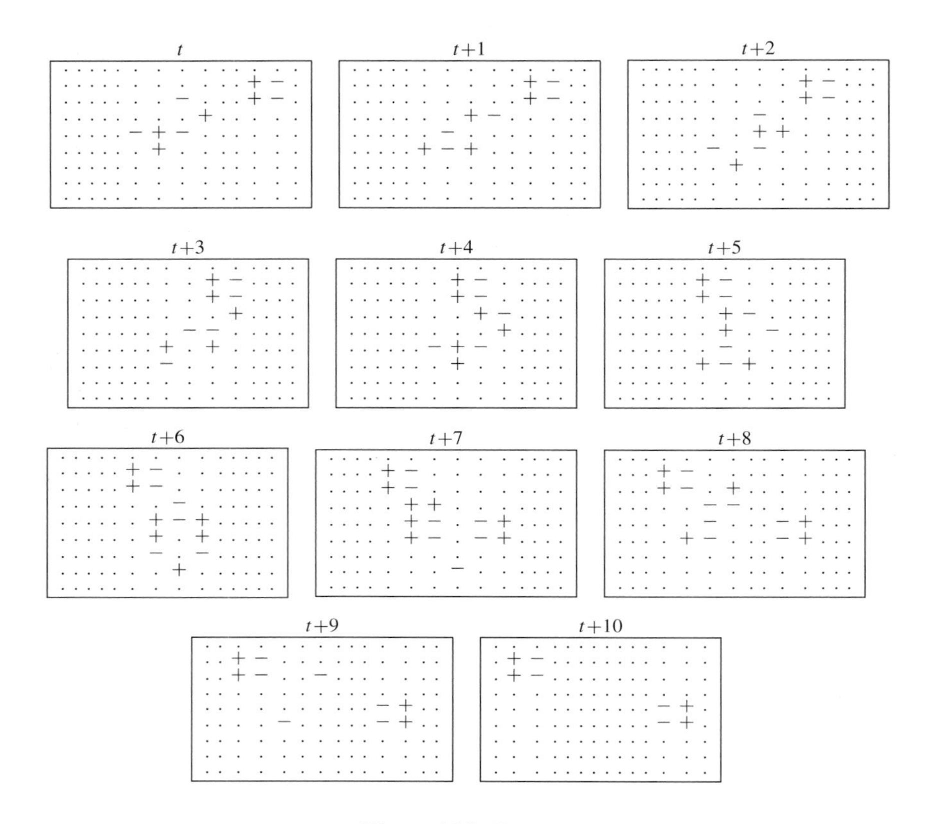

Figure 4.28. Gate g7.

The gate is excitation-conservative because

- the number of incoming particles equals the number of out-coming particles
- the number of overall excited states before the collision equals the number of overall excited states after the collision (figure 4.21).

The minimal gates $x \wedge y$ and $\overline{x} \wedge y$ in a 2^+-medium have two input and three output trajectories and a (5×2)-cell size.

Let $\gamma(\mathrm{g})$ be the size of a gate g, i.e. the maximum among the sizes of the input and output particles, and sizes of the collision and perturbation areas. Then we have the following space complexities of the gates: $\gamma(\mathrm{g1}) = (5 \times 5)$, $\gamma(\mathrm{g2}) = (4 \times 2)$, $\gamma(\mathrm{g3}) = (5 \times 2)$, $\gamma(\mathrm{g4}) = (4 \times 6)$, $\gamma(\mathrm{g5}) = (5 \times 3)$, $\gamma(\mathrm{g5a}) = (3 \times 6)$, $\gamma(\mathrm{g6}) = (4 \times 7)$, $\gamma(\mathrm{g6a}) = (7 \times 4)$, $\gamma(\mathrm{g8}) = (5 \times 6)$, $\gamma(\mathrm{g9}) = (7 \times 3)$ and $\gamma(\mathrm{g10}) = (4 \times 7)$.

The g2 gate has the minimal size. Unfortunately, it does alter inputs (figure 4.19). Gate g3 is slightly larger than gate g2 but it does not alter data. It covers a 10-cell area and computes both $x \wedge y$ and $\overline{x} \wedge y$ functions.

Re-usability of the computing device is one of the advantages of excitable-lattice computations. To start a new computational process on the same lattice we should clean up all results from previous computations. If the lattice has absorbed boundaries this will not be a problem at all. We can show that for a 2^+-medium an evolution started at any random configuration of excited and refractory states of the lattice nodes finishes in the entirely-at-rest lattice. In the case of periodic boundaries there is a possibility that several particle-like waves will travel along this discrete torus along the non-intersecting trajectories. Here we can use the so-called `erase` collision (figure 4.29) to clean up the lattice. To form the *cleaner* we collide the 2^+-particle with the 3^+-particle as shown in figure 4.29, $t, \ldots, t+3$. As a result of the collision a mobile growing pattern is formed (figure 4.29, $t+10, \ldots, t+16$). The pattern has two growth points. They move perpendicularly to each other. Therefore the pattern is stretched (figure 4.30). If any particle collides into the pattern's body the particle disappears. To kill the cleaner we crash 2^+-particles into the growth points of the cleaner.

The following types of interaction gates are realized in a 4^+*-medium:*

$\mathrm{h1}(x, y) = x\overline{y} \triangledown \overline{x}y$

$[\langle 4^+, 4^+\rangle \to 4^+ \triangledown 4^+]$ or $[\langle 6^+, 6^+\rangle \to 6^+ \triangledown 6^+]$ or $[\langle 4^+, 6^+\rangle \to 4^+ \triangledown 6^+]$

$\mathrm{h2}(x, y) = xy \triangledown x\overline{y} \triangledown \overline{x}y$

$[\langle 4^+, 6^+\rangle \to 4^+ \triangledown 4^+ \triangledown 4^+]$

$\mathrm{h3}(x, y) = x\overline{y} \triangledown y$

$[\langle 4^+, 6^+\rangle \to 4^+ \triangledown 6^+]$

$\mathrm{h4}(x, y) = (xy \vartriangle xy) \triangledown x\overline{y} \triangledown \overline{x}y$

$[\langle 4^+, 6^+\rangle \to (4^+ \vartriangle 4^+) \triangledown 4^+ \triangledown 6^+]$

$\mathrm{h5}(x, y) = 1$

$[\langle 6^+, 6^+\rangle \to G_4^2]$

$\mathrm{h6}(x, y, z) = (xyz \vartriangle xyz \vartriangle xyz \vartriangle xyz\vartriangle) \triangledown \overline{xy}z \triangledown \overline{x}y\overline{z} \triangledown x\overline{yz}$

$[\langle 6^+, 6^+, 6^+\rangle \to (4^+ \vartriangle 4^+ \vartriangle 4^+ \vartriangle 4^+) \triangledown 6^+ \triangledown 6^+ \triangledown 6^+ \triangledown 6^+]$.

The h1 and h2 gates are similar to the g1 and g10 gates in a 2^+-medium. The coordinates of the excited and refractory cells before the collision in an h2-gate

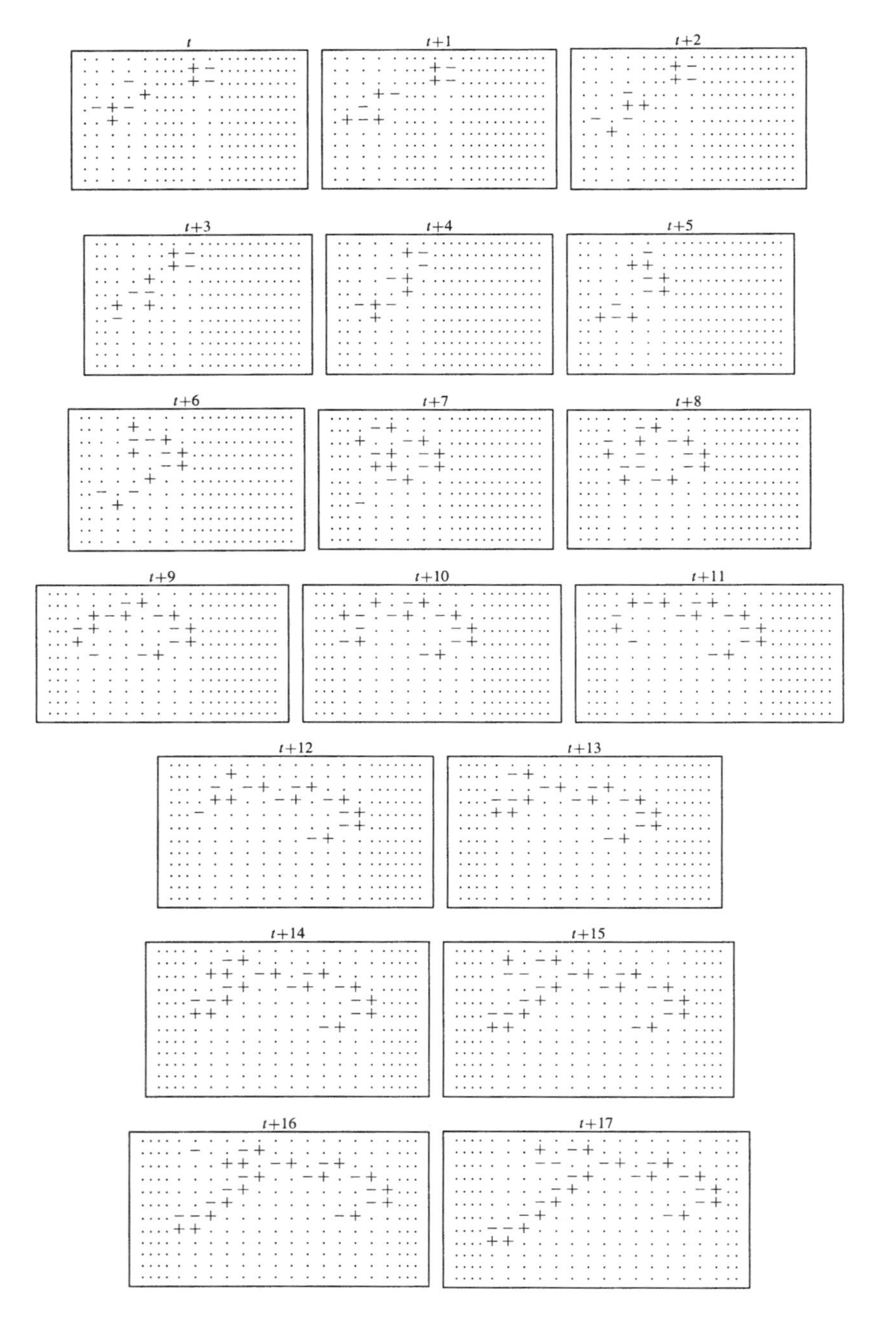

Figure 4.29. Dynamical building of a cleaner in the `erase` collision of two particles.

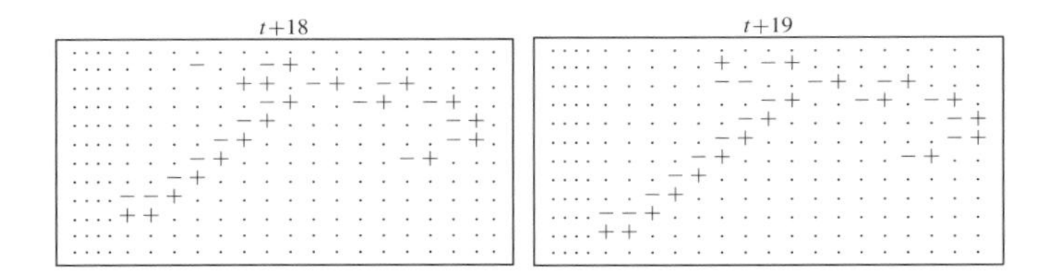

Figure 4.29. (Continued)

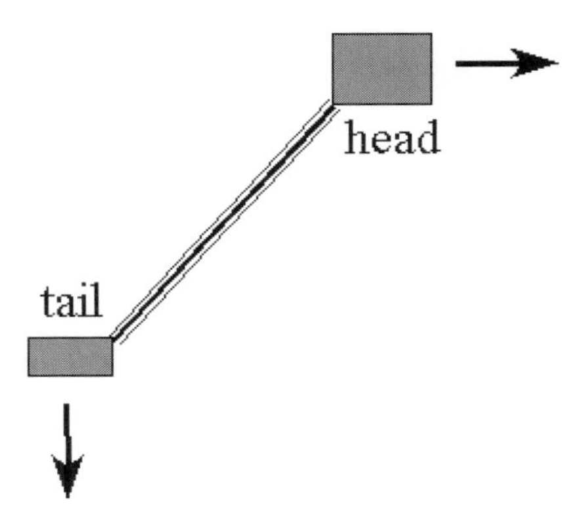

Figure 4.30. An erasion scheme.

are as follows:

$$\mathbb{C}_+ = (0,0,0), (0,0,1), (0,1,0), (0,1,1), (2,-1,0), (2,-1,1), (0,-9,0), (1,-9,0), (0,-9,1), (1,-9,1)$$

$$\mathbb{C}_- = (-1,0,0), (-1,0,1), (1,0,0), (1,0,1), (1,-2,0), (1,-2,1), (0,-10,0), (1,-10,0), (0,10,1), (1,-10,1).$$

An example of the collision is shown in figure 4.31.

In the collision between 4^+- and 6^+-particles in the h3 gate a new 6^+-particle is generated. It moves along the same trajectory as the previous 6^+-particle when approaching a collision site (figure 4.32).

When two particles interact with one another in a h4 gate they give birth to two 4^+-particles. These new 4^+-particles move perpendicularly to the collision plane. The collision skeleton is shown in figure 4.33; the coordinates of excited and refractory cells representing the positions of 6^+- and 4^+-particles before

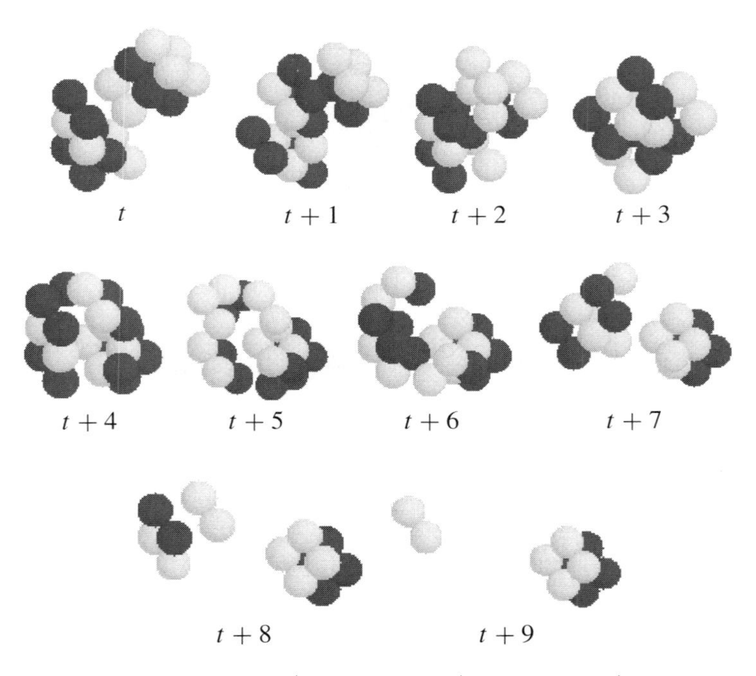

Figure 4.31. h2-gate: collision of 4^+-particle with 6^+-particle; the 4^+-particle is formed as a result of the collision.

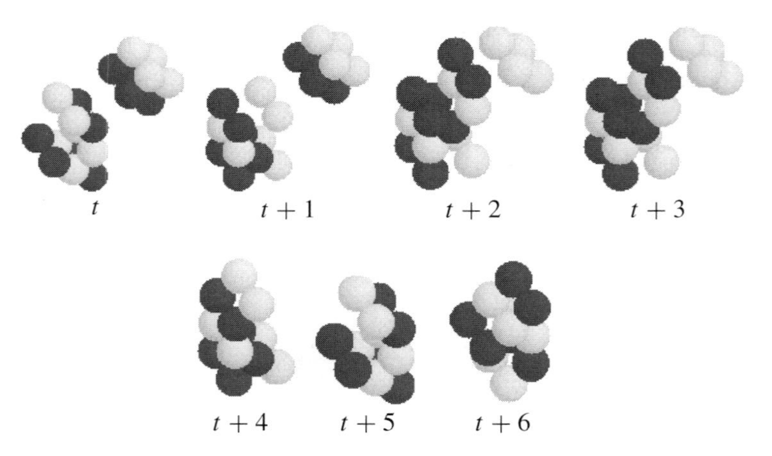

Figure 4.32. h3 gate: collision of a 4^+-particle with a 6^+-particle with the formation of a 6^+-particle.

Figure 4.33. Skeleton of the collision in a h4 gate.

collision are as follows:

$$\begin{aligned}\mathbb{C}_+ &= (0,0,0), (0,0,1), (0,1,0), (0,1,1), (2,-1,0), (2,-1,1), (0,-5,0),\\ &\quad (1,-5,0), (0,-5,1), (1,-5,1)\\ \mathbb{C}_- &= (-1,0,0), (-1,0,1), (1,0,0), (1,0,1), (1,-2,0), (1,-2,1),\\ &\quad (0,-6,0), (1,-6,0), (0,-6,1), (1,-6,1).\end{aligned}$$

A tetrad of 4^+-particles is formed when a ternary collision of 6^+-particles takes place in an h6 gate. The scheme and the skeleton of the collision are shown in figure 4.34.

The coordinate of three 6^+-particles colliding in h6 gates (a full history of the collision is shown in figure 4.35):

$$\begin{aligned}\mathbb{C}_+ &= (0,0,0), (0,0,1), (0,1,0), (0,1,1), (2,-1,0), (2,-1,1), (1,5,0),\\ &\quad (1,6,0), (3,7,0), (1,5,1), (1,6,1), (3,7,1), (4,1,0), (4,2,0),\\ &\quad (2,3,0), (4,1,1), (4,2,1), (2,3,1)\\ \mathbb{C}_- &= (-1,0,0), (-1,0,1), (1,0,0), (1,0,1), (1,-2,0), (1,-2,1), (0,6,0),\\ &\quad (2,6,0), (2,8,0), (0,6,1), (2,6,1), (2,8,1), (3,2,0), (5,2,0),\\ &\quad (3,4,0), (3,2,0), (5,2,1), (3,4,1).\end{aligned}$$

4.12 Reflectors, counters and registers

In the Fredkin–Toffoli model [243, 469] stationary mirrors are used to deflect the trajectory of a signal, to make a sideways shift, to delay a signal or to realize crossover of two signals. In cellular-automata models the stationary mirrors are usually represented by the stationary patterns of the non-resting cell states [469]. In this section we discuss how things stand with immobile self-localized patterns—patterns that do not change their size globally but only in an oscillating mode, do not translate themselves on the lattice and do not generate other

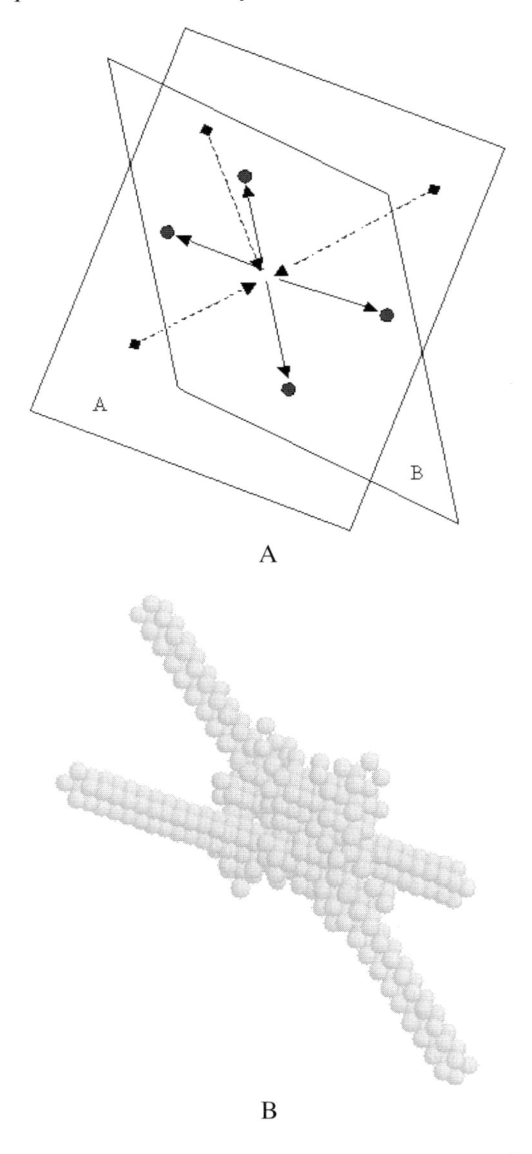

Figure 4.34. The scheme (A) and skeleton (B) of a collision of three 6^+-particles resulting in the formation of four 4^+-particles. Gate h6 is implemented in the collision.

patterns—in excitable lattices and illustrate possible ways of realizing deflections of the particle-like waves and the construction of counters and registers.

If there is an immobile pattern in an excitable lattice then it is an oscillator.

The transitions $+ \rightarrow -$ and $- \rightarrow \bullet$ are unconditional, therefore any configuration σ_1 of excited cells at time t transforms itself into a configuration of refractory cells at time $t+1$ and becomes stationary at $t+2$. Before the excited cells of σ_1 become refractory they excite some of the surrounding cells of σ_2, which, in their turn, excite the third pool σ_3 of the cells. If the cells of σ_1 lie in the neighbourhood of the cells from σ_3 they will be excited at time $t+3$.

Unfortunately, the following finding takes place for two-dimensional lattices:

There are no immobile patterns in a 2^+-medium.

Let such a pattern exist. Then it has at least two excited cells and at least two refractory cells. In any connected set of excited and refractory cells in a two-dimensional lattice there are pairs of neighbouring excited cells that have more than one resting neighbour. These resting neighbours will be excited and they also have more than one neighbour at rest. Hence, the diameter of the excited set grows. The outcome of the proposition states that there are no stationary reflectors nor stationary registers in two-dimensional excitable lattices.

The minimal immobile pattern in a 4^+-medium has 12 non-rest-state weight, a $3 \times 3 \times 3$-cell volume, and it oscillates with a period of three time steps.

This pattern, the so-called blinker B, has three configurations, which change in the loop $c_1 \rightarrow c_2 \rightarrow c_3 \rightarrow c_1$. In the $[(x, y, z), (x, y, z+1), (x, y, z+2)]$-slice the states of B change in the following manner:

$$c_1 = \left[\begin{pmatrix} \cdot & + & \cdot \\ - & + & - \\ \cdot & + & \cdot \end{pmatrix} \begin{pmatrix} \cdot & \cdot & \cdot \\ - & \cdot & - \\ \cdot & \cdot & \cdot \end{pmatrix} \begin{pmatrix} \cdot & + & \cdot \\ - & + & - \\ \cdot & + & \cdot \end{pmatrix} \right]$$

$$c_2 = \left[\begin{pmatrix} \cdot & - & \cdot \\ \cdot & - & \cdot \\ \cdot & - & \cdot \end{pmatrix} \begin{pmatrix} + & + & + \\ \cdot & \cdot & \cdot \\ + & + & + \end{pmatrix} \begin{pmatrix} \cdot & - & \cdot \\ \cdot & - & \cdot \\ \cdot & - & \cdot \end{pmatrix} \right]$$

$$c_3 = \left[\begin{pmatrix} \cdot & \cdot & \cdot \\ + & \cdot & + \\ \cdot & \cdot & \cdot \end{pmatrix} \begin{pmatrix} - & - & - \\ + & \cdot & + \\ - & - & - \end{pmatrix} \begin{pmatrix} \cdot & \cdot & \cdot \\ + & \cdot & + \\ \cdot & \cdot & \cdot \end{pmatrix} \right]$$

A period of three time steps is the minimal possible period of the oscillation. This follows from the cell state transition rule: a cell excited at time step t can only be excited again at a time step greater than $t+3$ because it takes a refractory state at step $t+1$ and the cell returns to the rest state at step $t+2$.

As a result of a collision with the blinker B a 4^+-particle is destroyed. Pattern B may be destroyed or not destroyed depending on the phase differences of the patterns at the moment preceding the collision.

This finding is the result of exhaustive search of all possible collisions between 4^+-particles and the blinker B. Arranging several copies of the blinker on a lattice and specifying the distances between them we can realize a register. We still do not know whether it is possible or not to generate the blinker B in

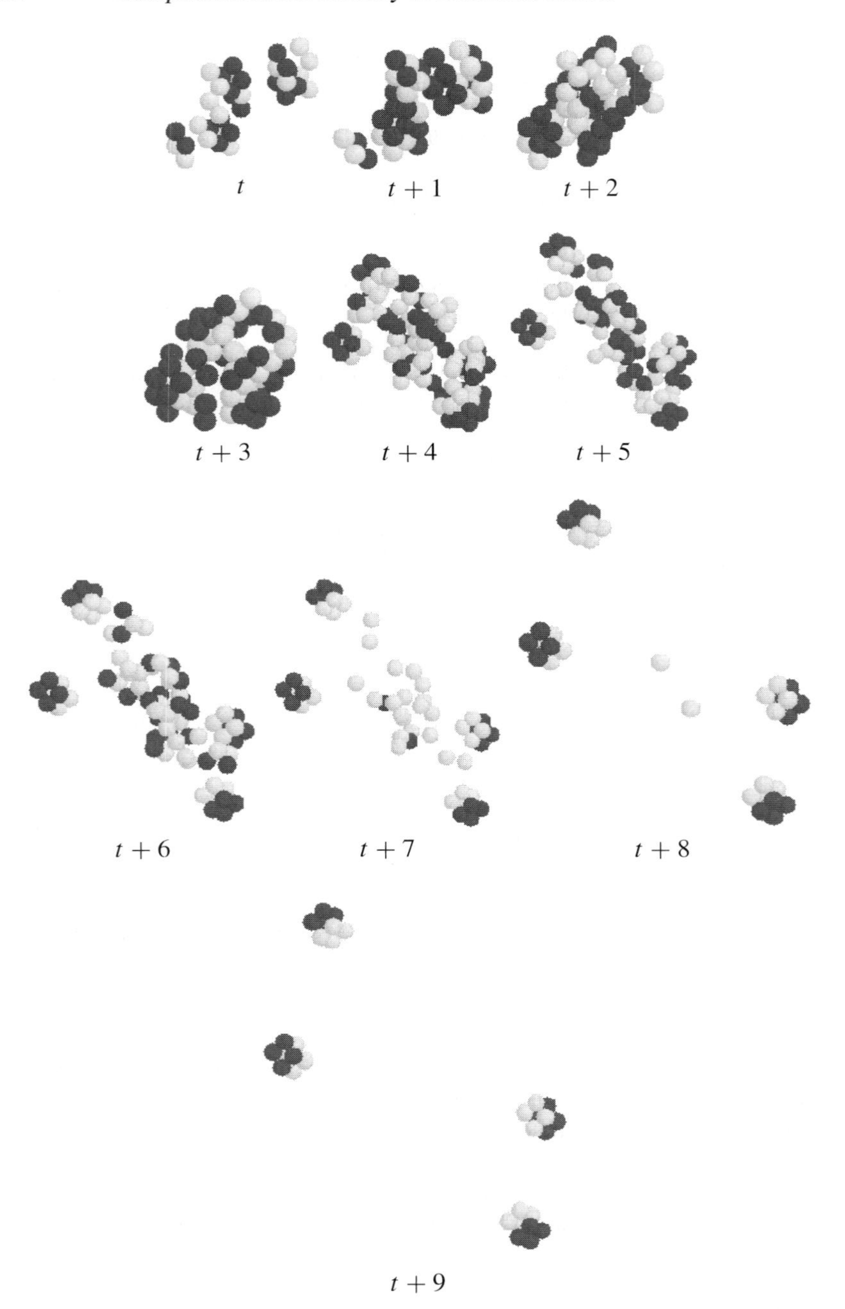

Figure 4.35. The collision of three 6^+-particles resulting in the formation of four 4^+-particles. Gate h6 is implemented in the collision.

a collision of 4^+- or 6^+-particles, therefore currently we are able to design the `read only` registers in the 4^+-medium.

There are no immobile reflectors in excitable lattices.

As we previously found every elementary particle may play the role of the mobile reflector.

In a 2^+*-medium the mobile counter representing* m *digits is the mobile pattern of* $2 + m$ *excited states. It has size of* $2(m + 1) \times (2 + m)$ *cells.*

The counter representing zero is the 2^+-particle itself. To increase the value of the counter we collide the 2^+-particle into the counter pattern, e.g.

$$\begin{pmatrix} \cdot & + & + & \cdot & \cdot & \cdot \\ \cdot & - & - & \cdot & + & - \\ \cdot & \cdot & \cdot & \cdot & + & - \end{pmatrix}.$$

As a result of such a collision the counter bearing the value 1 (this is just an arithmetical 1 not a logical value) is formed

$$\begin{pmatrix} \cdot & + & + & \cdot & \cdot & \cdot \\ \cdot & - & - & + & \cdot & \cdot \\ \cdot & \cdot & \cdot & - & \cdot & \cdot \\ \cdot & \cdot & \cdot & \cdot & + & \cdot \\ \cdot & \cdot & \cdot & \cdot & - & \cdot \end{pmatrix}.$$

Each increment is implemented via a collision of a 2^+-particle with the *tail* of the counter pattern in the following manner:

$$\begin{pmatrix} \cdot & + & + & \cdot & \cdot & \cdot & \cdot & \cdot \\ \cdot & - & - & + & \cdot & \cdot & \cdot & \cdot \\ \cdot & \cdot & \cdot & - & \cdot & \cdot & \cdot & \cdot \\ \cdot & \cdot & \cdot & \cdot & + & \cdot & \cdot & \cdot \\ \cdot & \cdot & \cdot & \cdot & - & \cdot & + & - \\ \cdot & \cdot & \cdot & \cdot & \cdot & \cdot & + & - \end{pmatrix}.$$

To decrease the value of the counter we crash a 3^+-particle into the end of the *tail*.

4.13 Excitation or Life?

What place do excitable lattices occupy in the hierarchy of universal computation models? Here we compare our models with some other non-standard computationally universal machines.

The first candidate—the *Banks computer* [65]—has a stationary architecture: predetermined wires and gates are represented in a combination of non-rest cell states. Therefore despite the fact that Banks computer has a minimal number of cell states, i.e. 2, it cannot qualify for comparison because it is not a dynamical computer.

The *Life without Death* [276] model—will be discussed later in great detail—has no predetermined architecture and simulates the interaction gates of the billiard ball model quite explicitly. However, all trajectories of the signals are

Table 4.1. Elementary parameters of the *Game of Life* (GL) and excitable lattices (EL). The complexity of the particle-like waves and stationary patterns is written in the format (*weight*, *volume*), where *weight* is the number of non-rest cell states. The complexity of the particle guns has the format (*weight*, *volume*, *period of particle generation*). Parameters of the three-dimensional *Game of Life* model were evaluated for the 5655 rule [73].

Parameters	2D GL	3D GL	2D EL	3D EL
Minimal stationary patterns	(4, 4)	(7, 8)	No	(12, 27)
Minimal mobile pattern (moving along columns or rows)	(8, 16)	(18, 40)	(4, 4)	(8, 8)
Minimal mobile pattern (moving along diagonals)	(5, 9)	No	(6, 16)	(16, 32)
Minimal stationary gun	(45, 180, 30)	No	No	(12, 48, 3)
Minimal mobile gun	No	No	(26, 54, 4)	(16, 48, 2)
One-collision logical gate	$x \wedge \overline{y}$	$x \wedge \overline{y}$	$x \wedge \overline{y}$ and $x \vee y$	$x \wedge \overline{y}$ and $x \vee y$

represented in non-rest cell states, which are the absorbing states in cell evolution. That is, a full history of the computation in *Life without Death* model can be extracted from the final stationary configuration. The model is not reusable. This is a small but significant disadvantage. Even if *Life without Death* is taken into consideration its space complexity is not satisfactory: the width of a signal trace is not less than eight lattice cells.

The interpretation of a billiard ball model in a two-dimensional *partitioning cellular automaton* [469] seems to be slightly artificial. In fact, those 16 states which are used in the cellular-automata model are quite enough to encode the motion of an abstract object in eight directions and to adjust the collision and reflections in a conventional cellular-automata model.

The *sand-pile* model [279] has at least three cell states and allows us to perform computation using one-dimensional gliders moving on linear graphs. The imaginary sand-pile computer should have a stationary quasi-one-dimensional architecture. Frankly speaking all implementations of sand-pile computing devices look rather like weird interpretations of the propagation of an impulse along neuronal dendrites, where logic operations take place at the site of merging of two or more dendrite branches.

Therefore we have only one candidate to compare with the excitable lattice—the *Game of Life* model [253, 81]. Both excitable lattice and *Game of Life* models exhibit bounded growth from the randomly chosen configuration, and represent binary signals by the translating patterns, called gliders and particle-like waves. The *Game of Life* is obviously minimal in the number of cell states. Both models have exactly the same cell neighbourhoods. The *Game of Life* is unquestionably the winner in the complexity of immobile patterns but it loses in all other parameters. Excitable lattices have smaller and lighter self-localized mobile patterns and stationary guns. *Game of Life* has no mobile guns at all. Excitable lattices do. The only gate realizable in a single binary collision of gliders in the *Game of Life* is $\overline{x} \wedge y$ whereas we can compute both $\overline{x} \wedge y$ and $x \wedge y$ in the single collision of two particle-like waves in excitable lattices. The exact parameters of the elementary objects in these two models are shown in table 4.1.

We should also note that in three dimensions only the 4555, 5766 and 5655 rules exhibit gliders and a bounded growth from the random configuration [73]. In his paper [73] Bays emphasises that gliders should occur naturally in the evolution, i.e. they have to appear reasonably often. However, gliders in the 5655 rule are extremely rare and glider guns have yet to be discovered. In contrast, the 4^+- and 6^+-particles are the stable solutions in every numerical experiment with excitable lattices. Moreover, particle guns emerge in every series of at most 100 experiments with a 4^+-medium.

Remarkably, there are no generators of the spaceships, i.e. particles moving along the coordinate axes, in the *Game of Life*; in excitable media there are no generators of 3^+- and 6^+-particles.

4.14 Search for universality

There are several hints that allow us to detect appropriate candidates in the real world. Thus, for discrete breathers we know at least two criteria that can be used in the experimental search: the energy threshold and scattering of planar waves by the breathers [233]. The viability of the first criterion is proved in numerical experiments with cellular-automata models of excitable media [25]: self-localized excitation in the lattice appears in the evolution when every cell of the lattice has the critical excitation threshold; increasing the threshold of cell excitation leads the system from spatial particle-like wave solutions through quasi-chaotic regimes to regimes with unidirectional and spiral waves [25]. Another confirmation of the 'subcritical threshold' is presented in [240]: in one-component systems breathers exist only below a critical coupling. In investigations of the spatially localized excitations in a lattice of coupled oscillators it was found that breather-like patterns are more typical for a small coupling between oscillators whereas a large coupling leads to a globally chaotic state [1].

Jakubowski *et al* [325] offer a realistic guide for searching for physics-based computation universality. The possible computationally universal system should satisfy the following requirements:

- self-localized waves must preserve their integrity after a sequence of collisions,

- self-localized waves must lose a negligible amount of energy through radiation [325].

Based on the results of numerical experiments with the cubic nonlinear Schrödinger equation Jakubowski *et al* [325] show that even simple measures of radiation may be very suitable in practical evaluations of the usefulness of wave collisions. They indicate the possible candidates: Kerr materials, media with laser beam propagation, spatio-temporal photoreactive optical solitons and optical solitons in some types of atomic crystals [325]. We will discuss some of the prospective candidates in the forthcoming section.

In abstract cellular-automata models (without any realistic constraints) we can simply build automata with the required behaviour. In this case we can use genetic algorithms (see e.g. [459]), the sculpturing of the basin attraction fields [648], reconstruction of the local transition rules from the given global configurations [9] and generation of the predetermined patterns using integer programming techniques [99].

> *Let $\mathfrak{A}$ be the d-dimensional three-state closest-neighbourhood cellular-automata model of lattice excitation, the cells of which update their states by the rule*
>
> $$x^{t+1} = \begin{cases} + & x^t = .\ \textit{and}\ |\{y \in u(x) : y^t = +\}| = \theta \\ - & x^t = + \\ \bullet & \textit{otherwise}. \end{cases}$$
>
> *Is this true for any $d > 0$ that $\mathfrak{A}$ is universal when $\theta = 2^{d-1}$.*

This criterion of universality takes place for the dimensions $d = 2$ and $d = 3$. The criterion also works for $d = 1$, one-dimensional lattice. To realize the $x \wedge \overline{y}$ gate in a one-dimensional lattice we launch waves of the excitation on the tips of the lattice. For $d > 3$ the answer is still uncertain. The computation abilities of various excitable lattices in the context of the phenomenology of excitation dynamics are discussed in chapter 5.

Nonlinear media that exhibit self-localized mobile patterns in their evolution are potential candidates for the role of universal dynamical computers. Here we tackle some of the models: breathers, solitons, light bullets and some exotic findings.

4.15 Solitons, light bullets and gaussons

4.15.1 Solitons

A soliton is a solitary wave pulse, originally observed in the numerical integration of the Korteweg–de Vries equation—a nonlinear partial differential equation [663, 245].

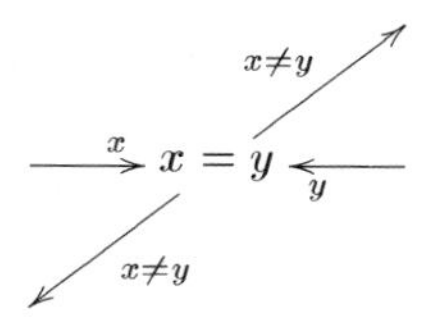

Figure 4.36. A comparison operator realized in a collision of two solitons. Variables x and y are represented by the solitons' amplitudes.

In general, we can say that solitons behave like effective particles when they interact with each other. In *bright solitons*, those described by a nonlinear Schödinger equation, the interaction of solitons depends upon the relative phase of the solitons at the moment of interaction. Solitons attract each other if they have the same phase; if phases are in opposition the solitons are repelled [241, 490]. Dark solitons repel each other unconditionally.

If we consider a compound soliton, i.e. a soliton consisting of both dark and bright components, the situation would be different. When a bright component (out of phase) is associated with each dark soliton the attractive interaction between the solitons forms a proper balance of attractive and repulsive forces and a two-soliton bound state is generated [490]. Unfortunately, bright components do not survive alone in a defocusing medium: they are diffracted when separated from dark components [490].

Makhankov [410] investigates the kinetics of solitons in a nonlinear Schrödinger equation model with a self-consistent potential and a wave equation:

$$\mathrm{i}\dot{\psi} + \psi_{xx} - U\psi = 0$$

and

$$(\partial_t^2 - \triangle)U = \partial_x^2|\psi|.$$

Amongst several types of soliton interactions discussed in [410] two phenomena may be interesting for us. When two solitons with similar amplitudes collide they fuse. Solitons with amplitudes that differ by at least two orders of magnitude undergo quasi-elastic collision. This means, ideally, a comparison operator is realizable in this collison (figure 4.36). Moreover, if three solitons collide a sound packet is created in the collision of two solitons. The sound packet may destroy the third soliton if its amplitude does not exceed half the amplitude of the colliding solitons.

From results reported by Islam *et al* in 1992 [321] we can speculate that at least three basic types of soliton interaction can be used in the design of collision-based logic gates:

(i) *Soliton collision*.

- Elastic collision. This happens when solitons (usually, orthogonally polarized) approach one another at relatively small angles. The solitons

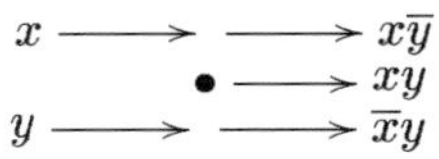

$$\begin{array}{lll} x \longrightarrow & \longrightarrow & x\overline{y} \\ & \bullet \longrightarrow & xy \\ y \longrightarrow & \longrightarrow & \overline{x}y \end{array}$$

Figure 4.37. A gate employing soliton attraction.

are deflected as a result of the interaction. In a one-dimensional case we can see temporal shifts.

- Inelastic collision. This type of collision happens when solitons collide at large angles. The solitons pass through each other. Shifts in the centre frequency form in the one-dimensional case. In the many-dimensional case we can see spatial shifts.

(ii) *Soliton attraction.* If two solitons move along parallel trajectories, an (un)stable bound state may be formed. This new double soliton propagates in the same direction as its two predecessors.

(iii) *Weak near-interaction*, dragging of solitons and formation of time delays. Soliton dragging is the formation of a bound pair and the propagation of two solitons of the pair at an angle close to the weighted mean of the initial angles.

A specific soliton collision can be used to build a standard billiard ball gate:

$$\langle x, y \rangle \rightarrow x\overline{y} \bigtriangledown \overline{x}y \bigtriangledown (xy \bigtriangleup xy).$$

The attraction of solitons gives us the chance to build the following gate:

$$\langle x, y \rangle \rightarrow x\overline{y} \bigtriangledown \overline{x}y \bigtriangledown xy$$

the graphical form of which is shown in figure 4.37.

An interesting finding is reported in [228]. This is the state known as Frenkel–Kontorova lattices, lattices which combine a harmonic interaction between neighbouring nodes and an anharmonic on-site potential and which support a special type of mobile self-localized excitation. When an external field is applied stable collective excitations, in the form of topological solitons (kinks), spread across the lattice. In fact, these dislocations are dragged by moving excitation waves [228].

Little can be said about the interaction of *more than two solitons* in two and three dimensions except that soliton complexes and phase shifts may be observed as a result of multiple soliton collisions [245]. Thus, for example, analytical expressions for an effective potential of interaction between two- and three-dimensional solitons are produced and discussed in [409] based on the model

$$\mathrm{i}u_t + \nabla^2 u + |u|^2 u - g|u|^4 u = 0$$

where g is the ratio of nonlinearity. It can be shown that if solitons have similar parameters they are attracted to each other, which may lead to the formation of the orbiting bound states of two- or three-dimensional solitons. This reminds us of the comparison gate.

Surprisingly, more findings have been obtained in experiments with automata models of solitary waves than in the numerical solutions of differential equations. One-dimensional cellular-automata models of solitons have become widely recognized since the papers by Park, Steiglitz and Papatheodorou *et al* [492, 493, 578]. Several classes of (ir)reversible cellular automata have also been characterized by soliton-like patterns in [3, 250].

One particularly interesting class of soliton automata on chemical graphs, based on the structure of (CH)-chains and closures, has been investigated by Dassow and Jurgensen [165, 166, 258].

Recent results on soliton automata include the transformation of discrete soliton equations into binary cellular automata [589, 90], automaton derivations of the generalized Toda equation [468, 433] and the construction of an integrable cellular automaton with many-soliton solution [599].

Regarding cellular-automata models of solitons we must certainly mention the very promising paradigm of particle machines and measures of the information transfer in soliton collisions [325] and the method of ultradiscretization used in [293] to derive a one-dimensional cellular automaton which simulates the dynamics of solitons in crystals.

4.15.2 Light bullets

Still widely considered as the three-dimensional solitons of a Schrödinger equation light bullets may not survive collision without loss of energy [87, 204]. Therefore, they are quite suitable for interaction gates.

The results of numerical simulations of the three-dimensional generalized nonlinear Schrödinger equation in a model with an alternative refractive index demonstrate the propagation of stable light bullets with soliton-like behaviour [204, 205].

There are several computationally useful outcomes of bullet binary collisions [206]:

- light bullets survive collisions;
- two bullets fuse into a stationary wave;
- bullets pass through each other but a third stationary wave is formed (figure 4.38);
- bullets rotate one around another, therefore their outcoming trajectories are changed (figure 4.39).

The g2 gate is realized in light bullet interactions.

This follows from the results of the simulation of light bullets [206, 204, 207, 205, 203]. In the g2 gate the input trajectories of x and y variables are coincident

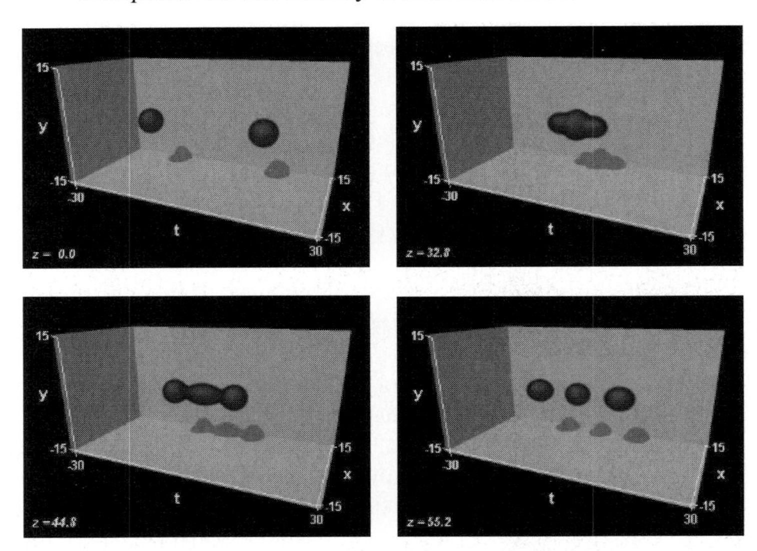

Figure 4.38. Light bullets pass through each other and also form a third stationary bullet as a result of this collision [206]. Published with kind permission of Darran Edmundson and Richard Enns.

with the output trajectories of the $\overline{x}y$ and $x\overline{y}$ terms, respectively. The value of xy can be registered from two different outward trajectories (figure 4.21). One of the examples of the g2 gate is the spiralling light bullet [206]. When two light bullets move towards one another and collide side by side with a certain offset they rotate around one another without capture or fusion, and later escape the interaction zone with transverse velocities (see e.g. figure 4.39) [206].

Let λ^+ be a moving light bullet and λ^0 be a stationary solution (non-moving bullet). Then two following interaction gates are realizable in a light bullet collision:

$$h_1(x, y) = \overline{x}y \bigtriangledown x\overline{y} \bigtriangledown (xy \bigtriangleup xy \bigtriangleup xy)$$
$$[\langle \lambda^+, \lambda^+ \rangle \to \lambda^+ \bigtriangledown \lambda^+ \bigtriangledown (\lambda^+ \bigtriangleup \lambda^0 \bigtriangleup \lambda^+)].$$

This is a fission situation [206]. The bullets move towards each other, pass through one another and continue their motion but a third stationary wave is also formed. The newly born wave remains at the collision site. There are two waves before the collision and three waves after it (figure 4.38). The stationary wave represents xy:

$$h_2(x, y) = \overline{x}y \bigtriangledown xy \bigtriangledown x\overline{y}$$
$$[\langle \lambda^+, \lambda^+ \rangle \to \lambda^+ \bigtriangledown \lambda^0 \bigtriangledown \lambda^+].$$

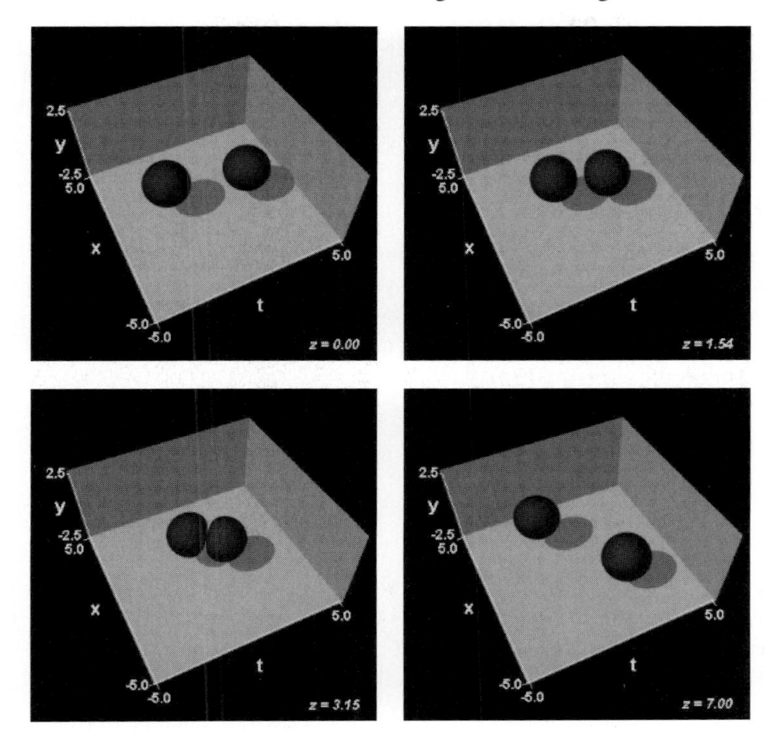

Figure 4.39. Elastic collision, or repulsion, of two light bullets [206]. Published with kind permission of Darran Edmundson and Richard Enns.

Bullet fusion is the keystone feature of the gate [206]. If the velocities of the bullets are below a certain threshold the colliding bullets do not overcome the binding forces and fuse into a single stationary soliton.

The formation of a stable stationary soliton may be used to build a counter in the light bullet universal computer. This allows us to organize the `write` operation. To `read` from the counter we should crash a light bullet into the stationary soliton. Sadly, we do not know all the possible outcomes of such collisions. Applying our experience with breathers we can propose that, depending on the velocity of the moving bullet and the degree of offset in a collision, there are several possible products of the collision:

- reflection of the moving bullet;
- a previously stationary soliton is set into motion;
- the moving bullet disappears.

There is also one possibly promising phenomenon of soliton tunnelling which involves bistable solitons, i.e. light bullets with the same energy but different radial profiles [393]. If the sizes of two colliding bullets differ significantly the smaller one can easily penetrate through the middle of the larger bullet. This fact may be used to reduce the complexity of collision-based circuits.

In fact, almost all outcomes of light bullet collisions have their analogues in gausson collision phenomenology [482]. So, let us tackle gaussons.

4.15.3 Gaussons

We discuss gaussons as they were originally presented by Oficialski *et al* [482], and studied by Makhankov [410]. When dealing with the equations we take into account the corrections made by Jakubowski [327].

Gausson collisions are investigated using the following equation (we discuss here Jakubowski's version [327]):

$$-\mathrm{i}\frac{\partial u}{\partial t} = \frac{h}{2m}\frac{\partial^2 u}{\partial^2 x} + \frac{b}{h}\ln(a|u|^2)u$$

where a and b are constants

$$a = \mathrm{e}^{1-E_0/b} \qquad \text{and} \qquad b = \frac{h^2}{2ml^2}$$

computed from the gausson length l and the gausson rest energy E_0.

The equation has a solitary wave solution, which is called a gausson (again from [327]):

$$u(x,t) = \left(h\sqrt{\frac{\pi}{2mb}}\right)^{-1/2} \times \exp\left\{-\frac{\mathrm{i}u_\mathrm{e}^2}{2mh}t + \frac{\mathrm{i}u_\mathrm{e}}{h}x - \frac{mb}{h^2}\left(x - \frac{u_\mathrm{e}t}{m}\right)^2 + \mathrm{i}\phi_0 + \frac{E_0}{2b}\right\}$$

where θ_0 is an initial arbitrary phase, u_e is an envelope velocity.

From [482] we can classify the binary interactions of gaussons into the following types (the parameters of colliding gaussons can be found in [482]):

- Gaussons either pass through each other or collide elastically (figure 4.40). No sensible logic operation can be realized using such a collision because both data and results share the same trajectory.
- The colliding gaussons produce a third stationary gausson (figure 4.41). This stationary localization can be detected via a collision with another gausson. Therefore we can design an AND gate using this type of collision.
- The small initial space shift changes the outcome of a gausson collision impressively (figure 4.42). After the collision the gaussons change their trajectories. So, we can design the gate shown in figure 4.42(C).
- For a particular set of parameters a third mobile pulse is formed (figure 4.43). This mobile daughter gausson follows the trajectory of one of the initial

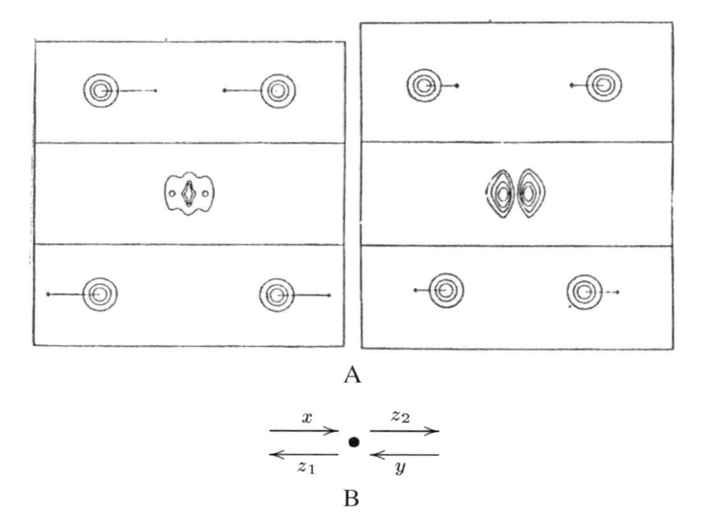

Figure 4.40. Two gaussons collide and either pass through each other or undergo an elastic collision: (A) snapshots of non-elastic (on the left) and elastic (on the right) gausson collisions (from [482]); (B) a trivial gate realized in the collision: $z_1 = y$ and $z_2 = x$.

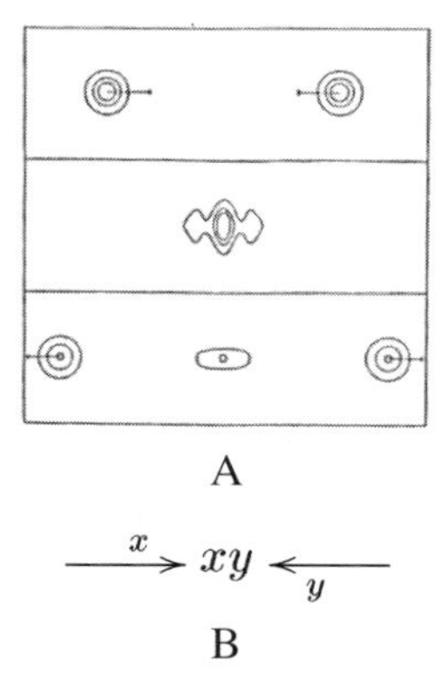

Figure 4.41. Two gaussons collide, form an unstable bound state and split again hence generating a stationary solution: (A) snapshots of the gausson collision (from [482]); (B) the gate realized in the collision.

gaussons. The newly formed gausson represents the result of the logic operation xy.

It is also possible to observe a hybrid collision when the gaussons' velocity vectors are changed as a result of the collision and a third, stationary, localization is formed (figure 4.44).

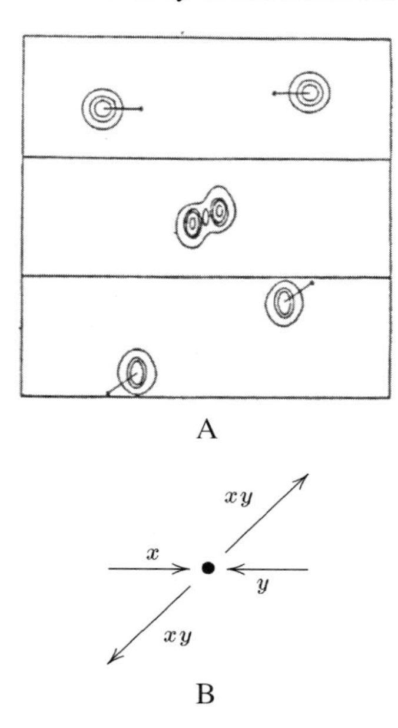

Figure 4.42. Two gaussons collide, form an unstable bound state, split again hence the angles of their velocity vectors are changed: (A) snapshots of the gausson collision (from [482]); (B) the gate realized in the collision.

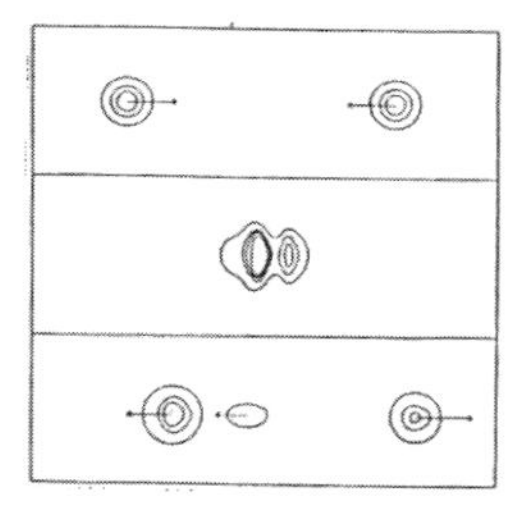

Figure 4.43. Two gaussons collide, form an unstable bound state, split again and hence a third mobile gausson is generated: snapshots of the collision (from [482]).

4.16 Breathers

A discrete breather is a time-periodic, spatially localized solution of the equations of motion for classical degrees of freedom interacting on a lattice [233]. A breather is a multisoliton solution in the sense of inverse scattering transformations [117, 118].

Sometimes breathers are called bions because they were considered to be dual states of kinks and antikinks [410]. Let us look at an example of a one-

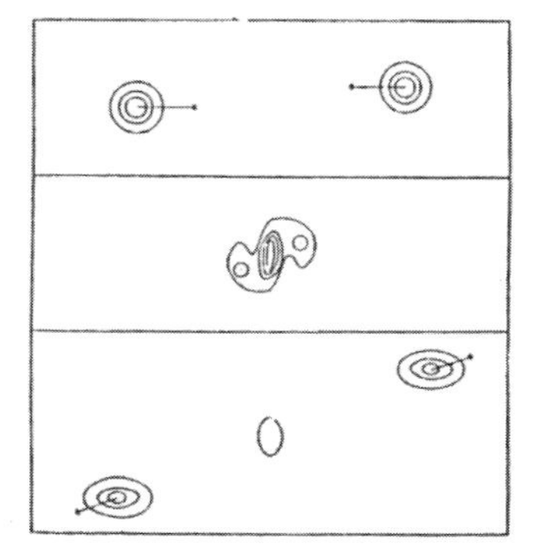

Figure 4.44. Two gaussons collide, form an unstable bound state, split again with changed velocity vectors and a third stationary gausson is generated: snapshots of the collision (from [482]).

dimensional sine–Gordon model offered by Chung [133]. The model has a boson Hamiltonian

$$H = \int \mathrm{d}x \left[\frac{1}{2\tau_0} [\phi_t^2 + \phi_x^2 + 2m^2(1 - \cos\phi)] \right]$$

where τ_0 is a coupling constant and m is the mass of a fundamental linear mode, i.e. a phonone, furthermore $m = 1$. The system describes numerous real-life phenomena such as the electronic gas, chains of spins or organic molecules, superlattices and liquid crystals [133]. The equation of motion has the form

$$\phi_{tt} + \phi_{xx} - \sin\phi = 0.$$

There is a localized mode of a soliton:

$$\phi_\pm = 4\tan^{-1}\{\mathrm{e}^{\pm x}\}$$

where '+' represents the soliton and '−' the antisoliton. The bound state of a soliton and antisoliton is assumed to be a breather:

$$\phi(x, t) = 4\tan^{-1}\left\{ \frac{\sqrt{1-\omega}\sin(\omega t)}{\omega\cosh(\sqrt{(1-\omega^2 x)})} \right\}$$

where $0 \le \omega \le 1$ [133]. We can produce solitons and breathers moving with arbitrary speed with the help of the Lorentz transformation. These three states are characterized by the topological charge [133]:

$$Q = \frac{1}{2\pi}\int_{-\infty}^{\infty} \phi_x \,\mathrm{d}x = \begin{cases} 1 & \text{for soliton} \\ 0 & \text{for breather} \\ -1 & \text{for antisoliton.} \end{cases}$$

Recall also that in the early 1970s three-dimensional spherically symmetric long-lived bubbles were discovered in ϕ^4 models by Bogolubsky and Makhankov [410]. These coherent configurations were called pulsons. Only later did [265] it become clear that pulsons were the first ever examples of the now popular breathers and oscillons.

One of the common models of breathers supporting nonlinear systems is a chain of harmonically coupled particles. Each particle situated at position u_x is under the influence of the substrate potential

$$v(u_x) = \omega_d^2 \left(\frac{u_x^2}{2} - \frac{u_x^3}{3} \right)$$

where ω_d^2 is a parameter measuring the amplitude of the substrate potential (actually the parameter controls the discreteness of the system) [201, 233, 232]. The model is represented by Hamiltonian

$$H = \sum_x [\tfrac{1}{2}\dot{u}_x^2 + \phi(u_x - x_{x-1}) + v(u_x)]$$

where x is a space variable, i.e. the index of a lattice node, $\dot{u}$ is a time derivative [201, 233]. The system has the following equation of motion:

$$\ddot{u}_x = -\frac{\partial H}{\partial u_x}.$$

The potential functions $v(\cdot)$ and $\phi(\cdot)$ can be expanded around an energy minimum as follows:

$$v(z) = \sum_{i=2,3,\ldots} \frac{v_i}{i} z^i \qquad \text{and} \qquad \phi(z) = \sum_{i=2,3,\ldots} \frac{\phi_i}{i} z^i$$

and the linear spectrum is

$$\omega_q^2 = v - 2 + 4\phi_2 \sin^2(\pi q/2)$$

where ω is the frequency and q is the wavenumber.

Spatially localized and time-periodic solutions of the previously described system are usually called discrete breathers.

Under certain conditions, when for example $k\omega \neq \omega_q$, $k\omega$ are the frequencies of the higher harmonics and k is the Fourier number, *the discrete breather is structurally stable* [233].

The so-called DNA breather is another remarkable example of the presence everywhere of breathers, which will be discussed in one of the following sections.

Christiansen *et al* [135] studied the dynamics of breathers on a two-dimensional lattice via the numerical simulation of a two-dimensional generalized discrete nonlinear Schrödinger equation:

$$\begin{aligned} &\mathrm{i}\dot{u}_{i,j} - 4u_{i,j} + 2(1-\alpha)u_{i,j}|u_{i,j}|^2 \\ &\qquad + (u_{i+1,j} + u_{i-1,j} + u_{i,j-1} + u_{i,j+1})\left(1 + \frac{\alpha}{2}|u_{i,j}|^2\right) = 0 \end{aligned}$$

where $\alpha = 1$ stands for a discrete self-trapping equation and $\alpha = 0$ induces Ablowitz–Ladik discretization [135, 136]. These authors show that stationary excitation will either collapse or spread all over the lattice in finite time. As for mobile excitation the pulse contracts as it moves on the lattice. Eventually it slows down and becomes pinned. For $\alpha = 1$ the moving breather can leave a trail of standing spikes; for $\alpha = 0$ only one stationary breather is formed from the moving breather.

4.16.1 When are breathers born?

Flach *et al* [232] obtained a reliable numerical solution of discrete breathers in three-dimensional lattices and predicted a positive energy threshold for real-life three-dimensional lattices, e.g. the dynamics of atoms in the crystals. They demonstrated that breathers are generated on a lattice if the coupling between the lattice nodes is weak. Thus, for example, considering a Hamiltonian with two variables describing the transverse displacement of two bases belonging to the base pairs in the DNA molecule under different values for the coupling constants between two nucleotides along the same strand, Forinash *et al* [239, 240] show that intrinsic local modes can be accidentally formed due to the localization of thermal fluctuations and one of their growth mechanisms is the exchange of energy.

Breather solutions are calculated for conjugated polymers, polyacetelen, ionic crystals and electrical lattices; they can also be generated optically [233]. In fact, the mechanism of discreteness-induced energy localization works in a large variety of physical lattice systems [201].

Some issues related to the breather excitation threshold for nonlinear lattices are raised in [636]; as no exact thresholds are given this approach may be useful in future investigations.

Through analysis of the stability of band edge phonons Flach [233] shows that a breather appears due to the bifurcation of phonon orbits, i.e. the phonon orbit becomes unstable at the bifurcation energy $E_c \sim N^{-2/d}$, where N is the size of the lattice and d is the dimensionality. The newly formed orbit is not invariant under translation symmetry and a new discrete breather is born.

Breathers are born when the energy localizes itself into large amplitude excitations. One of the first factors in the creation of a localized excitation may possibly be a modulation instability [201]. The energy localizes itself because of discreteness. If the discreteness of a system is strong enough then very stable large amplitude breathers co-exist in the model [201].

Extremely encouraging findings are published in [529], where a one-dimensional discrete non-integrable thermalized (i.e. in a phonone bath) Schödinger equation is studied numerically. That is, the one-dimensional equation

$$\mathrm{i}\dot{\phi}_x + (\phi_{x+1} + \phi_{x-1}) + a|\phi_x|^2\phi_x = 0$$

is considered, where x is the index of the lattice node and a is a real coefficient used to tune the nonlinear term [529]. The equation has the following Hamiltonian form:

$$H(\{\phi_x, \mathrm{i}\phi_x^*\}) = \sum_x \left(-\phi_x\phi_{x+1}^* + \phi_x^*\phi_{x-1} - \frac{a}{2}|\phi_x|^4\right).$$

The system is initialized by giving each node x of the lattice a random initial value v with probability $p(x)$ distributed as here:

$$p(x) = \frac{k}{\pi}\frac{1}{v^2 + k^2}.$$

When evolution starts we see [529] that high-amplitude excitations arise from random fluctuations. They are long-lived breathers. The following results of [529] may be useful.

- *Discrete breathers seem to cause larger phonon fluctuations in their neighbourhood, which increase the probability of new breather generation in the vicinity of an old one* [529].
- *The newly born breather may be ejected as a propagating breather by the effect of phonon noise* [529].
- *A stationary 'old' breather may recover after the generation of a new moving breather and repeat the generation* [529].

This may be considered to be an explicit description of the *stationary particle gun* found in our experiments with cellular-automata models of an excitable lattice.

> *The findings of [529] prove that particle guns, patterns which periodically emit mobile self-localizations, can, in principle, exist in a physical reality.*

4.16.2 How do breathers collide?

Collision subsumes motion. There is still no rigorous proof of breather mobility [232, 233]. However, moving breathers with a long lifetime have been found in numerical experiments [127, 135]. For example, Chen and Aubry [127] demonstrate that a breather moves as a classical particle in the presence of an appropriate perturbation.

What happens if we perturb a moving breather? The numerical study provided in [233] shows that the perturbed breather can either radiate some energy and become time periodic again or disappear in chaotic patterns when the internal resonance is involved.

Many results have been obtained on the effects of internal degrees of freedom on the mobility properties of localized excitations on nonlinear lattices [233]. Thus it was shown that interactions between breathers and impurity modes may lead to the fusion of two breathers and the generation of a larger excitation [239, 240]. In the presence of an external potential a discrete breather could be broken into two spatially separate, coherent structures with individual motions [117, 118]. And finally, perturbation of the pinning mode in a non-continuous limit gives us a method for constructing moving breathers with minimum shape alteration [127].

When two breathers collide a new pattern may be formed. This new pattern remains localized or generates two new breathers depending on the perturbations and phase differences [127]. Bang and Peyrard [63] and Forinash *et al* [239, 240] also report the survival of breathers after collisions and that their energy exchange depends on phase differences.

Gates g2 and h2 are realized in breather interactions.

Breathers may survive collisions or they may generate a new localized excitation, which, in its turn, forms two new breathers [127, 63, 239, 240]. Therefore gate g2 is realizable. In [127] it is claimed that the new pattern formed in the collision of two breathers can remain localized. This gives us an opportunity to construct an h2 gate.

It has been shown that the result of a collision depends strongly on the precise conditions of the collision [201]. The next finding is not only important but it also builds the strongest ever bridge between our cellular-automata models of an excitable lattice and theoretical physics results on localized excitations.

The result of a collision between several breathers depends on the relative phase of the breathers when they collide. [201]

Dauxois and Peyrard [201] simulated a chain of harmonically coupled particles with a substrate potential:

$$V(u_x) = \omega_d^2 \left(\frac{u_x^2}{2} - \frac{u_x^3}{3} \right)$$

where ω_d^2 is a parameter, which measures the amplitude of the substrate potential and controls discreteness. The Hamiltonian looks as follows:

$$H = \sum_x \left[\frac{\dot{u}_n^2}{2} + \frac{(u_x - u_{x-1})^2}{2} + V(u_x) \right].$$

The fact that the discrete media energy of a breather is not conserved is also discussed in [201]. When two breathers collide the energy exchange tends to increase the larger breather. Moreover, an approximate relation between breather amplitude and its frequency (see [201]) demonstrates the following result:

The smaller the amplitude of a breather is the more freely it moves in the lattice.

The larger breather 'feels' the discreteness more strongly.

Finally we summarize all information about breather collisions in the following list of possible outcomes of a binary collision of breathers:

(i) The result of a collision depends on the relative phases of the breathers when they collide.
(ii) Energy exchange between two collided breathers is proportional to the difference in their amplitudes.
(iii) A large breather increases its energy as a result of the collision; when two breathers collide with one another only the breather with a large amplitude remains recognizable, however it loses some energy; if the amplitude of the larger breather increases, this means that the breather collects some energy from thermal fluctuations.
(iv) The larger breather is often set into motion by the collision.
(v) Multiple collisions can prevent a weakly unstable breather from decaying.
(vi) If a breather grows in amplitude the minimum energy barrier increases and this breather will be trapped by discreteness.
(vii) Local defects on the lattice, e.g. regions with different coupling constants, may cause breather reflection, temporarily trapping or even acceleration depending on the parameters.

4.16.3 Collision of breather with impurity

The following findings are adopted from [239]. The model studied in [239] is a chain of harmonically coupled particles influenced by an on-site asymmetric potential. The model has the Hamiltonian

$$H = \sum_x \left[\frac{\dot{u}_x^2}{2} + \frac{(u_x - u_{x-1})^2}{2} + \omega_d^2 \left(\frac{u_x^2}{2} - \frac{u_x^3}{3} \right) \right]$$

where u_x is the position of the particle x and ω_d is a relative measure of the on-site potential and coupling energy; the larger ω_d is the weaker the coupling is and the larger the effects of lattice discreteness are. Forinash *et al* [239] use $\omega_d = 2.2$ in their experiments. The interaction of a breather with a mass impurity is studied by numerical solution of the equation of motion derived from the Hamiltonian, for a chain of 100 particles.

When a breather collides with an *unexcited* impurity there are four outcomes depending on the breather amplitude and mass of the impurity (figure 4.45).

When a breather collides with an unexcited impurity, four outcomes may be expected:

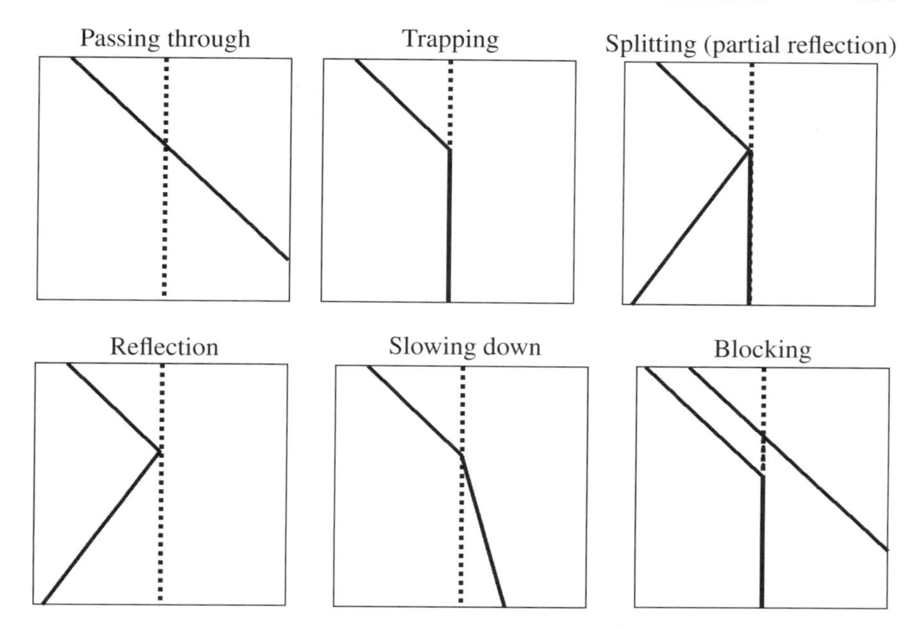

Figure 4.45. The interaction of a breather with an impurity. The breather (full line) moves on a one-dimensional lattice and interacts with the impurity (dotted line). The time runs from top down.

- *Passing through. The breather passes through the impurity without any noticeable change in its form and energy.*
- *The breather is trapped by the impurity. The breather stays at the impurity site.*
- *Splitting (partial reflection): part of the breather excitation becomes trapped by the impurity mode, the other part of the breather excitation is reflected.*
- *Reflection. The breather is almost reflected* [239].

The express parametrizations of breather collisions with stationary impurity is shown in figure 4.46. In some cases it is also possible to slow down the breather (figure 4.45). As demonstrated in numerical experiments by Forinash *et al* [239] the breather may be trapped at the impurity site for a long time and escapes later with a reduced velocity.

What happens when a breather collides with an *excited* impurity? The collision outcomes are determined by the relative phases of the breather and the impurity mode [239], i.e. by the distance in lattice nodes between breather and impurity.

When colliding with an excited impurity the breather is trapped by the

	1.025	1.05	1.1	1.2	1.4
0.55	P	P	S	R	R
0.61	P	P	T	R	R
0.66	P	P	T	R	R
0.72	P	T	T	S	R
0.77	P	T	T	S	R

Figure 4.46. Outcomes of a breather colliding with a stationary impurity depends on breather amplitude (first column) and impurity mass (first row). The cases of passing, trapping, reflection and splitting are shown by symbols P, T, R and S, respectively. Adapted from [239].

impurity if the distance between the initial position of the breather and impurity site is odd and the breather passes through the impurity if the distance is even. [239]

Let us imagine that one breather follows another breather running towards a stationary impurity. The first breather passes through the impurity but excites it. The second breather therefore collides not with the stationary but with the excited impurity. Depending on the distance between two breathers the second breather may be captured by the impurity (figure 4.45) or passed along it [239].

For certain sets of parameters it is possible to fuse two breathers into a single (im)mobile breather using a stationary impurity [239]. When two breathers follow each other towards an impurity a situation may develop when the first breather is trapped. However, when the second breather arrives at the impurity site it also becomes trapped, however only temporarily, because both captured breathers give rise to a larger breather which can start moving.

A stationary impurity can play the role of a summator.

If we set up parameters so that only the energy of n breathers, trapped by the impurity, is enough to start a large mobile breather, the impurity may be considered to be an elementary threshold summator which takes the value 1 if it accumulates at least n breathers and it takes the value 0 otherwise.

We could note that when a breather travels along DNA strands, defects in the DNA molecule can cause breather reflections or breather deceleration or acceleration [240]. Let us discuss this particular example in detail.

4.17 Collision gates in DNA

If we decide to compute with DNA molecules we have three options. The first is to use DNA molecules as 'conventional' wires or semiconductors (see experimental evidence on charge transport in [229, 631]). The second is to use recombination techniques, i.e. computing via assembling, [35, 532, 495]. The third option, discussed in this section, is to exploit the localization modes emerging in DNA molecules to implement practically the ideas of ballistic computing.

Recall that travelling defects can be found in numerical experiments in the spacetime collective dynamics of nonlinear discrete systems, for example chains of point-like masses, which are coupled by a nonlinear on-site potential or nonlinear forces. Thus, breather solutions are common for conjugated polymers, polyacetelen, ionic crystals and electrical lattices. They can also be generated optically [233]. In fact, the mechanism of discreteness-induced energy localization works in a large variety of physical lattice systems [201]. In this part of the book we consider one-dimensional biopolymers—DNA molecules—and discuss what types of logic gates can be realized in collisions of localizations travelling along the DNA chains. We mainly exploit the results from [239] and [240] to extract architectures of possible logic gates from the phenomenology of breather collisions, and substantiate our implications using cellular-automata models.

Forinash *et al* [239, 240] considered a Hamiltonian with two variables, describing the transverse displacement of two bases belonging to the base pairs of a DNA molecule under different values for the coupling between nucleotides along the same strand. They found that the intrinsic local modes are spontaneously formed. The modes emerge because of the localization of thermal fluctuations. The modes grow because of an exchange of energy, i.e. the local modes are derived from the Hamiltonian

$$H = \sum_{x} \frac{m}{2}[\dot{u_x}^2 + \dot{v_x}^2] + \frac{k}{2}(u_{x+1} - u_x)^2 + \frac{k'}{2}(v_{x+1} - v_x)^2 \\ + D(\mathrm{e}^{-a(u_x - v_x)} - 1)^2$$

where the degrees of freedom u_x and v_x at site x describe transverse displacements of the bases belonging to the base pair x of the DNA molecule [240]; k and k' are the coupling constants of two nucleotides in the same strand. In simulation experiments we may use a numerical solution of the equations of motion

$$\frac{\mathrm{d}^2 u_x}{\mathrm{d}t^2} = k(u_{x+1} + u_{x-1} - 2u_x) + 2\mathrm{e}^{v_x - u_x}(\mathrm{e}^{v_x - u_x} - 1)$$
$$\frac{\mathrm{d}^2 v_x}{\mathrm{d}t^2} = k'(v_{x+1} + v_{x-1} - 2v_x) - 2\mathrm{e}^{v_x - u_x}(\mathrm{e}^{v_x - u_x} - 1).$$

Forinash *et al* demonstrated that this self-localization—a breather—is mobile. Once excited the breathers travel along DNA strands. They interact with one another. The breathers also interact with defects, both natural and artificial, in a DNA molecule [240].

What types of breather-based gates can be implemented in a DNA molecule? A modest catalogue of DNA breather gates, extracted from the results published in [239], is displayed in figure 4.47.

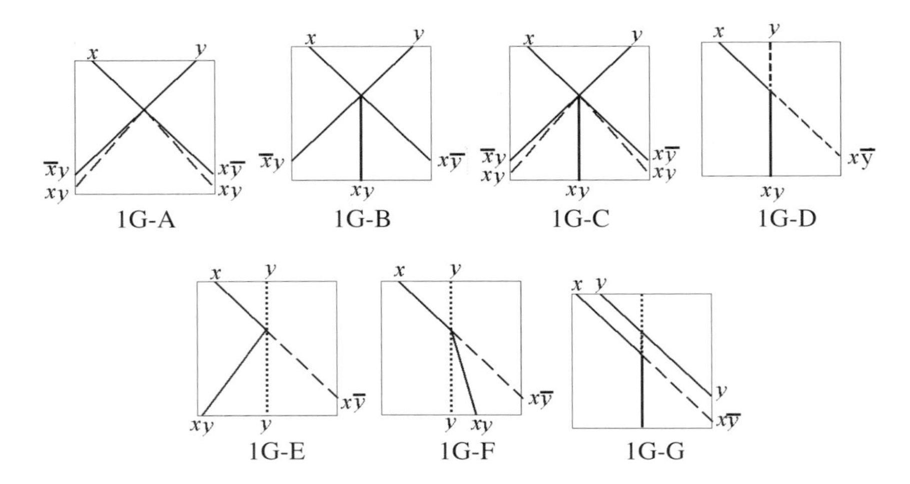

Figure 4.47. Catalogue of gates implemented in a DNA molecule in collisions between two breathers or between a breather and an impurity. Time runs down the page.

The gate 1G-A is typical amongst collision-based gates. When two breathers collide they undergo phase shifts and thus arrive at check points later than non-colliding breathers. This is similar to the original billiard ball model, where balls slightly deviate from their original trajectories as a result of a collision.

Sometimes a third localization is formed in a binary breather collision. Usually it remains immobile. This stationary localization symbolizes a logical conjunction of the variables represented by the colliding breathers. Thus gate 1G-B is designed.

In contrast, gate 1G-C is an amalgam of gates 1G-A and 1G-B. Colliding breathers are delayed or accelerated and a third stationary localization is formed.

The next four gates—1G-D, 1G-E, 1G-F and 1G-G—are derived from interactions of breathers with impurities [239], as previously discussed.

The trapping of a breather by an impurity offers us the means to construct gate 1G-D. We assume a stationary impurity is not detected. Thus when a breather collides with the impurity and becomes trapped the conjunction of two Boolean variables, one represented by the breather and the other by the impurity, is calculated. Gate 1G-E is designed from splitting and reflection types of breather–impurity collisions.

In some cases it is also possible to slow down a breather with an impurity. It has been demonstrated in numerical experiments [239] that a breather may be trapped at the impurity site for quite a long time. It escapes later with a reduced velocity. This gives us a realization for gate 1G-F.

Recall the case when two breathers are headed for a stationary impurity. We assume the initial positions of the breathers are adjusted to make the first breather pass through the impurity. The impurity is excited by the first breather. The second breather collides with the excited impurity. The breather may be captured by the impurity or pass through it [239]. Thus we obtain the structure of gate 1G-G.

Can all logic gates found in experiments with DNA breathers be implemented in cellular-automata models? Are there any new gates that were not found in DNA models? To answer these questions we employ a standard model of a one-dimensional cellular automaton with binary cell states. Each cell x_i of the automaton takes two states, 1 and 0, and updates its state in discrete time depending on its own state and the states of its four closest neighbours:

$$x_i^{t+1} = f(x_{i-2}^t, x_{i-1}^t, x_i^t, x_{i+1}^t, x_{i+2}^t).$$

We represent the cell state transition rules in the format

$$x_{i-2}^t x_{i-1}^t x_i^t x_{i+1}^t x_{i+2}^t \rightarrow x_i^{t+1}.$$

Such a representation of the cell state transition function consists of 32 rules, from $00000 \rightarrow x_i^{t+1}$ to $11111 \rightarrow x_i^{t+1}$. We use this format everywhere in the rest of this section, mostly in the examples of collision-exhibiting automaton evolution (figures 4.48, 4.49, 4.51–4.58).

The cell state 0 is a quiescent state, that is, the rule $00000 \rightarrow 0$ is obligatory for all functions of cell state transitions. We will not go into the full details of how to identify the rules that support mobile self-localized excitation, this can be found elsewhere (see e.g. [382, 315, 289, 650]). Here we present the rules found by a combination of identification algorithms and the brute force of exhaustive search.

Let us consider the evolution of the cellular automata in figure 4.48. On the lefthand part of the spacetime configuration we can recognize a localization 11101 that travels from left to right. This localization collides wih the stationary excitation, which oscillates between 01110 and 00100. The second localization, also oscillating as $01110 \leftrightarrow 00100$, is formed as a result of the collision (figure 4.48(A)). This collision realizes a gate similar to gate 1G-D, due to the rough equivalence between an impurity and a stationary localization (figure 4.47).

An attractive sequence of collisions happens in the righthand side of the spacetime configuration shown in figure 4.48. The localization 01011 travels from right to left. It collides with the stationary localization, which oscillates as $01110 \leftrightarrow 00100$. The mobile localization 01011 is destroyed as a result of the collision (figure 4.48(B)). Again, an operation similar to gate 1G-D is implemented. Localization 01011, travelling from right to left, collides with the immobile localization $01110 \leftrightarrow 00100$. The mobile localization 01011 is reflected as a result of the collision, while the stationary localization is destroyed (figure 4.48(C)). Thus, an analogue of gate 1G-B (figure 4.47) is constructed.

Two logic gates are implemented in collisions of localizations in the evolution of the cellular automaton shown in figure 4.49, that is, the localization

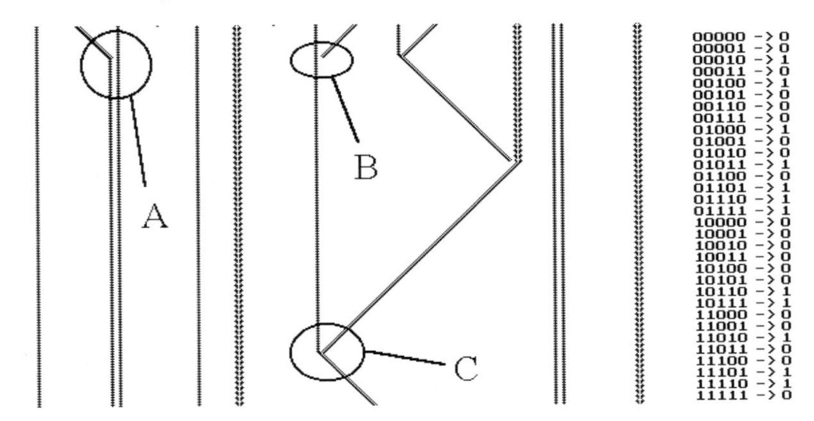

Figure 4.48. Spacetime evolution of a one-dimensional cellular automaton, which exhibits mobile self-localizations. Time runs down the page. The rules are shown on the right. The results of three particular collisions of mobile and stationary localizations are encircled: formation of a stationary localization (A), destruction of a mobile localization (B), reflection of a mobile localization (C).

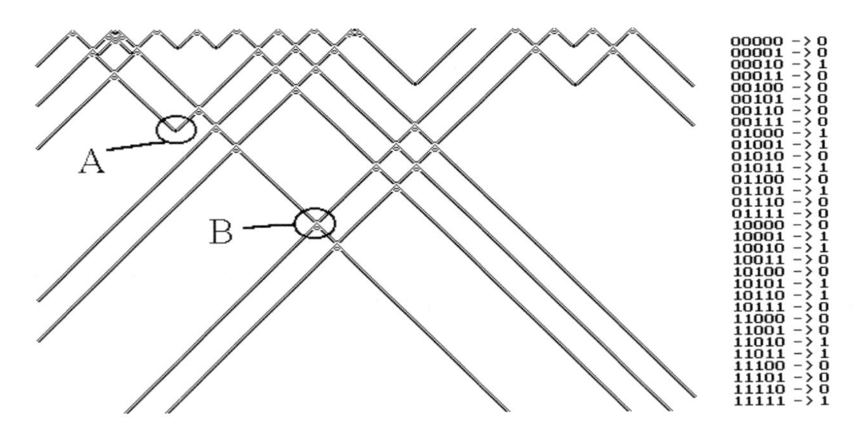

Figure 4.49. Spacetime evolution of a one-dimensional cellular automaton, which exhibits mobile self-localizations. Time runs down the page. The rules are shown on the right. The results of three particular collisions of two mobile localizations are encircled: destruction of both localizations (A); and phase shifts of both localizations (B).

11010 travels from left to right and the localization 01011 from right to left. They are destroyed when they collide with one another (figure 4.49(A)). Therefore, we obtain gate 1G-H (figure 4.50). Localizations 11010 and 01011 undergo phase shifts as a result of the collision, their trajectories are therefore translated (figure 4.49(B)). Such a collision gives us an analogue of gate 1G-A (figure 4.47).

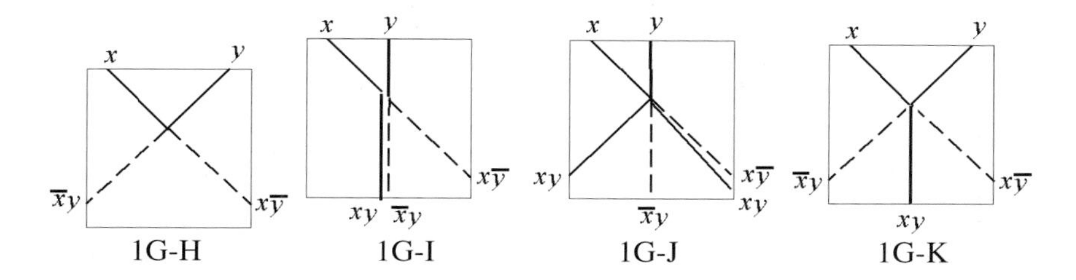

Figure 4.50. Catalogue of gates implemented in a one-dimensional cellular automaton in collisions between two (im)mobile localizations. These gates are not found in DNA breathers models. Time runs down the page.

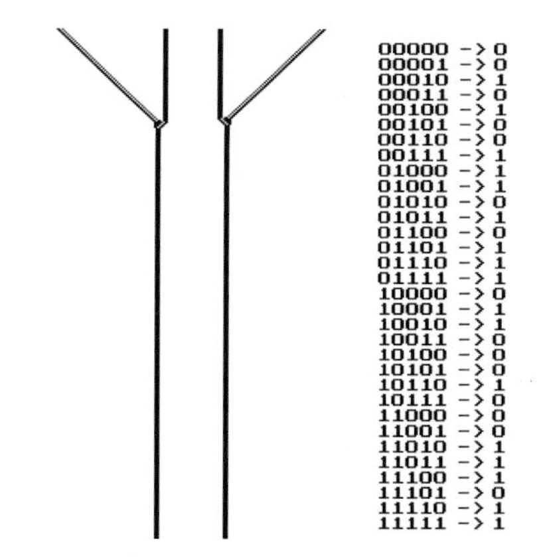

Figure 4.51. Spacetime evolution of a one-dimensional cellular automaton, which exhibits mobile self-localizations. Time runs down the page. The rules are shown on the right. In the picture we can see how mobile localizations collide with stationary localizations, and therefore implement shifts in the stationary localization at some fixed distance.

In the evolution of the automaton shown in figure 4.51 we can observe how two mobile localizations collide with stationary localizations. The mobile localizations are destroyed as a result of the collision. At the same time, the stationary localization is shifted toward the side from which the mobile localization comes. Thus, gate 1G-I (figure 4.50) is implemented.

The splitting of a mobile localization as a result of a collision with a stationary localization is shown in figure 4.52; this collision realizes gate 1G-J (figure 4.50).

A typical fusion of two mobile localizations is shown in figure 4.53. A stationary localization produced in the collision of two mobile localizations, x and y, represents the operation xy. Therefore gate 1G-K (figure 4.50) is implemented.

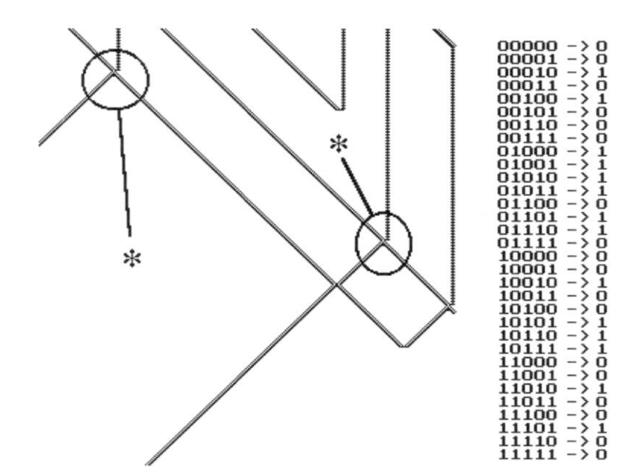

Figure 4.52. Spacetime evolution of a one-dimensional cellular automaton, which exhibits mobile self-localizations. Time runs down the page. The rules are shown on the right. In the picture we can see how mobile localizations collide with stationary localizations (sites of collision are encircled and marked by asterisks), continues its travel, however another mobile localization is formed as a result of the collision.

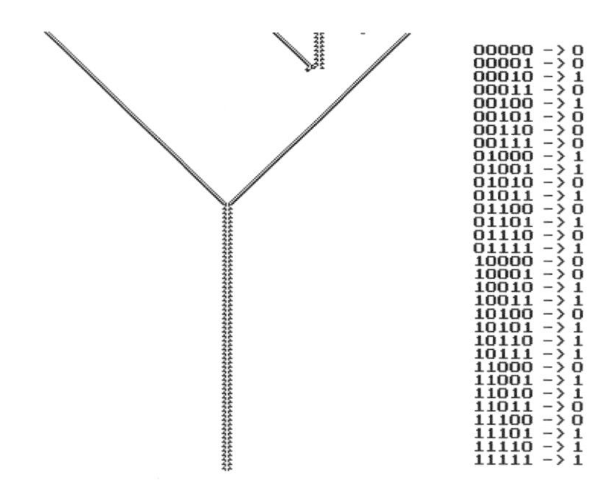

Figure 4.53. Spacetime evolution of a one-dimensional cellular automaton, which demonstrates the formation of a stationary localization in a collision of two mobile localizations. Time runs down the page. The rules are shown on the right.

What about collisions between mobile localizations and impurities, discussed previously? In a cellular-automaton model an unexcited impurity is represented by a cell that takes the stationary state, 0, at any time step of the evolution; the cell x is an unexcited impurity if $x^t = 0$ for any t. By analogy, the cell x is an excited impurity if $x^t = 1$ for any t. In words, for an impurity we fix the state of the cell and do not let it obey the rules which all other cells obey.

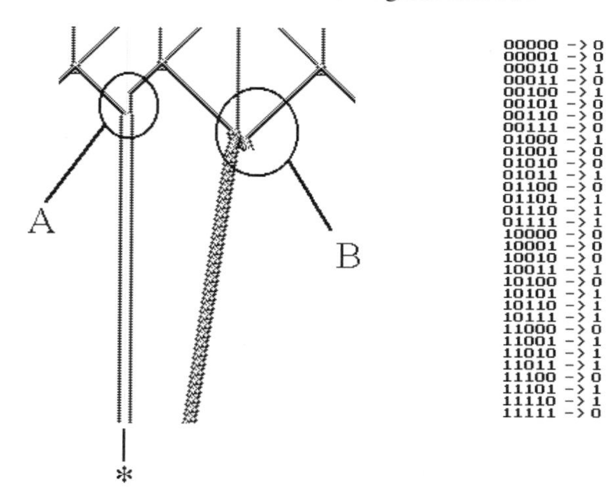

Figure 4.54. Spacetime evolution of a one-dimensional cellular automaton, which demonstrates two phenomena: (A) trapping of two mobile localizations by an unexcited impurity (the position of the impurity is marked by an asterisk); and (B) the collision of two mobile localizations with one stationary localization with the subsequent formation of one large slowly moving localization. No impurity is involved in collision (B). Time runs down the page. The rules are shown on the right.

We have found that in general, an excited impurity behaves as a stationary localization, therefore we now tackle an unexcited impurity.

Quite typically, as we see in figure 4.54(A), mobile localizations are trapped by an unexcited impurity.

A very interesting chain of collisions between an impurity and different types of localizations is shown in figure 4.55. A mobile localization a collides with an impurity *. The localization is trapped by the impurity. A stationary localization is formed. Then the stationary localization is shifted to the right by colliding with mobile localization b. After that a localization c crashes into the stationary localization and destroys it. At the same time, localization c is reflected as a result of the collision.

To be able to implement the negation operation we need a constant 1, or constant `Truth`. The constant may be represented by a stream of mobile localizations, emitted periodically. If a localization, representing `Truth`, collides with one of the localizations of the stream both localizations, involved in the

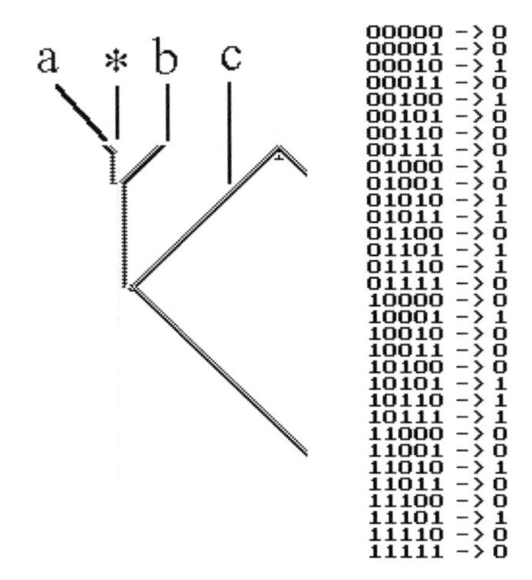

Figure 4.55. Spacetime evolution of a one-dimensional cellular automaton, which shows the interaction of localizations with an impurity. The position of a single-cell impurity is marked by an asterisk; mobile localizations are shown by symbols a, b and c. Time runs down the page. The rules are shown on the right.

collision, may be destroyed. The 'empty' member of the stream represents `False`, thus the negation is implemented.

In cellular-automata models, the localizations are emitted by either stationary or mobile guns, generators of mobile localizations. An example of three stationary guns, emerging in the evolution of a cellular automaton, is presented in figure 4.56. In figure 4.57 we can see how several mobile generators of mobile localizations are formed as a result of collisions between mobile localizations and stationary localizations.

Could we calculate Boolean functions with several variables? Certainly yes. An example of a collision of three localizations is shown in figure 4.54(B). Two mobile localizations, x and y, crash into stationary localization, z (figure 4.54(B)). The large, slowly drifting localization, w, is formed as a result of the collision. This represents $w = xyz$, i.e. a conjunction of three variables, x, y and z.

From the collision shown in figure 4.51 we can also count mobile localizations. Assume that all localizations travel from one direction, say from left to right, and collide with a stationary localization. Each mobile localization translates the stationary localization at a fixed distance, let it be Δ. Then the difference, $\Lambda^* - \Lambda^0$, between the final position, Λ^*, of the stationary localization on the Δ-stepped lattice and the initial position, Λ^0, of the stationary localization represents the number of mobile localizations that have crashed into the stationary localization.

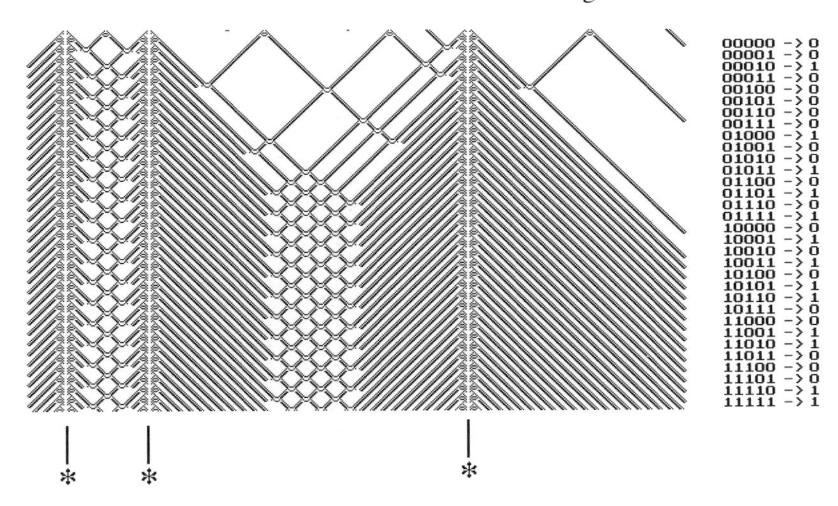

Figure 4.56. Spacetime evolution of a one-dimensional cellular automaton, which exhibits stationary generators, or guns, of mobile localizations. Time runs down the page. The rules are shown on the right. The positions of several generators are marked by asterisks. These guns emit mobile localizations on both sides.

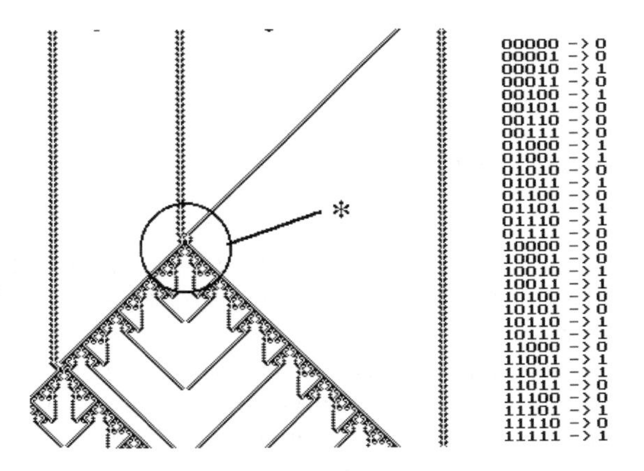

Figure 4.57. Spacetime evolution of a one-dimensional cellular automaton, which demonstrates the formation of mobile generators of localizations. Time runs down the page. The rules are shown on the right. Two mobile guns are formed in the collision between a stationary localization and mobile localization. The collision site is encircled and marked by an asterisk.

Using localizations travelling from different ends of the automaton lattice we can also solve simple equations. Let there be an equation $x - y = z$,

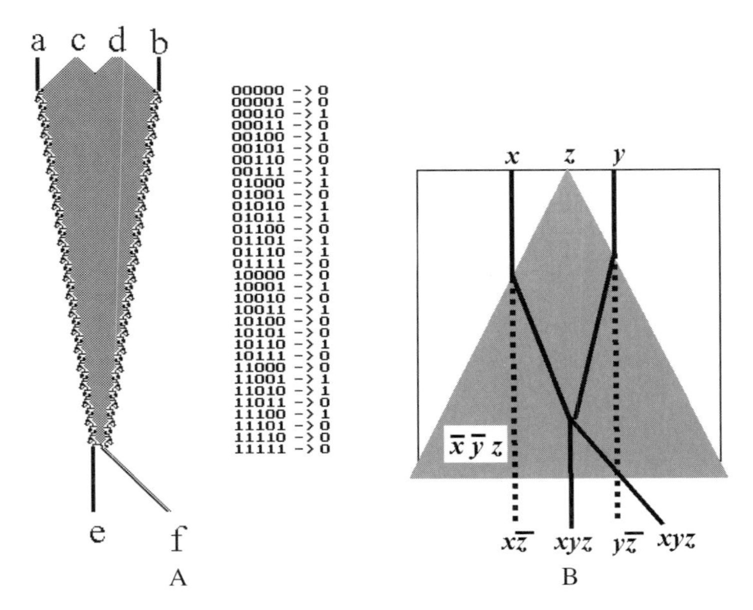

Figure 4.58. Interaction of stationary localizations with non-localized waves. (A) Spacetime evolution of a one-dimensional cellular automaton, which demonstrates the interaction of localizations and non-localized excitations. Time runs down the page. The rules are shown on the right. The initial positions of two localizations and two excitation sources are marked by letters a and b (localizations) and c and d (excitation sources). The interaction results in the generation of one stationary localization (e) and one mobile localization (f). (B) A possible gate built as a result of the interaction.

z = unknown. Assume the distance on the lattice increases from left to right. Therefore, we represent positive numbers by localizations travelling from right to left and negative numbers from left to right, and obtain the solution $z = \Lambda^* - \lambda^0$.

Spreading non-localized waves are more common in the evolution of excitable media than localized waves. Algorithms of computation with excitation waves are extensively discussed elsewhere (for a brief overview see [31]). However, there are few, if any, reports on interactions between localizations and non-localized waves. The example in figure 4.58 demonstrates the potential of such interactions. A cellular automaton starts the evolution with two stationary localizations, a and b, and two sources of non-localized waves, c and d. The waves c and d fuse into a single wave, the fronts of which collide with the localizations a and b. As a result of this collision, two defects are formed. The defects drift slowly towards each other and destroy any non-localized excitations (figure 4.58(A)). When the two defects collide they produce one stationary localization e and one mobile localization f. The possible gate realized by these mutual collisions is shown in figure 4.58(B).

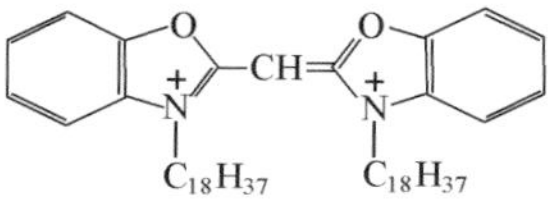

Figure 4.59. The structure of the donor molecule [463]. The acceptor molecule is sterically similar to the donor molecule: only it has sulphur instead of oxygen in its molecular structure.

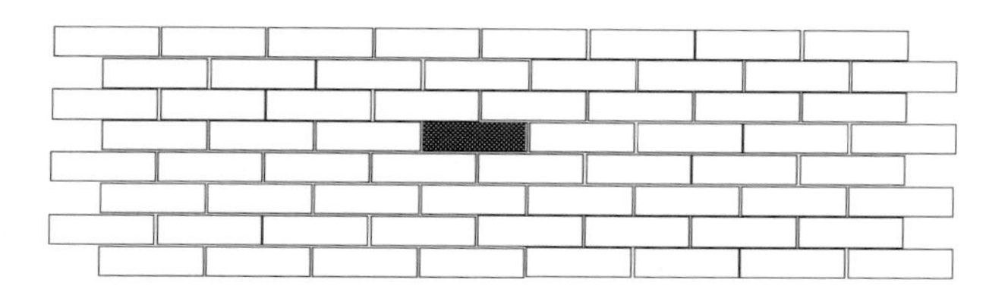

Figure 4.60. A rough diagram of a Scheibe monolayer. Few, usually one in 10^4, donor molecules of the oxycyanine group are replaced by acceptor molecules (dark filled) of the thiacyanine group. See [463] for details.

4.18 Scheibe aggregates and excitons

Monomolecular layers which exhibit a localized transfer of energy may be appropriate candidates for real-life collision-based computers.

What happens when we optically excite an electronic system in a molecular aggregate? Two competitive mechanisms are activated. An interplay between these two mechanisms leads to the localization of the excitation. The first mechanism—delocalization—is the transfer of excitation energy from site to site of the molecular aggregate, because of intermolecular resonance [603]. The second mechanism—localization—employs the imbalance of forces activated by electronic charge redistribution and acts on the relevant atoms to find new equilibria for the molecular configuration coordinates [603]. This mechanism destroys the resonance of electronic excitation on neighbouring sites and thus stabilizes the excitation. Scheibe aggregates give us the best example of localized energy transfer in monomolecular layers.

To make a Scheibe aggregate we mix oxacyanine dye molecules (figure 4.59) with inert molecules (usually in a molar ratio 1:1) to form a monomolecular planar array and add thiacyanine dye molecules. The fabricated array can be transferred to a solid substrate by various techniques [463]. If the array contains at least one acceptor molecule (thiacyanine dye) to 10^4 donor molecules (oxacyanine dye), see figure 4.60, it absorbs the energy of photons and transfers this energy over long distances without noticeable losses [463, 71].

On exciting a Scheibe aggregate a coherent exciton—a domain occupying a certain number of dye molecules oscillating in phase at

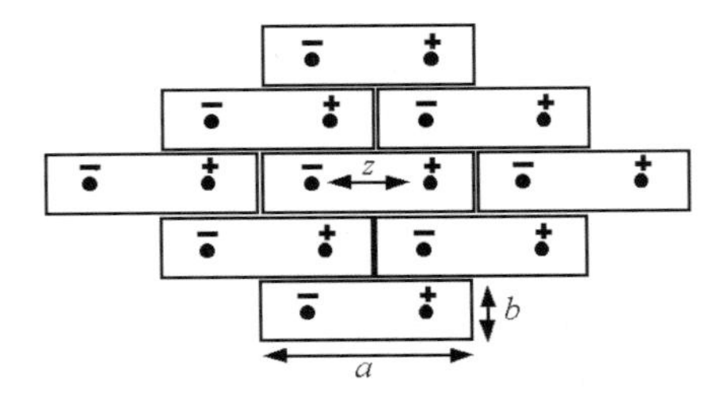

Figure 4.61. The architecture of an exciton. Each dye molecule is represented as a dipole. The dipole length is $l = 8.9$ Å; $a = 15.5$ Å and $b = 4$ Å [463].

> *the frequency of the fluorescence—is generated and it moves across the monomolecular array. The size of the exciton remains constant while the exciton moves. The exciton seems to move with supersonic speed in the aggregate.* [463]

A diagram of an exciton is shown in figure 4.61. In their investigations [463] Möbius and Kuhn proposed that

> *a coherent exciton is equally distributed over the eight molecules nearest to the acceptor molecule.*

This means we can simulate these localized excitation in a two-dimensional lattice with eight-node connections, or a cellular automaton with an eight-cell neighbourhood. The physical size of the exciton is about 27 on 20 Å which follows from the numerical evaluations in [463]. In general, the size of the exciton and the probability of energy transfer between the sites of the array is inversely proportional to the temperature, while the lifetime of the exciton is proportional to the temperature [463].

Investigating, by means of numerical simulation, the dynamics of excitons in Scheibe aggregates with thermal fluctuations of the phonons Bang *et al* [64] also demonstrate that excitons remain coherent during their lifetime in a noisy environment if the temperature is lower than 3 K [64]. They also confirmed that the lifetime of an exciton, found from experimental observations, increases with temperature [136].

The mobility of excitons is proved by Bartnik *et al* [69, 70, 71] who developed a quantum-mechanical model of energy transfer in Scheibe aggregates and studied the propagation of excitons resulting from the photon absorption process.

> *Moving excitons may undergo mutual annihilation when within each other's influence.* [352]

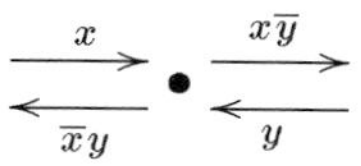

x $x\overline{y}$

$\overline{x}y$ y

Figure 4.62. The gate realized in the collision of two excitons.

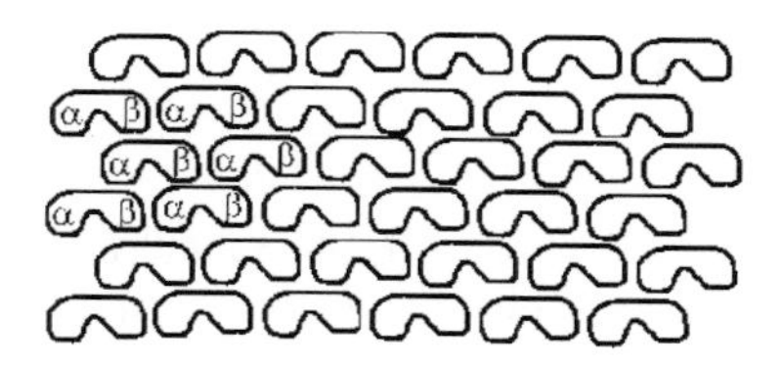

Figure 4.63. Array of tubulin dimers.

This conjecture, although proved experimentally [352], tells us that gates $\overline{x}y$ and $x\overline{y}$ (figure 4.62) are implemented in a collision, or interaction, of two excitons.

4.19 Microtubule computing

Mobile localizations can also be observed in tubulin microtubules. A microtubule's localization 'exists' at room temperature in contrast to the nearly absolute zero comfort of excitons in Scheibe aggregates.

The possibility of information-processing in microtubules has attracted the attention of researchers for a long time [288, 287, 498]. Most published results consider the dynamics of the activity in a microtubule, however they do not usually specify the particular type of dynamic. Could we do collision-based computing in microtubules? We can generate mobile signals and even control their behaviour but we know nothing about the outcomes of signal collisions.

A microtubule is an array of tubulin dimers (figure 4.63) rolled into a tube. Each dimer has an electric dipole. Thus, the array can be thought of as an array of dipole moments, which interact with their closest neighbours by dipole–dipole forces (see e.g. [111]). In cellular-automata models [287] and even more realistic computer simulations [110] it has been shown that the orientations of the dipoles of tubulin dimers in a microtubule are ordered.

In general, a microtubule undergoes a continuous process of assembling and disassembling. The kinetics of the processes can be partially controlled by temperature, concentration of tubulin and GTP (see a review in [608]). When GTP is hydrolized to GDP because of the addition of a new tubulin subunit it releases a certain amount of energy. This energy can be employed to change the configuration of the molecule and the electric dipoles. Thus, a defect, or a group of oppositely aligned dipoles, emerges [110]. The defect remains coherent if it covers several tubulin dimers.

	Scheibe aggregate	Microtubule
Neighbourhood size	6	6
Mobile localizations	Excitons	Defects
Size of localizations	8	Unknown
Triggering event	Collision of a photon	Addition of a tubulin unit

Figure 4.64. Comparison of Scheibe aggregates and microtubules.

Ideally, the defect may represent a signal that travels along the tubulin lattice. Unfortunately, we cannot guarantee that the signal moves autonomously in the initially determined direction. Probably, the defect just wanders at random [110]. However, applying an external field may certainly bias signal propagation.

Another possibility of achieving mobile signals is to consider coupling dipole states to tubulin elastic states. This coupling may result in the formation of solitons that travel along the microtubule [609]. The great advantage of microtubule machines is that, in contrast to Scheibe aggregates, mobile localized patterns might emerge in microtubules at physiological temperatures.

While talking about computing with microtubules we are tempted to touch a little bit on a more exotic problem. The problem is how consciousness may be derived from the dynamics of excitation. Hameroff and Penrose [286] claim that consciousness occurs when a system develops and maintains a coherent superposition until some specific criterion is reached [286]. They also point to intracellular microtubules as a physical substrate for consciousness. Simplifying the matter we can say that consciousness may be associated with mobile excitation patterns, such as dipole defects and solitons, travelling in active nonlinear media. In this situation all events of consciousness correspond to the collision sites between the travelling excitations.

We have found no results on how localizations in microtubules do collide and what happens as a result of the collisions. Hopefully, they behave like excitons. The next section presents some speculations on the collisions of two-dimensional localizations derived from cellular-automata models of Scheibe aggregates and microtubules.

4.20 Automata models of Scheibe aggregates and microtubules

As we can see in figure 4.64 there is a striking similarity between Scheibe aggregates and microtubules.

In both structures a non-edge unit—a molecule in Scheibe aggregates and a dimer in microtubules—is in contact with six neighbouring units. Therefore it would be reasonable to represent both systems by a two-dimensional hexagonal lattice, the cells of which obey the simple rules of finite automata. Such a representation is not new for microtubules (see e.g. [288, 287, 121]), although we discuss this in a slightly different context.

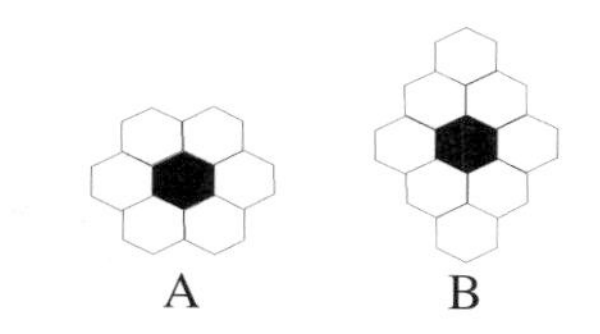

Figure 4.65. Cell neighbourhood templates of a two-dimensional hexagonal lattice: (A) six-cell localizations; (B) eight-cell neighbourhood. The central cell of the neighbourhood is darkened.

Thus, at the beginning we are dealing with a hexagonal lattice, the cells of which take three states: stationary, excited and refractory. A resting cell becomes excited depending on the excitation dynamic in its local neighbourhood. An excited cell becomes a refractory cell. A cell in the refractory state takes a resting state. All cell state transitions other than excitation are unconditional, i.e. a cell of the lattice unconditionally changes its state from excited to refractory and from refractory to stationary. There are myriads of ways for a lattice cell to measure the local excitation dynamic when it is deciding whether to become excited or not. Threshold excitation is a typical one. A look-up transition table is a comprehensive one. In a series of previous papers (see e.g. [23]) I discovered yet another representation of the excitation rules, the so-called excitation interval. I demonstrated [23] that an excitation interval function gives a comprehensive tool for analysing the spacetime dynamics of excitation in two- and three-dimensional lattices. The function assumes that each resting cell becomes excited if the number of excited neighbours belongs to the interval $[\theta_1, \theta_2]$, where $1 \leq \theta_1 \leq \theta_2 \leq k$, where k is the size of the cell neighbourhood or the number of the cell's neighbours.

The neighbourhood size k equals six in a hexagonal lattice with an immediate neighbourhood template (figure 4.65(A)).

In computer experiments we investigated all possible interval sensitivity functions of lattice excitation. We found that for the excitation intervals $[\theta_1, \theta_2]$, where $1 \leq \theta_1 \leq \theta_2 \leq 6$, the hexagonal lattice exhibits either a 'classical' (subject to model discreteness) wave dynamic (figure 4.66(A) and (E)) or replication-like behaviour (figure 4.66(C)) or disorganized chaotic-like behaviour (figure 4.66(B) and (D)). Unfortunately, we found no mobile self-localizations with lifetimes longer than three discrete time steps. Why? Possibly because we have chosen the wrong neighbourhood. If we look back at the section 4.18 about excitons we find that an exciton might occupy the eight molecules, closest to the central one. Let us update the neighbourhood template and inspect changes in the spacetime dynamic of lattice excitation.

We modify the neighbourhood template to make it identical to the molecular template occupied by an exciton in Scheibe aggregates. Thus, we add two additional neighbours to the lattice cell neighbourhood. From this moment each non-edge cell of the lattice has exactly eight neighbours as shown in figure 4.65(B)

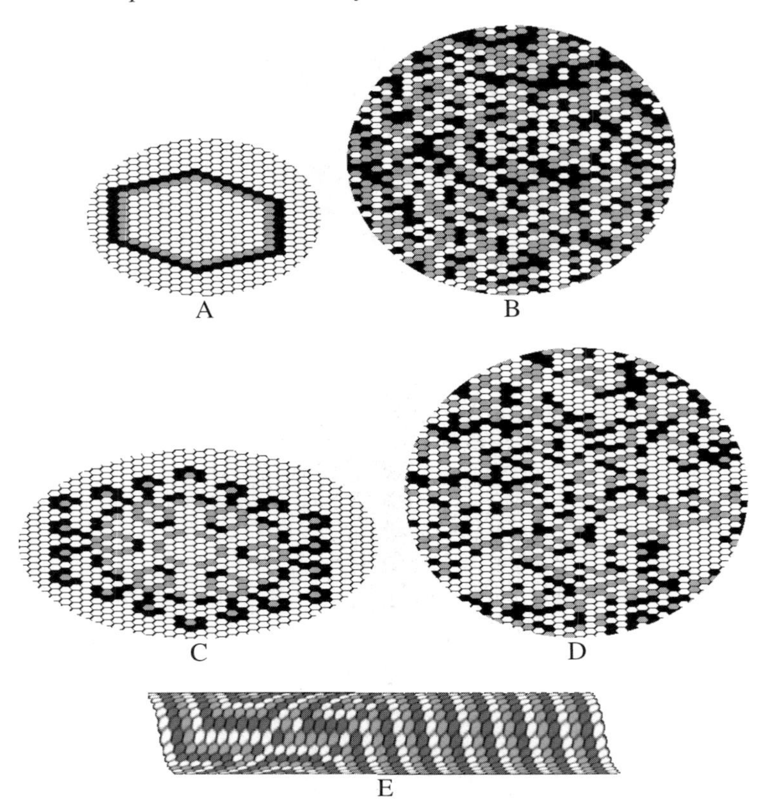

Figure 4.66. Examples of the excitation dynamic in hexagonal lattices; the cell neighbourhood is shown in figure 4.65 (A) with excitation intervals [1, 8] ((A) and (B)) and the excitation interval [1, 1] ((C) and (D)). The lattice has evolved from two initial configurations: (A) and (C), only one cell is excited initially; and (B) and (D), each cell is initially excited with the probability 0.3. The dynamics of the target of waves on a microtubule cylinder is shown in (E).

and the neighbourhood becomes asymmetric. Now the modified excitable lattice exhibits all regimes of spacetime excitation that have been previously found in rectangular lattices [13, 19, 24, 26] with eight-cell neighbourhoods.

They include a chaotic-like dynamic (figure 4.67(A)), a 'classical' wave dynamic (figure 4.67(B)) and mobile self-localizations (figure 4.67(C)). In lattices with excitation interval [2, 2] we detect all the types of self-localization that have already been found in rectangular excitable lattices with eight-cell neighbourhoods, the so-called 2^+-media (see the extensive analysis in [26]). One example of the collision implementation of a logical gate (figure 4.68) is shown in figure 4.69. The `Truth` value of a variable x is defined by the presence

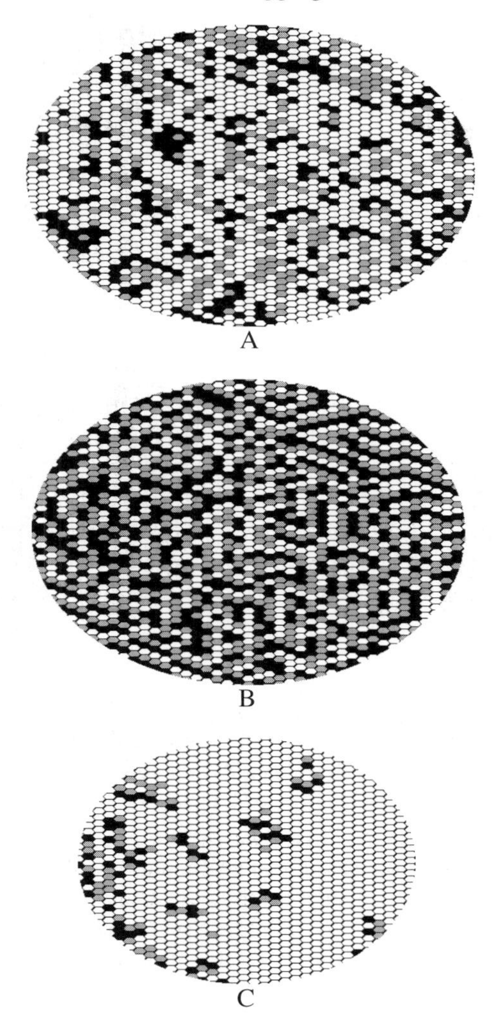

Figure 4.67. Excitation dynamic in a two-dimensional excitable lattice with an eight-cell neighbourhood (as shown in figure 4.65(A)). Each cell is initially excited with probability 0.3. (A) The excitation interval is [1, 1]; (B) the excitation interval is [2, 8]; (C) the excitation interval is [2, 2].

of a localization built from two excitation and two refractory cell states. The `Truth` value of a variable y is given by the presence of a localization formed from three excited and three refractory states. The localizations move toward each other along intersecting trajectories (figure 4.69, $t, \dots, t+4$). Eventually they collide (figure 4.69, $t+5$). As a result of the collision and subsequent local

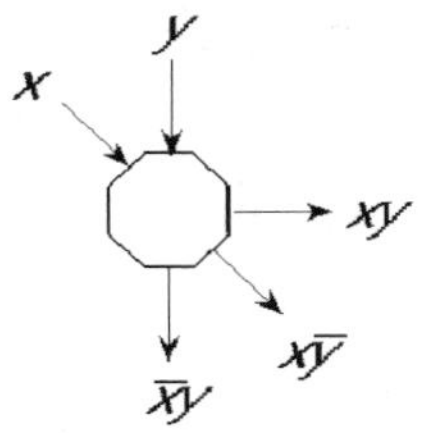

Figure 4.68. An example of a logical gate realized in a collision of two mobile localizations.

perturbations at the collision site one mobile localization is formed (figure 4.69, $t+8, \ldots, t+11$). This localization represents the value of the conjunction xy. The localization travels along a trajectory perpendicular to the trajectory of the signal y.

These findings confirm that all collision-based logic gates designed earlier in two-dimensional excitable lattices [26, 31] can, at least in the distant future, be implemented in experiments with Scheibe aggregates and microtubules.

4.21 Localizations in granular materials

Granular materials exhibit the properties of both solids and liquids. A vertically vibrated granular material exhibits a real variety of collective behaviour from standing waves and hexagonal structures to localized defects such as *oscillons* and *one-dimensional worms*.

Let us start with the phenomenological model offered by Tsimring and Aranson [605]:

$$\partial_t \psi = \gamma \psi^* - (1 - \mathrm{i}\omega)\psi + (1 + \mathrm{i}b)\nabla^2 \psi - |\psi|^2 \psi - \rho \psi$$
$$\partial_t \rho = \alpha \nabla \cdot (\rho \nabla |\psi|^2) + \beta \nabla^2 \rho.$$

The first equation is an amplitude equation for the order parameter ψ. The parameter $\psi(x, y, t)$ characterizes the local complex amplitude of granular oscillations at the frequency $\omega = \Omega/2$ at the site (x, y); Ω is the frequency of the flat surface oscillation. Parametric drive and excitation of standing waves is guaranteed by the term $\gamma \psi^*$. Nonlinear saturation of a granular material is simulated by the term $|\psi|^2 \psi$. The last term, $-\rho\psi$, of the first equation deals with the coupling of the order parameter ψ to the local average density ρ. The second equation deals with the conservation of granular material: ρ is the mass of granular material per unit square averaged over the period of vibrations. The first term of the second equation reflects the tendency of the granules to diffuse out of the regions with high-frequency oscillations; the second term tackles the diffusive

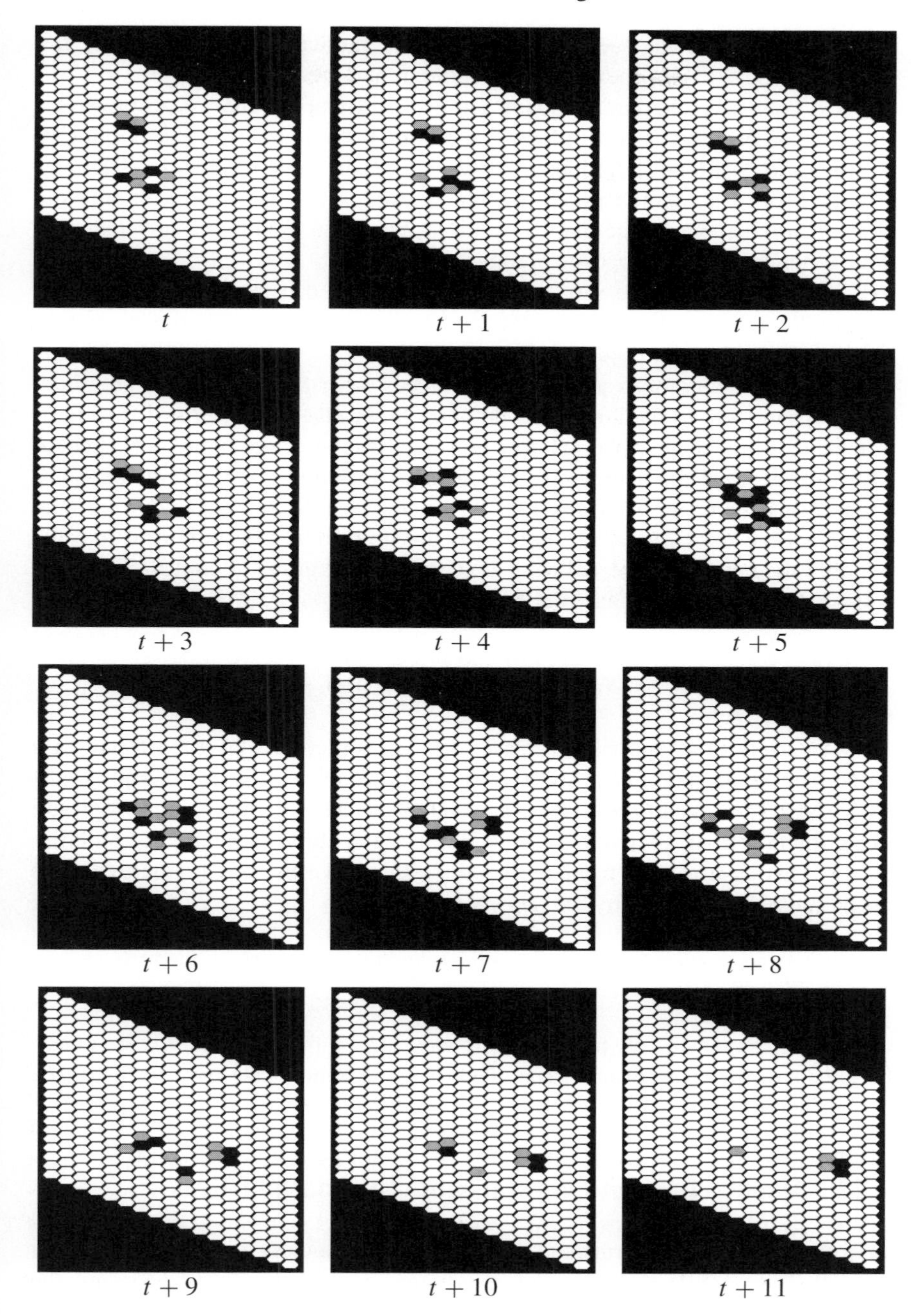

Figure 4.69. A collision between two mobile localizations in a hexagonal excitable lattice with an eight-cell neighbourhood.

relaxation [605]. It has been demonstrated in numerical experiments [605] that oscillons appear slightly below the threshold of parametric instability.

Do oscillons interact with one another? Tsimring and Aransonc [605] suggest that, because of the oscillon's oscillatory behaviour, the interaction forces between the oscillons should also be oscillatory. In [605, 622] some examples of bound states are demonstrated: the bound state of two oscillons with opposite phases, and the bound state of four oscillons, where an oscillon is surrounded by three other oscillons in opposite phases (figure 4.70).

Small granular noise can unbind weakly coupled pairs of oscillons [605]. The oscillons can duplicate themselves and organize some type of growing pattern that may imitate signal trajectories.

In certain conditions an oscillon produces other oscillons on its periphery [605]. The newly formed oscillons organize themselves, due to the diffusion of granules towards the periphery, into worm-like structures (see figure 4.71). These worms grow until the average density of the surrounding flat regions becomes so high that the creation of new oscillons is impossible [605]. If we were able to control the direction of oscillon-created worm growth accurately it may be possible to implement the collision of two or more growing cones of oscillon worms. In this imaginary case the gate

$$\langle x, y \rangle \rightarrow [\overline{x}y, x\overline{y}, xy]$$

may be constructed.

A variety of bound states of oscillon-type structures has been found in a numerical investigation of a Swift–Hohenberg model with strong damping of a basic state [158], see figure 4.72. Almost all bound states are immobile. However, one particular triad, shown in figure 4.72(E), may exhibit some kind of mobility [158]. This triad is built from an oscillon pair of like-polarity bound to a third oscillon of opposite polarity. As suggested in [158],

> *the oscillon triad breaks the spatial reflection symmetry and travels in the direction of two leading oscillons.*

What happens if two triads collide with one another? Possibly, a stationary bound state would be formed. The newly formed bound state may represent the $x \wedge y$ operation.

4.22 Worms in liquid crystals and automata lattices

In experiments with electroconvection in nematic liquid crystals [383], Dennin *et al* [179, 178, 180, 177] discovered two-dimensional mobile localized states, the so-called *worm* states. Such worm states are localized in a direction perpendicular to the direction of the liquid crystal and they extend in a parallel direction [606]. The convective amplitude is large at one end of the worm, called the head,

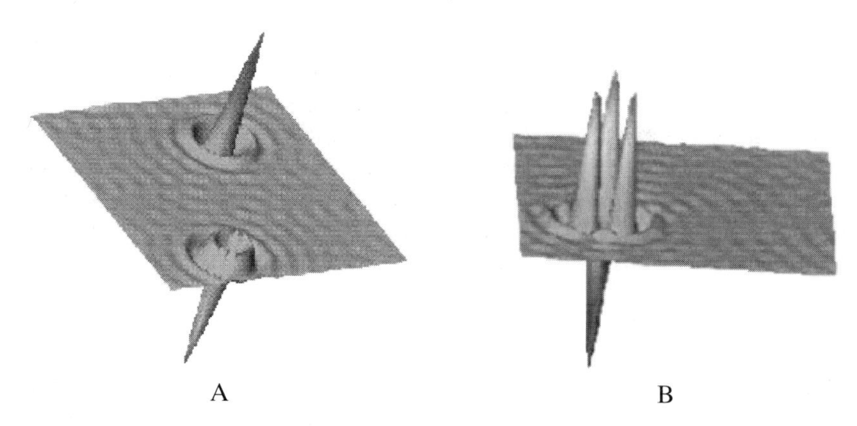

Figure 4.70. Examples of bound states of oscillons [622]; (A) two oscillons in opposing oscillating phases; (B) a tetrad of oscillons. Published with kind permission of Shankar Venkataramani.

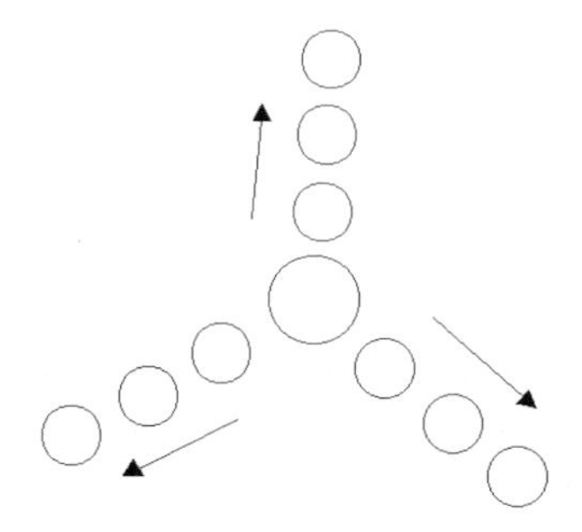

Figure 4.71. Growth of worm-like structures from a single central oscillon.

and decays gradually towards the other end [538]. The worms propagate head forward.

The liquid crystals have axial anisotropy (let it be the x-axis). There are four directions along which the realized waves propagate [541]. The worm moves along the x-axis if not perturbed [541] and it has a very distinctive shape. In the x-axis it rises steeply to its maximum at one end (head) and then decays gradually towards the end (tail). The worm is narrow along the y-axis. The widths of the worms are more or less equal whereas their lengths can vary significantly from exemplar to exemplar (figure 4.73).

The phenomenology of worms can be studied via numerical imitation of the system described by a modified Swift–Hohenberg equation for electroconvection

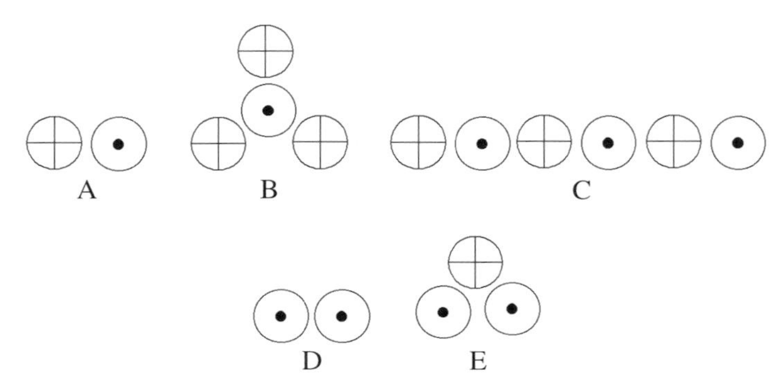

Figure 4.72. Bound states of oscillons. Oscillon polarities are shown by $\odot$ and $\oplus$. The bound states may consist of dimers (A) and tetramers (B) oscillons of opposing polarities; chains of alternating polarities (C); dimers of oscillons of like polarity (D); triads (E). Modified from [158].

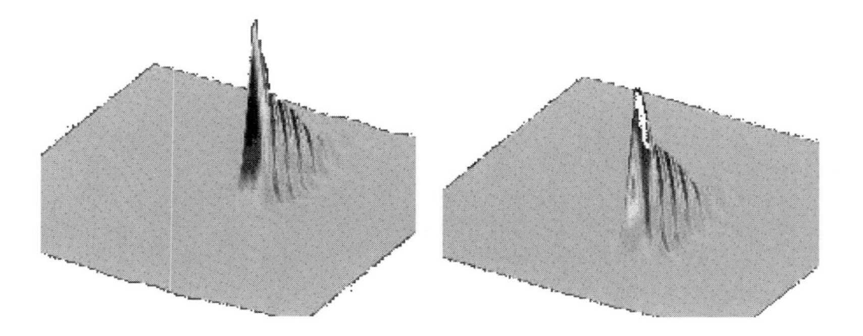

Figure 4.73. Two snapshots of a travelling steady-worm solution [539]. Published with kind permission of Hermann Riecke.

[606]:

$$\frac{\partial \phi}{\partial t} = (\epsilon + \mathrm{i}\omega)\phi - \sigma((\partial_x^2 + q_x^2)^2 + b(\partial_x^2 + q_x^2)(\partial_y^2 + q_y^2) + (\partial_y^2 + q_y^2)^2)\phi$$
$$+ \mathrm{i}v_g((\partial_x^2 + q_x^2) + a(\partial_y^2 + q_y^2))\phi + g_0|\phi|^2\phi + g_1|\phi|^4\phi$$

where $\phi(\boldsymbol{x}, \boldsymbol{t})$ is a complex order parameter, ϵ is the reduced Rayleigh number, ω is the Hopf frequency, $\boldsymbol{q} = (q_x, q_y) = (\cos\theta, \sin\theta)$ is a linearly unstable wavevector, b is an anisotropic parameter and σ is a complex constant [606]. The equation is simulated for a particular set of parameters in [606]. The simulation results show that there are two species of worm states [606]:

- Short worms. Their length (which is about three basic wavelengths) does not change with time. Short worms travel with a constant velocity proportional to the group velocity. Short worms are formed because of the spatial extension in both directions.

- Long worms. Their length grows with time. A long worm is the combined structure of a one-dimensional state in one direction and harmonic behaviour in the other direction.

A spectrum of binary collisions between worms is analysed in [606]. Three basic types of collisions may be highlighted:

- If two short worms collide they do not change their characteristics.
- If a short worm collides with a long worm the short worm disappears.
- If two long worms collide they stop spreading and form a boundary state.

The phenomenology of worm collisions is still not complete. Thus we will not even forecast what gates may be implemented with worms but make some references to the emergence of worms.

In a classical model of electroconvection the reduced Rayleigh number is an analogue of the excitation threshold (interval) in excitable lattices or coupling coefficient in coupled map lattices. The Rayleigh number determines the phenomenology of space and time evolution of the system. Thus, in the model designed in [606] the interval $[-0.1, -0.25]$ is responsible for the formation of worms (figure 4.74). For small numbers no pattern exists and for large numbers the patterns are extended and spatio-temporal chaos is more typical.

It has also been reported in [180] that chaotic localized states are found near the onset of electroconvection in nematic liquid crystals. These states are travelling waves, which have irregular location, small width and varying lengths.

Localized structures are observed in experiments in a binary mixture with Rayleigh–Bernard convection. It is reported that the localization may behave like solitons in integrable systems, thus, for example, they can pass through each other without changing their structure. The necessary conditions for the formation of localized structures are linear bistability (to stabilize the peak and tail of an impulse) and nonlinear dissipation (to stabilize the front connecting the peak and tails of the impulse) [606, 68].

In a one-dimensional model of binary mixture convection [537] we can observe pulses, i.e. stable self-localized waves propagating throughout the system like particles. The interaction of the pulses may be investigated numerically with the help of an extended version of the Ginzburg–Landau equation [537]. Thus, it is shown in [537] that when two impulses undergo head-to-head collision a stable bound state is formed only if the drift velocity of the pulses pushes them together. When two impulses interact only by their tails we can observe a repulsive interaction at short distances and a weak attraction at longer distances.

To warm up a reader's imagination on how collision-based computing with worms may be done we study *Life without Death*—a two-dimensional cellular-automata model of growing worm-like patterns [276]. The analogy between worms of liquid crystals and worms of automata lattices will be even stronger

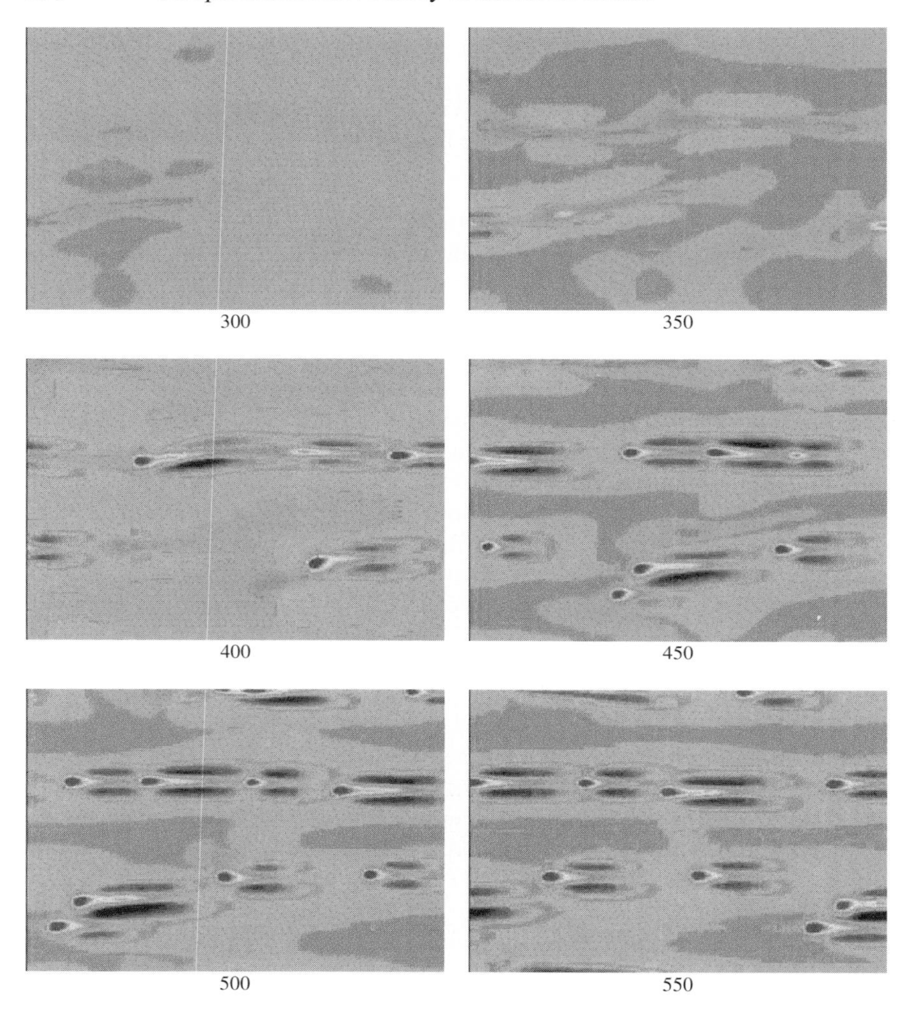

Figure 4.74. Development of worm solutions in two dimensions from small random initial conditions [539]. Published with kind permission of Hermann Riecke.

if we allow worms to grow: they become longer and longer [541]—just the same as the so-called ladders during the evolution in the cellular-automata model of Griffeath and Moore.

Griffeath and Moore [276] designed a model that realizes interaction logic gates in the collision and interaction of the worm-like patterns, the so-called *ladders*, which grow in the evolution of a two-dimensional cellular automaton with an eight-cell neighbourhood and binary cell state set. Every cell of the automaton takes states 0 and 1 and updates its state by the following simple rule:

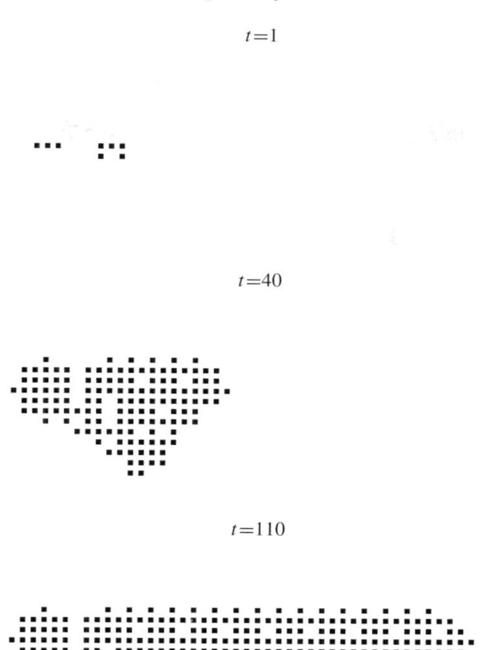

Figure 4.75. Growth of a ladder in a two-dimensional cellular-automaton model *Life without Death*. The initial configuration, a seed, of the ladder, is borrowed from [276].

> *a cell in state 0 takes state 1 if exactly three of its neighbours have state 1.*

The state 1 is absorbing: once a cell takes state 1 it never changes to state 0.

The signal trajectories, or impulses, representing Boolean values, are simulated by quasi-one-dimensional ladders that grow in one of the four directions, i.e. along the columns or rows of the cellular array. The ladders can be turned, blocked and delayed. *Immortality* is the only significant disadvantage of the model: the trajectories of the ladders cannot intersect each other.

Let us discuss the phenomenology in more detail.

The ladder is generated by the configuration of four cells in state 1:

$$\begin{array}{ccc} 1 & 1 & 1 \\ 1 & & 1 \end{array}$$

and grows towards the east because it is intentionally blocked on the west by the configuration 111 of three cells in state 1 (figure 4.75).

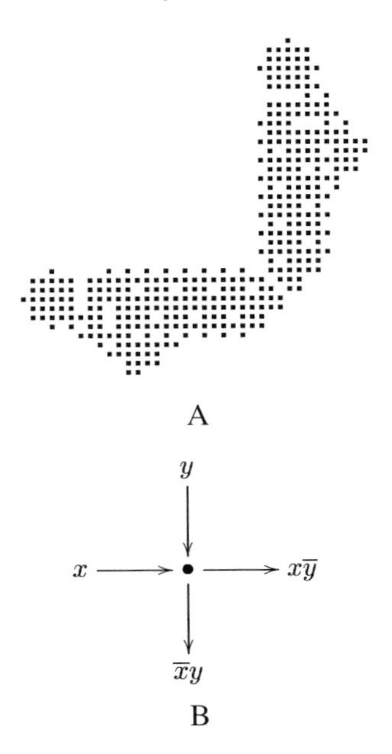

Figure 4.76. Collision of two ladders at an angle of 90^0. The ladders started their growth simultaneously and at the same distance from the collision site. One ladder grows from west to east another from north to south. (A) Configuration of two-dimensional cellular automaton representing one of the final steps of ladders collision. (B) The gate realized in the collision.

The outcome of ladder collisions strongly depends on the relative phase difference and the orientations of the colliding patterns. Our simulation experiments on ladder scattering show that, in most cases, state 1 patterns grow 'uncontrollably' and eventually the lattice is almost covered by 1s. This means that even if logic gates were to be constructed they will be very sensitive to local perturbations and noise.

Two types of ladder collision can be straightforwardly utilized in the construction of logic gates.

If two ladders collide at an angle of 90° and the collision site is at the same distance from the ladder origination sites both ladders stop growing (figure 4.76(A)). This means the gate

$$\langle x, y \rangle \to [x\overline{y}, \overline{x}y]$$

is implemented in the ladder collision (figure 4.76(B)).

If two ladders start their growth in different time steps, i.e. one starts to grow before the other, one of the ladders blocks the growth of the other (figure 4.77(A)).

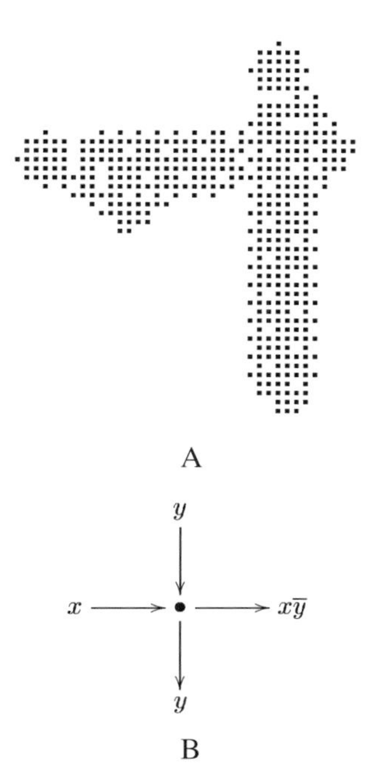

Figure 4.77. Collision of two ladders which started their development at different time steps. The ladder moving southward is born earlier than the ladder growing eastward. This first ladder blocks the growth of the second ladder. (A) Configuration of two-dimensional cellular automaton represents final step of ladders collision. (B) Gate realized in the collision.

In this situation the gate

$$\langle x, y \rangle \to [x\overline{y}, y]$$

is implemented (figure 4.77(B)).

When organizing complex computation in a *Life without Death* cellular automaton we could also shift (figure 4.78(A)) and turn (figure 4.78(B)) ladders; that is, we are able to control the signal trajectories sensibly.

4.23 Reaction–diffusion and active particles

Surprisingly, we do not have particularly many results relating to the formation and interaction of mobile self-localized states in reaction–diffusion media. Hence, we consider here just a couple of examples, possibly they are a bit artificial but nevertheless they are intuitively attractive. The first example is very small and deals with the division of a spot in a reaction–diffusion medium; the second is more lengthy—it tackles active Brownian particles.

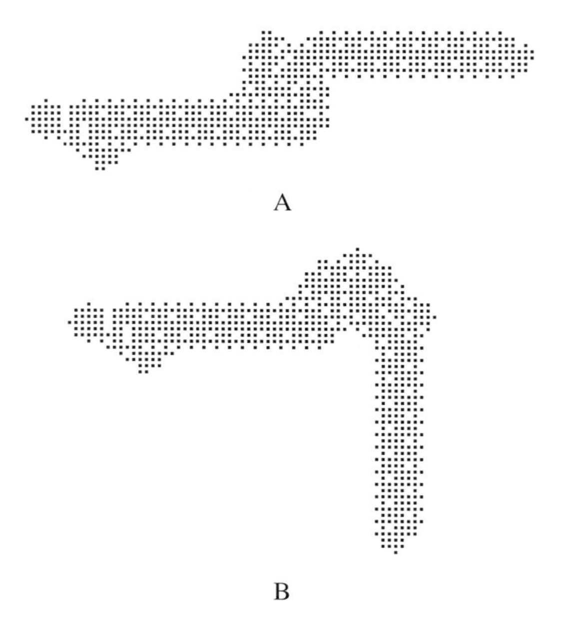

Figure 4.78. Configurations of a two-dimensional *Life without Death* cellular automaton which represent snapshots of shifted (A) and turned (B) ladders.

Various models of reaction–diffusion systems have proved to be computationally universal. Thus, for example, a lattice of chaotic coupled oscillators can realize NOR gates as shown in [565]. However, as in the overwhelming majority of other examples, the construction of the logic gates is artificial and they are closer to traditional implementations of Boolean circuits via neural networks. The example discussed here does not provide us with an explicit construction for logic gates but demonstrates how a signal can be multiplied in a reaction–diffusion medium.

Hagberg [284] numerically studied the following system of reaction–diffusion equations:

$$u_t = u - u^3 - v + \nabla^2 u$$

and

$$v_t = \epsilon(u - a_1 v - a_0) + \delta\nabla^2 v$$

for the parameters $a_1 = 2.0$, $a_0 = -0.15$, $\epsilon = 0.014$ and $\delta = 3.5$. This two-dimensional reaction–diffusion medium starts its evolution in a disc-shaped pattern. The disc expands. Some time after the start of the evolution the expanding disc is perturbed to an oval-shaped domain. When the domain expands to a certain size some parts of its boundary undergo front transitions. They annihilate and split into two vortices. The vortices split later into a couple of domains. The

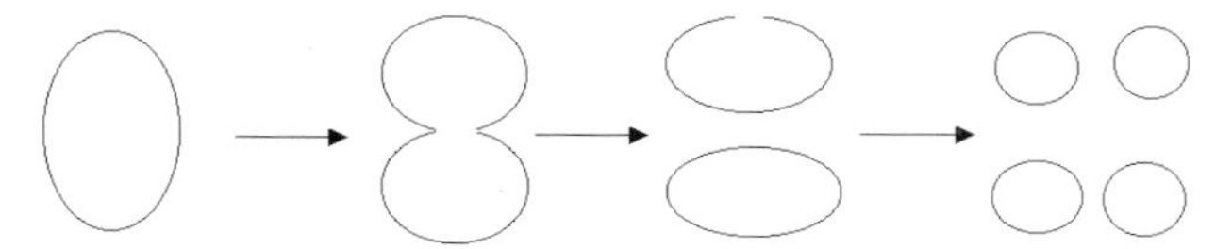

Figure 4.79. An example of spot pattern division in a reaction–diffusion model.

transitions are shown schematically in figure 4.79. These domains themselves undergo similar transformations.

The model studied by Schimansky-Geier *et al* [557] consists of N particles moving in a two-dimensional space. Every particle generates an intensity variable, which is similar to a field interaction. In the experiments [557] the intensity variable was identified with either a potential field, where particles move, or with an inhibitor density, damping the growth of the replicating activator, responsible for the particle generation rate. The intensity variable $h(r, t)$ generated by the particles with constant rate q, diffusing with coefficient D_h, and decaying with the rate k_h, is simulated by the equation

$$\frac{\partial}{\partial t}h(r,t) = q \sum_{i=1,N} \delta(r - x_i(t)) - k_h h(r,t) + D_h \Delta h(r,t)$$

where $x_i(t)$ is the coordinate of particle i at time t. The dynamics of the active particles themselves is described by the Langevin equations

$$\dot{x}_i = v_i$$

$$\dot{v}_i = -\gamma v_i + \alpha \nabla h(x_i, t) + \zeta_i(t)\sqrt{2\epsilon\gamma}$$

where γ is the friction coefficient, $\zeta(t)$ is a Gaussian white noise with intensity ϵ, α is the strength with which a particle searches for the maximum on intensity $h(r, t)$ [557].

In their experiments Schimansky-Geier *et al* [557] also assumed that the particles can be desorbed and absorbed at the surface; rates of desorbtion and absorption depend on the local intensity $h(r, t)$. Moreover, they considered a threshold-triggered replication of the particles: the particles replicate if their number per unit volume exceeds some threshold a_c. Thus we also have an absorption dependence on intensity $h(r, t)$:

$$\frac{\partial}{\partial t}a(x,t) = -k_a a(x,t) + \phi_0 \Theta(a(x,t) - a_c)(1 - h(x,t)) + D_a \Delta a$$

where $a(x, t)$ is the density of the particles at (x, t), k_a is the rate of particle desorption, $\phi_0(1 - h(x, t))\Delta t$ is the probability of the occurrence of a new particle at site x in the time interval $[t, t + \Delta t]$ [557].

When various intensities of global coupling of the excitation threshold are considered quite impressive localized mobile excitations are observed in the

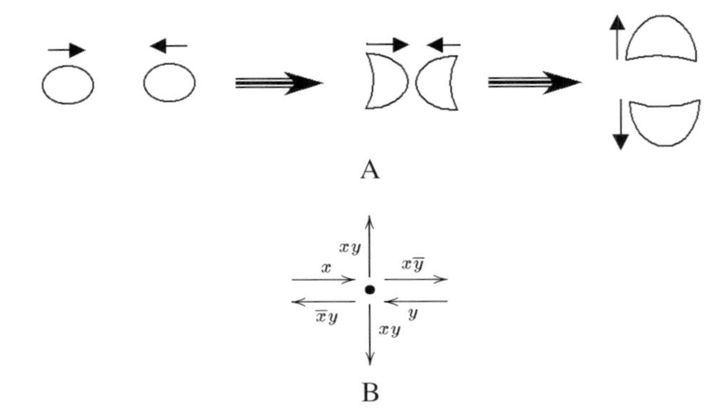

Figure 4.80. Head-on collision of two mobile regions of excited particles (A) and the logical gate constructed (B).

simulated system. For example, it has been demonstrated in [557] that in a simulated system of thousands of active particles, a spot of active particles, which travel with constant velocity as a bounded excitation region, is formed. This finding encourages us to think about collision-based computing not only in models with discrete diffusion but also in swarms of social insects and pedestrians.

What gates may be implemented in collision of two bounded regions of mobile excited particles (self-localizations)? Two examples from [557] may give us a partial answer.

The first example deals with the collision of two mobile spots of excitation (figure 4.80(A)). In the head-on collision the spots are reflected perpendicularly to their former direction of motion. We can question whether the spots are truly reflected or whether two new spots are formed and the particles are redistributed between the new spots; that is, we have the typical gate xy (figure 4.80(B)).

The second example demonstrates that a spot, or a region of excited particles, divides itself into four mobile excited spots. This finding may be used to realize the operation of branching when the trajectory of a signal, an analogue of a wire, is split into several trajectories of the same signal. In fact, such a self-replication is quite typical for many real-world reaction–diffusion systems; see the examples in figure 4.81(A) and (B).

There are also other exotic candidates for collision-based computing. They have been moderately investigated and no published results on the outcomes of real or numerically simulated interactions of these species are known.

Fluctuons, which are chains of short-lived pairs of creation–annihilation fermions [148], and *informions*, which arise from the self-interference of diffusional processes and are massless quanta of information [209], are abstract phenomena. They are certainly mobile but what happens when two of them

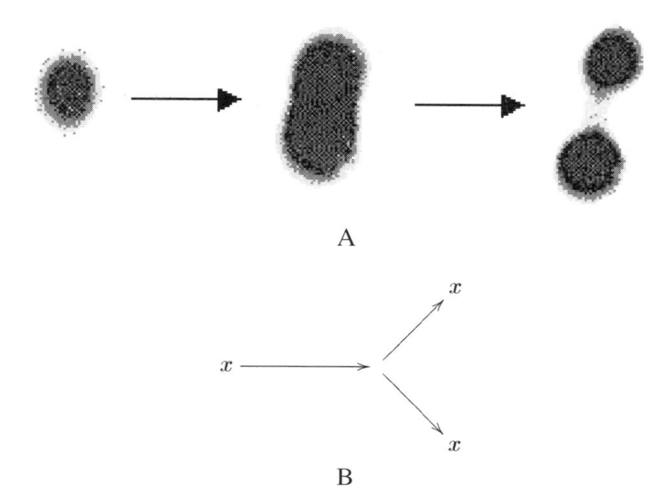

Figure 4.81. Self-replication of a chemical spot, observed in a real reaction–diffusion medium employing a ferrocyanide–iodate–sulphite reaction [387] (A) and a respective operation of branching (B).

collide? The spin waves in ferromagnetics form another family of exotic species; in this case we can appeal to *magnons* as elementary quasi-particle excitations [451].

Chapter 5

Emergence of computation in excitable lattices

How does the local determine the global? How do we find the most suitable medium for a particular computational task? Are there any parameters that govern the characteristic transitions of a medium between various domains of phase space? All these questions have intrigued scientists for a long time. The connection between the structure of a single automaton, like a neighbourhood, state set, cell state transition rule, boundary conditions, and the resultant behaviour of the whole automaton collective has been a persistent topic in the exploration and understanding of complex systems represented in terms of cellular automata and automata networks [382, 648, 491, 645, 647, 161, 459, 460, 461].

The automaton's behaviour itself has also been subjected to examination, in respect of its morphology, its utility, its complexity and, this is of particular interest to us, its computational abilities. One of the main concerns has been to discover a simple parameter which, when applied to the structural characteristics, will produce similar values for structures which produce similar behaviour.

Langton's λ-parameter [382] and Wolfram's classification of one-dimensional cellular automata [645, 647] triggered reseach into the emerging properties of a complex system and cellular automata in particular.

The λ-parameter is still a symbol of the progress in the emergent research. We do not fix on this parameter alone, however it is worth discussing immediately. The λ-parameter is structural in the sense that it is calculated from the cell state transition rules for cellular automata and the parameter itself does not take into account any kind of dynamical computation or analysis of the space and time configurations of the evolving systems. The λ-parameter measures the number of 'activation' entries in the cell state transition tables, i.e. the number of cell neighbourhood configurations that cause the central cell of the neighbourhood to take a non-resting state. Lower values of λ mean the more probable it is that the system will finish its evolution in an entirely resting state; a higher value of

λ the more probable it is that the whole lattice will be filled with non-resting states at the end of the evolution. By arranging cell state transition rules by the values of λ we can achieve a transition between ordered, complex and chaotic dynamics. Somewhere between the regions of ordered and chaotic behaviour an island of glider-based dynamics may be found [382, 648]. Due to the ability of gliders to perform both specialized and universal computations this region with complex dynamics is also considered to be an area of great computing potential. Unfortunately, when dealing with models of natural systems, the situation with the λ-parameter is not so clear as we would wish and many pitfalls wait for those who try to apply it straightforwardly. Usually, some additional parameters must be employed to characterize the transition surface between the useful and non-useful domains of the rule space efficiently (see e.g. [637, 161]).

In this book we have discussed mostly automata models of excitable media. In this case an excitation threshold seems to be the most obvious parameter, and this is quite realistic from a commonsense point of view. However, as we shall find later, the threshold poorly characterizes the spacetime dynamic: no 'interesting' phenomena are disclosed in this way. To discover richer regimes of the excitation dynamic and thereby finding some new computing abilities, we enhance the excitation threshold to the so-called excitation interval. In a lattice with local transition rules based on the excitation interval a cell is excited if the number of excited neighbours belongs to a certain interval. Using excitation intervals we are able to classify the lattice excitation regimes and to select those suitable for wave-based computing.

Even more amazing results are obtained when the so-called vector-based excitation rule is employed. In lattices with vector-based excitation rules a cell is excited only by certain numbers of excited neighbours. In this chapter we examine all possible regimes of cell excitation and subdivide the excitation rules into several classes as a function of the morphological characteristics of the resultant excitation configurations far beyond the transient period. Both spatial factors, such as number of clusters of excited states, size diversity of clusters, maximum size of clusters etc, and dynamic characteristics, such as length of transient period, activity level etc, are examined. Finally we show how to classify the potential computational abilities of excitable lattices using a combination of the morphological, dynamic and parametric classifications.

5.1 Three types of excitation rules

Typical cellular-automata models of an excitable media have the form of a two-dimensional cellular automaton, every cell of which has eight closest neighbours, takes three states and changes its states in discrete time. All cells have the same neighbourhood and the same state transition function. The cells change their states simultaneously.

The cell state set $\mathbf{Q} = \{\circ, +, -\}$ consists of three elements which represent the stationary, $\circ$, excited, $+$, and refractory, $-$, states of a cell. Usually, for a two-dimensional integer lattice $\mathbf{L}$ we can define a neighbourhood mapping $u : \mathbf{L}^k \to \mathbf{L}$. This associates each cell of the lattice with its k neighbours. In this chapter we consider eight-cell neighbourhood $u = \begin{smallmatrix}\bullet & \bullet & \bullet\\ \bullet & \diamond & \bullet\\ \bullet & \bullet & \bullet\end{smallmatrix}$, where the $\bullet$s are the neighbours of cell $\diamond$.

The cell state transition function $f : \mathbf{Q}^k \times \mathbf{Q} \to \mathbf{Q}$ determines the cell state transition rule $x^{t+1} = f(x^t, u(x)^t)$, where x^{t+1} and x^t are the states of cell $x \in \mathbf{L}$ at time steps $t+1$ and t, and $u(x)^t$ is the state of the neighbourhood $u(x)$ at time step t. The transition rule can be defined in many different ways; each is useful in a particular context.

> *Any excitation rule amongst those considered in this chapter has one overriding characteristic: the relative and absolute positions of excited cells in the neighbourhood are ignored; it is only the number which affects the state transition.*

5.1.1 Threshold excitation rule

The threshold function is a widely used cell state transition function. If the number of excited neighbours of a resting cell exceeds some specified integer $\theta = 1, \ldots, k$ then the cell becomes excited:

$$x^{t+1} = \begin{cases} + & (x^t = \circ) \wedge (\sum_{y \in u(x)} \chi(y^t, +) \geq \theta) \\ - & x^t = + \\ \circ & \text{otherwise} \end{cases}$$

where $\chi(y^t, +) = 1$ if $y^t = +$, and is zero otherwise; and θ is the excitation threshold.

5.1.2 Interval excitation rule

The so-called interval rule is developed in [23]. There are lower, θ_1, and upper, θ_2, limits of excitation, $8 \geq \theta_2 \geq \theta_1 \geq 1$ and a cell at rest becomes excited if the number of excited neighbours lies between these two limits θ_1 and θ_2:

$$x^{t+1} = \begin{cases} + & (x^t = \circ) \wedge (\sum_{y \in u(x)} \chi(y^t, +) \in [\theta_1, \theta_2]) \\ - & x^t = + \\ \circ & \text{otherwise} \end{cases}$$

where $\chi(y^t, +) = 1$ if $y^t = +$, and is zero otherwise; and θ_1 and θ_2 are the boundaries of the excitation interval.

5.1.3 Vector excitation rule

$$x^{t+1} = \begin{cases} + & (x^t = \circ) \wedge (\sum_{y \in u(x)} \chi(y^t, +) \in \mathbf{S}) \\ - & x^t = + \\ \circ & \text{otherwise} \end{cases}$$

where $\chi(y^t, +) = 1$ if $y^t = +$, and is zero otherwise; $\mathbf{S}$ is the set of all possible sums of excited neighbours, $\mathbf{S} = 2^{\{1,\dots,8\}}$.

We call this rule a vector rule. The vector rule is unusual because it allows us to consider *all* possible sensitivity modes of the cell.

In the forthcoming sections we analyse the full spectrum of vector excitation rules. Namely, basing on the eight-cell neighbourhood we examine all 256 possible vector excitation functions, represented by the vectors $s = (s_j)_{1 \leq j \leq 8}$ and ordered naturally:

$$\begin{aligned} &f_1 : s_1 = (00000000) \\ &f_2 : s_2 = (00000001) \\ &\dots \\ &f_{256} : s_{256} = (11111111). \end{aligned}$$

Let us explain the vector representation of the excitation rules. If vector s has entry 1 at position j, $s_j = 1$, then a resting cell becomes excited if it has j excited neighbours in its neighbourhood.

Thus, if the cells update their states by the function represented by the vector (01110010) then every rest cell will become excited if it has two, three, four or seven excited neighbours in its neighbourhood; the rest cell does not become excited if it has one, five, six or eight excited neighbours. Let the neighbourhood $u(x)$ of the resting cell x, $x^t = \circ$, be in the state $u(x)^t = \begin{smallmatrix} + & - & \circ \\ \circ & \circ & + \\ + & \circ & - \end{smallmatrix}$ and its transition function is represented by the vector (01110010) then the cell takes the excited state at the next discrete time step, $x^{t+1} = +$. This is because the vector (01110010) means that the cell will become excited if either two, three, four or seven of its neighbours are excited. In this particular example three neighbours are excited, therefore the cell is excited as well, $x^{t+1} = +$.

5.1.4 Relations between the rules

In spite of its triviality it is good to note the following fact: if M_{vc}, M_{th}, M_{in} are families of excitable lattices, the cells of which update their states by vector, threshold and interval rules of excitation, then

$$M_{\text{vc}} \supset M_{\text{in}} \supset M_{\text{th}}.$$

Taking θ as the excitation threshold and k as the neighbourhood size we have $s_j = 1$, if $j \in [\theta, k]$, and $s_j = 0$, otherwise, we prove $M_{\text{vc}} \supset M_{\text{th}}$. Vector

(10000001) tells us that $M_{vc} - M_{th} \neq \emptyset$. By analogy, $s_j = 1$, if $j \in [\theta_1, \theta_2]$, and $s_j = 0$ otherwise, it implies $M_{vc} \supset M_{in}$. And vector (10001001) indicates that $M_{vc} - M_{in} \neq \emptyset$. Assuming $\theta_1 = \theta$ and $\theta_2 = k$ we obtain $M_{in} \supset M_{th}$. Moreover, we have $M_{in} - M_{th} \neq \emptyset$.

5.2 Lattices with interval excitation

There are 36 regimes of lattice excitation, or 36 rules, determined by sensible ranges of lattice excitation intervals $[\theta_1, \theta_2]$, $1 \leq \theta_1 \leq \theta_2 \leq 8$. For each of the 36 rules we provide an analysis of the space and time excitation dynamic. This results in a morphological classification, which is also described here.

5.2.1 Basics of classification

We would like to start by discussing the global behaviour of excitable lattices by making some analogies with Wolfram's classification of cellular-automata spacetime dynamics [645, 647]. An interpretation of the excitation dynamic in terms of Wolfram's classes is as follows.

I The lattice evolves toward a uniform resting state. The excitation is damped in the result.

II The lattice falls into either the uniform resting state or an oscillatory state, depending on the initial distribution of excited states. All oscillations are quite simple; their periods are short.

III The behaviour of the lattices is characterized by aperiodic and less predictable oscillations.

IV The lattices exhibit propagating localized patterns (self-localized excitations) in their spacetime evolution.

5.2.2 Findings

- There is only one excitable lattice that belongs to class IV. This is the lattice with excitation interval [2,2], where a resting cell of the lattice becomes excited if exactly two of its neighbours are excited.
- The lattices with excitation intervals $\theta_1 = 1$ and $\theta_2 = 1, 2$ and $\theta_1 = 2$ and $\theta_2 = 3$ belong to class III.
- The lattices with excitation intervals $\theta_1 = 1$ and $\theta_2 = 3, \ldots, 8$; and $\theta_1 = 4, \ldots, 8$ and $8 \geq \theta_2 \geq \theta_1$ lies in class II.
- The lattices lie in class I if the lower boundaries of cell excitation intervals exceed 3, $\theta_1 \geq 3$.

The actual distribution of lattices with interval excitation rules among the various classes is as follows: 58.3% (I), 30.6% (II), 8.3% (III) and 2.8% (IV).

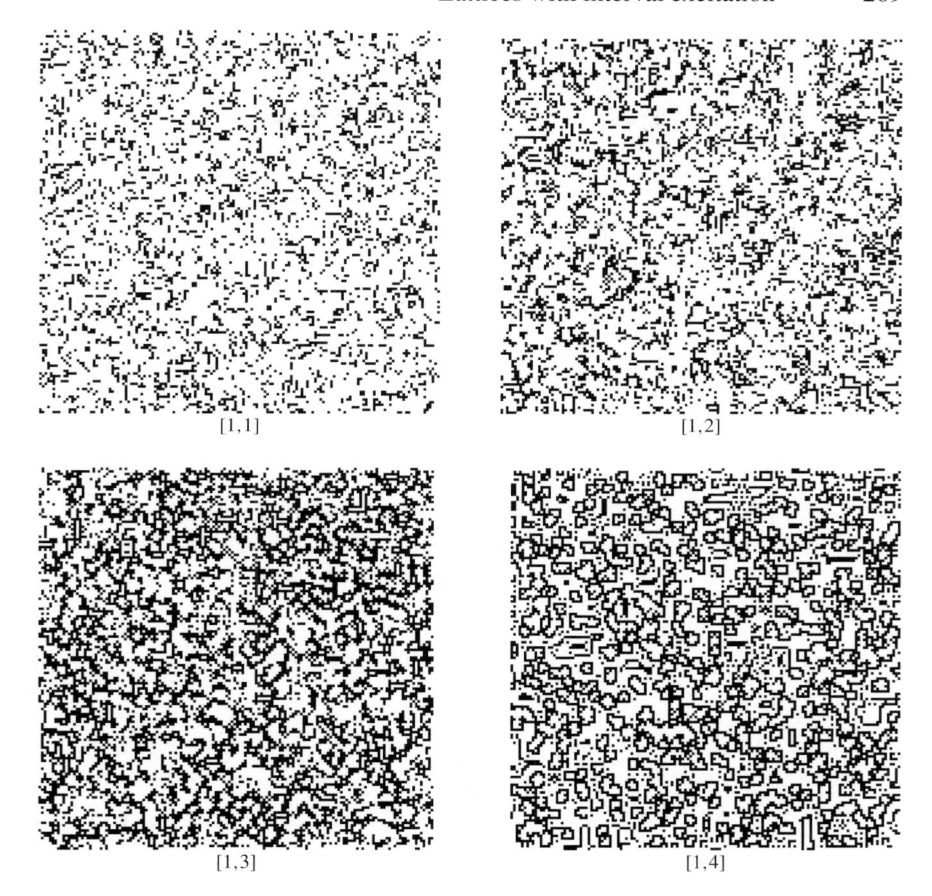

Figure 5.1. Configuration of the excited cells of two-dimensional excitable lattices recorded at the 100th step of the evolution. The cell excitation intervals are [1, 1], [1, 2], [1, 3] and [1, 4]; they are shown below the configurations. At the beginning of the evolution one-third of all cells, chosen at random, are excited.

5.2.3 Analysis of excitation dynamic

Now we can summarize the results of examining the spacetime (figures 5.1, 5.2, 5.3 and 5.4) and integral dynamics (figures 5.5 and 5.6) of activities in excitable lattices.

We see that most types of lattice lie in class I except for the lattices with excitation intervals [1, 1], [1, 2] and [2, 3], which belong to the class III, and the lattice with cell excitation interval [2, 2], and periodic boundaries, which are in class IV. Looking at the behaviour of those lattices with excitation intervals [2, 3], [2, 4] and [2, 5], it is apparent that in the early stages of development they

Figure 5.2. Configuration of excited cells of two-dimensional excitable lattices recorded at the 100th step of the evolution. The cell excitation intervals are [1, 5] and [1, 6]. At the beginning of the evolution one-third of all cells, chosen at random, are excited.

may be wrongly classified as elements of class IV. However, after the transient period they exhibit behaviour which is undoubtedly in class III. When the lower boundary of the excitation interval exceeds three, $\theta_1 \geq 3$, then, regardless of the upper boundary θ_2 of the excitation interval, and independently of the boundary conditions, the lattices belong to class I, where the uniform resting states are global attractors of the lattice evolution. We therefore restrict our investigation to those lattices for which $\theta_1 = 1, 2$ and $\theta_1 \leq \theta_2 \leq 8$.

In the evolution of the lattices with cell excitation interval [1, 1], when every resting cell becomes excited only if exactly one of its neighbours is excited, connected sets of excited states are generated. They spread over the lattice and are transformed into groups of small clusters as a result of collisions with other spreading excitations. With time, they may again combine into powerful sets, then dissolve into small groups, and so on. A very typical excitation configuration is shown in figure 5.1 [1, 1].

The 'springs' of connected excited states are more visible in the lattice excitation configuration for the cell excitation interval [1, 2] (figure 5.1, [1, 2]).

Due to an increase in the number of possible situations leading to cell excitations for $\theta_1 = 1$ and $\theta_2 = 3$, islands of excited states are formed in an ocean of rest and refractory states, and *vice versa* (figure 5.1 [1, 3]). These islands expand on the lattice and are characterized by wavefronts with a 'toothed' appearance due to the alternation of excited and refractory states along the outline of a normal wavefront (figure 5.1, [1, 3]).

The overall spacetime structure of the excitation dynamics is preserved for $\theta_1 = 1$ and $\theta_2 = 4$, but the lattice is increasingly filled by excitation states. The most notable fact is that classical waves first appear in the lattice with the

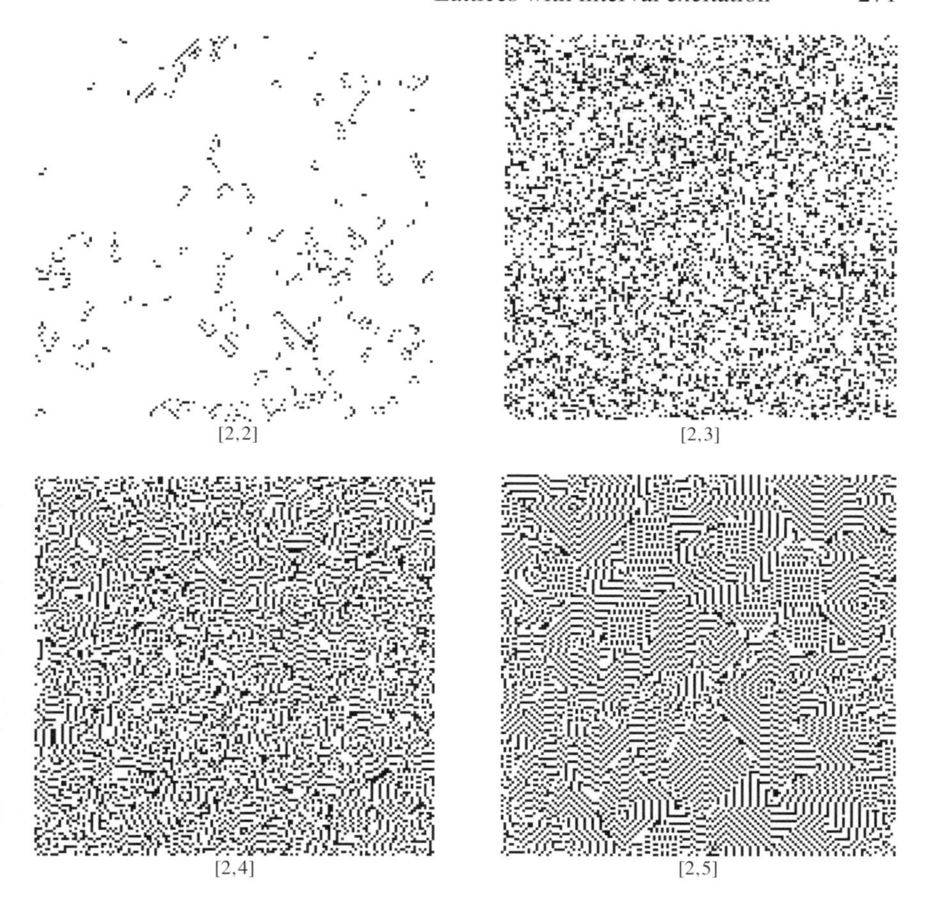

Figure 5.3. Configuration of excited cells of two-dimensional excitable lattices recorded at the 100th step of the evolution. The cell excitation intervals are [2, 2], [2, 3], [2, 4] and [2, 5]; they are shown below the configurations. At the beginning of the evolution one-third of all cells, chosen at random, are excited.

excitation interval [1, 4]. In a snapshot of lattice activity (figure 5.1, [1, 4]) not only are the typical waves, but also the generators of the waves are visible in the ocean of dissociated excitations; the generators appear in the snapshot as small clusters of black pixels (excited states) encircled by 'full' wavefronts.

In lattices with the excitation interval [1, 5] 'full' waves are generated around the small clusters, usually consisting of between two and five cells. These waves spread on the lattice and annihilate one another as a result of collisions (figure 5.2, [1, 5]). However, the excitation activity remains practically stationary because the generators are not destroyed and so continue to produce waves.

In the evolution of lattice with cell excitation interval [1, 6] we can observe

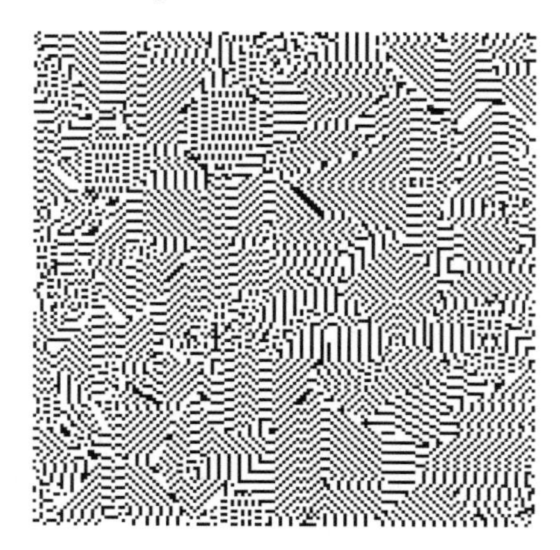

Figure 5.4. Configuration of excited cells of two-dimensional excitable lattices recorded at the 100th step of the evolution. Such configurations are usually found in the evolution of lattices with cell excitation intervals [2, 6], [2, 7] and [2, 8]. At the beginning of the evolution one-third of all cells, chosen at random, are excited.

a decrease in the number of wave generators, and a relative straightening of the wavefronts or boundaries of excited waves (figure 5.2, [1, 6]).

For excitation intervals [1, 7] and [1, 8] no long-term activity exists. This is because quite classical excitation waves emerge at the beginning of the evolution, they travel along the lattice and destroy each other when they collide. Therefore, the lattice finishes its evolution in an entirely resting state. So here we do not provide the spacetime dynamics or integral activity illustrations for the excitation intervals [1, 7] and [1, 8].

Let us now increase the lower limit of the cell excitation interval: $\theta_1 = 2$. Remember that $\theta_2 \geq \theta_1$. The first excitation interval to consider is [2, 2]. In lattices with cell excitation interval [2, 2] small groups of excited states move quite freely on the lattice because the neighbouring excited elements of a moving cluster can mutually stabilize one another (figure 5.3, [2, 2]). For excitation intervals [2, 3] and [2, 4] we can also observe extended but unstable patterns (figure 5.3, [2, 3] and [2, 4]). These patterns easily emerge but their collisions generate rather dispersed excitation states and the lattice becomes filled with numerous small clusters very similar to those observed in lattices with excitation interval [1, 1].

On increasing the upper boundary of excitation interval, θ_2, from three to eight, we see a 'solidification' of the wavefronts. The waves, completely disordered in the case of interval [2, 3], are sparse for interval [2, 4] and they present a finely dispersed labyrinthine structures in the lattices with excitation intervals from [2, 5] to [2, 8] (figures 5.3 and 5.4).

In the examples of integral excitation dynamics (figures 5.5 and 5.6) we see that for fixed θ_1 the increase in θ_2 leads to an increase in the aggregate excitation level ($\theta_2 = 1, \ldots, 3$) that stabilizes after $\theta_2 = 4$. The amplitude of the oscillations increases somewhat when θ_2 changes from one to three, and decreases after $\theta_2 = 4$. The lower limit $\theta_1 = 2$ gives the most rapid growth of the excitation level (figure 5.6).

The only lattices with cell excitation interval [2, 2] *are strongly affected by boundary conditions: the level of integral activity stabilizes with time for periodic boundaries and all excitation completely vanishes when the boundaries are absorbing.*

This is because, after a certain period of the evolution, only small patterns of mobile self-localized excitations remain on the lattice. If the boundaries are periodic the patterns travel around the lattice torus along the non-intersecting trajectories; otherwise the patterns move away from the boundaries, when they are absorbing, of the lattice.

5.2.4 Morphological diversity

In addition to the dynamic characteristics we could try to employ some of the morphological characteristics of excitation patterns. Morphological diversity is one of the most appropriate and this is easily calculated.

The morphological diversity of a global configuration of an excitable lattice is the number of clusters of excited states counted in the excitable lattice configuration when the transition period has passed, and the level of activity has become quasi-stable. Our calculation technique for morphological diversity is discussed in [23]. To compare lattices with different excitation modes by the morphological diversity of the excitation patterns we compute the distributions of the sizes of the connected subsets of excited states.

For excitable lattices with periodic boundaries and $\theta_1 = 1$, increasing θ_2 from one to four leads to an increase in morphological diversity. However, after $\theta_2 = 4$ the diversity increases considerably and only stabilizes when the upper limit of cell excitation equals $6, \ldots, 8$. The situation is similar for the $\theta_1 = 2$ case, but stabilization occurs at the maximum level.

We could assume that the morphological diversity characterizes, to some degree, the complexity of the excitation patterns. Thus, for example, for $\theta_1 = 1$ an increase in morphological diversity reflects an increase in the size of the excited clusters, and the corresponding diversity of activity patterns. For $\theta_2 = 4$ (and $\theta_1 = 2$) the first embryonic waves are generated but the clusters still persist. This marks the maximum range of morphological diversity in the transition from a 'particle' system to a 'wave' system.

Excitable lattices with the cell excitation intervals [1, 4] *and* [2, 4] *have the maximum morphological diversity amongst the excitable lattices considered in the section.*

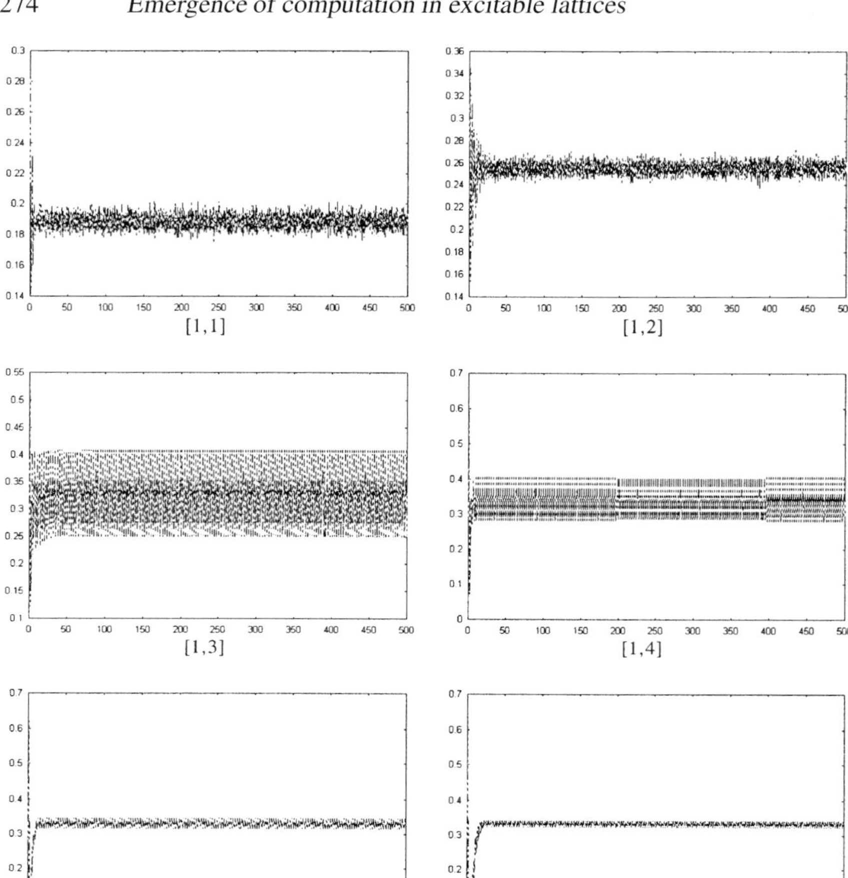

Figure 5.5. Integral activity of an excitable lattice of 100×100 cells, in 10 runs, each of 500 time steps; the boundaries are periodic. Cell excitation intervals [1, 1]–[1, 6] are shown below the graphs. At the beginning of the evolution cells are excited with probability 0.3. The ratio of excited cells for every step of the evolution is shown in the vertical axis, the time arrow is the horizontal axis.

It appears that lattices of this kind occupy a critical position intermediate between the systems with mobile self-localizations and systems with typical target waves.

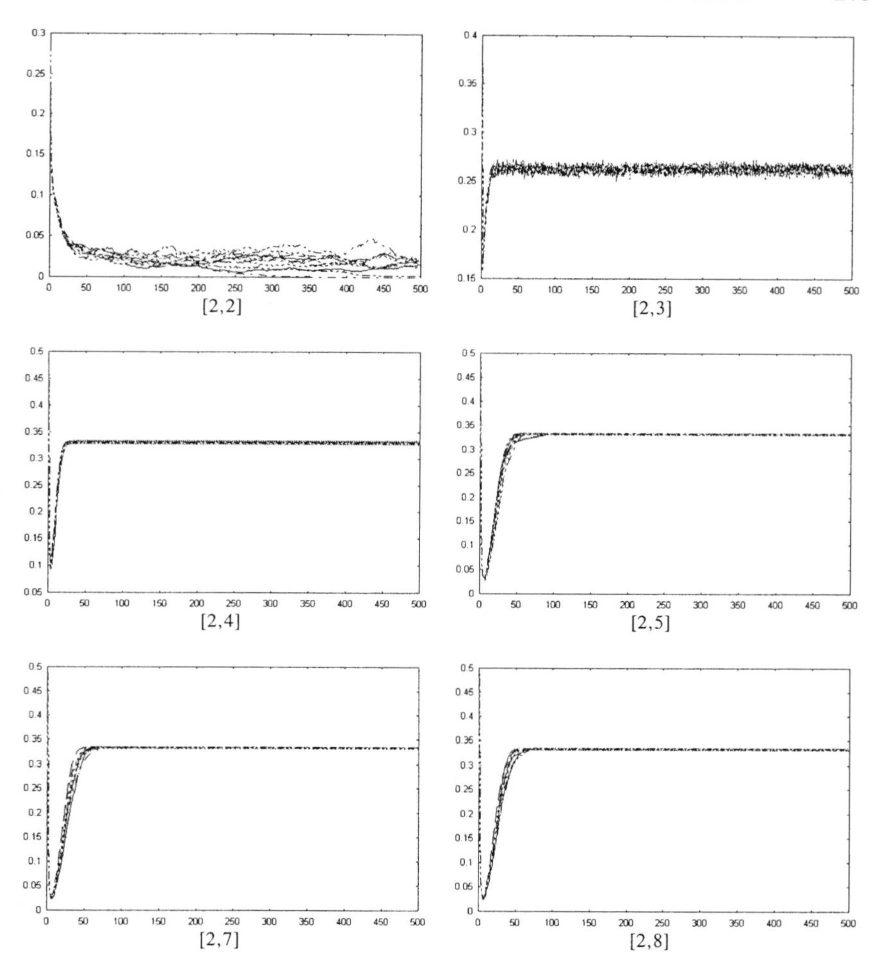

Figure 5.6. Integral activity of an excitable lattice of 100 $\times$ 100 cells, in 10 runs, each of 500 time steps; the boundaries are periodic. The cell excitation intervals [2, 2]–[2, 8] are shown below the graphs. At the beginning of the evolution the cells are excited with probability 0.3. The ratio of excited cells for every step of the evolution is shown in the vertical axis, the time arrow is the horizontal axis.

5.3 Interval excitation controllers for mobile robots

In the previous section we found that there are regions of excitation rules responsible for the universal computation capabilities of excitable media. If we want to implement collision-based logic gates then the lattice with the excitation interval [2, 2] may be employed. Thus, the domain of universality is quite narrow.

Excitable lattices with other intervals of cell sensitivity may certainly be used in image-processing and probably computational geometry, however a manual design of excitation rules must be involved.

To make a fair comparison between the different domains of an excitation rule space we must choose a problem that can, in principle, be solved by any excitable lattice. Using such a test problem we can talk about the performances of excitable lattices with different regimes of node excitation. Here we choose the problem of robot navigation, or a primitive phototaxis, because its affinity to the real-life behaviour of excitable media is obvious. Our solutions of the problems are discussed in [27, 28]; in this chapter we give a brief overview of the results.

> *Given an excitable lattice, a robot and a light source, how can we employ the spatio-temporal dynamics of light-induced excitations on a lattice to guide the robot toward the light source?*

The answer is rooted in the following three ideas.

The first one is a proof of the efficiency of two-dimensional excitable lattices with local vectors for computing the shortest paths on locally connected graphs [16]. The computation of shortest paths and field-like balancing of communication networks (as described in chapter 3) also benefits from the theory of a programmable force field developed in [93]: having a force field surrounding a given object causes the object to move.

The second wave of inspiration may come from the artificial amoeba project (see e.g. [661, 662]), where virtual and real collectives of interconnected robots are designed to form an elastic mobile system. The artificial amoeba supports our understanding of the necessity of a diffusive component in the control of distributed systems.

Metabolic controllers for robots [672] form the third part of our inspirational results. In general, a metabolic controller can be imagined as a reactor with several chemical reagents [672, 66]. The reactor is connected with light sensors. Changes in the values of the light sensors invoke changes in the concentrations of the reagents in the reactor. The chemical reactions between the metabolites result in the corresponding changes in robot behaviour. Simulated robots with simulated metabolic networks can solve a range of simple tasks, e.g. obstacle avoidance and phototaxis [672].

5.3.1 Architecture of the excitable controller

We base the proposed controller on a two-dimensional finite lattice **L** with the following characteristics. Every node of the lattice is coupled with the eight closest neighbours and it can be in one of three states: rest ($\circ$), excited ($+$) and refractory ($-$). The nodes of the lattice update their states simultaneously in discrete time. All nodes of the lattice are subdivided into two groups: edge nodes and internal nodes.

All *internal nodes* behave in the following manner. A node calculates its next state depending on its current state and the states of its eight closest neighbours. The node switches from an excited state to the refractory state and from a refractory state to the rest state unconditionally, i.e. independently of the states of its neighbours. Being in the rest state the node takes an excited state only if the sum of its excited neighbours lies in the interval $[\theta_1, \theta_2]$, where $1 \leq \theta_1 \leq \theta_2 \leq 8$.

All *edge nodes* are excited only by light. At every step of the simulation an edge node x becomes excited when illuminated with the probability $p_r(x)$ which is determined by the position of the node in light rays relative to other edge nodes. That is, for every node x of $\mathbf{L}$ the light-dependent probability of excitation is calculated as $p_r(x) = \xi(1 - \zeta(x) - \epsilon)$, where $\xi(a) = a$ if $a > 0$, and $\xi(a) = 0$, otherwise; $\zeta(x) = (d(x) - m)/s$, where $m = \min_{z \in \mathbf{L}}\{d(z)\}$, $d(z)$ is the distance from the point source of light to the node z and s is the maximal distance between two nodes of the lattice $\mathbf{L}$; ϵ is a very small positive real number.

When a certain proportion of neighbouring edge nodes are illuminated they are excited by light and excite their neighbours; their neighbours excite their neighbours etc. Thus an excitation wave travels across the lattice from the site relatively closest to the source of light. The velocity vector of the wavefront is directed away from the light source. Thus if a node of the lattice wants to know where the light is at a given time step the node detects the maximum excitation density in its neighbourhood. The vector from the node to the maximum density indicates the relative position of the distant light source. So, every internal node of the lattice must be supplied with a local vector to indicate the relative position of the light.

Every node x of the lattice $\mathbf{L}$ is assigned a so-called vector state $v_x = (v_{ix}, v_{jx})$ consisting of two components v_{ix} and v_{jx}, which relate to the lattice coordinate axis i and j. Both components of the local vector state take values from the set $\{-1, 0, 1\}$. The vector states of the nodes are updated every step of the evolution by the following rules:

- $v_x^{t+1} = v_x^t$ if $x^{t+1} \neq +$
- $v_x^{t+1} = (\sigma(\sum_{y \in u(x): y^t = +}(x_i - y_i))\ \sigma(\sum_{y \in u(x): y^t = +}(x_j - y_j)))$
  $\sigma(a) = a/|a|$, if $x^{t+1} = +$.

At every time step the configuration of local vectors represents the distributed 'belief' of the lattice nodes in the direction towards the light source. How can we transform this configuration into robot movement?

Our aim is to show that light-generated excitation waves can be transformed into commands which can be sent to the robot motors, which subsequently orient the robot toward the source of light. Therefore we introduce the so-called global vector $V = (V_i, V_j)$. The vector represents the estimated direction of the light source with respect to the robot. It is updated at each step of the simulation as follows:

$$V^t = \left(\sigma\left(\sum_{x \in \mathbf{L}} v_{xi}^t\right), \sigma\left(\sum_{x \in \mathbf{L}} v_{xj}^t\right)\right)$$

where $\sigma(\cdot)$ acts as above. The components of the global vector are transformed into the rotation angles as shown in the following table.

		V_j^t		
		-1	0	1
	-1	$-\alpha$	-3α	-6α
V_i^t	0	0	0	18α
	1	α	3α	6α

For ease of use the rotations are represented as multiples of a basic unit of rotation—the angle α. In experiments α was in the range 0–30°.

The following algorithm may be employed in both virtual and real robots in some primitive experiments:

```
do
 Evolve Lattice and Calculate Local Vectors;
 Calculate Global Vector;
 Rotate Robot;
 Move robot at fixed distance;
until Robot Reaches Source of Light.
```

5.3.2 Activity patterns and vector fields

At every trial and for every set of parameters the initial distance between the robot, set at a random orientation, and the light source is measured. All further measurements are taken relative to that distance. The two main parameters of robot motion are the basic angle of rotation $\alpha = 0.5$ and the motion increment $\delta = 0.001$.

The spacetime dynamics of lattice controllers is fairly rich (see figures 5.7 and 5.8 for sample excitation patterns to sense the overall dynamics). For node sensitivity intervals of either $[1, 1]$ or $[1, 2]$, the excitation patterns are highly dispersed over the lattice with small clusters of excitation. When the node excitation interval is set at $[1, 3]$, long and thin 'primordial' waves spread coherently across the lattice.

When the upper bound of the excitation interval exceeds three, i.e. $\theta_1 = 1$ and $\theta_2 \geq 4$, typically centripetal circular and spiral waves of excitation are formed. The waves are generated repeatedly. This is because certain combinations of rest, refractory and excited states are formed in the early stages of the simulation when external nodes are excited for the first time. Such configurations of lattice node states are called generators. The generators become stable and some of them cannot even be destroyed by further excitation of edge nodes.

Lattices with the $[2, 2]$ interval of node excitation are extremely attractive because they possess several features of dynamical computational universality (see [26]). Their dynamics may be considered as complex because of their

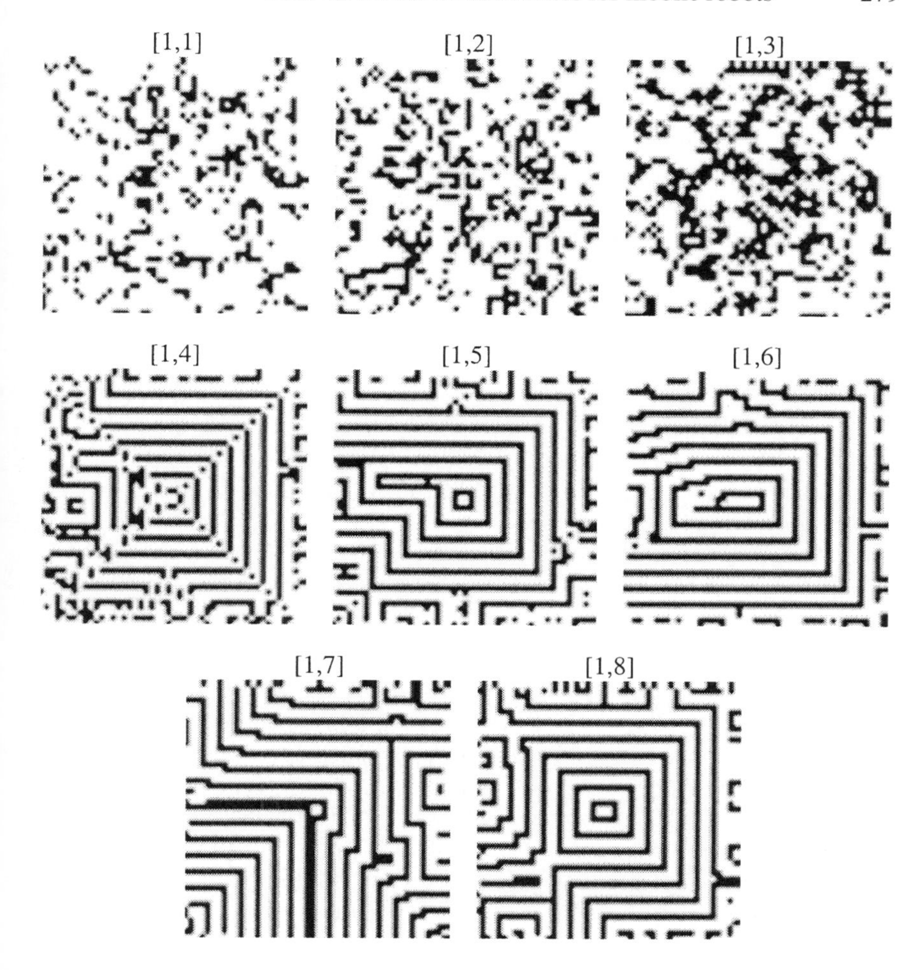

Figure 5.7. Typical excitation patterns in excitable-lattice controllers for $[\theta_1, \theta_2]$ excitation intervals with internal nodes: $\theta_1 = 1$ and $\theta_2 \in [1 \ldots 8]$. The excited nodes of the lattice are shown in black, other nodes are white. The patterns are recorded on the 200th step of the simulation.

sensitivity to initial conditions, long transient periods (in the absence of external stimuli) and bounded growth of excitation clusters. Particle-like waves, built from at least two or three excited states, spread all over the lattice and often collide with one another. The collisions produce other types of particle waves.

If we widen the node excitation interval to $[2, 3]$ we obtain waves of excitation with spacetime compositions which range between single particles to classical waves. In this case a collision between minimal particle waves may lead to the formation of relatively long strings of excitation.

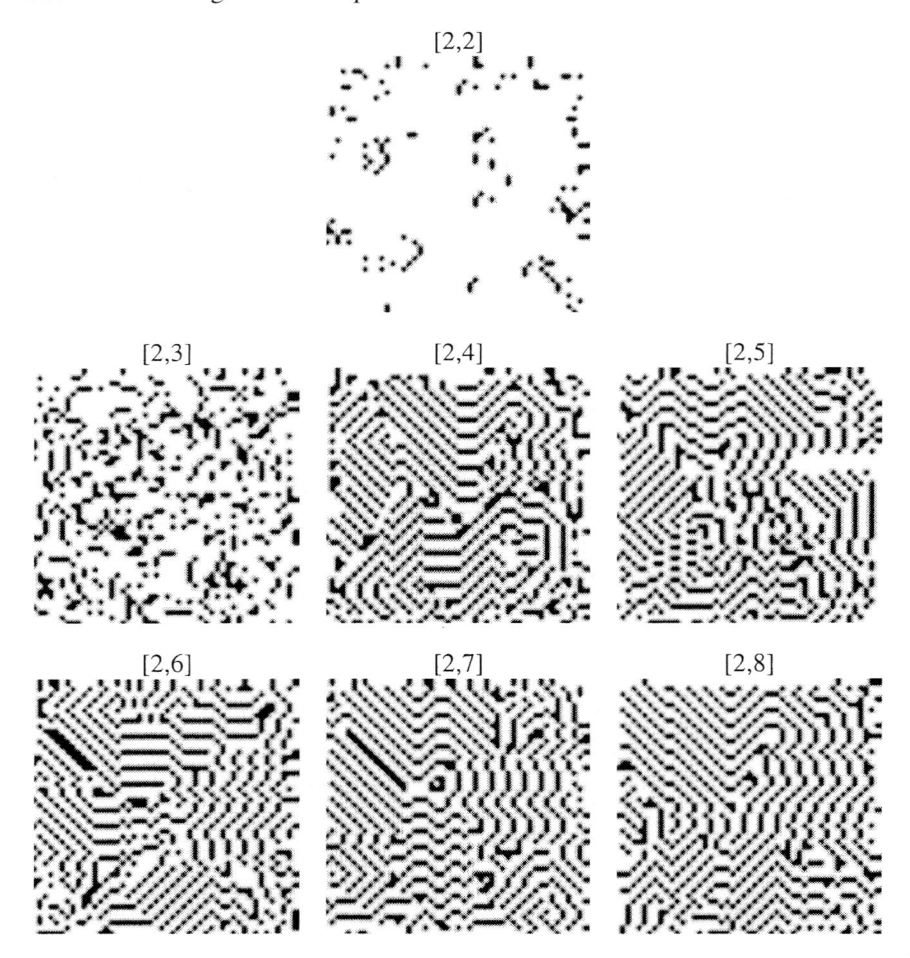

Figure 5.8. Typical excitation patterns in excitable-lattice controllers for $[\theta_1, \theta_2]$ excitation intervals of internal nodes: $\theta_1 = 2$ and $\theta_2 \in [2 \ldots 8]$. The excited nodes of the lattice are shown in black, other nodes are white. The patterns are recorded on the 200th simulation step.

As soon as θ_2 exceeds four labyrinthine excitation patterns are formed. The waves are usually centripetal; however, quite often newly formed wave generators emit short-range centrifugal waves.

It is quite obvious from observation of the excitation patterns that the resulting vector field reflects the excitation dynamics. For example, the excitation patterns shown in figures 5.7 and 5.8 correspond to the local vector configurations shown in figures 5.9 and 5.10 (the small vector arrows show the orientation),

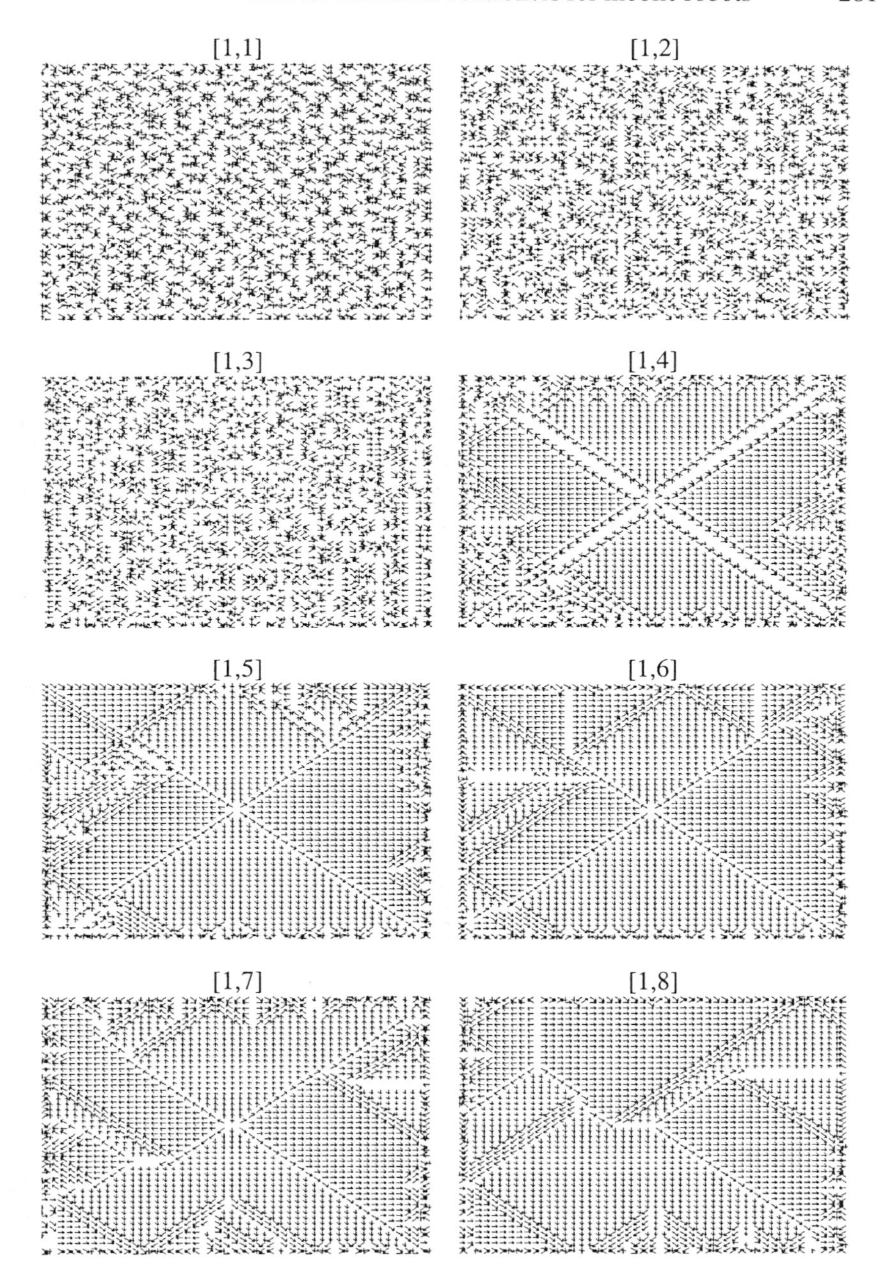

Figure 5.9. Typical vector 'fields' of lattice controllers for $[1, \theta_2]$ excitation intervals of internal nodes, $\theta_2 \in [1 \ldots 8]$. The vector 'fields' are recorded after the 200th simulation step to allow all lattice nodes to update their local vectors.

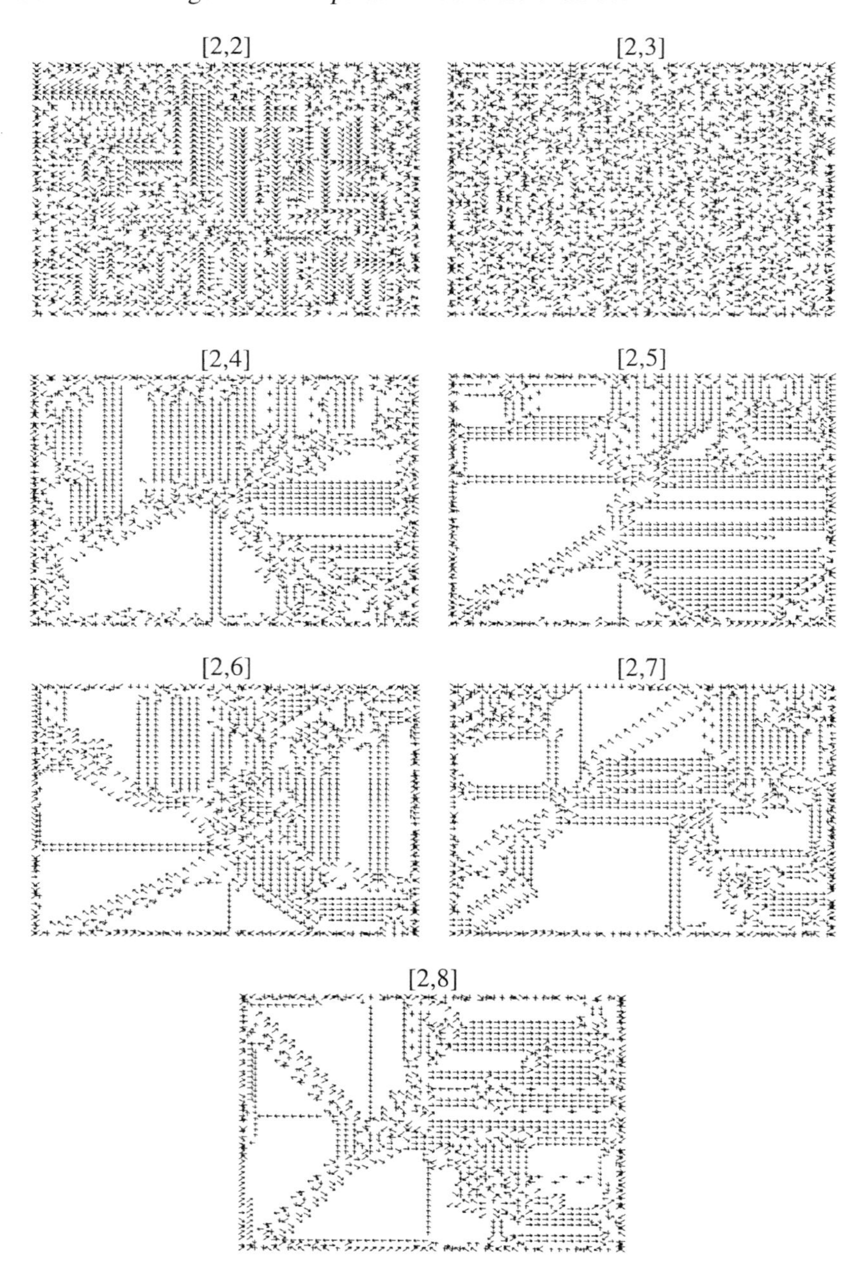

Figure 5.10. Typical vector 'fields' of lattice controllers for $[2, \theta_2]$ excitation intervals of internal nodes, $\theta_2 \in [2 \ldots 8]$. The vector 'fields' are recorded after the 200th simulation step to allow all lattice nodes to update their local vectors.

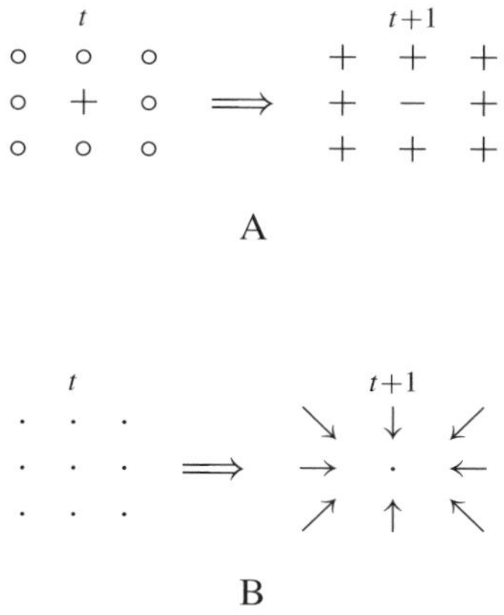

$$\begin{array}{ccc} & t & \\ \circ & \circ & \circ \\ \circ & + & \circ \\ \circ & \circ & \circ \end{array} \quad \Longrightarrow \quad \begin{array}{ccc} & t+1 & \\ + & + & + \\ + & - & + \\ + & + & + \end{array}$$

A

$$\begin{array}{ccc} & t & \\ \cdot & \cdot & \cdot \\ \cdot & \cdot & \cdot \\ \cdot & \cdot & \cdot \end{array} \quad \Longrightarrow \quad \begin{array}{ccc} & t+1 & \\ \searrow & \downarrow & \swarrow \\ \rightarrow & \cdot & \leftarrow \\ \nearrow & \uparrow & \nwarrow \end{array}$$

B

Figure 5.11. Example of the formation of local vector attractors when a single node excites all its neighbours: (A) excitation dynamic at two subsequent time steps; (B) orientation of the local vector.

which gives us a clear idea of what is happening in the lattice controllers.

Thus in lattices with a [1, 1] node excitation interval, where a node can be transformed from rest to an excited state if it has only one excited neighbour, small rectangular clumps can be generated (see figure 5.11).

Increasing θ_2 from one to eight enlarges the areas with the same local vector orientation. The lattice becomes virtually subdivided into homogeneous domains of collinear vectors. The boundaries between homogeneous domains are not necessarily fixed.

Local vector configurations generated in the evolution of lattices with a [2, 2] interval of node excitations remarkably combine chaotic subconfigurations with straight routes formed by co-oriented vectors. These straight routes are formed by the minimal particle waves that move along the columns and rows of the lattice as well as along the diagonals. In particle collisions some disordered patterns can be formed. This fact is reflected in the chaotic subconfigurations which may be found on the intersections of the particle traces. Increasing the interval from [2, 2] to [2, 3] gives an increased probability of node excitation.

Lattices with the node excitation intervals [2, 4], [2, 5], [2, 6], [2, 7] and [2, 8] have sufficient ‘attracting’ neighbours in positions which cancel out their individual effects so that a null local vector is often generated. This often creates a sufficient number of null local vectors which can be seen as blank regions in figure 5.9. Since the lattice, therefore, has a small number of oriented vectors (not null vectors) the global vector may be considered to be sensitive to an extreme change in orientation which can affect the robot trajectory. However, as we will see later, this is not always the case.

5.3.3 Taxonomy of robot trajectories

What happens when we release a robot into a simulated space and switch on a light source?

The edge nodes of the lattice controllers are excited, excitation patterns move inwards; their movements modify the local vectors of the nodes they pass through. The modification of the local vectors leads to certain changes in the orientation of the global vector which then results in rotations of the robot. Persistent increments or decrements of the angle of the global vector, in the absolute coordinate system, is seen as cycloidal rotation of the robot. In contrast, small adjustments of the global vector may result in a more direct trajectory toward the light source.

Several typical types of robot trajectories are shown in figures 5.12 and 5.13. These sample trajectories are provided to give the reader an overall impression of the modes of excitation-induced robot motion. The following basic types of robot locomotive behaviour are highlighted: graceful motion, pirouette motion and cycloidal motion [28].

(1) *Graceful motion.* Initially the robot positions itself toward the source of light and then makes full rotations around its centre no longer. The trajectory is quite straight with only a few turns. The robots controlled by lattices with node excitation intervals [1, 1] and [2, 3] provide the best examples. If controllers have cell excitation intervals [1, 2] and [2, 2] the robot behaviour still lies in this class; however, the trajectories are not so direct. Sometimes a robot can circle around the target when it is in the vicinity of the light source (e.g. [2, 2); see figures 5.12, [1, 1], [1, 2], and 5.13, [2, 2], [2, 3].
(2) *Pirouette motion.* This class only includes robots employing a [1, 3] node excitation interval in the lattice controller. A robot reaches the target quite directly, often only two times slower than the robots from the class of graceful motion. However, they may exhibit repetitive patterns of cyclic motion around the light source. The radii of such cycles can vary extensively. Often the robots take a very long time to actually make contact with the target; see, e.g., figure 5.12, [1, 3].
(3) *Cycloidal motion.* The class of cycloidal motion includes robots in which lattice controllers employ the node excitation interval $[\theta_1, \theta_2]$, where θ_1 is as usual and $\theta_2 \geq 4$. The outcomes of robot evolution can be classified onto two groups. A robot either eventually hits the target or find itself locked into a closed trajectory which prevents it from ever reaching the target.

In the first case the robot approaches the target by implementing many subsequent cycles, as shown in figures 5.12, [1, 4], [1, 5], [1, 8], and 5.13, [2, 4], [2, 6], [2, 7]. Finally the robot hits the target. In the second case the robot executes a circular stable motion. This lock-in can be explained by the creation of stable generators of excitation waves which are formed at the early stages of evolution. These generators can be destroyed by the excitation waves from the edge nodes excited by light. Therefore the robot does not react to light at all. Figure 5.12,

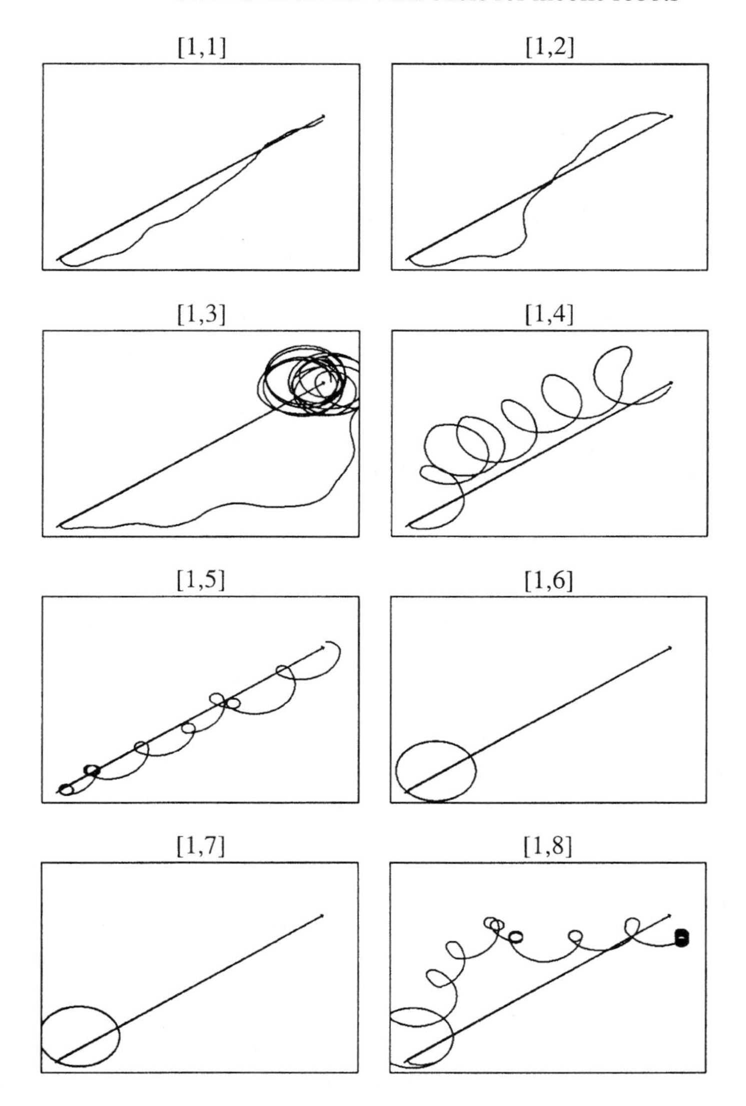

Figure 5.12. Typical trajectories of a robot controlled by two-dimensional excitable lattices with $[\theta_1, \theta_2]$ excitation intervals of internal nodes: $\theta_1 = 1$ and $\theta_2 \in [1 \ldots 8]$. The initial position of the robot is connected by the arrow to the position of the light source. The controllers do not have any defects or noise and $\epsilon = 0.1$.

[1, 6] shows a snapshot of the cyclic rings of excitation expanding from the centre of the lattice. This has the effect of suppressing any waves created by external

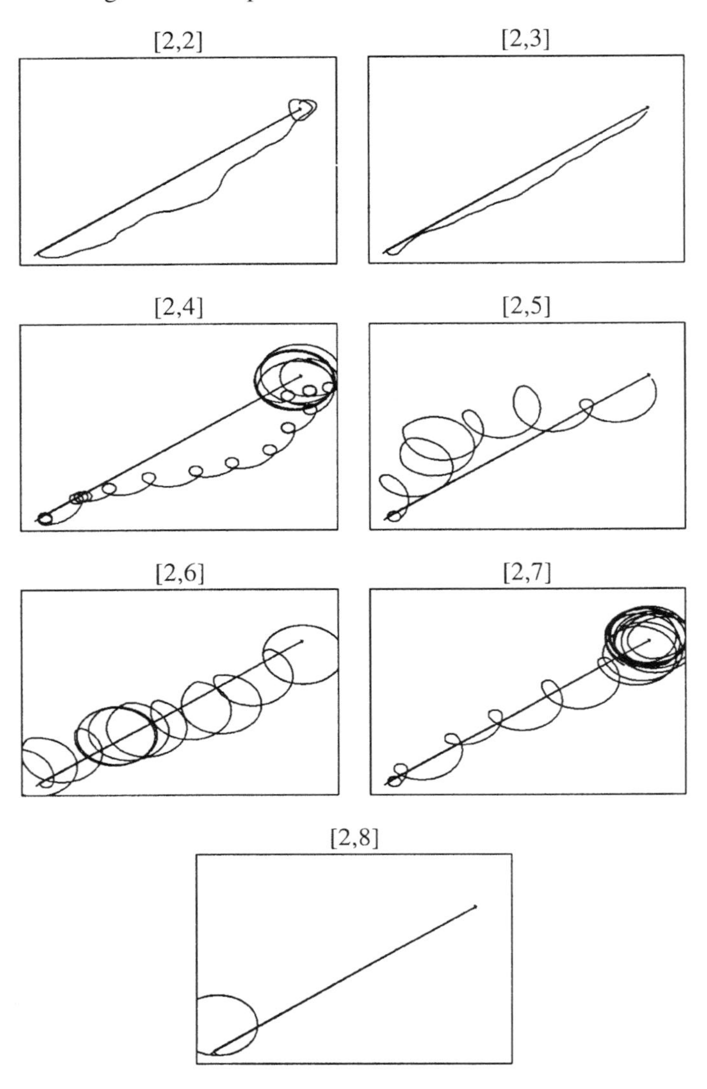

Figure 5.13. Typical trajectories of a robot controlled by two-dimensional excitable lattices with $[\theta_1, \theta_2]$ excitation intervals of internal nodes: $\theta_1 = 1$ and $\theta_2 \in [2 \ldots 8]$. The initial position of the robot is connected by an arrow to the position of the light source. The controllers do not have any defects or noise and $\epsilon = 0.1$.

stimulation at the boundaries of the lattice. Three excellent examples of cyclic behaviour are shown in figure 5.12, [1, 6], [1, 7], and 5.13, [2, 8].

5.3.4 Robot's performance and excitation intervals

Time-dependency relationships, expressed in time-to-target, are explored in simulations. For each node excitation interval 10 trials are conducted. In each trial the simulated robot is placed one unit from the light source at a random orientation. Figures 5.14 and 5.15 illustrate the time–distance graphs for each of the 10 trials. They show that there is a large variance in the time-to-target distance. However, the general trajectory trends are easily recognizable. A trajectory of the robot in two-dimensional space depends sufficiently on a chain of external and internal events which is initiated on the edges of the lattice as well as internally inside the lattice. It should be noted that the initial few steps, in terms of such parameters as orientation, excitation on the edges and time steps of the experiment temporal coincidence of the events with physical orientation of the robot (and controller, respectively) relatively to the source of light also plays an important role.

Our main result states that

Robots in which the controllers have node excitation intervals [1, 1], [1, 2], [2, 2] *and* [2, 3] *are the best performers in chasing a light target.*

5.3.5 Fault tolerance of interval excitation lattice controllers

When dealing with distributed controllers it is worth considering how good excitable lattice are when they are partly destroyed or are subject to arbitrary external influences unrelated to the task in hand. We consider only two types of lattice-controller defects: (i) silent defects, when some nodes of the lattice are removed initially with probability p_s; and (ii) noisy defects, when some nodes of the lattice are assigned 'spontaneous excitation' status at the beginning of simulation with probability p_n.

Once a node is assigned a defective state, it remains defective throughout the trial. The behaviour of the defective nodes reflects their nicknames: silent nodes (or absent nodes) do not react on excitation of their neighbours and stay in the rest state throughout the simulation. The noisy nodes burst into excitation independently of the states of their neighbours. A silent node simply remains in the rest state independently of an excitation in its neighbourhood.

The behaviour of a robot with a defective controller, in terms of finding the target, is considered unreliable if the time taken to reach the target is several orders of magnitude greater than for a robot with no defects in its lattice controller. Ten simulation trials were conducted, as described earlier, for the following node excitation interval: [1, 1], [1, 2], [2, 2] and [2, 3]. It was found that the best performers are reliable when either the probability of silent defects is less than 0.05 or the probability of noisy defects is less than 0.02.

In general, regarding the size of lattice controllers (50×50; 2500 nodes) the probabilistic thresholds 0.05 (for a silent defect) and 0.02 (for a noisy defect) of fault tolerance mean that there should be no more than 125 silent defects or

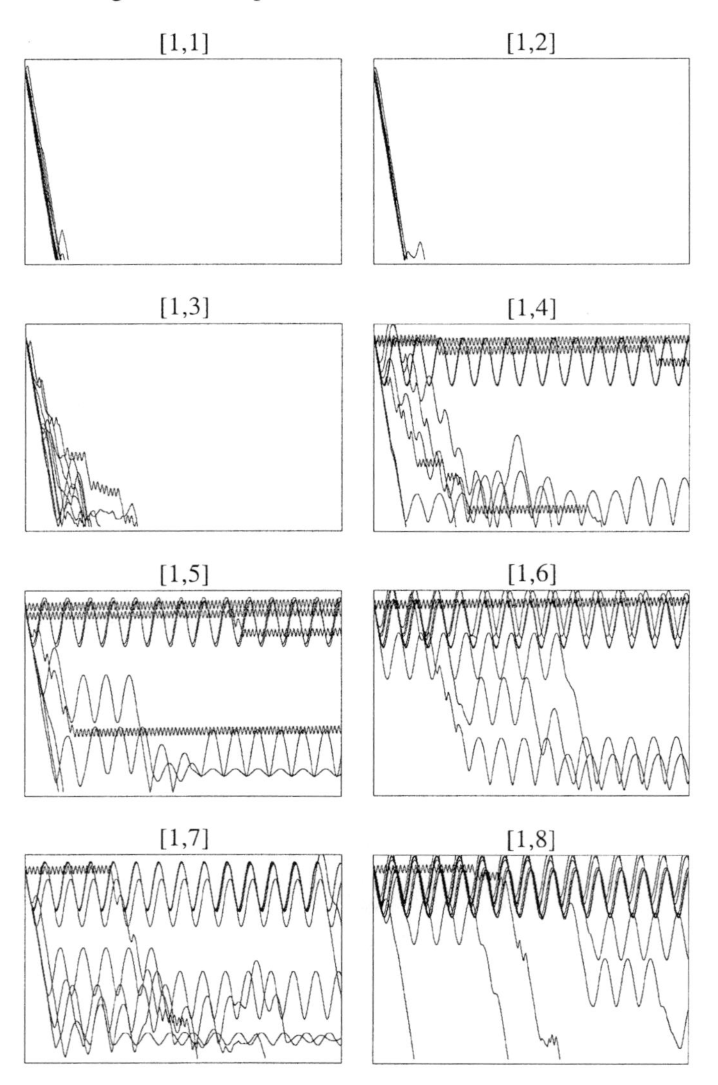

Figure 5.14. Distance-to-target (vertical axis) versus time (horizontal axis) diagrams merged in 10 trials. The virtual robots are controlled by two-dimensional excitable lattices with $[\theta_1, \theta_2]$ excitation intervals of internal nodes: $\theta_1 = 1$ and $\theta_2 \in [1, \ldots, 8]$. The controllers have no defects or noise and $\epsilon = 0.1$. Every 'distance–time' box has size 500×10000 units, where distance is real valued and time is discrete.

50 noisy defects in an effective lattice controller. The probabilities 0.05 and 0.02 are guaranteed thresholds of reliability for any interval of [1, 1], [1, 2], [2, 2] and

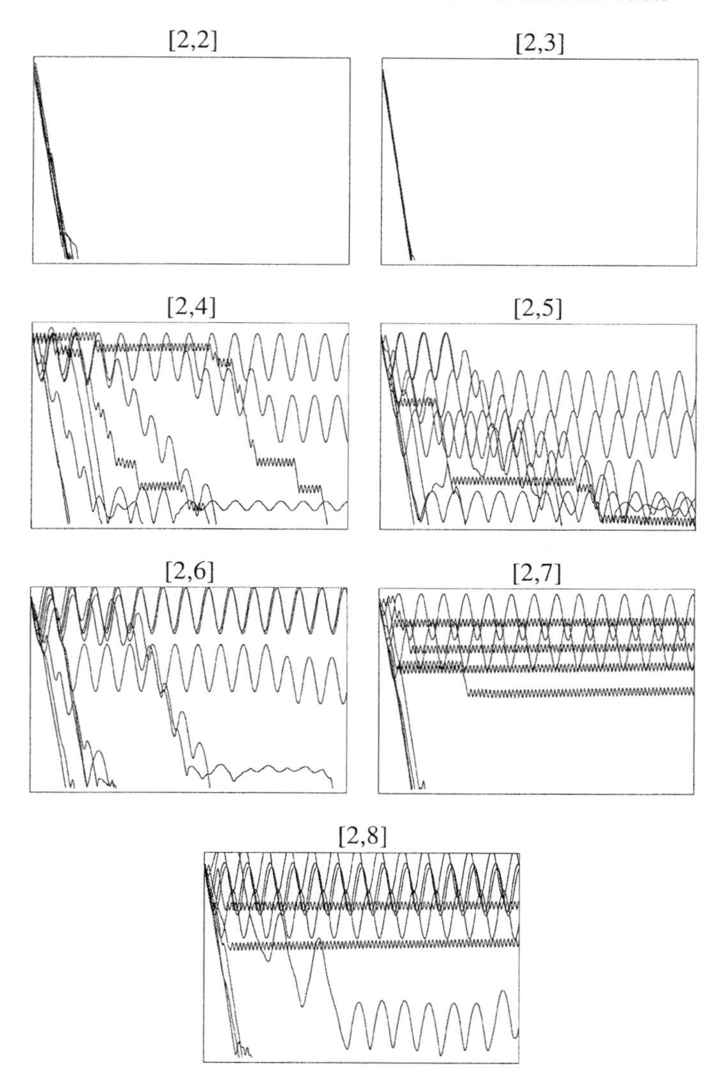

Figure 5.15. Distance-to-target (vertical axis) versus time (horizontal axis) diagrams merged in 10 trials. The virtual robots are controlled by two-dimensional excitable lattices with $[\theta_1, \theta_2]$ excitation intervals of internal nodes: $\theta_1 = 2$ and $\theta_2 \in [2, \ldots, 8]$. The controllers have no defects or noise and $\epsilon = 0.1$. Every 'distance–time' box has size $500 \times 10\,000$ units, where distance is real valued and time is discrete.

[2, 3]. However, we may expect that precise thresholds of reliability will vary from one interval to another.

The most robust controller with respect to noisy defects uses a node excitation interval of [2, 3]. *The most robust controller with respect to silent defects uses a node excitation interval* [1, 1]. *Their thresholds of reliability are 0.07 and almost 0.2, respectively.*

5.3.6 Simulated controllers in real robots

In laboratory experiments we use a modified mobile robot in an experimental environment which is intentionally designed to investigate the principles of the collective behaviour of minimalist mobile entities (see e.g. [447]). The robots are small in size (approximately 23 cm in diameter), portable, able to turn on the spot and carry sensors. An excitable lattice controller was simulated on a Motorola 68332 on-board processor, which is programmed in C. In the experiments a robot carries two optical sensors on the front and one optical sensor on the back. The robot runs in a large arena which is 1760 times the area of the robot.

Ideally, every edge node of the lattice **L** should be supplied with its own sensor. This is a bit complicated technically. Therefore, to verify simply the principles of excitation-based navigation we restricted the robot to only three light sensors: a left-hand sensor, a right-hand sensor and a rear sensor. The connection between the sensors and edge nodes is reduced to a simple one: the left-hand sensor influences the left-hand part of the front edge of the lattice **L**, the right-hand sensor excites the right-hand part of the lattice's front edge, and the rear sensor sends excitation signals to the rear edge of the lattice. The basic scheme is shown in figure 5.16.

In experiments with real robots we demonstrated that excitable lattice controllers can be employed for practical tasks [28]. The behaviour of the real robots matched the behaviour of their simulated counterparts.

Here we present only sample trajectories of the robot with an excitable lattice controller employing node excitation interval [2, 2]; all other parameters are the same as the parameters used in computer experiments. The initial configuration of the experimental setup is shown in figure 5.17. At the beginning the robot performs a typical U-turn, often observed in computer experiments. After that it moves towards the light target. Several serial snapshots of the robot moving towards light are shown in figure 5.18; the robot trajectories are also indicated there.

Finally we see (figure 5.19) that even at the macro-level an excitable lattice controller can be successfully used as a specialized processor for phototaxis.

5.3.7 Real-life candidates for excitable controllers

In computer and laboratory experiments we have demonstrated that the phototactic behaviour of both virtual and real robots can be implemented with the help of excitable light-sensitive lattices. We also found that excitation domains

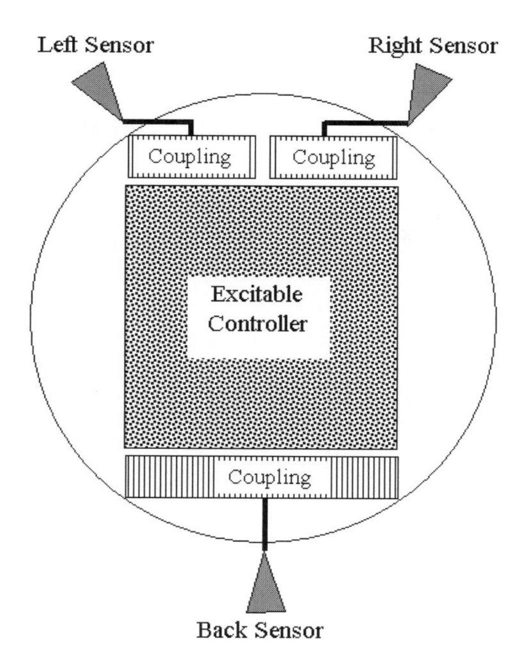

Figure 5.16. Basic diagram integrating the edges of an excitable lattice with light sensors.

rule the space which makes the corresponding robots good performers. Are there any natural materials that can be used to fabricate real-life controllers?

At least two groups of appropriate candidates might be considered. The first group may consist of active chemical media. Light-sensitive modification of the Belousov–Zhabotinsky reaction, particularly its printed realization [581], is probably the most obvious candidate. To couple excitation waves in a Belousov–Zhabotinsky reaction with actuators we can employ the fact that the reaction may participate in a nonlinear polymerization reaction [213]. Ideally, by coupling oscillations in a reaction layer with polymerization of a substrate we can obtain some kind of undulating motions in the substrate film, thus making it swim.

The implementation of the Belousov–Zhabotinsky reaction at the nano-level is somewhat dubious. However, this form of reaction may give us some indication of the direction in which to search for light-sensitive planar reactors.

Molecular two-dimensional arrays form another group of prospective candidates. Perhaps the best example is Scheibe aggregates manufactured by the Langumir–Blodgett technique. Such molecular arrays support mobile self-localized excitations, called excitons (see e.g. [463]). Excitons are induced by light and travel along monomolecular arrays. Remarkably Scheibe aggregates

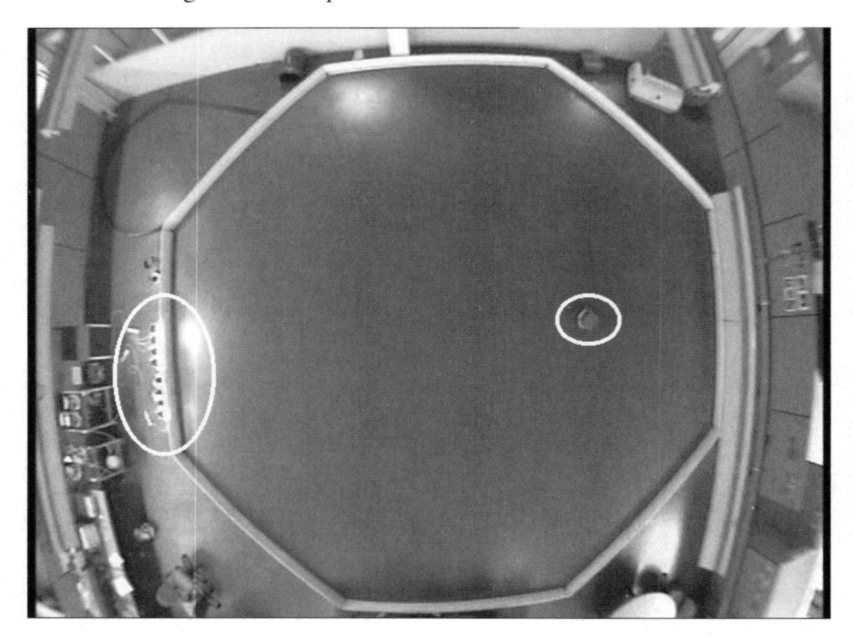

Figure 5.17. Initial configuration of an experiment with a real robot. Several lamps are put in a line to form a light source (ellipse on the left). The robot is placed at some distance and oriented away from the light source (circle on the right).

are analogous to excitable lattices with the node excitation interval $[\theta_1, \theta_2]$ where $\theta_1 = 2$ and $\theta_2 = 2, 3$.

Let us imagine that every molecule of the monomolecular array could incorporate an actuator. Then the distributed controller for the distributed robot may be designed from the monomolecular array in figure 5.20). The arrays of propulsive actuators may look like arrays of cilia of *Paramecium*. We assume the molecules are excited by photons. When the edge molecules become excited they transmit excitation/energy to the internal molecules of the array. We can restrict an actuator's position so that it takes no more than eight possible orientations, i.e. orientations corresponding to the eight neighbours of the molecule. If the actuator positions itself away from the direction from which the excitation arrives, then the actuator could produce a local propulsive force. The combined propulsive forces of the array actuators give a robot an overall propulsive vector toward the light source. In other words, we simply substitute local vectors by real actuators. In fact, the interaction between the excitation patterns and actuators is more complicated than we could imagine. However, it is reasonably correct to say that the interaction between the local forces generated by each actuator and the environment implicitly create a form of vector integration which causes the rotation and translation motion in the robot.

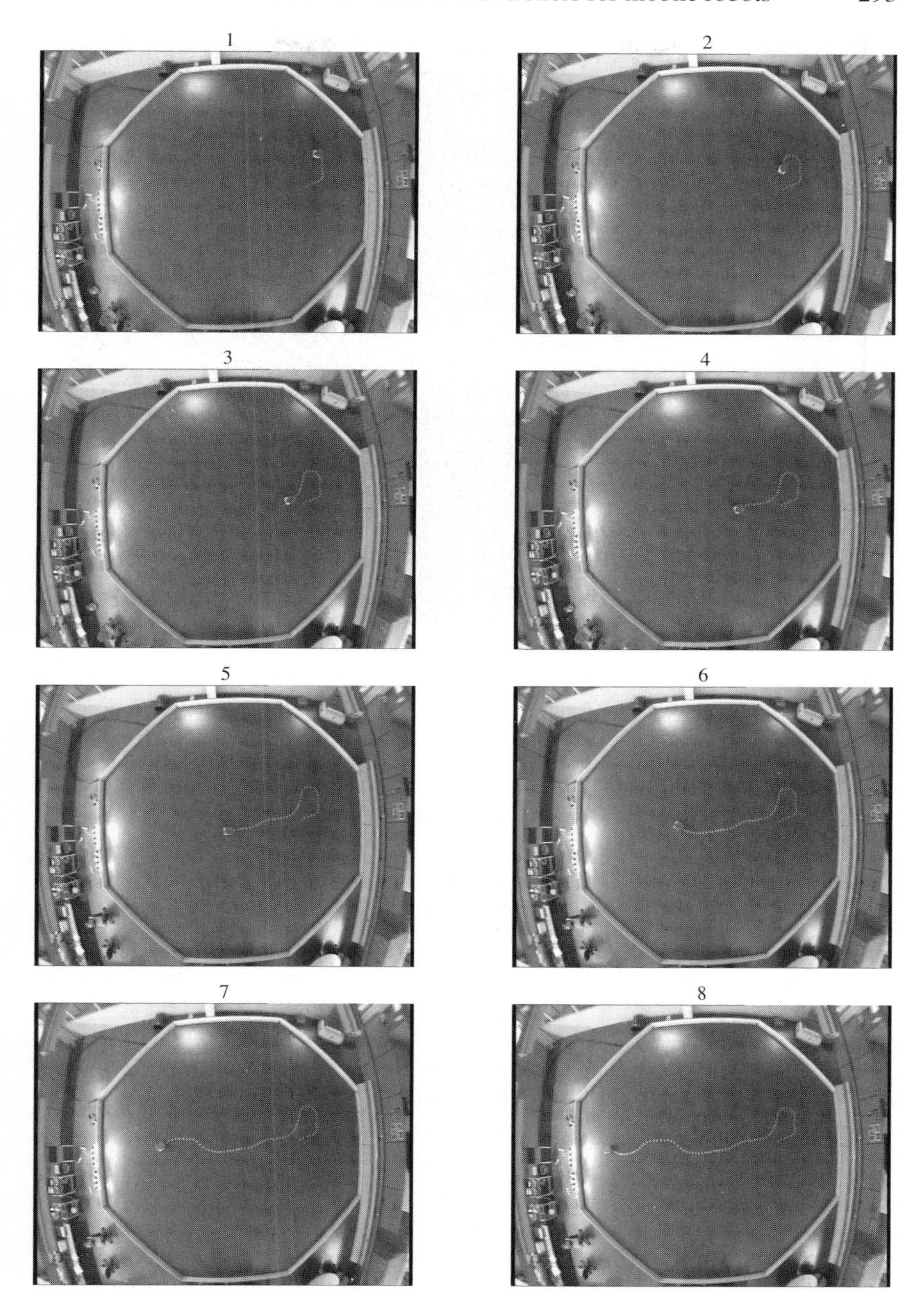

Figure 5.18. Snapshots of a real mobile robot and its trajectory. The robot is guided toward the light target by an onboard excitable controller.

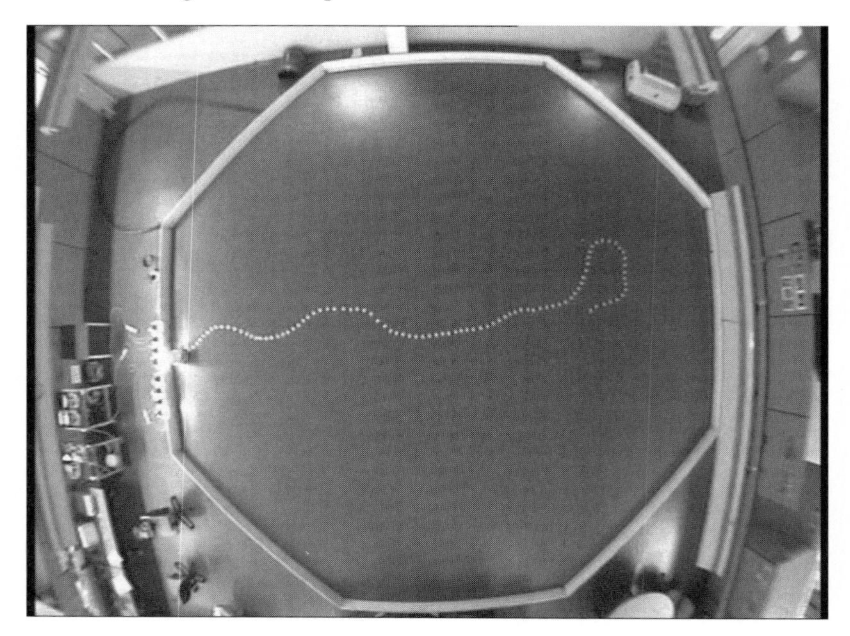

Figure 5.19. A trajectory of the robot obtained in the real experiments. The trajectory starts on the left and finishes near the light source on the right.

If we were to fabricate distributed robots with the graceful type of motion we should choose an underlying molecular array that supports self-localized mobile excitation. In this case Scheibe aggregates again give us the best opportunity. A real single self-localized mobile excitation in a Scheibe aggregate has a diameter of 30 Å. Then we can expect the minimal size of the two-dimensional controller to be approximately 0.2 μm by 0.2 μm.

Luckily, we could also mention here a living analogue of cilia arrays, which exhibit a positive taxis. This is a phylum *Ciliophora*, or ciliates—protists, the bodies of which are covered with cilia. We could refer to *Paramecium caudatum* (figure 5.21) as a typical example [418, 639]. Each cilium, attached to a membrane of a protist, sweeps with a power stroke in the direction opposite to the intended direction of organism movement. The cilia are usually arranged in rows; this arrangement forms a longitudinal axis of beating. Cilia beat coherently in waves and propel the individual forward. The physical integration of the strokes of many cilia propel the protist in the opposite direction to the direction of beating.

There are still limited data on the coordination of cilia. Control may be implemented by either information transfer via a submembrane network of microtubules or coordinated contractions of the membrane travelling along the protist's body. In any case, we can speculate about direct analogies between the waves of cilia beating, caused by excitation waves of the lattice, and waves of co-

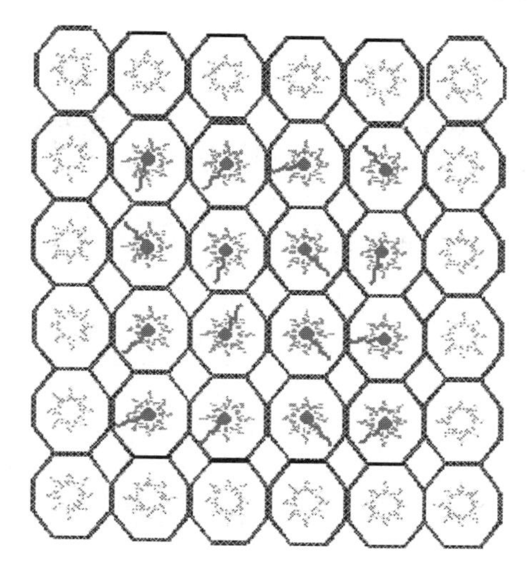

Figure 5.20. A scheme for an excitable molecular array with incorporated actuators. Molecules exchange excitation locally. Actuators are shown explicitly.

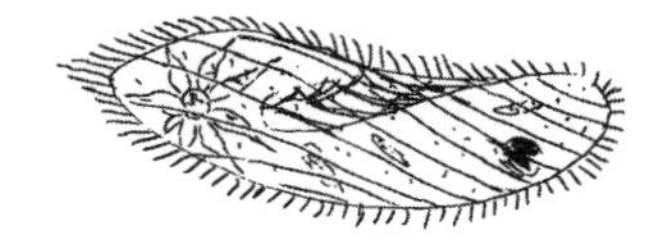

Figure 5.21. *Paramecium caudatum.*

beating of cilia in protists, that move from the front to the rear of the organism.

The absence of any directional photoreceptors in ciliates [416] is even a closer match with our theoretical models. Thus, for example, it has been proved in real-life experiments [417, 416] that only the frequency of directional changes together with the frequency distribution of the angle of directional changes are affected by light; the speed is completely unaffected. This was an intuitive assumption in our model of mobile excitable lattices. So, both excitable lattices and real protists are oriented by steering towards the direction of maximal light intensity [416].

The rest of the chapter deals with the phenomenology of the vector excitation rules defined in section 5.1.3.

5.4 Lattices with vector excitation rules

When investigating the spacetime dynamics of two-dimensional excitable lattices with vector excitation rules we employed several characteristics to classify lattices with various specific cell state transitions rules. That is, for each excitation rule

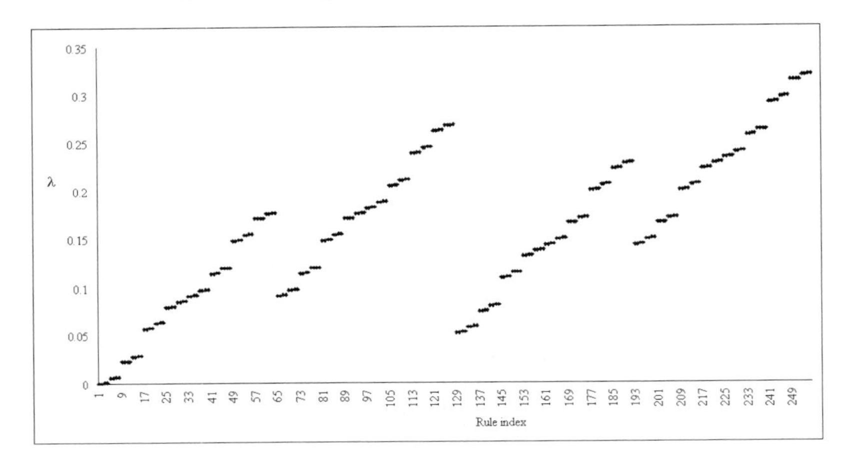

Figure 5.22. Values of the λ^+-parameter for the excitation rules.

we calculated a static parameter, which represented a unique attribute of the rule; for each set of evolution trials we calculated the dynamical and morphological characteristics of the excitation patterns.

5.4.1 λ-like parameter of excitation rules

For every function f we consider a static parameter

$$\lambda^+ = \frac{\sum_{w \in \mathbf{Q}^8} \chi(f(w), +)}{3^8}. \tag{5.1}$$

Given excitation rule f the parameter $\lambda^+(f)$ indicates the ratio of cell neighbourhood configurations (every configuration is represented as a string of eight symbols of the ternary alphabet) which may excite the central cell (owner of the neighbourhood).

This parameter is inspired by the λ-parameter, which was developed in [382] to link the intrinsic characteristics of cell state transition rules with the spatio-temporal behaviour of certain classes of cellular automata. In spite of some uncertainties about its applicability in some contexts, the λ-parameter remains a useful way to predict the behaviour likely to be produced by various cellular-automata rules, and has been influential in the development of cellular-automata theory. In our case the λ^+-parameter can be readily interpreted as the intrinsic probability of excitation. Furthermore, since only a cell at rest can become excited, we have $0 \leq \lambda^+ \leq \frac{1}{3}$ (figure 5.22).

The distribution of the 'number of functions' mapping each value of the parameter (5) is shown in figure 5.23.

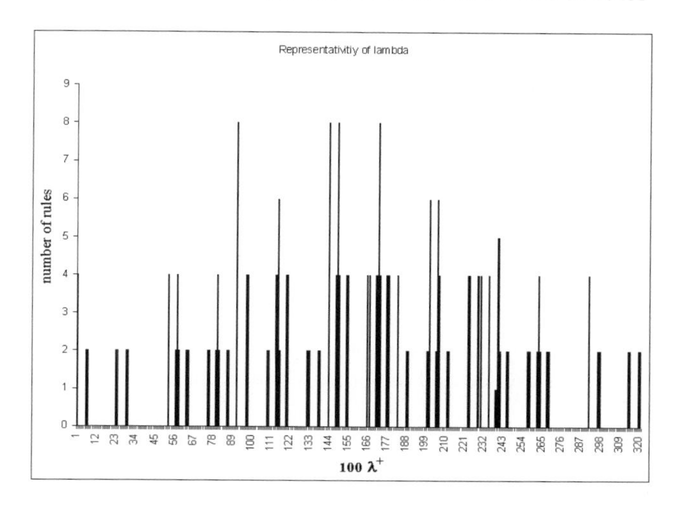

Figure 5.23. Distribution of excitation rules with values of λ^+.

5.4.2 Initial conditions

To investigate the behaviour of excitable lattices under different initial conditions we either excite every lattice cell with some fixed probability p, or assign to every cell one of three states from $\mathbf{Q}$ with probability $p = \frac{1}{3}$. The spacetime dynamics of an excitable lattice is usually analysed in absorbing and periodic boundary conditions.

To classify the lattices with different excitation rules we use both the morphological and dynamic characteristics: number and size of clusters, activity level and transient period.

5.4.3 Characteristics of the global dynamic

Let $A_{[t_i,t_j]}$ be an interval activity level of a two-dimensional excitable lattice measured in the time interval $[t_i, t_j]$:

$$A_{[t_i,t_j]} = \frac{\sum_{x\in\mathbf{Z}^2;t\in[t_i,t_j]} \chi(x^t, +)}{|\mathbf{Z}^2| \cdot |t_i - t_j|}.$$

An *activity level*, which can also be thought of as the extrinsic probability of excitation, for the function f is defined as $\alpha = A_{[t_1,t_2]}$, where $t_2 > t_1 \gg \tau$ and $|t_2 - t_1|$ is sufficiently large.

A *transient period* is computed as an integer τ such that for any $t \geq \tau$ the trajectory of the model in global state space spins on a fixed point or cycle, i.e. for any $t_4 > t_3 > t_2 > t_1 \geq \tau$ and for very small ϵ we have

$$|A_{[t_1,t_2]} - A_{[t_2,t_3]}| \leq \epsilon.$$

It is not difficult to measure this difference over large time intervals, larger than any oscillation period.

5.4.4 Morphological characteristics

We did not resist the temptation to describe the lattice configurations using such subjective phrases as 'homogeneous', 'heterogeneous', 'finely grained', 'coarsely grained' etc. To support the verbal descriptions with some objective numerical evidence, we also counted the numbers and sizes of clusters of excited states in the configurations far beyond the transient period.

A *cluster* of excited states is defined as the connected subset of the lattice, every cell of which is excited and has at least one excited neighbour, i.e. $\eta \subset \mathbf{L}$ is a cluster at time t if

$$\forall x \in \eta : (x^t = +) \wedge (\exists y \in u(x) : y^t = +).$$

We derive the following measures from analysis of the clusters present in the configurations:

- *number of clusters* κ: $\kappa = |\mathbf{K}|$, where $\mathbf{K} = \{\eta \subset \mathbf{L} : \eta \text{ is a cluster}\}$ and $\forall \eta', \eta'' \in \mathbf{K}, \eta' \neq \eta'' : \eta' \cap \eta'' = \emptyset$
- *maximum cluster size* $\omega_{\max}$: $\omega_{\max} = \max_{\eta \in \mathbf{K}} |\eta|$;
- *typical cluster size* ω: $\omega : d_\omega = \max_{1 \leq i \leq |\mathbf{L}|} d_i$, where $d_i = |\{\eta : |\eta| = i\}|$, $1 \leq i \leq |\mathbf{L}|$;
- *size diversity of clusters* υ: $\upsilon = \sum_{1 \leq i \leq |\mathbf{L}|} sg(d_i)$, $sg(a) = 1$ if $a > 0$ and is zero otherwise.

Another useful measure is derived from the concept of richness, and relates to what are called complete configurations of cellular automata. The configuration of a cellular automaton is called complete if it contains all the possible states of the cell neighbourhood. We can reconstruct, or identify, the cell state transition function of an arbitrary cellular automaton from a given configuration only if the previous configuration is complete [9].

In the excitable lattice model, a cell calculates its next state not from the configuration of its neighbourhood but from an integral characteristic to it—the number of excited neighbours.

We therefore say cell x is r-rich, $r = 1, 2, \ldots$, at time step t if its neighbours from the neighbourhood $u_r(x) = \{y \in \mathbf{L} : |x - y| \leq r\}$ have all possible sums of their excited neighbours; that is, cell x is r-rich at time step t if

$$\left| \bigcup_{y \in u_r(x)} \left\{ \sum_{z \in u(y)} \chi(z^t, +) \right\} \right| = 9.$$

The *richness* ρ of an excitation rule is defined as the number of r-rich cells in the configurations far from the transient time at time $t \gg \tau$, $r = 1, 2, \ldots$. Realistic values of r will be discussed later.

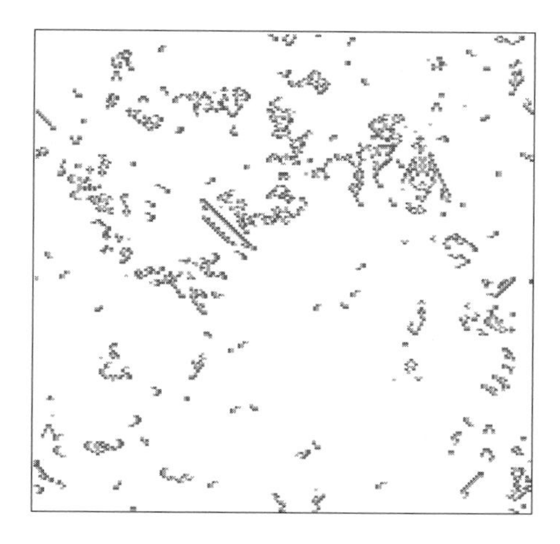

Figure 5.24. Typical configuration of an excitable lattice from the 2^+-class.

5.4.5 Morphological classification

After examining the structure of configurations far beyond the transient period, we conclude that all functions can be assigned to one of 11 morphological classes:

0: any initial configuration evolves to a state of uniform rest;

2^+: in the case of periodic boundaries and any initial conditions the lattice evolves into configurations comprising particle-like waves or localized excitations, which travel around the lattice, collide with each other, and generate new moving patterns as a result of collisions (figure 5.24);

H: any initial conditions produce a homogeneous activity pattern (figure 5.25);

F: fingerprint-like patterns appear in the configurations together with a small number of wave generators (figure 5.26);

L: labyrinth-like patterns emerge with a large number of wave generators (figure 5.27);

GW: configurations far from the transient period are characterized by islands of strip-like waves surrounded by almost homogeneous patterns (figure 5.28);

FW: 'focused waves' appear;

CGFG: the members of the class exhibit a transition from coarsely grained to finely grained patterns (figure 5.29);

HCGFG: the members of the class exhibit a transition from more coarsely grained to finely grained patterns (figure 5.30);

CGSW: the members of the class pass from coarsely grained patterns to sawtooth wave patterns (figure 5.31);

SW: the lattice behaviour ranges from disordered wave patterns to the formation of spiral waves (figure 5.32).

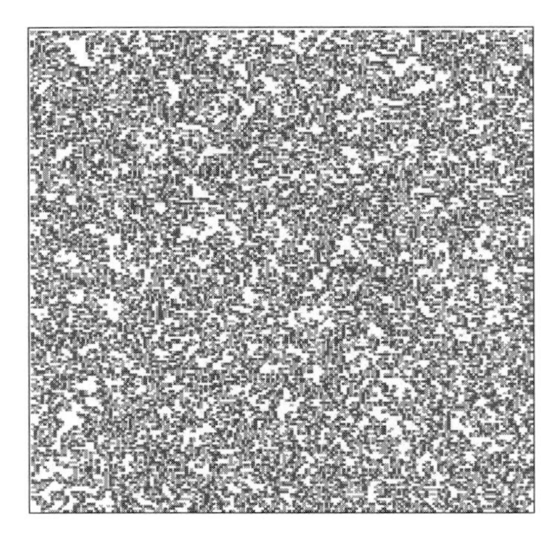

Figure 5.25. Typical configuration of an excitable lattice from the H-class.

Figure 5.26. Typical configuration of an excitable lattice from the F-class.

The class memberships are characterized in table 5.1. Typical 'space + time' configurations of excitable lattices with representative excitation rules from most classes are shown in figures 5.33–5.37. These activity patterns are calculated by the following standard procedure. Each cell of a 30×30 cell lattice with absorbing boundaries in an $\mathbf{L} \times \mathbf{T} = (x_{ij}^t)_{1 \leq i,j \leq 30}$, $t = 0, 1, 2, \ldots, 30$ space was initially excited with probability $p = \frac{1}{3}$. The localized excitation patterns in the 2^+-class can be clearly seen in figure 5.33. The formation of wave generators and the filling of the lattice with fingerprint-like patterns of waves is an attribute of the F-class (figure 5.34). The spacetime homogeneity of the evolution of the H-class model

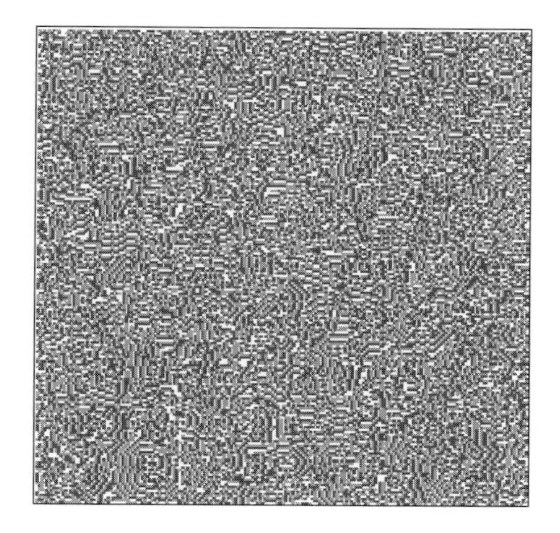

Figure 5.27. Typical configuration of an excitable lattice from the L-class.

Figure 5.28. Typical configuration of an excitable lattice from the GW-class.

is clearly visible in figure 5.35. Figure 5.36 reveals the formation of sawtooth-like waves in SW lattices. The mechanism for the generation of focused waves (FW-class) is revealed in figure 5.37.

We can generalize the results by merging the observed classes into four typical groups:

(1) *Damping group*: class 0
(2) *Wave group*: classes L, F, SW and FW
(3) *Particle group*: The 2^+-class
(4) *Transitional group*: classes CGFG, HCGFG and CGSW

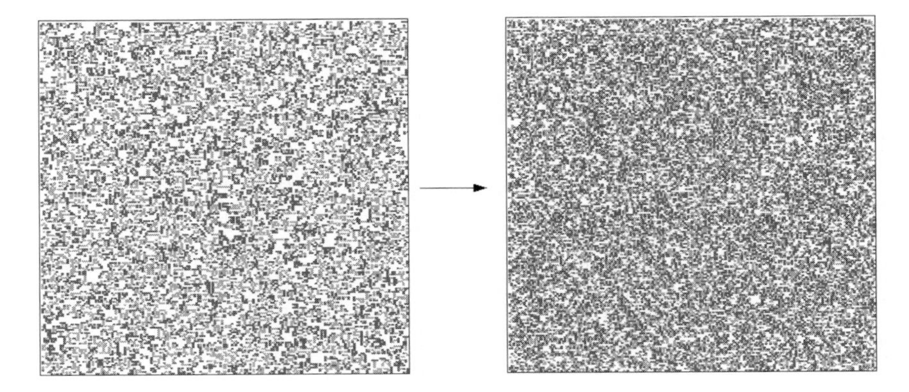

Figure 5.29. Typical configurations of an excitable lattice from the CGFG-class. The left-hand configuration presents more coarsely grained patterns and the right-hand the most finely grained ones.

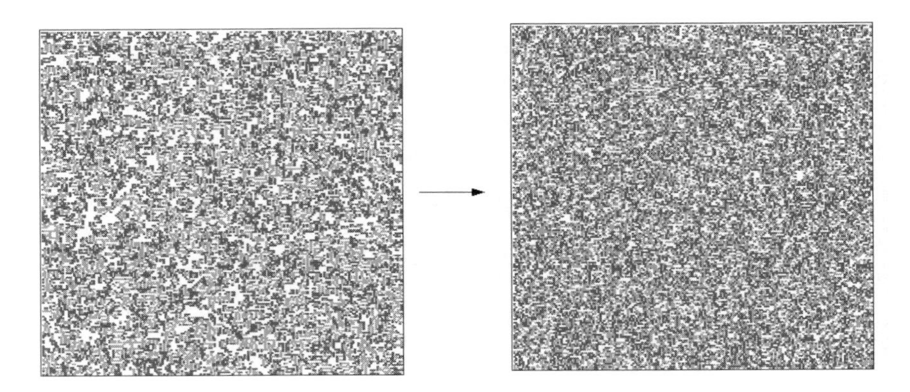

Figure 5.30. Typical configurations of an excitable lattice from the HCGFG-class. The left-hand configuration presents more coarsely grained patterns and the right-hand one finely grained.

Solid–liquid transition: classes CGFG and HCGFG
Incipient waves: CGSW class

5.4.6 How initial conditions influence the structure of the classes

Does the class structure alter with a change in the initial conditions of lattice evolution? Let us start with a discussion of boundary conditions. Most classes, except for 2^+ and FW, preserve their structure and membership when the boundaries change from absorbing to periodic, or *vice versa*. Class FW exists only

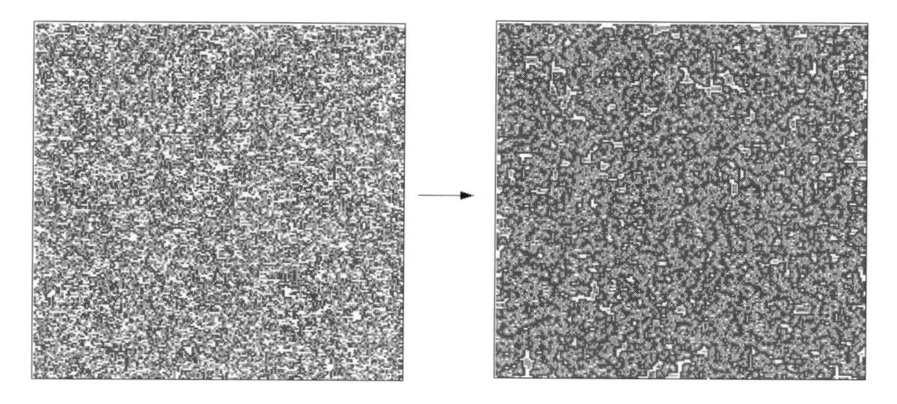

Figure 5.31. Typical configurations of an excitable lattice from the CGSW-class. The left-hand configuration presents more coarsely grained patterns and the right-hand one sawtooth wave patterns.

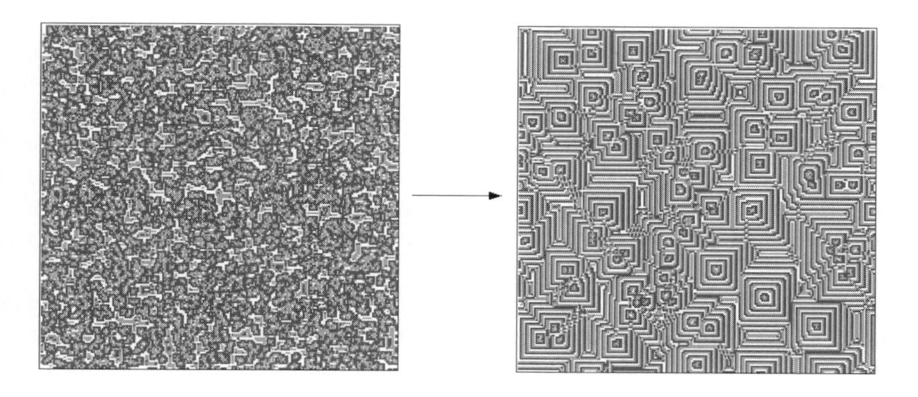

Figure 5.32. Typical configurations of an excitable lattice from the SW-class. The left-hand configuration presents incipient wave patterns and the right-hand one circular and spiral waves.

for periodic boundaries; when the boundaries become absorbing it degenerates into the class 0. Similarly, the class 2^+ is typical for periodic boundaries, but it merges with class 0 for absorbing boundaries. This happens because there are no stationary patterns in the evolution of 2^+ lattices. All patterns move on the lattice and inevitably encounter the boundaries, disappearing when the boundaries are absorbing.

The structural changes occurring after an increase in the probability p of initial excitation are shown in figure 5.38. For every rule, each cell of the initial

Table 5.1. Specification of morphological classes. Classes GW and SW emerge only when the probability for a cell to be excited in an initial configuration is greater than 0.6.

Class	Rule vector s	Rules indices	Number of rules
0	(00000000) ... (00111111)	1 ... 64	64
2^+	(01000000) ... (01011111)	65 ... 96	34
H	(01100000) ... (01101111)	97 ... 112	15
L	(01110000) ... (01110111)	113 ... 120	7
F	(01111000) ... (01111111)	121 ... 128	7
CGFG	(10000000) ... (10111111)	129 ... 192	63
HCGFG	(11000000) ... (11011111)	193 ... 224	31
CGSW	(11100000) ... (11110111)	225 ... 248	23
GW	(11100000) ... (11101001)	225 ... 234*	9
SW	(11111000) ... (11111101)	249 ... 254*	5
FW	(11111110), (11111111)	255, 256	2

configuration becomes an excited state with probability $p = 0.1, 0.2, \ldots, 0.9$, and the rest state with probability $1 - p$.

It is clear that classes 0, 2^+, CGFG, HCGFG and FW are stable when the probability of initial excitation increases, and that classes H, L and CGSW are not stable. Class L branches into L and F classes when p exceeds 0.2, and loses all its members to class F when $p \geq 0.6$. The members of class H move to class F when $p \geq 0.7$. Class F is also vulnerable to increasing p and loses some of its members to class 0 for $p \geq 0.5$ (figure 5.38).

Consider the mechanisms underlying these structural changes. In excitable lattices, as in any cellular automaton, the cells interact locally and, therefore, the spatio-temporal dynamics will change catastrophically when the boundary conditions are changed only when we have a more or less stable transmission of 'influence' between distant parts of the lattice. The spiral waves in the lattices from the classes CGSW and SW cannot achieve this because with the initial conditions used here, there are many waves, which undergo many collisions and consequent annihilations, when the number of wave generators is increased with increasing p or as a result of collisions. On the other hand, the unidirectional waves in class FW may be responsible for a sort of 'broadcasting'; they cover the lattice reasonably fast, and the number of wave generators is typically reduced as a result of wave competition. We can therefore expect class FW to be vulnerable to a change in boundary conditions, and this has been confirmed by numerical experiments. The particle-like waves in class 2^+ are much more specific than the waves in class FW, and can transmit information directly across the lattice. Again, not surprisingly, this class is unstable when the boundary conditions change.

The observed dependence on the probability p of initial excitation can possibly be explained by considering the critical probability p_c of percolation,

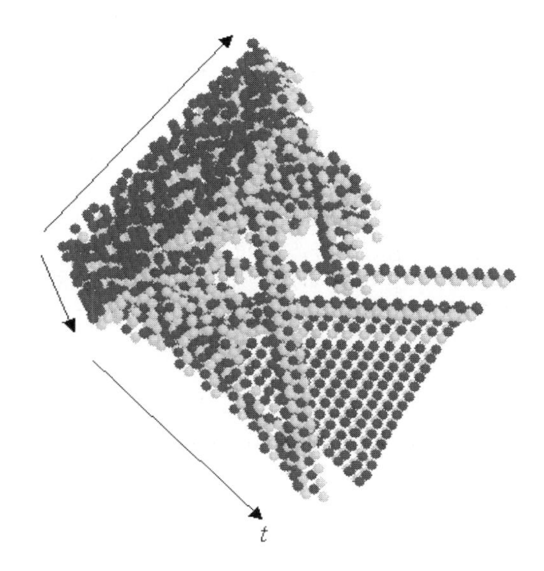

Figure 5.33. Evolution of the lattice from class 2^+ in $(2+1)D$ space (discrete space plus time). The lattice has 30×30 cells with periodic boundaries. The evolution takes 30 time steps. Initially every cell of the rest lattice is excited with probability 0.3. One of the octants is removed so the internal structure can be seen. Excited cells are black and refractory cells are grey.

i.e. the ratio of excited sites to the total size of the lattice at which an infinite cluster of excited states appears on the lattice (see e.g. [572]). The values of the probability p which force one or more classes to suffer structural changes are $0.15, 0.6, 0.9$, with most changes happening when p is about 0.6. Could there be a connection with the critical probability of percolation? The following bounds for p_c are known: $0.54 \leq p_c \leq 0.68$ for a square lattice [449, 673] and $0.65 \leq p_c \leq 0.81$ for a hexagonal lattice [400]. Although it is not possible to make anything approaching a strong case for a connection between p_c and the probability p of initial excitation, we can at least note that the value $p \geq 0.66$ appears to be significant.

If we look at the diagram showing the number of clusters κ in a random initial configuration where each cell becomes excited with probability p (figure 5.39) we find that the number of clusters κ decreases for $p \geq 0.15$ and it is equal to one with a very high probability for $p \geq 0.66$. The first bound determines the branching of class L into classes L and F (figure 5.38); decreasing the number of potential generators reduces the number of wavefronts and transforms

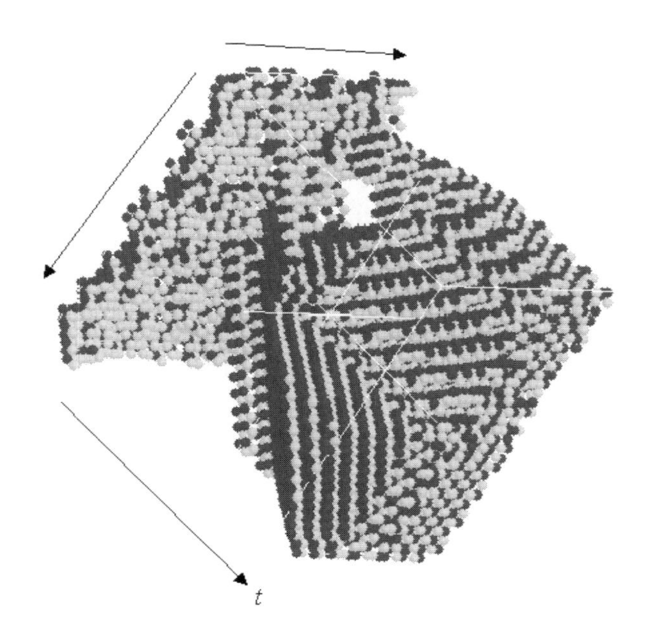

Figure 5.34. Evolution of the lattice from class F in $(2+1)D$ space (discrete space plus time).

the dynamical labyrinth-like excitation patterns into fingerprint-like excitation patterns. Class F acquires all members from H and L once $p \geq 0.66$ (figure 5.38); in other words, the expected length of the shortest path out of the labyrinth patterns catastrophically increases for $p \geq 0.66$. Obvious wave patterns also appear in class CGSW for $p \geq 0.66$ (figure 5.38).

What is the dependence of the activity level α on lattice size and initial probability of excitation? Increasing the number of excited cells in the initial configuration up to $\frac{1}{3}$ of all lattice cells leads to an increase in the extrinsic probability of excitation in classes L, F, CGSW, SW and FW (figure 5.40). Classes CGFG and HCGFG are much more stable with respect to the initial conditions and this could be due to their 'stochastic' behaviour. When the initial excitation probability exceeds 0.5 the members of classes L, CGFG, HCGFG and CGSW exhibit damping of the excitation dynamic (figure 5.40). We found that the extrinsic probability of excitation decreases when lattice size increases (figure 5.41); this happens for all classes except 0 and 2^+. The situation with class 0 is trivial whereas the relative invariance of class 2^+ to lattice size can be explained by the formation of particle-like waves.

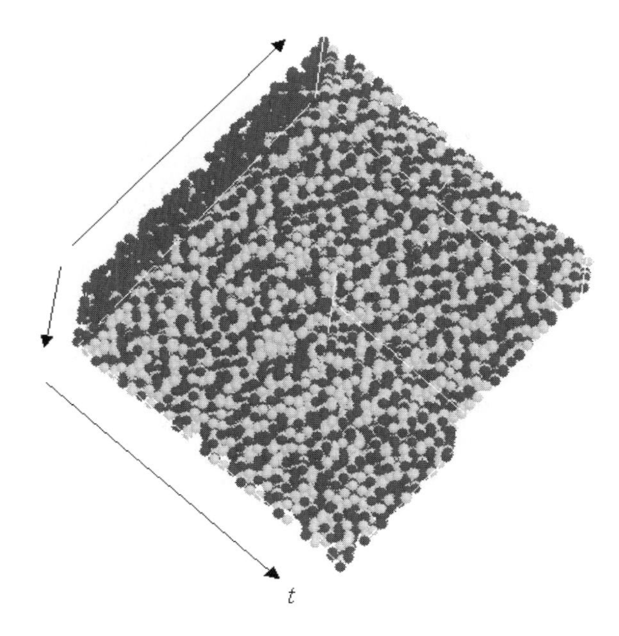

Figure 5.35. Evolution of a lattice from class H in $(2+1)D$ space (discrete space plus time).

5.4.7 Morphological diversity

The morphological diversity associated with the excitation rules can be evaluated by studying clusters of excited states in configurations far beyond the transient period, and analysing the richness ρ of these configurations. In the numerical experiments we obtained values for the number of clusters κ (figure 5.42), the maximum cluster size $\omega_{\max}$ (figure 5.43), the typical cluster size ω (figure 5.44) and the size diversity of clusters υ (figure 5.45). They are computed for lattices of 100×100 cells with periodic boundary conditions; at time step $t = 0$ every cell becomes excited with probability 0.3 or remains in the rest state with probability 0.7. The results of the computation are averaged over 50 experiments independently of each of 192 excitation rules (from rules with indexes 65 to 256). Since any configuration in class 0 evolves into the configuration with all cells in the rest state, we did not consider the rules from this class. Parameters such as the number of clusters κ and the size diversity of clusters υ are the most representative characteristics. They clearly separate the rules into the morphological classes defined in table 5.1. An ascending hierarchy of classes (obtained from figure 5.40) derived from the number of clusters κ is presented in table 5.2.

The size diversity of clusters υ changes dramatically from class to class

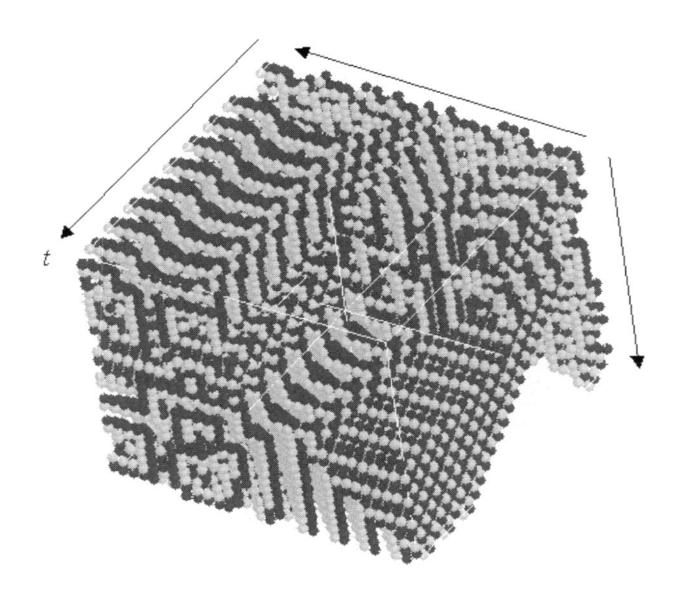

Figure 5.36. Evolution of a lattice from class SW in $(2+1)D$ space.

before vector rule (10111111) with index 192, and increases very smoothly after this rule (figure 5.45). Nevertheless, we are still able to discriminate between the classes (table 5.2).

For all classes, except 2^+, L, F, SW and FW, $\omega = 1$ (figure 5.44). However, solitary waves in class 2^+, and spiral and unidirectional waves in classes L, F, SW and FW, increase the typical cluster size ω. In class 2^+ most of the patterns that survive long-term evolution are the so-called 2^+-particles [13, 14]. These particle-like waves travel around the discrete torus along non-intersecting trajectories. In classes L and F, the typical long-wave fibres are clearly recognizable, but they are outnumbered by the small disordered groups of 'dying' clusters which are formed after the collision of two or more waves. The long waves in class FW determine the relatively large typical cluster size ω. In the diagram showing the maximum cluster size $\omega_{\max}$ versus rule index (figure 5.43) we can detect two rather messy peaks. They belong to classes L and F ($\omega_{\max} = 150$–250, respectively), SW ($\omega = 500$), FW ($\omega = 300$) and CGSW. The value of this parameter in the last class displays very high variability. In this transitional class, the development may extend from coarsely grained patterns to the onset of saw-tooth waves; but once the waves are established (SW and FW) $\omega_{\max}$ decreases. It may be that these transitional features are, in some way, responsible for the very high variability, but we are still far from an explanation. The discussion will

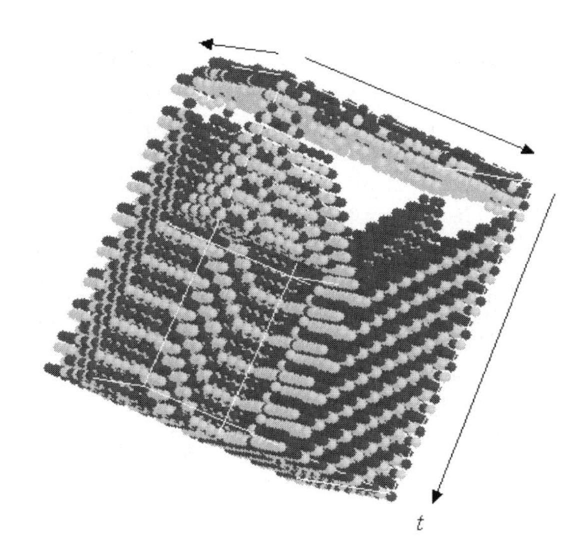

Figure 5.37. Evolution of an excitable lattice from class FW in $(2+1)$-dimensional space.

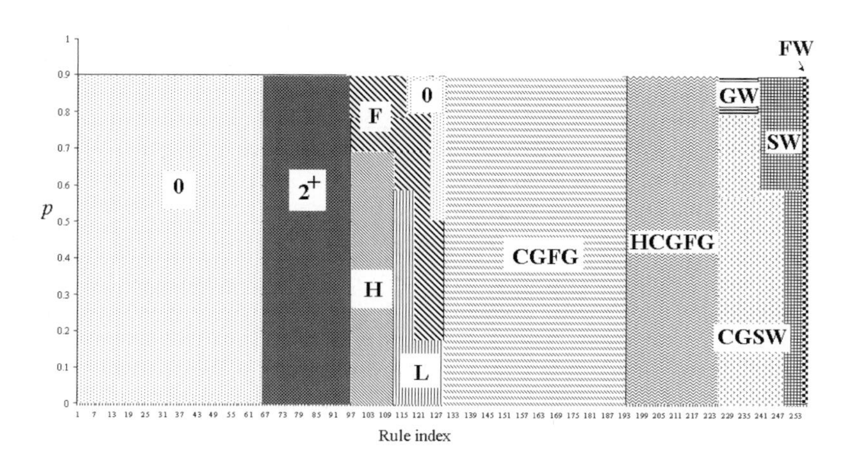

Figure 5.38. Heredity of the morphological classes when the probability p (vertical axis) of initial excitation increases. The rule indices are arranged on the horizontal axis.

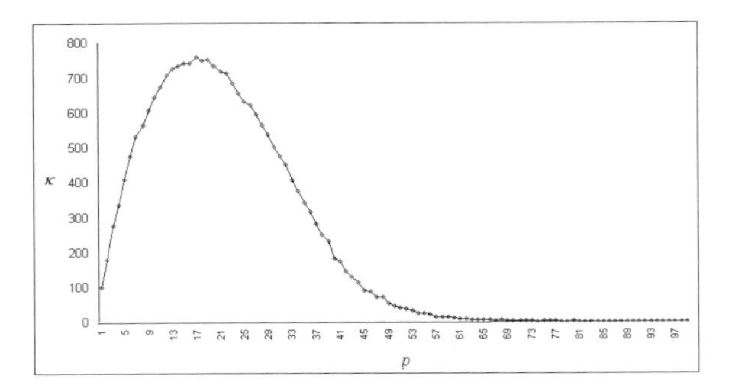

Figure 5.39. The number of clusters κ in random configurations of only excited and rest states with different probabilities of excitation. For every probability the results are averaged over 100 runs.

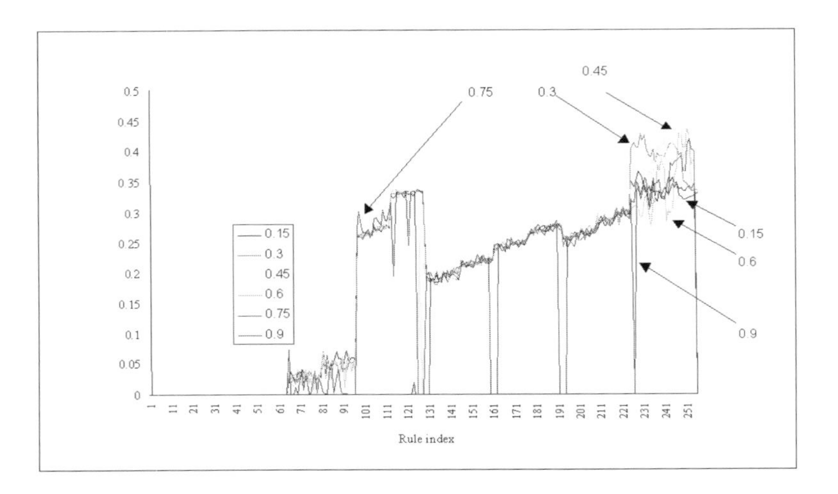

Figure 5.40. Dependence of the activity level α (vertical axis) on the initial excitation probability p (graph labels). Every cell of the lattice at the initial time step becomes excited with probability p. The rule indices are arranged along the horizontal axis.

show that interactions between chaotic dynamics, wave formation and λ^+ play a key role in this phenomenon: chaotic spacetime dynamics together with high λ^+ increase the probability of cluster formation, whereas wave generation decreases it.

The richness ρ of the configurations generated by the excitation rules is another indicator of morphological diversity. In the numerical experiments we found that if a configuration generated by a given rule has rich cells, then it has them for $r \geq 2$. When discussing ρ we therefore imply $r = 2$, i.e. second-order

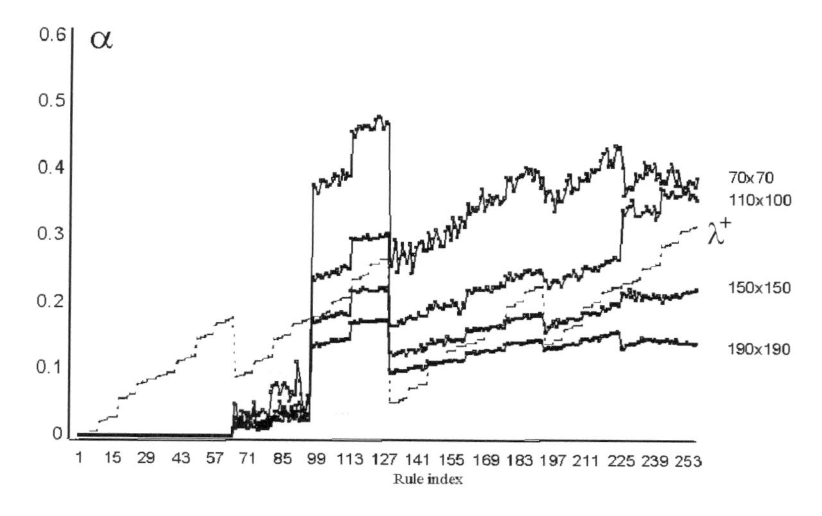

Figure 5.41. Dependence of the activity level α on the size $n \times n$ of the excitable lattice. Four graphs are labelled with lattice sizes and the fifth graph is λ^+.

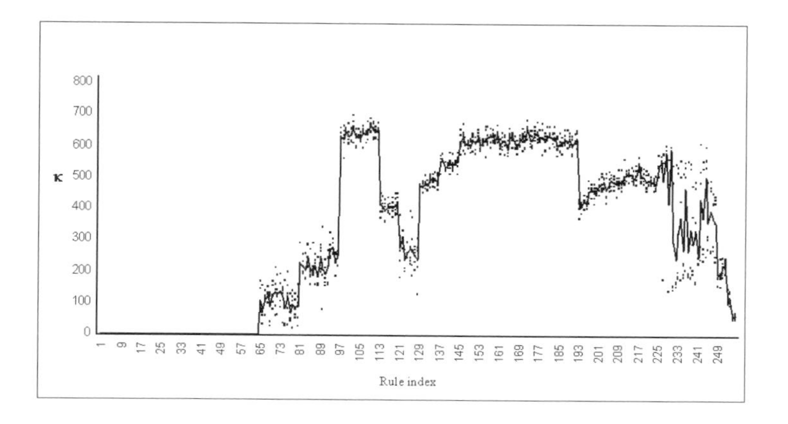

Figure 5.42. The number of clusters κ (vertical axis) in sample configurations determined long after the transient period. The data for every excitation rule are averaged over 50 independent runs. Initially every cell of the lattice is excited with probability 0.3. The lattice has 100×100 cells and periodic boundaries.

neighbours are included in the consideration when calculating ρ. The average value of ρ for the excitation rules is shown in figure 5.46. Classes 0, 2^+, L, F and FW have $\rho = 0$, and the ρ of class H is near zero. The remaining classes show a maximum (minimum) ρ for members with a maximum (minimum) λ^+. The rate of decrease increases very rapidly from CGFG through HCGFG to CGSW. On

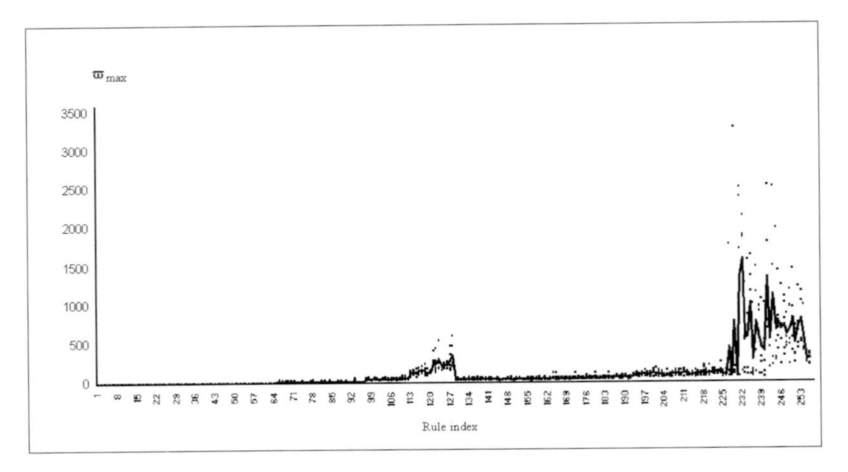

Figure 5.43. The maximum cluster size ω_{max} in sample configurations determined long after the transition period. The data for every excitation rule are averaged over 50 independent runs. Initially every cell of the lattice is excited with probability 0.3. The lattice has 100×100 cells and periodic boundaries.

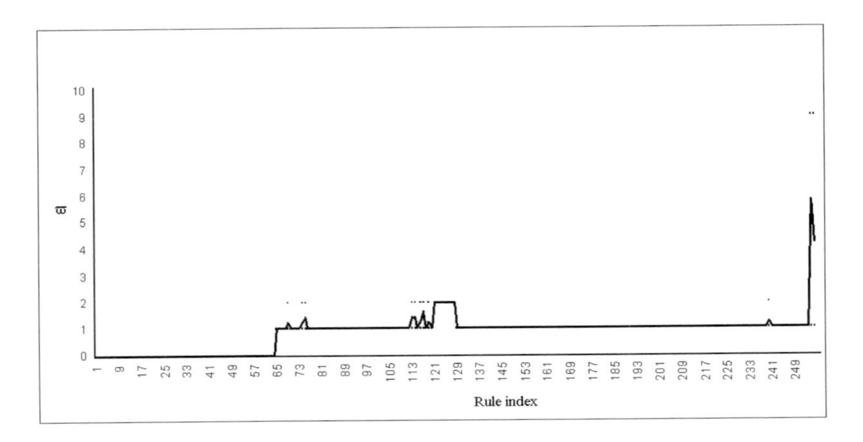

Figure 5.44. The typical cluster size ω (vertical axis) in the sample configurations determined long after the transition period. The data for every excitation rule are averaged over 50 independent runs. Initially every cell of the lattice is excited with probability 0.3. The lattice has 100×100 cells and periodic boundaries.

the richness ρ' averaged over all members of a class, the classes can be ordered as

$$\rho'(\text{CGSW}) > \rho'(\text{HCGFG}) > \rho'(\text{SW}) > \rho'(\text{CGFG}).$$

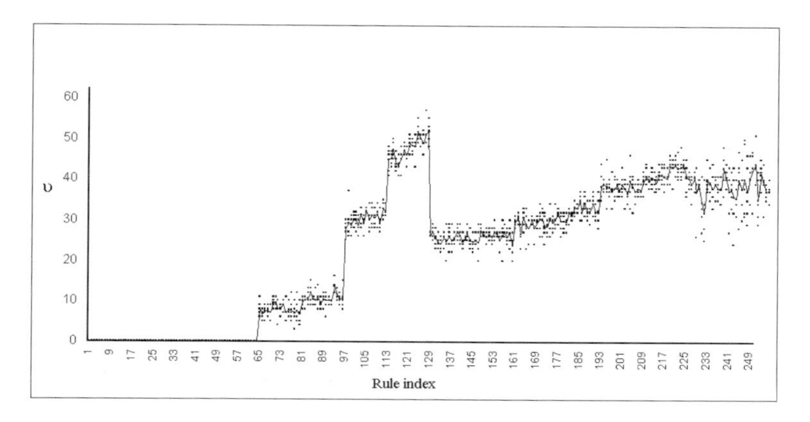

Figure 5.45. The size diversity of clusters υ in the sample configurations determined long after the transition period. The data for every excitation rule are averaged over 50 independent runs. Initially every cell of the lattice is excited with probability 0.3. The lattice has 100×100 cells and periodic boundaries.

Table 5.2. Values of the most important parameters: number of clusters κ, activity level α and transient period τ, computed in 100 experiments for every rule and averaged over the members in each class.

Class	κ	υ	α	τ
0	—	—	—	3
2^+	150	10	0.04	32
H	605	30	0.27	10
L	400	45	0.28	16
F	250	50	0.34	42
CGFG	650	30	0.24	1
HCGFG	500	40	0.26	1
CGSW	350	40	0.33	10
SW	200	40	0.34	9
FW	60	35	0.34	40

Remarkably, ρ decreases in CGSW when incipient waves are present. It is worth noting that only the ρ of CGSW exceeds the maximum possible ρ of a random configuration (figure 5.47); it is about four times greater than the ρ of the random configuration in which every cell becomes excited with probability 0.48–0.53.

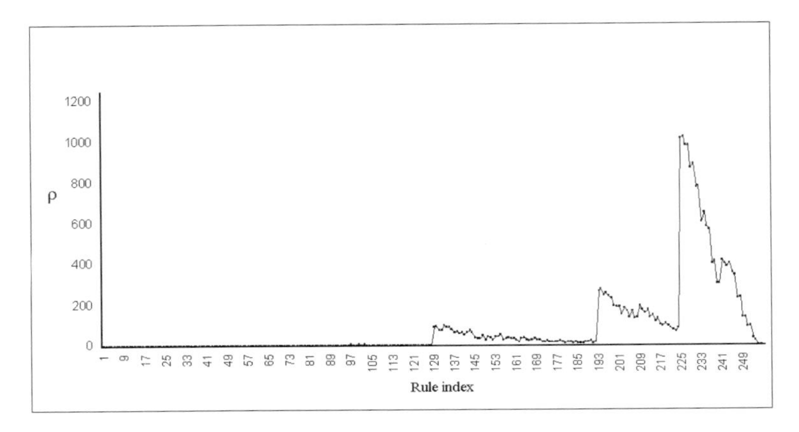

Figure 5.46. The richness ρ (vertical axis) of the sample configurations determined long after the transition period. Results for every excitation rule is averaged on 50 independent runs. Initially every cell of the lattice is excited with probability 0.3. The lattice has 100×100 cells.

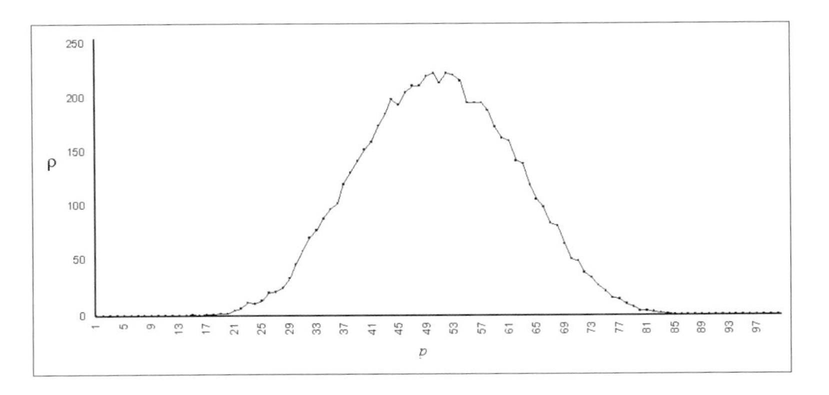

Figure 5.47. The richness ρ (vertical axis) of a random configuration of only excited and rest states with different probabilities p (horizontal axis) for a cell to become excited. For every probability p the results are averaged over 100 runs. The boundaries are periodic.

5.4.8 Excitation generators

Generators are blocks of cells in excited, rest and refractory states that generate excitation patterns which grow without limit on an infinite lattice. A finite fragment of an infinite lattice, or a finite lattice with absorbed boundaries, including initially excited cells, may be at rest after some steps even if the activity is unbounded because the excitation waves go beyond the boundaries. In some cases, even a finite fragment will retain the excitation at any step of the evolution if the excitation rules support a recurrent dynamic.

Let σ^t and σ_-^t be sets of cells, which are excited and refractory at time step t:

$$\sigma^t = \{x \in \mathbf{L} : x^t = +\}$$

and

$$\sigma_-^t = \{x \in \mathbf{L} : x^t = -\}.$$

A set σ is called a minimal ∞-generator if the following conditions are satisfied:

- $\sigma_-^0 = \emptyset$,
- $|\sigma^t| \neq \emptyset$, for $t \geq 0$,
- $|\sigma| = \min_{\sigma^0 \in \{\circ,+\}} \{|\sigma^0|: \forall t \geq 0, \sigma^t \neq \emptyset\}$.

A set σ is called a minimal generator if it is a minimal ∞-generator and there is a positive integer p such that for every time step t such that $t = mp$, $m = 1, 2, \ldots$: $\sigma^0 \subset \sigma^t$.

In words this is as follows.

> *An ∞-generator gives rise to a non-degrading excitation activity at least once; the activity may persist by spreading across an infinite lattice.*
>
> *A generator does not simply give birth to a non-degrading excitation activity but is self-reconstructed periodically during lattice evolution.*

The existence of minimal generators globally determines the non-decreasing activity on a finite lattice. Sites excited at the initial step of the evolution will become excited again in the evolution. If a local excitation function has a minimal generator then the function has a minimal ∞-generator but not *vice versa*.

A set σ of lattice cells is called a trivial generator if it is a minimal generator and consist of a compact group of excited, refractory and resting cells: $\sigma_-^0 \neq \emptyset$ and $\sigma_-^0 \subset \mathrm{hull}(\sigma^0)$, where hull$(\cdot)$ is the hull of a discrete set. The set of excited, rest and refractory sites is a trivial generator if there is a spacetime excitation loop in the region covered by the initial configuration of the generator. This means that at every step of the evolution every resting cell of the generator has enough excited cells in its neighbourhood to become excited at the next step of the evolution. Therefore, we have $\sigma_\circ^t \subset \sigma_-^{t-1}$, $\sigma_-^t \subset \sigma_+^t$ and $\sigma_+^t \subset \sigma_\circ^{t-1}$.

Let us look at an example. In the example we reduce vector representation of the excitation rule to $s = (s_i)_{1 \leq i \leq 8}$, where $s_i = 1$ if $i \geq \theta$. This gives us the threshold case. In this way the minimal trivial generator for $\theta = 1$ consists of three cells and evolves as follows:

$$\begin{pmatrix} + & \circ \\ * & - \end{pmatrix} \to \begin{pmatrix} - & + \\ * & \circ \end{pmatrix} \to \begin{pmatrix} \circ & - \\ * & + \end{pmatrix} \to \begin{pmatrix} + & \circ \\ * & - \end{pmatrix} \to \cdots$$

the cell state marked $*$ may be ignored.

The evolution of the minimal trivial generator for $\theta = 2$ is as follows:

$$\begin{pmatrix} - & \circ & - \\ + & \circ & + \\ - & \circ & - \end{pmatrix} \to \begin{pmatrix} \circ & + & \circ \\ - & + & - \\ \circ & + & \circ \end{pmatrix} \to \begin{pmatrix} + & - & + \\ \circ & - & \circ \\ + & - & + \end{pmatrix} \to \begin{pmatrix} - & \circ & - \\ + & \circ & + \\ - & \circ & - \end{pmatrix} \to \cdots .$$

Let $w \in \mathbf{Q}^{m^2}$ be an arbitrary block of $m \times m$ sites. We say a block w is naturally non-constructible for cell state transition function f if there is no such block $w' \in \{+, \circ\}^{(m+2)^2}$ of $(m+2) \times (m+2)$ sites that by applying f in parallel to m^2 inner sites of w' we obtain w. The difference in block sizes is chosen to avoid specifying boundary conditions.

Trivial generators are naturally non-constructible.

The usual way to detect the non-constructibility of a given block for an arbitrary cell state transition function exactly is to make an exhaustive search over all possible blocks that can be transformed into the given one (pre-image construction). Fortunately, in the case of local excitation rules we can avoid this procedure thanks to the unconditional transition $+ \to -$. Let us assume a trivial generator σ^t has a predecessor σ^{t-1} then for any cell $x \in \sigma^t \cap \sigma^{t-1}$ and $x^t = \circ$ there is a neighbour $y \in u(x)$ such that $y^t = -$, i.e. $y^{t-1} = +$ and $x^{t-1} \in \{\circ, +\}$. If $x^{t-1} = +$ then $x^t = -$ and if $x^{t-1} = \circ$ then $x^t = +$. This contradicts $x^t = \circ$.

Class 0 has no generators.

We say a class of excitable rules has generators if it has at least an ∞-generator. Hence, we should show that, for any $f \in$ class 0, an infinite lattice and $|\sigma^0| < \infty$, there is such a $t \geq 0$ that $\sigma^t = \emptyset$. The situation is trivial if $\sigma^0 = \emptyset$; so, assume that it does not take place. Any rule of class 0 is presented by such a vector s, where $s_i = 0$ for $i \leq 2$. It is enough to consider f_{64}, $s = (00111111)$. The rule implies $\sigma^{t+1} \neq \emptyset \Rightarrow |\sigma^t| \geq 3$. Let $\sigma^t = \underbrace{(+ \cdots +)}_{h}$, $h \geq 3$, then σ^{t+1} consists of two strings $\underbrace{+ \cdots +}_{h+2}$ and $\sigma^{t+O(h/2)} = \emptyset$. Correspondingly,

$$\sigma^t = \begin{pmatrix} & & + & & \\ & & \vdots & h_1 & \\ & h_3 & \vdots & & \\ + & \cdots & + & \cdots & + \\ & & \vdots & h_4 & \\ & h_2 & \vdots & & \\ & & + & & \end{pmatrix}$$

is transformed into

$$\sigma^{t+1} = \begin{pmatrix} & & + & + & & \\ & & \vdots & \vdots & h_1-2 & \\ & h_3-2 & \vdots & \vdots & & \\ + & \cdots & + & + & \cdots & + \\ + & \cdots & + & + & \cdots & + \\ & & \vdots & \vdots & h_4-2 & \\ & h_2-2 & \vdots & \vdots & & \\ & & + & + & & \end{pmatrix}$$

and $\sigma^{t'} = \emptyset$ for $t' \geq \max(h_1, \ldots, h_4)$. Any connected σ^t can be presented in this form and any connected subset of disconnected σ^t can be analysed separately. Therefore, we have $|\sigma^{t+1}| < |\sigma^t|$.

As soon as it is impossible to construct even the trivial generator, in an infinite lattice, or a lattice with absorbing boundaries, we can do so on the torus. In the worst case the configuration evolves as follows (do not forget about the periodic boundaries)

$$\begin{pmatrix} \circ & - & \circ & - \\ + & \circ & + & \circ \\ - & + & - & + \end{pmatrix} \rightarrow \begin{pmatrix} + & \circ & + & \circ \\ - & + & - & + \\ \circ & - & \circ & - \end{pmatrix} \rightarrow \begin{pmatrix} - & + & - & + \\ \circ & - & \circ & - \\ + & \circ & + & \circ \end{pmatrix} \rightarrow \begin{pmatrix} \circ & - & \circ & - \\ + & \circ & + & \circ \\ - & + & - & + \end{pmatrix} \rightarrow \cdots.$$

Classes CGFG, HCGFG and CGSW, SW and FW have minimal ∞-generators of size 1.

The proposition states that for all functions $f_{129} \ldots f_{256}$, i.e. vector excitation functions indexed by $129 \ldots 256$, any rest configuration on an infinite lattice with only one excited cell has excited cells at any step of its evolution. Let $x^0 = +$ then for any neighbour $y \in u(x)$ of cell x we have $y^1 = +$ and there are four such neighbours $z \in u(y)$ of cell y such that $z^1 = \circ$ and $\sum_{v \in u(x)} \chi(v^t, +1) = 1$. For any function $f_{129} \ldots f_{256}$ we have $s_1 = 1$. Therefore there are at least four such cells $z \in u(y)$ that $z^2 = +$. Let there be four cells $z \in u(y)$ such that $z^t = +$ and the cells are the proximal elements of the set σ^t; then at time step $t + 1$ we have at least four excited cells.

Using arithmetical representation of the vector function, where the rest, excited and refractory states are represented by the integers 0, 1 and 2, and $sg(a) = 1$ if $a > 0$ and is zero otherwise,

$$x^{t+1} - sg(x^t)((x^t + 1) \mod 3) = (1 - sg(x^t)) s_{\sum_{y \in u(x)} y^t \mod 2}$$

it is possible to demonstrate that the minimal ∞-generator is simply a minimal generator in classes CGFG and HCFG and that classes CGSW, SW and FW have no minimal generators of size 1.

The hints may include an indication about the fractals dynamic of the lattices with the rules from CGFG and HCFG (see figure 5.48) and the fact that members of CGSW, SW and FW exhibit omnidirectional waves after singular excitation of an entirely at rest lattice. Here we make two preliminary statements about the recurrent dynamics in CGFG and HCFG and omnidirectional waves in CGSW, SW or FW.

If we compare the classification of computational abilities with the results on generators we find the following.

Excitable lattice processors capable of image filtration should be subject to a fractal dynamic; both universal and wave processors are not.

Does the fractal dynamic itself imply a capability for image-processing?

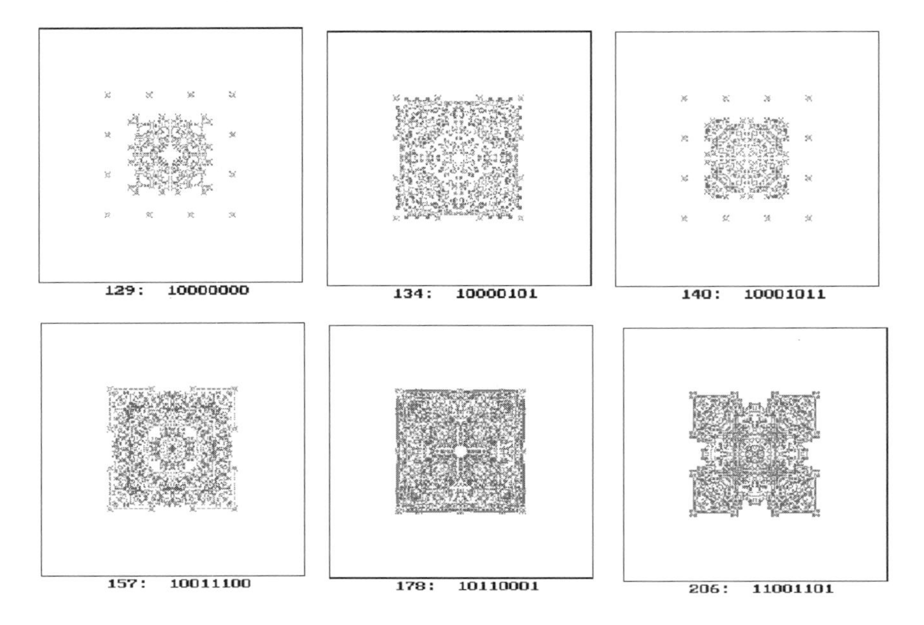

Figure 5.48. Fractal excitation patterns generated in lattices from the classes CGFG and HCFG.

Let function f of classes CGFG or HCFG act on an infinite lattice and $\sigma_{-}^{0} = \emptyset$ and $|\sigma^{0}| = 1$ then there are a t and $x \in \mathbf{L}$ such that $x^{t} \in \sigma^{t-2}$.

The first element of a vector representing any function from these classes equals 1, so we have

$$\sigma^{1} \cup \sigma_{-}^{1} = \begin{pmatrix} + & + & + \\ + & - & + \\ + & + & + \end{pmatrix}.$$

Using the structure of $\sigma^{1} \cup \sigma_{-}^{2}$ the functions can be subdivided into the following three groups (the first two lie in class CGFG and the third is, in fact, HCFG):

$$\text{(i)} \quad f_{129} \ldots f_{160}\colon \sigma^{1} \cup \sigma_{-}^{2} = \begin{pmatrix} + & & & & + \\ & - & - & - & \\ & - & \circ & - & \\ & - & - & - & \\ + & & & & + \end{pmatrix}$$

$$\text{(ii)} \quad f_{161} \ldots f_{192}\colon \sigma^{1} \cup \sigma_{-}^{2} = \begin{pmatrix} + & & + & & + \\ & - & - & - & \\ + & - & \circ & - & + \\ & - & - & - & \\ + & & + & & + \end{pmatrix}$$

$$\text{(iii)} \quad f_{193} \ldots f_{224}\colon \sigma^{1} \cup \sigma_{-}^{2} = \begin{pmatrix} + & + & & + & + \\ + & - & - & - & + \\ & - & \circ & - & \\ + & - & - & - & + \\ + & + & & + & + \end{pmatrix}.$$

The situation with the third group is trivial. For the first group we have that

$\sigma^3 \cup \sigma^3_-$ has at least four blocks of the form:

$$\begin{pmatrix} + & + & \circ \\ - & \circ & \circ \\ \circ & \circ & \circ \end{pmatrix}$$

the excited sites of which are proximal elements of $\sigma^3 \cup \sigma^3_-$; central site of the block will be excited at next step of time. A block

$$\begin{pmatrix} + & \circ & \circ \\ - & \circ & - \\ \circ & \circ & \circ \end{pmatrix}$$

with the same properties exists in $\sigma^4 \cup \sigma^4_-$ from the second group.

Let function f from class FW act on an infinite lattice, $\sigma^0_- = \emptyset$, $\sigma^0 \neq \emptyset$ and $|\sigma^0| < \infty$. Then $x^t = \circ$ for $x \in \text{hull}(\sigma^0)$ and $\forall x \in \sigma^t : \sum_{y \in u(x)} \chi(y^t, +) = 2$, where $t = O(d(\sigma^0))$, $d(\cdot)$ is the diameter of the set.

For some t let set σ^t be disconnected and consist of two eight-connected sets σ^t_1 and σ^t_2. Due to the form of functions f_{255} and f_{256} we have for any non-empty σ^t_i: $\text{hull}(\sigma^{t+1}_i) \supset \text{hull}(\sigma^t_i)$, therefore $\text{hull}(\sigma^{t+p}_1) \cap \text{hull}(\sigma^{t+p}) \neq \emptyset$ for some $p \geq |\sigma^t_1 - \sigma^t_2|$.

Consider a finite non-empty σ^0 consisting of two disconnected subsets σ^0_1 and σ^0_2 such that $\text{hull}(\sigma^0_1) \supset \text{hull}(\sigma^0_2)$. Then σ^1 will consists of four subsets that $\text{hull}(\sigma^1_1) \supset \text{hull}(\sigma^1_2) \supset \text{hull}(\sigma^1_3) \supset \text{hull}(\sigma^1_4)$. And for $t \geq 2$ we have $\text{hull}(\sigma^{t+1}_i) \supset \text{hull}(\sigma^t_i)$, $i = 1, \ldots, 4$.

Diameter $d(\sigma^0)$ is finite, therefore for $t > \max(|\sigma^1_3, \sigma^1_2|, d(\sigma^1_4))$ the set σ^t is a connected expanding set: $\text{hull}(\sigma^t) \supset \text{hull}(\sigma^{t-1})$. To prove the second part of the proposition we can consider an expanding connected set σ^t where some excited cells have more than two excited neighbours. The elements of this set can be excited (due to the eight-cell neighbourhood) only the rest sites closest to sites of σ^t; that is, σ^t has rest shell of width 1 and σ^{t+1} satisfies the second condition of the proposition.

A stronger result on the formation of omnidirectional waves in class FW follows.

Class FW has no minimal generators.

This is a straightforward consequence of the previous proposition. The propositions imply that any finite lattice evolving from a configuration with only excited or rest cells will be entirely at rest after a time proportional to the lattice size independently of boundary conditions.

Let function f from classes CGSW, SW or FW act on an infinite lattice and $\sigma^0_- = \emptyset$ and σ^0 is a singleton. Then $x^t = \circ$ for $x \in \text{hull}(\sigma^0)$ and

$\forall x \in \sigma^t : \sum_{y \in u(x)} \chi(y^t, +) = 2$, *where* $t = O(d(\sigma^0))$, $d(\cdot)$ *is the diameter of the set.*

Classes H, L and F have minimal generators of size 2.

We should consider functions $f_{97} \ldots f_{128}$; therefore, two excited cells lying in the same row or column form the minimal ∞-generator. The configuration starting with these two excited cells will evolve as follows (the $\circ$ state is indicated when necessarily only the core states of growing configurations are presented):

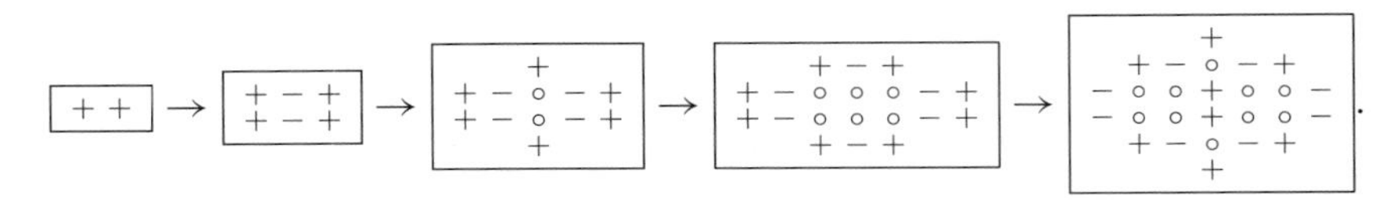

Therefore at the fourth step of the evolution we have $\sigma^0 \subset \sigma^4$. Moreover cells belonging to σ^0 have exactly four rest neighbours that can be excited. This means that for period $p = 4$ we can find such a t, $t = mp$, $m = 1, 2, \ldots$, that $\sigma^0 \subset \sigma^t$. Thus, σ^0 is a minimal generator. Moreover, we can say that every cell x, such that $|x - \sigma^0| > 2t$, becomes excited with period 4.

Class 2^+ *has minimal* ∞*-generators of size 2.*

Differing from the classes discussed previously, for the vector s, representing a function from class 2^+, we have $s_2 = 1$ and $s_3 = 0$. Therefore until the fourth step of the evolution the excitation continues in the same way but the rest cells neighbouring the cells of σ^0 are not excited because they have either fewer or more than two excited neighbours. This means that the configuration will expand but the excitation will not necessarily return to the sites of σ^0.

Some examples of generator-induced patterns in 2^+ are shown in figure 5.49.

It is also possible to demonstrate that $\sigma^0 \subset \sigma^4$, for $t = mp$, $m = 1, 2, \ldots$, when p is not simply a positive integer but some integer-valued function in which the arguments are t and f. The collision dynamics near opposite angles of growing rhombi of excited sites leads to the generation of 2^+-particles:

$$\begin{array}{cc} - & - \\ + & + \end{array} \quad \begin{array}{cc} + & + \\ - & - \end{array} \quad \begin{array}{cc} - & + \\ - & + \end{array} \quad \begin{array}{cc} + & - \\ + & - \end{array}$$

that move towards each other and disappear after collision [14] but before disappearing they excite the sites of σ^0.

5.4.9 Integral dynamics

Both activity level α and transient period τ are integral characteristics because they do not take into account any spatial features of the excitation. Nevertheless they are useful for examining the system from a dynamical point of view. The

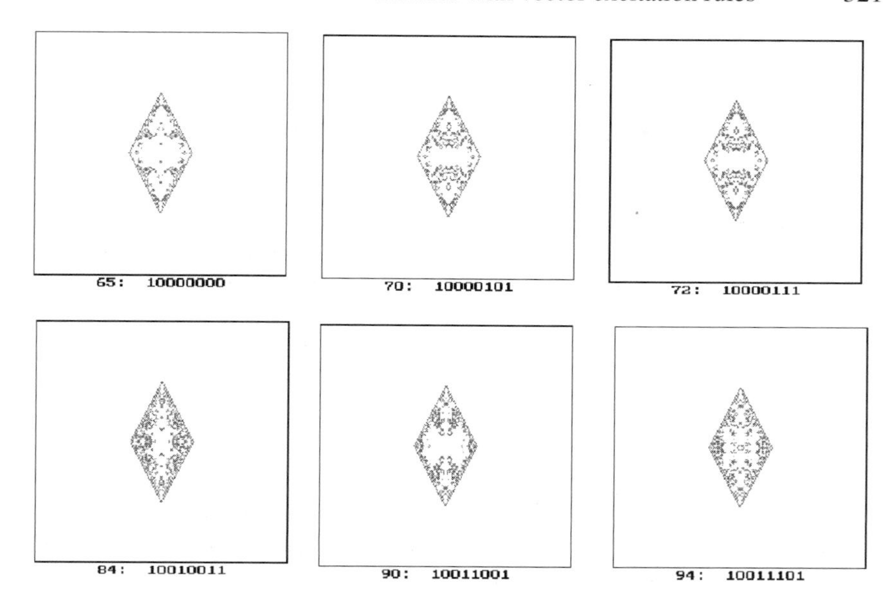

Figure 5.49. Generator-induced excitation patterns on lattices from class 2^+.

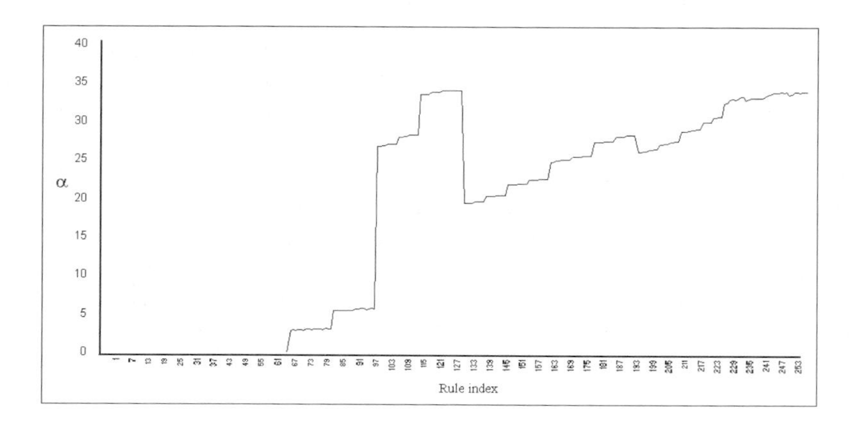

Figure 5.50. The activity level α (vertical axis) calculated on the sample configurations determined long after the transition period. The results for every excitation rule are averaged over 50 independent runs. Initially every cell of the lattice becomes excited with probability 0.3. The lattice has 100×100 cells and periodic boundaries.

activity level α of the sample configurations generated by the excitation rules is shown in figure 5.50.

Class 2^+ is at the lower level of the hierarchy. The upper levels are occupied

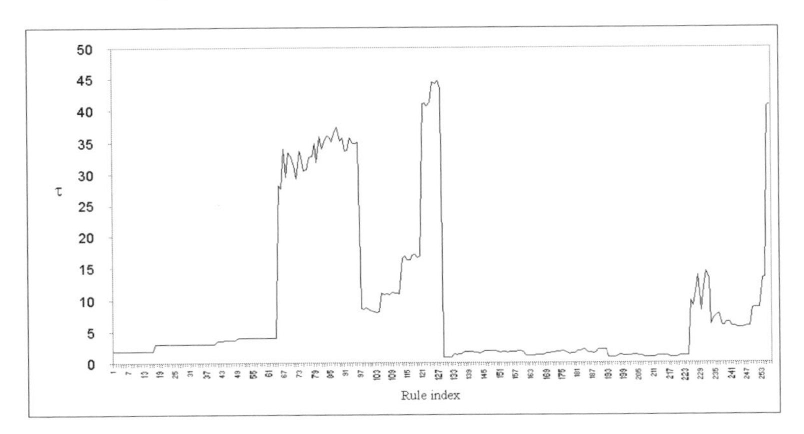

Figure 5.51. The transient period τ (vertical axis) calculated on the sample configurations determined long after the transition period. The results for every excitation rule is averaged over 50 independent runs. Initially every cell of the lattice becomes excited with probability 0.3. The lattice has 100×100 cells and periodic boundaries.

by class members which exhibit wave dynamics in their evolution. Classes with disordered spatial dynamics lie between them. As will be demonstrated later, the activity level α is proportional to the λ^+ of the excitation rule. The situation in respect of the transient period τ is much more interesting (figure 5.51). The smallest values of the transient period τ are found in classes 0, CGFG and HCGFG.

All classes with wave-like spatial dynamics are situated at the upper levels of the τ-hierarchy, except for SW, which is close to classes with disordered dynamics. With the exception of CGSW, the dispersion of the transient period τ values for members within classes is not as high as it is for the activity level α. The τ *versus* α classification is presented in figure 5.52. The position of class 0 is trivial. Most classes are concentrated in the region with high values of the transient period τ and a low level of activity α; the exceptional classes include FW, F and 2^+. The members of 2^+ have a low level of activity α and longest transient periods τ. The classes FW and F are also characterized by long transient periods τ but have a very high level of α (figure 5.52).

5.4.10 Parametric classification

Let us explore three static parameters of excitation rules: the excitation threshold θ, the excitation interval $[\theta_1, \theta_2]$ and λ^+.

The model using the rule of excitation threshold can be represented by eight rules of vector excitation. For $\theta = 3, \ldots, 8$ all models belong to class 0. Class F includes models with $\theta = 2$ (rule 128 is represented by vector (01111111)). The

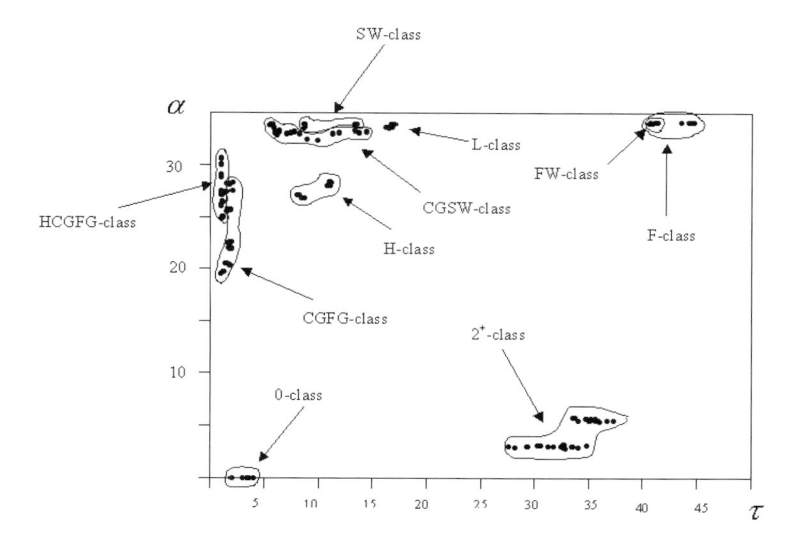

Figure 5.52. Distribution of the vector excitation rules, each rule is represented by a dot, on the (α–τ)-plane.

rule expressed via the excitation threshold $\theta = 1$ lies in class FW (this is rule 256 with vector (11111111)). Therefore, by using the threshold θ alone we can only subdivide the rules into three classes:

- damping of any excitation,
- complicated wave dynamics and
- very simple dynamics.

> *The excitation threshold θ cannot be considered to be an adequate parameter for an exhaustive classification of the possibilities inherent in the excitable lattices.*

Let us tackle the rules of the excitation interval. The relation of the interval rules to the current morphological classification is shown in table 5.3. The interval parametrizations which is, of course, two component, covers all morphological classes very well, and all the phenomena obtained for the interval classification in [23] are in strong agreement with the current results. It is possible to summarize the findings by noting that increasing the upper excitation limit leads the system from a region of 'chaotic' dynamics to the emergence of wave dynamics. To find an excitable system with localized excitations, the lower excitation limit must be increased. A careful combination of the lower and upper excitation limits allows us to produce any of the possible regimes of spatio-temporal activity.

In numerical experiments we refer to classical cellular-automata models with finite lattices, finite neighbourhoods and a finite number of cell states. This allows

Table 5.3. Correspondence between interval and morphological classifications. Note, that for $\theta_1 > 2$ all rules fall into class 0, so here we do not indicate rules employing $\theta_1 > 2$.

	θ_2							
θ_1	1	2	3	4	5	6	7	8
1	CGFG	HCGFG	CGSW	CGSW	SW	SW	FW	FW
2		2^+	H	L	F	F	F	F

us to compare our findings with Wolfram's classification of cell state transition rules [645] and the associated parametrizations due to Langton [382]. Let us take a closer look at the Wolfram classes again [645, 491, 382]:

I Evolves to a unique homogeneous state from almost all initial states. Evolves to a simple limit point in phase space.

II Evolves to a limit cycle with short period. The separated simple structures arise from particular initial patterns; they are stable or oscillate with a small period. The number of different structures generated in this class is sufficiently small. The automata may work as filters and preserve certain initial patterns, therefore they can be used in image-processing. In two-dimensional cellular automata we observe some type of crystallization of an initially homogeneous pattern into several stable domains (each region evolves towards a particular phase).

III Evolves to aperiodic chaotic patterns independently of the initial state. Several rules exhibit particle-like patterns. Evolution from a rest configuration with only one non-rest site generates fractal-like patterns. They are irreversible in the sense of decreasing entropy with time. The transition period is sufficiently long when irregular patterns are generated. Circular expansion of the patterns is typical in the two-dimensional case.

IV Propagating particle-like structures are generated in the evolution; this indicates the ability to implement universal computation. Their behaviour is rather unpredictable. The evolution is highly irreversible. Long transient periods are typical.

First of all we will try to identify the correspondences between our morphological classes and those of the Wolfram classification, using the same informal notions as Wolfram did. Class 0 obviously belongs to Wolfram's class I. Class 2^+ should belong to class IV on the basis of its morphological characteristics, but it has a short transition period. Classes H, CGFG and HCGFG correspond to class III, but as the coarseness of the homogeneous patterns increases, members migrate to the class II. Rules exhibiting wave-like excitation patterns were not discussed in Wolfram's classification because it was mainly based on investigations of one-dimensional cellular automata. One of the uncertain points in making this correspondence is that in one dimension we cannot separate particle-like waves and unidirectional waves, especially when a wavefront is formed (i.e. the velocity vector of a wave is determined by the spatial disposition of states analogous to excited and refractory states). Therefore class IV may properly include class 2^+ and classes F, L, SW and FW. Here we propose that classes SW, L, F and FW should lie between classes II and III if particle-like waves and unidirectional waves are, in fact, distinct in the one-dimensional case, or between classes III and IV if they are not distinct. The second situation seems much more realistic. We can then regard class CGSW as lying in class III with its 'coarsely grained' tail and in class II or III with its 'sawtooth wave' head. Some curious results on the existence of class IV in nature are given in [430].

Now we are ready to tackle the static parameter derived from the cell state transition rules. Let us look on what is happening to the spacetime dynamic of a *one-dimensional* cellular automaton when λ increases. The following parametrizations of the automata behaviour are obtained in [382] for one-dimensional cellular automata with binary cell states and a ternary cell neighbourhood (sometime we use wave terminology to facilitate interpretation; the corresponding values of λ are shown at the beginning of each description):

- $\lambda = 0.00$–0.15. Evolves to a uniform rest state. The length of the transition period increases proportionally with λ.
- $\lambda = 0.20$. Stable structures appear in the evolution. The transition period increases.
- $\lambda = 0.25$–0.30. Evolution depends on initial conditions. Homogeneous and heterogeneous fixed points with islands of non-rest states oscillating in time are the most frequent outcomes of the evolution. The transition period increases.
- $\lambda = 0.35$. Dynamical variability increases. The transient period is very long. Primordial propagating structures appear in the evolution.
- $\lambda = 0.40$. Patterns propagating unidirectionally become recognizable.
- $\lambda = 0.45$. Catastrophic increase in length of transient period. Directional waves appear.
- $\lambda = 0.50$. Spreading waves interacting one with other generate new waves. Large fluctuations in the dynamics lead to the collapse of activity (wave collisions and annihilation). Another enormous increase in the length of the transient period occurrs.
- $\lambda = 0.55$–0.65. Area of dynamical activity expands rapidly with time. Behaviour is chaotic.
- $\lambda = 0.70$–0.75. Behaviour is chaotic after a very short transient period.

In his work, Langton found that $\lambda = 0.75$ produces maximal dynamical disorder and $\lambda = 0.45$ corresponds to patterns with particle-like behaviour. This last value of λ was assumed to be critical [382]. Combining his results with those of Wolfram's classification, Langton proposed that an increase in λ from 0 to 0.45

leads to an increase in the complexity of the cell state transition rules, taking the system past Wolfram's classes I and II to class IV, which has maximal complexity. Increasing λ further leads to class III and a decrease in complexity [382]. Class IV, of course, is known to be able to support universal computation [645].

From a historical perspective we should mention that λ-like parametrizations were also used more than 30 years ago in the paper by Ashby and Walker [630] on the complex dynamics of automata networks and we devote the next paragraph to briefly discussing Walker–Ashby parametrizations.

In [630] networks of finite automata with a binary state set were investigated. In the Walker–Ashby model every automaton receives inputs from two other automata of the network. The network evolves in discrete time and every finite automaton calculates its next state at $t + 1$ on its internal state and the states of its two inputs at time t according to the state transition function $f : \{0, 1\}^3 \rightarrow \{0, 1\}$. The so-called internal homogeneity, designated here by the symbol γ, is introduced in [630] to

> *reflect the tendency for ... to output the same state on successive occasions.*

The internal homogeneity γ is calculated as follows [630]:

$$\gamma = \max(|\{w \in \{0, 1\}^3 : f(w) = 0\}|, |\{w \in \{0, 1\}^3 : f(w) = 0\}|).$$

Let m be the number of 1s entered in the state transition tables for Walker–Ashby automata with binary states. Then, in the case of a three-cell (three-node) neighbourhood and binary cell (node) state set the Walker–Ashby and Langton parameters relate to each other in the following manner:

$$\gamma = \max((\lambda^{-1} - 1)m, m).$$

Obviously, $4 \leq \gamma \leq 8$, and the maximal homogeneity $\gamma = 4$ corresponds to $\lambda = \frac{1}{2}$ (for binary states only).

As a result of an empirical investigation of all 256 local evolution functions, Walker and Ashby [630] found that increasing γ tends to decrease the lengths of attracting cycles in the global evolution, the length of the transient period ('disclosure lengths') and to increase the 'number of equilibrium states'. Despite the similarity in the measures the results of the parametrizations obtained in [630] and [382] are very different which may be the result of the specific relation (5.4.10) between the parameters and differences in the architectures of the investigated systems (random regular graphs in [630] and linear graphs in [382]).

In our models of excitable lattices we tried to set up a correspondence between the λ^+-parameter and our morphological classes (figures 5.53 and 5.54). Membership of the morphological classes is determined long after the transient period. Two types of initial conditions are tackled.

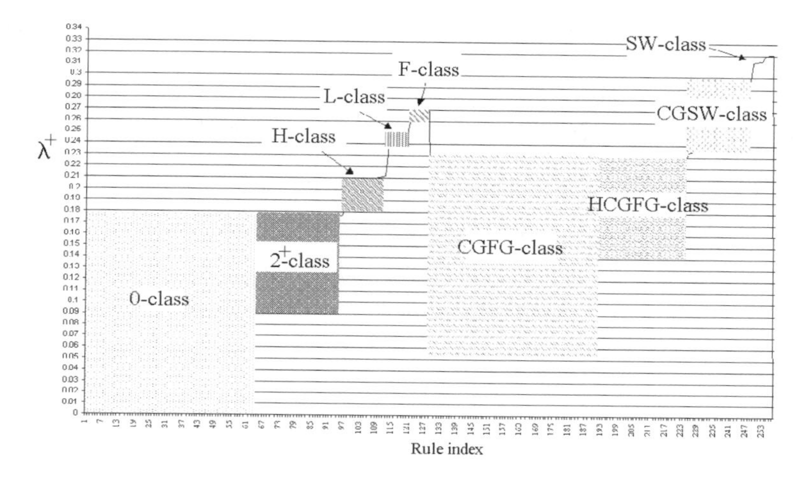

Figure 5.53. Correspondence between the values of λ^+ (vertical axis) and membership of the morphological classes in the case of 'natural' initial conditions, i.e. when initially every cell is excited with probability 0.3 and remains at rest otherwise.

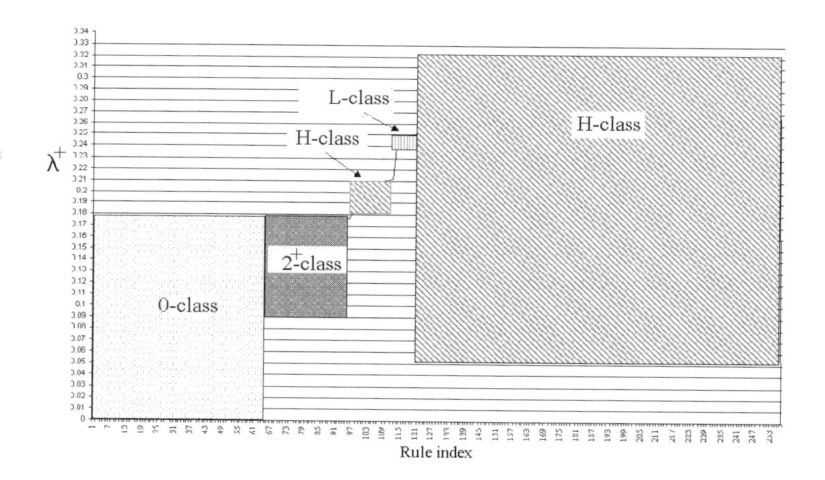

Figure 5.54. Correspondence between the values of λ^+ (vertical axis) and membership of morphological classes in 'artificial' initial conditions, i.e. when initially every cell takes rest, refractory or excited state with the probability 0.3.

In the first case (figure 5.53) we initially excited the resting lattice in such a way that every cell became an excited state with $p = 0.3$ and remained at rest with probability $p = 0.7$. The situation in which a cell may be initially at rest or excited is called the natural condition because this is typical for

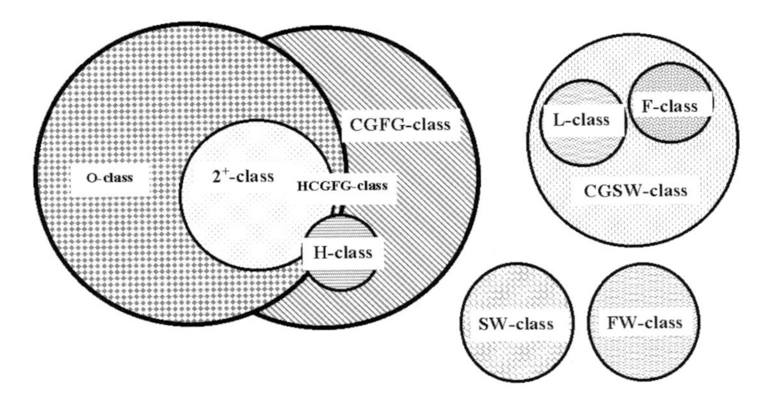

Figure 5.55. Membership relation between the morphological classes based purely on the value of λ^+ of the excitation functions.

electrophysiological experiments. Such initial conditions allow us to distinguish all classes easily.

In the second situation ('artificial initial conditions') every cell takes one of three possible states with probability $\frac{1}{3}$ (figure 5.54). This fails to discriminate between the classes: all classes except 0, 2^+, and L fall into class H. Class H itself is made up of two sections, one very large and the other very small. We can therefore see (figure 5.53 and 5.54) that it is quite difficult to discriminate between the classes using the λ^+-parameter under 'artificial initial conditions'. This becomes particularly evident when we look at the interrelationships of the classes indexed by the λ^+ parameter alone (figure 5.55).

We now consider whether it is possible to use λ^+ to predict any other characteristics of excitable media behaviour. We deal first with the two dynamic characteristics: the transient period τ and the activity level α.

In the (λ^+–τ)-plane (figure 5.56) we can see that some classes appear to be grouped together. In describing their attributes, we shall use the notions 'high', 'medium' and 'low' to indicate the positions within the non-empty range of λ^+ and the transient period τ ($\lambda^+ \in [0, 0.32]$ and $\tau \in [0, 45]$). The following groups of classes can be observed.

(i) Short transient period τ and wide range of λ^+: 0, CGFG and HCGFG.
(ii) Moderate transient period τ and medium-to-high λ^+: H, CFSW, SW and L.
(iii) Long transient period τ and medium λ^+: 2^+.
(iv) Long transient period τ and high λ^+: F and FW.

This means that we cannot use the λ^+-parameter alone to predict the transient period τ.

In the (λ^+–α)-plane (figure 5.57) we see two groups:

(i) Low level of activity α and wide range of λ^+: 0 and 2^+.
(ii) Medium-to-high level of activity α and small to high λ^+: all other classes.

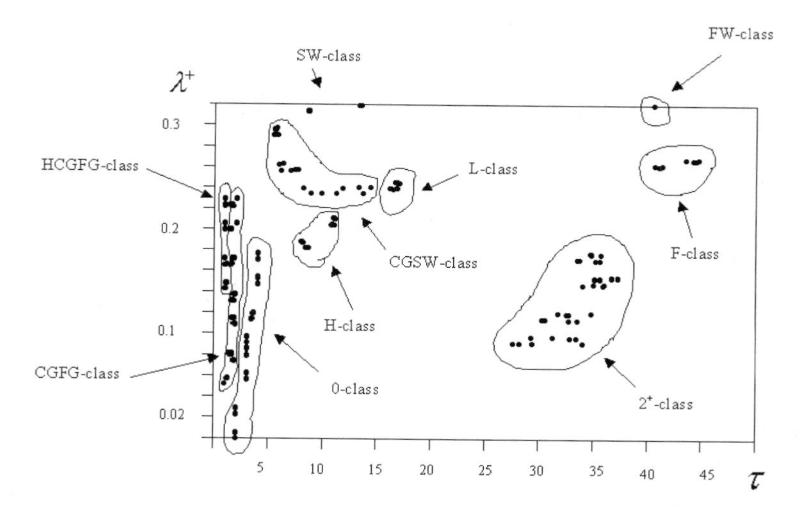

Figure 5.56. Distribution of the vector excitation rules (each rule is represented by a dot) on the (λ^+–τ)-plane.

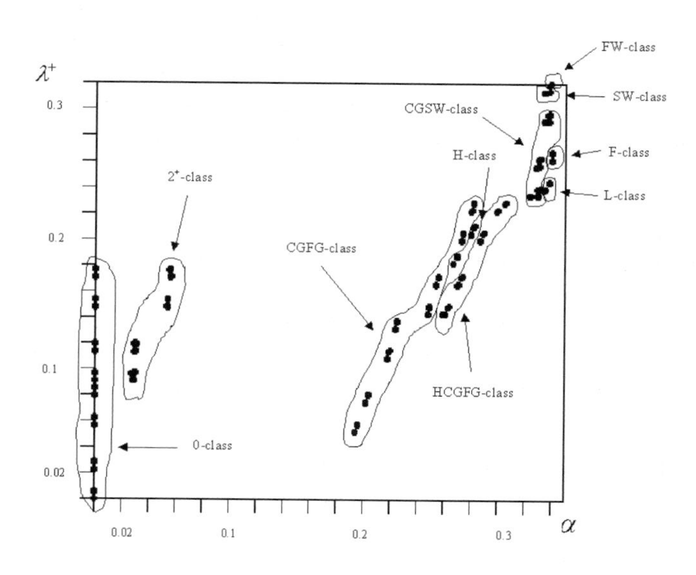

Figure 5.57. Distribution of the vector excitation rules (each rule is represented by a dot) on the (λ^+–α)-plane.

Figure 5.57 demonstrates that for all classes except 0 and 2^+ the activity level α increases with λ^+. This means that λ^+ may be useful in the prediction of the activity level α.

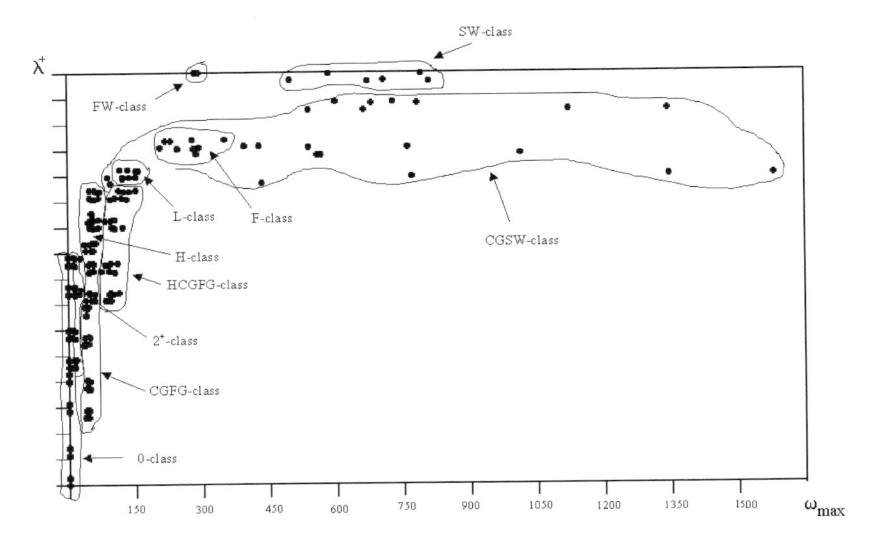

Figure 5.58. Distribution of the vector excitation rules (each rule is represented by a dot) on the (λ^+–$\omega_{\max}$)-plane.

Table 5.4. Differences between expected α' computed from the distribution of average number of clusters in a random configuration, where every cell takes an excited state with probability α' and the activity level α computed from the sample configurations in the evolution of members of the morphological classes.

Class	α'_1	α'_2	$\min_{i=1,2} \|\alpha - \alpha'_i\|$
FW	0.0	0.50	0.16
SW	0.02	0.40	0.06
2^+	0.02	0.40	0.02
F	0.03	0.39	0.05
CGSW	0.05	0.37	0.04
L	0.06	0.36	0.08
HCGFG	0.07	0.33	0.07
CGFG	0.10	0.30	0.06
H	0.12	0.28	0.01

The plots of the class membership against λ^+ and the other parameters are also interesting. The maximum cluster size $\omega_{\max}$ grows exponentially when λ^+ increases (figure 5.58), as would be expected in view of the α–λ^+ dependency. What is really surprising here is the dominance of the transitional class CGSW over all other classes in the maximum cluster size $\omega_{\max}$.

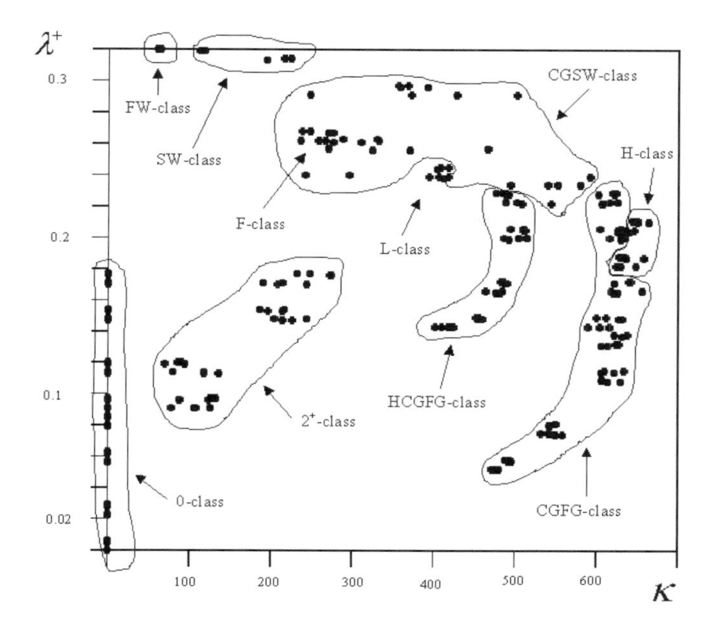

Figure 5.59. Distribution of the vector excitation rules (each rule is represented by a dot) on the $(\lambda^+–\kappa)$-plane.

In the $(\lambda^+–\kappa)$-plane (figure 5.59) the classes are much more distinct. We compare how the number of clusters κ matches the activity level α. Using κ calculated from a random configuration (figure 5.39) and the average number of clusters κ calculated on the sample configurations (table 5.2) we can extract the expected probability of the excitation (table 5.4) and compare it with the activity level α (table 5.2). Function $\kappa(\alpha)$ is parabolic with maximal value 14–16, therefore for some values of κ we have two possible values of expected excitation probability. Comparing table 5.2 with table 5.4 we find that only classes 2^+ and H do not correspond to the descending order in the tables. But at the same time they have the smallest differences between these two probabilities (0.01 for H and 0.02 for 2^+). The next best classes in ascending order are CGSW and F. The formation of clusters in class H is stochastic. At the same time the highly ordered space dynamics in class 2^+ gives us the same results. The correspondence between highly ordered and highly disordered systems may be typical. CGSW and F are also far from chaotic in space and time.

Class 2^+ is also separated from all other classes on the $(\lambda^+–\upsilon)$-plane (figure 5.60). The transitional classes CGFG, HCGFG and CGSW occupy the region $\upsilon = 25–45$ and $\lambda^+ = 0.022–0.30$. The wave classes FW, SW, and

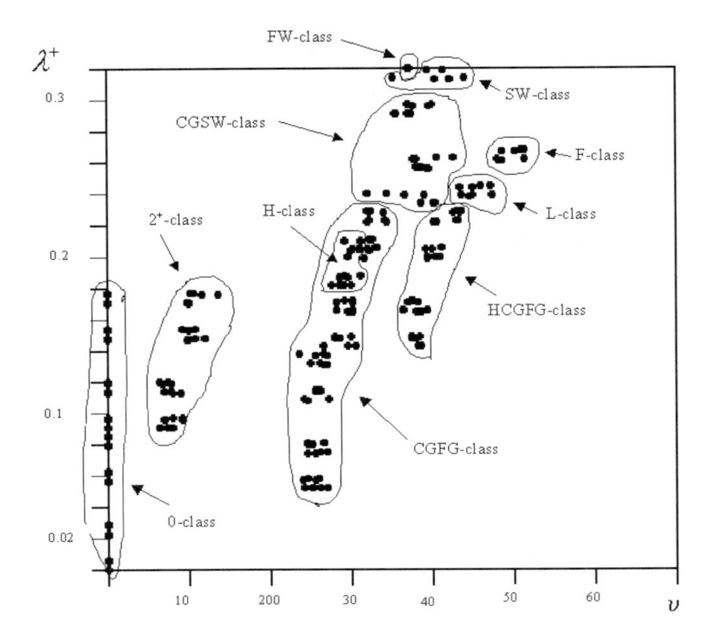

Figure 5.60. Distribution of the vector excitation rules (each rule is represented by a dot) on the $(\lambda^+ - \upsilon)$-plane.

especially F and L tend to be in regions with relatively high values of λ^+ and great diversity in the size of clusters υ (figure 5.60).

Maximal ρ is typical for the transitional classes CGSW and HCGFG (figure 5.61). Class SW has a moderate ρ for relatively high values of λ^+. All other classes can be considered to be very poor. It is remarkable that all classes except CGFG ($\rho = 1$–100) have much richer configurations than it is possible to reach by randomly assigning excited states to the lattice cells. This means that from a richness point of view all the space dynamics are not chaotic and the systems are readily identifiable [9] because any configuration which has evolved from a random initial configuration has at least one rich block that represents the complete subconfiguration of the excitable lattice; and using two consecutive configurations with rich blocks we can reconstruct the cell state transition table completely.

5.5 Computation in excitable lattices

There are three main computation classes in cellular automata and automata models of reaction–diffusion and excitation.

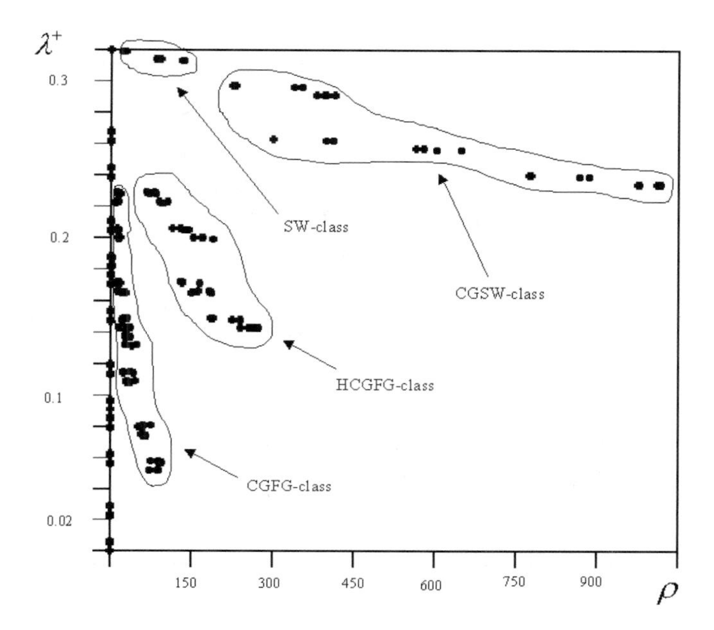

Figure 5.61. Distribution of the vector excitation rules (each rule is represented by a dot) on the $(\lambda^{+}-\rho)$-plane.

(1) *Wave computation or long-distance computation.* The excitation waves spread around the lattice and computation is implemented when waves interact with one another. The main feature of this class is that the wave, in general, should pass at a distance proportional to the size of the computational space before it comes into collision with another wave. The communication is distant but, sometimes, similar to broadcasting.

(2) *Filtering or short-distance computation.* The filters, also known as transducers, work on short time scales. The data configuration is transformed into the result configuration after the first few steps of the evolution. All communications are short distance.

(3) *Universal computation.* This is the class which implies the ability to realize basic logic gates. The logic gates are the interaction gates, i.e. the logic operation is realized at the collision sites of the quanta of information, or the bits, which are represented by movable self-localized patterns, gliders, solitons, breathers etc.

Sometimes, the boundaries between the classes are very doubtful, and we can demonstrate this in three examples. The first deals with the boundary between wave and universal computations. If we consider a one-dimensional cellular automaton it would be difficult to differentiate between gliders or solitons and trivial waves. Therefore, being in one dimension it is impossible to discern whether an automaton implements computation using gliders or breathers or with unidirectional waves. The second example shows that there is no clear difference between filters and wave computers: allowing the implementation of only one or two steps of the evolution the automaton acts as a filter but if we consider long-term evolution the initial pattern can exhibit unlimited growth and it spreads like a target wave. So, we are unable to differentiate between filters and wave computers. This is true for all dimensionalities. If we consider the Banks model of a two-dimensional cellular automaton [65] with binary cell states that has been proved to be universal we realize that actually this automaton works like a filter at every step of its evolution.

The computational capabilities of different classes of cellular automata were first discussed in Wolfram's papers [645, 491]. In his papers, Wolfram clearly underlined the capacity for universal computation in class IV, but also mentioned the possibility of using automata from class II for image-processing. Now this sort of image-processing is clearly a type of computation, and we feel that we should perhaps approach the topic of computation in cellular automata by asking what varieties of computation can be supported by various automata models, rather than by confining our inquiry solely to the subject of universal computation. We claim that cellular automata can support three basic types of computation: filtering, wave computation and universal computation. We discuss each of these in turn, with examples, and then consider the extent to which the existence of these computational regimes can be predicted in our classification.

5.5.1 Specialized computation: image transformation

Filters typically implement the standard operations of image-processing, such as erosion, dilution or contouring. In contrast to wave processors (where waves of reagents travel across or around the lattice, and information is broadcast) and universal excitable processors (where travelling self-localized excitations representing quanta of information interact far from their sites of origination), filter processors analyse and transform input data (typically an image) in a perfectly local manner, i.e. each element of the computing medium changes its state (which may correspond to some grey-scale encoding) as a function of the current configuration of the states of its closest neighbours, or of some integral characteristic calculated from the neighbourhood configuration. The obvious examples are such things as median filtration, and contour detection and enhancement. There is no need for long-term evolution when a computing medium implements one of these image-processing procedures because most of the basic operations can be done within a few time steps. Thus, for example, a cellular automaton with an eight-cell neighbourhood and binary cell states can filter noise (in the form of occasional black pixels) from a black and white image if its cells update their states in the following way: let the states of the cells represent white and black pixels, and let every black cell change its state to white if it has fewer than θ black neighbours, where θ represents the threshold of the filtration. The examples of cellular-automata-based image-processing can be found in [506].

Cellular-automata models of wave computation in reaction–diffusion processors are applied to several problems of computational geometry in [10, 12, 15, 16, 22, 21]. Working prototypes of chemical processors are designed and tested in [600, 17, 38]. Even some optimization problems, e.g. the detection of motion direction [510, 511], which involves the search for the minimum or maximum of some function in a parametric space, can be solved by wave-based algorithms in excitable media. Wave computation in excitable media structured as homogeneous neural networks was discussed in [310]. The image-processors were implemented using the Belousov–Zhabotinsky reaction in chemical media in various experimental approaches [376, 514, 515, 516, 517, 519, 520, 522, 523, 524, 525, 526, 527].

To finish this subsection logically we discuss which image-processing operations are implemented in the excitable lattices from different morphological classes.

> *Among all classes of lattice excitation the only classes F, L and SW preserve information on an image and, therefore, can be straightforwardly used in image-processing.*

> *The lattice from classes F, L and SW lose information on an image in various different ways but they preserve it analogously.*

In all experiments the image is presented in the following way: black pixels are the excited nodes of the lattice, white pixels are the rest nodes.

The situation with class FW is the clearest (figure 5.62). Consistently with the results from generators the waves that move the contour of the image inwards collide with each other ($t = 4$–10) and disappear as a result of collisions ($t = 12$). The wave representing the contour travels outward ($t = 12$–22). At every evolution step of this wave is a cluster of excited cells every one of which has exactly two excited neighbours. On a finite lattice with absorbing boundaries the wave goes outside the boundaries. On a torus-like lattice it collides with itself and disappears; that is the final configuration will be entirely at rest in any case.

The contour of the image will spread outward in lattices from class CGFG (figure 5.63) but regions of the lattice inside the spreading contour are covered by homogeneous excitation patterns after a period proportional to the initial diameter of the image ($t = 10$–22). As a result of the evolution the entire lattice is filled with homogeneous excitation pattern.

The same situation is typical for class HCGFG (figure 5.64) except that some elements of the expanding contour (namely, point-like singularities) give rise to growing fractal-like excitation patterns ($t = 18, 20$). The homogeneous activity pattern will cover the lattice in a time proportional to the lattice size.

In the evolution of CGSW (figure 5.65) and H (figure 5.66) the shell of recursive waves and core of the finely or coarsely grained patterns are formed.

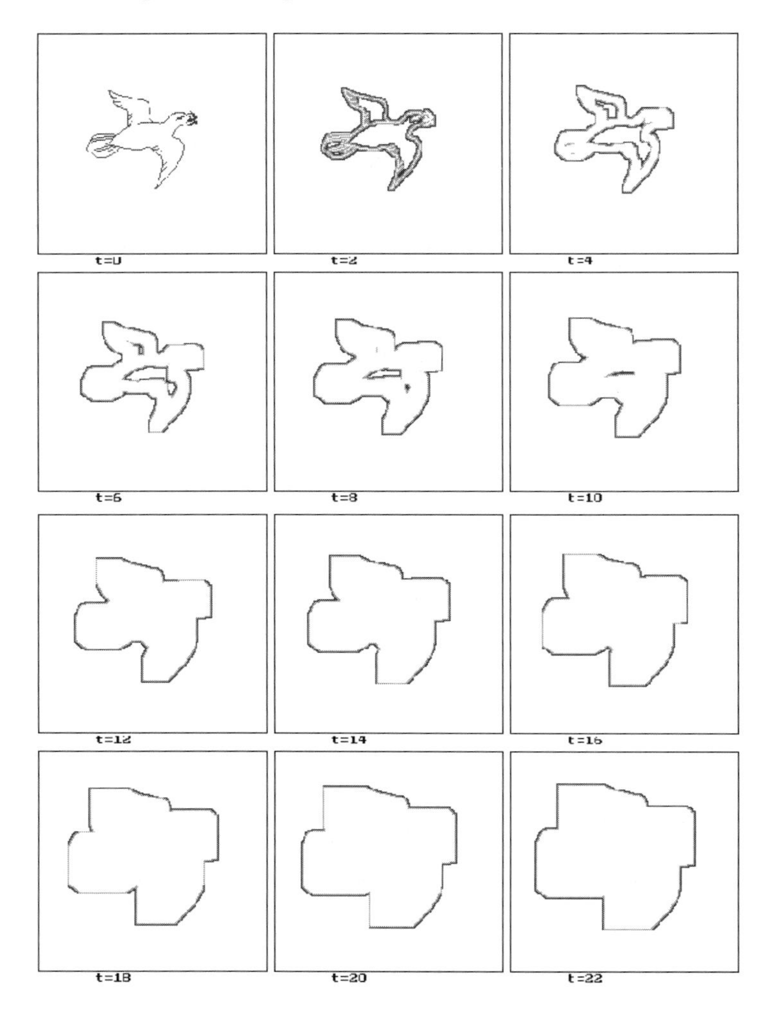

Figure 5.62. Image-processing in an excitable lattice of class FW.

The disordered core expands and covers the whole lattice in a finite time. In the evolution of F (figure 5.67) the waves travelling inside the contour of the image form spiral wave generators as a result of mutual collisions. The positions of these generators reflect in one way or another the structure of the image contour ($t = 34$).

The recognition capabilities of lattices from SW are more promising. Generators of unidirectional waves are formed because of the internal features

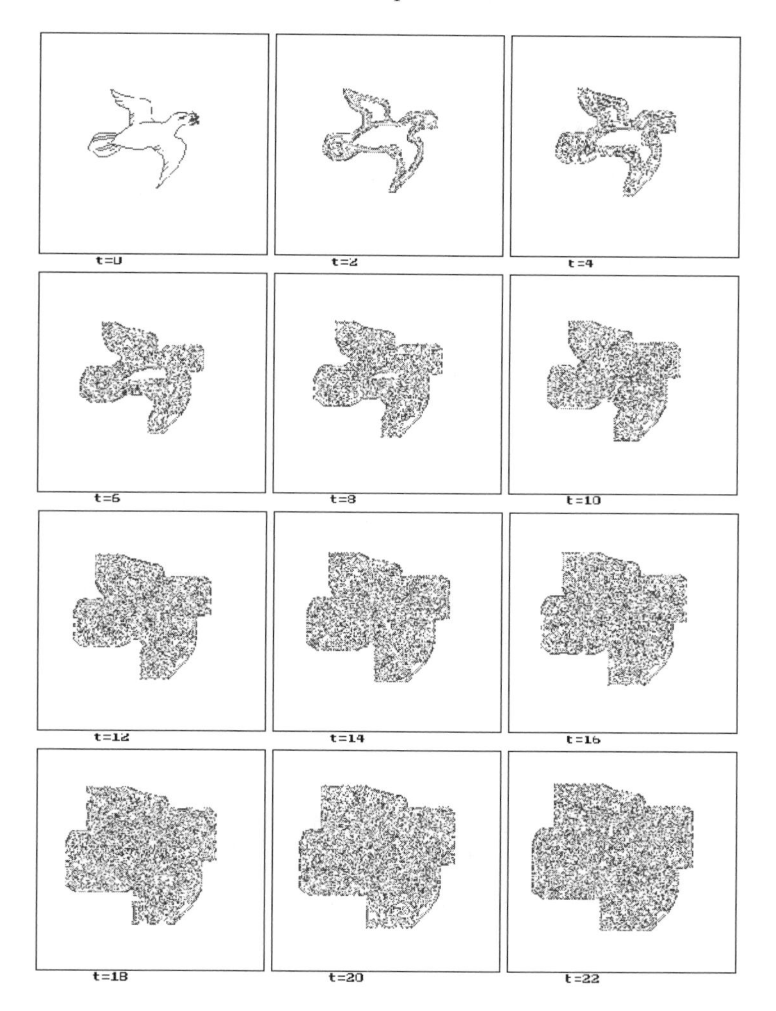

Figure 5.63. Image-processing in an excitable lattice of class CGFG.

of the image (figure 5.68). These generators do not move and, therefore, give some kind of structural representation of the image. The features of an image can be preserved in the disposition of the generators of spiral or unidirectional waves.

And, as in all previous discussions, class 2^+ is outstanding (figure 5.69). The integral structure of the image becomes broken relatively soon after the start of the evolution ($t = 4$) but various self-localized patterns with particle-like behaviour are formed. They travel on the lattice, collide each with each other and give birth to other particle-like patterns. Not surprisingly, all of the mobile patterns leave the

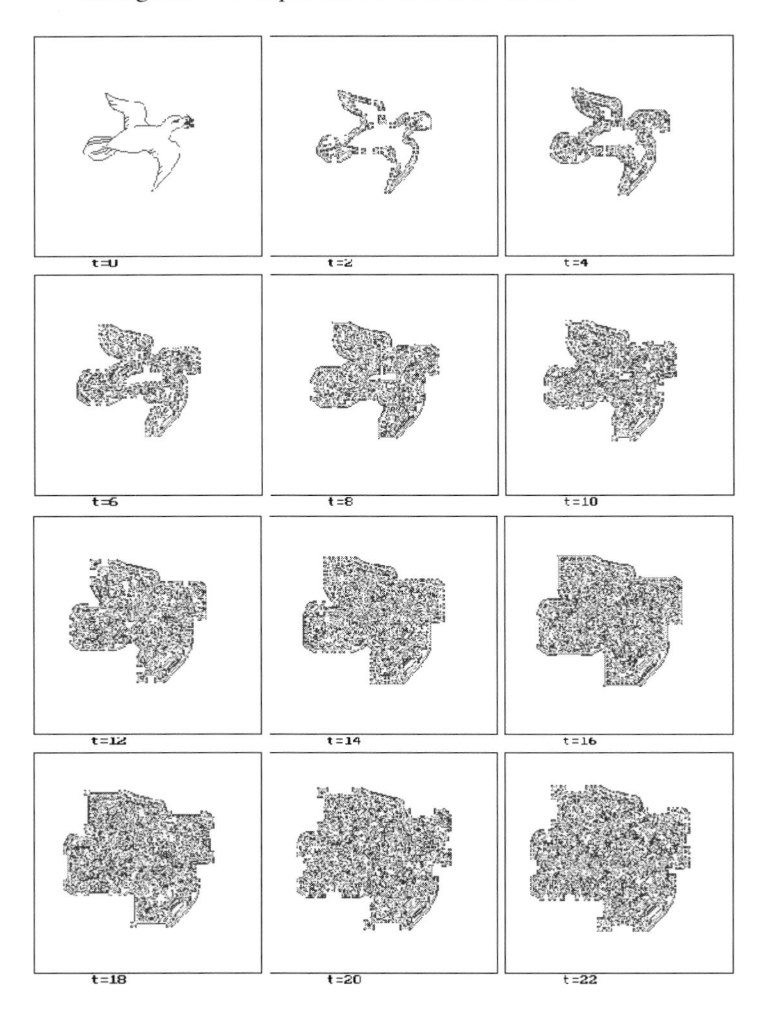

Figure 5.64. Image-processing in an excitable lattice of class HCGFG.

lattice boundaries with time. As demonstrated in [13, 14, 26] most self-localized patterns (which are still alive long after the start of the evolution) can be dissected into elementary particles (the so-called 2^+- and 3^+-particles). Assuming that the particles present bits of information we can imagine the so-called 2^+-image-processor, where an image is applied on the lattice (and presented in the excited-cell configuration) but information about the image is encoded into the 2^+- and 3^+-particles and can be detected from the boundary edges of the lattice.

At the present moment only Belousov–Zhabotinsky-reaction-based chemical

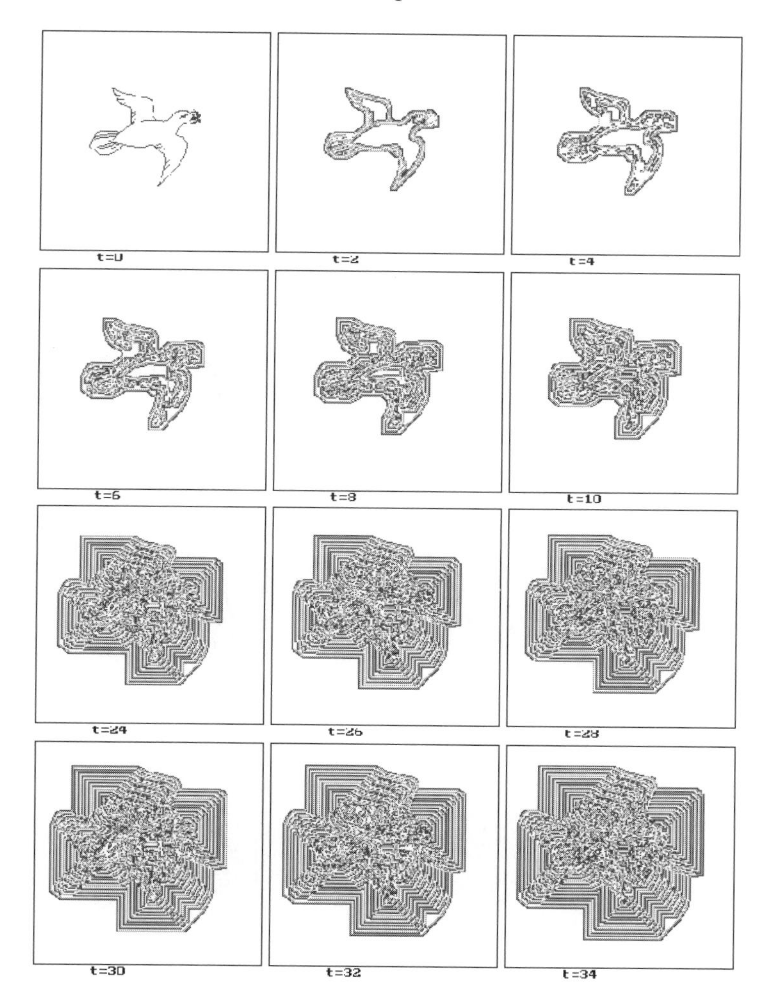

Figure 5.65. Image-processing in an excitable lattice of class CGSW.

processors have been tested for the purposes of image-processing. How do our models match the basic operations realized in Rambidi's processors? The basic operations realized experimentally in [514, 515, 516, 517, 519, 520, 522, 523, 524, 525, 526] include:

- contour enhancement;
- segmentation of an image, quasi-skeletonization;
- partial reconstruction of an image;
- filtration (removal or enhancement of small features which is implemented via dilution and erosion operations);

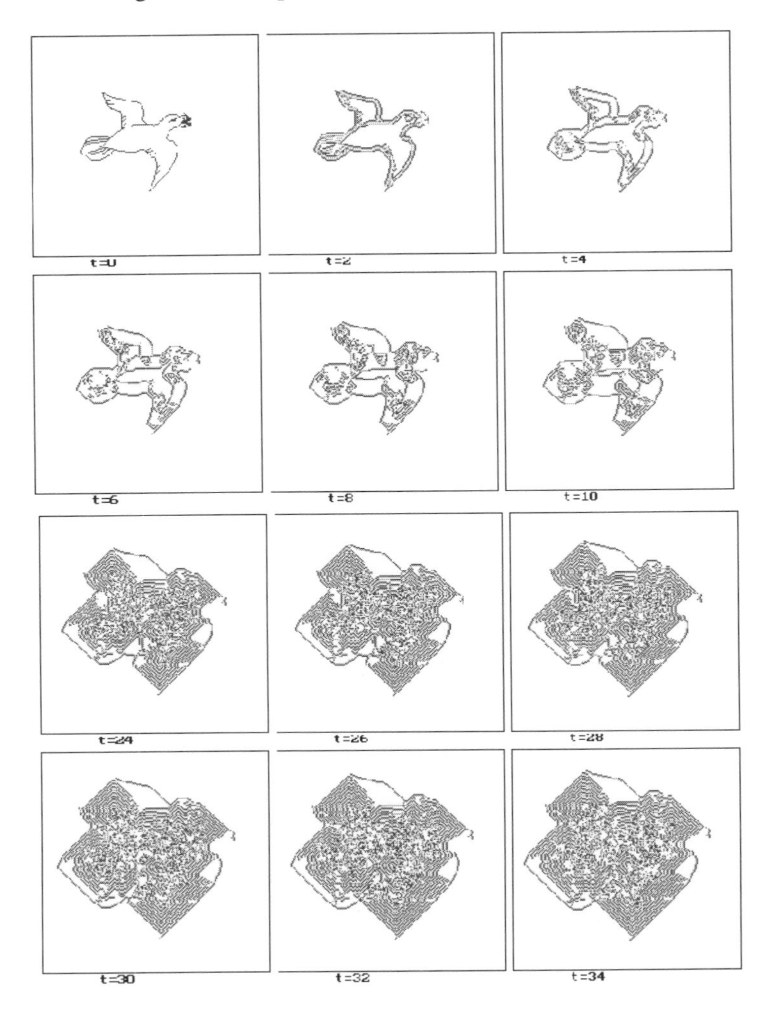

Figure 5.66. Image-processing in an excitable lattice of class H.

- synthesis of the general outline of an image.

We could also add the seventh operation—'memorizing'—the key features of an image are preserved by one or other configurations of the wave generators in the computing media, and the wave generators lie inside the convex hull of the initial image. The operation has not been discussed in experimental papers but we could predict its potential.

To analyse the overall productivity of excitable lattices for image-processing

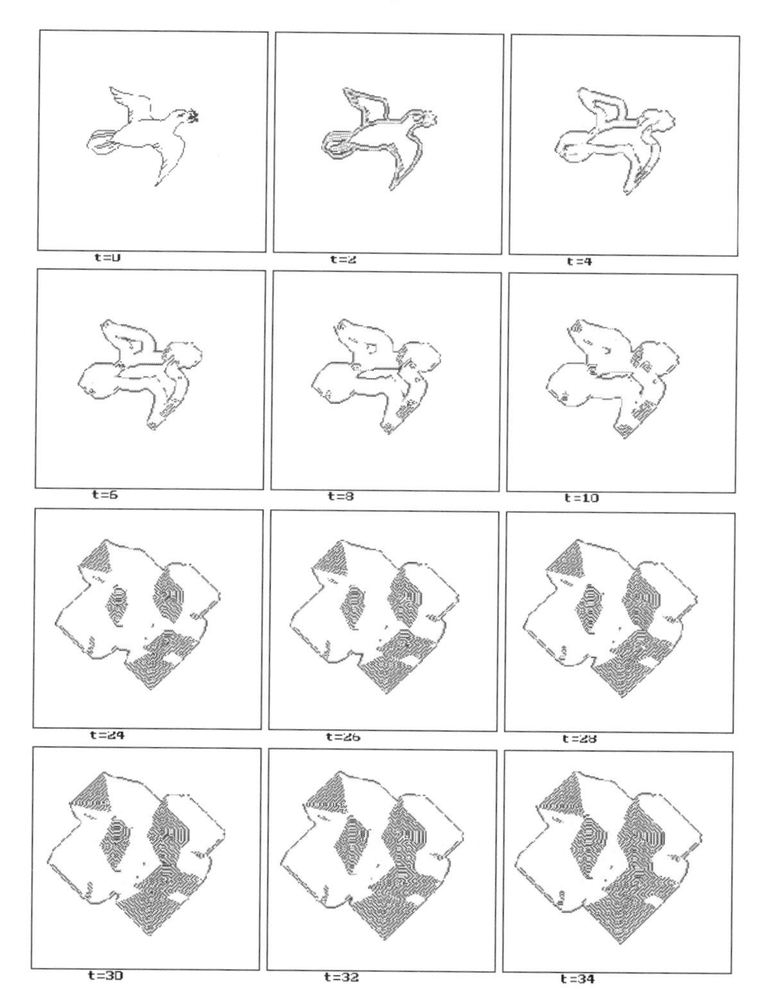

Figure 5.67. Image-processing in an excitable lattice of class F.

we allow operations to be implemented in sequences. This means that when the first operation, e.g. contour enhancement, is finished, the computation is not stopped, but the next operation is started, e.g. skeletonization, which may be followed by a third operation, e.g. synthesis of the general outline. During the implementation of these operations the wave generators can be formed and the positions of the generators may reflect the spatial features of the initial image. So, the image will be memorized. The latter notion means that some of the operations (which are logically consistent with one another) can be implemented in the same time intervals.

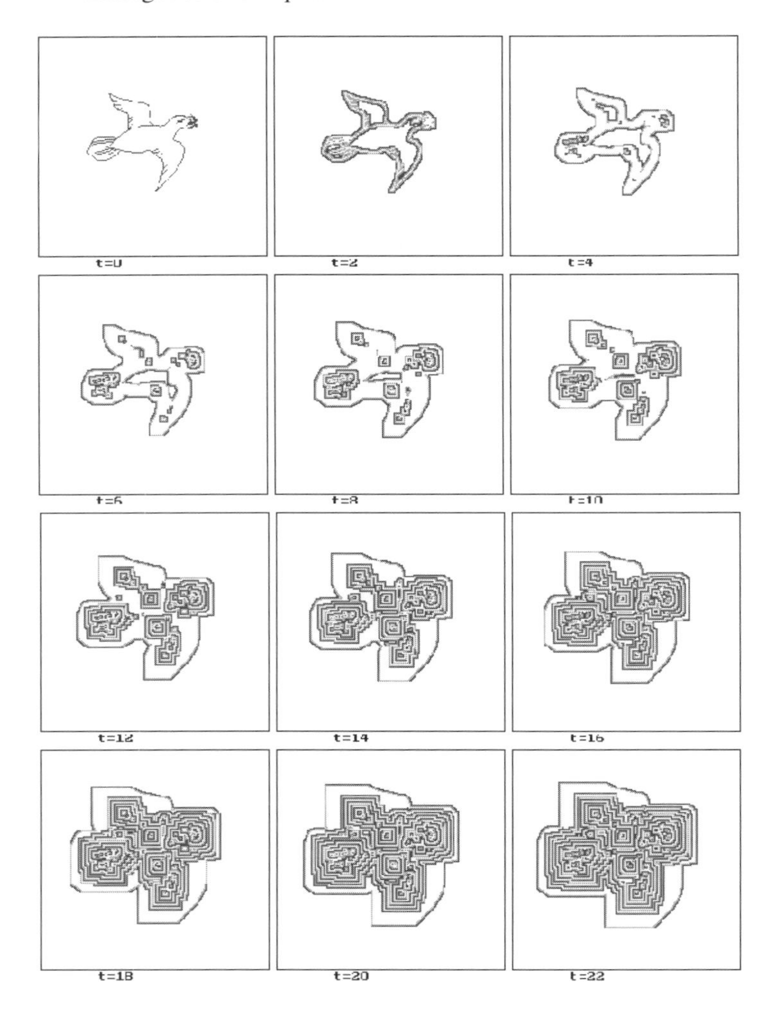

Figure 5.68. Image-processing in an excitable lattice of class SW.

We group the morphological classes into the several following operational clusters.

(i) *Contourers*: H, L, F, CGFG, CGSW, SW, FW
(ii) *Segmenters*: 2^+, HCGFG
(iii) *Skeletoners*: L, F, SW, FW
(iv) *Filters*: 2^+, H, L, F, CGSW, SW, FW

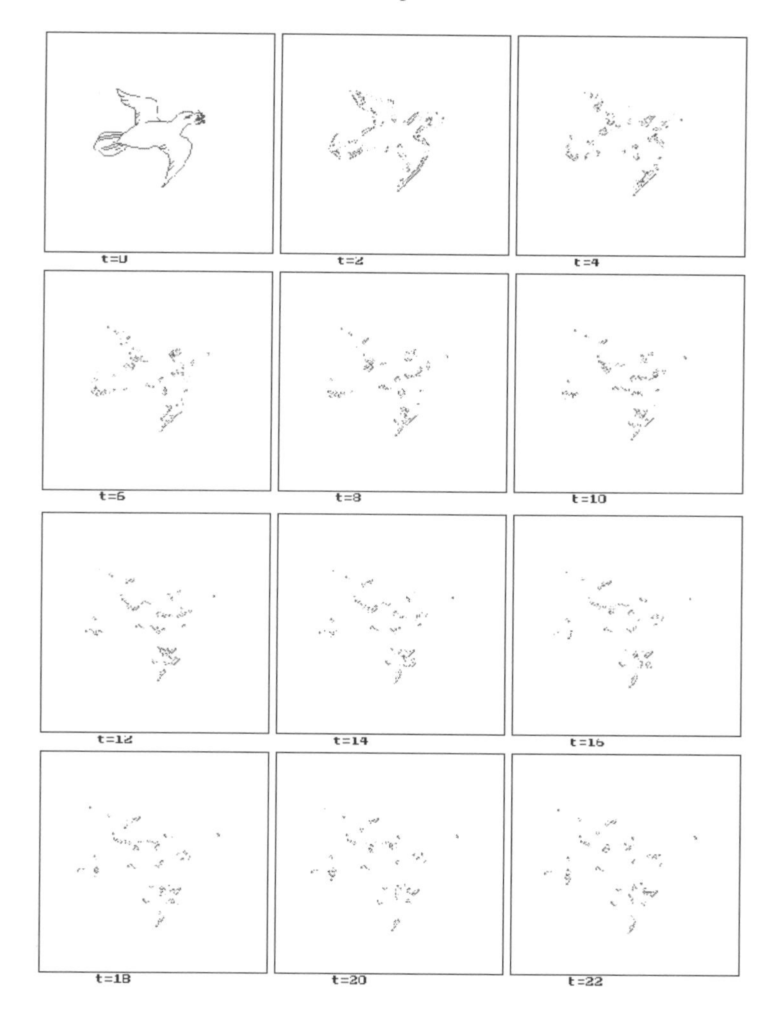

Figure 5.69. Image-processing in an excitable lattice of class 2^+.

(v) *Outliners*: CGFG, HCGFG, CGSW, FW
(vi) *Memorizers*: 2^+, L, F, SW.

The position of contourers is quite obvious. Two excitation waves appear at the second computation step around the boundaries of the original image. These operations are performed at the first computation step. Later these two contouring waves move in and out of the image. The segmentation is strongly related to the results on minimal generators: some segments of the image boundaries do not excite the outlying regions of rest lattice and, therefore, the contours become broken. The skeletonization is not absolute: the waves, generated by the image boundaries, travel inside the image contour and die as a result of collisions. Therefore, the configuration, determined at the last time step before the internal waves disappeared, reflects the skeleton of the image. Note that to compute the true skeleton of the stationary structure we need the fourth state and excitation function depending on the exact configuration of the neighbourhood (not only sums of the excited neighbours) [17]. The filters are only capable of detecting space-invariant features. As previously stated, the spatial features of an image can be represented in excitable lattices in the spatial disposition of the wave generators. The generators occur at the first evolution steps and do not change their position with time. Therefore, they represent some features of an image correctly.

5.5.2 Universal computation

Implementations of filter and wave computation can be regarded as specialized processors or 'instance machines' [668, 526]; in contrast, universal processors can compute any function. The term 'universal computation' is essentially reduced to the ability to realize any Boolean function, which in turn can be reduced to the ability to support one (or more) basic logic gates. Logic gates can be implemented in excitable media in many different ways. Advanced results on universal computing in class 2^+ were presented in chapter 4.

5.5.3 Parametrizations of computational abilities

In his key paper [382], Langton announced that he had derived a general theory about where to locate rules likely to support computation in any cellular-automata-rule space.

For his chosen example, binary automata with ternary neighbourhoods, such rules are located where the λ-parameter is 0.35–0.50. The viability of this hypothesis has also been shown using the methods of mean field theory [282] and by the analysis of the global transition graphs of one-dimensional cellular automata [190]. However, different computational abilities, as in table 5.5, have not been previously discriminated. It is generally understood that the computational abilities of cellular automata are only concerned with computation with particle-like patterns. We prefer to retain a much broader view of computation, and to think that by increasing λ it is possible to change the class of possible computations which can be implemented in a cellular automaton (table 5.5), and their models of excitable media.

The simple parametrizations of computational abilities found for one-dimensional cellular automata (table 5.5) works sufficiently well. However, it does not work as *we would like it to* in two-dimensional cellular-automata models of lattice excitation: as we have shown, the rules can be subdivided into classes with different computational abilities, but the λ^+-parameter, or intrinsic excitation probability, does not allow us to predict this subdivision in detail. As far as we know, only a two-component interval parametrization using a combination of the lower and upper excitation limits allows us to do this (table 5.6).

5.6 Motility-based classification of excitable lattices with vector excitation rules

In the previous sections we briefly discussed the computation abilities of excitable lattices with various forms of vector excitation rules. Some correspondence between morphological classes and domains of problems has been offered. Excitable lattices have been mostly separated on the basis of those which are suitable for collision-based universal computation and those that may be utilized in image-processing and, at least partially, computational geometry. On finishing the previous section we had a feeling that one more example of specialized excitable processor, even arbitrarily selected from a wide range of case studies, must be considered. We satisfy ourselves in the present section, which presents the results of computer simulation of mobile excitable lattices, the cells of which obey vector excitation rules, as defined in section 5.1.3. In the experiments we used the same design of mobile lattices, as that employed in excitable-lattice controllers for mobile robots discussed in section 5.3, with the only exception that the internal nodes of the lattices obey the rule defined in section 5.1.3. The results, presented in this section, are based mainly on our publication [32].

For each function we calculated a lattice travel time in the following manner. We chose a 'ceiling travel time' τ^*. To give lattices plenty of time to fulfil their tasks we take τ^* equal 2×10^4 in our experiments, which is several orders greater than the time (this equals 930 time units) to travel straightforwardly from a starting point to a light source. If a lattice does not reach the target in τ^* steps the attempt is deemed unsuccessful. For each rule we conducted 30 trials and recorded the number of failures and successes. In each trial the lattice was put at the same site but rotated randomly; the sequence of initial random rotations of a lattice was the same for each tested lattice excitation rule. Only the travel times of successful trials are taken into account.

From the distribution of success rates, calculated in computer experiments, we classified the excitation rules into the four following classes.

- C0: The lattices, in which the cells are governed by the functions of the group, never exhibit excitation patterns, and, therefore, do not move at all.
- C1: The majority of the lattices, namely 98% of all excitation functions, hit the target in every trial (the success rate 1.0). The remaining 2% of rules cause the lattice to reach the target with the success rate 0.99.
- C2: Almost all rules have success rates either 1.0 (44%) or 0.99 (37%). The tail of the distribution represents functions with success rates 0.97 (12%) and 0.96 (7%).

Table 5.5. λ-parametrizations of cellular automata due to their computational abilities (one-dimensional case). The classification is based on Wolfram's phenomenology [645, 491], Langton's parametrization [382] and our results on the phenomenology of excitation [23, 25].

Behaviour of λ	Computational abilities	Class of processors
	Not capable to computation because of dull dynamic.	No
	Purpose oriented computation, e.g. image filtration, because of phase transitions	Filters (no long distance communications)
Increasing downward	Universal computation because of particle-like patterns	Universal computers (dynamic networks of logic gates)
	Purpose oriented computation, e.g. geometric algorithms, because of spreading and interacting waves	Wave computers (long distance communication)
	Not capable to computation because of chaotic dynamic	No

Table 5.6. The possible computational abilities of excitable lattices in interval parametrizations; 'F' means filters, 'W' represents wave processors, and 'U' are the universal computers.

	θ_2							
θ_1	1	2	3	4	5	6	7	8
1	F	F	F	F	W	W	W	W
2	—	U	—	W	W	W	W	W

C3: The distribution of the success rates in this class is shown in figure 5.70. As we see neither of the functions has a success rate exceeding 0.74; 12% of all the functions have the rate 0.39. Other functions have rates 0, 0.3 or 0.6.

To describe rule classes compactly we recall that set **S** may be represented by a vector $\boldsymbol{v}$ such that the cell x, resting at time step t, becomes excited at time

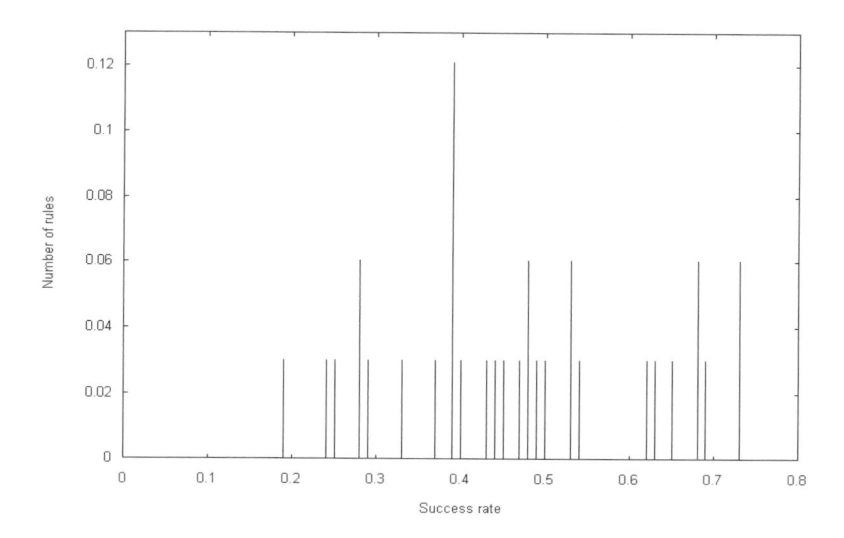

Figure 5.70. Distribution of success rates among functions of class C3.

step $t+1$ if

$$v_{\sum_{y \in u(x)} \chi(y^t, +)} = 1.$$

This means the indexes of non-zero entries of vector $\boldsymbol{v}$ represent the number of excited neighbours of cell x that cause excitation of cell x. Using this vector representation we are able to represent groups of excitation functions by Boolean expressions, which employ the notation $x \vee y$ for disjunction and xy for conjunction. Consider, for example, set $\mathbf{S} = \{2, 5, 7, 8\}$. This is represented by the vector $\boldsymbol{v} = (01001011)$ or in Boolean notation by the following expression $\overline{v_1}\, v_2 \overline{v_3}\, \overline{v_4}\, v_5\, \overline{v_6}\, v_7\, v_8$.

Thus, we can represent class C by a minimal Boolean formula $P(\mathrm{C})$ variables of which are entries of the vector $\boldsymbol{v}$. These representations are shown here.

$$P(\mathrm{C0}) = \overline{v_1}\,\overline{v_2}\,\overline{v_3}\,\overline{v_4}\,\overline{v_5}\,\overline{v_6}\,\overline{v_7}\,\overline{v_8}$$

$$P(\mathrm{C1}) = \overline{v_1}\, v_2\, (\overline{v_3} \vee x_3\, \overline{v_4}\,\overline{v_5}\,\overline{v_6}\,\overline{v_7}\,\overline{v_8}) \vee v_1\, (\overline{v_2} \vee v_2\, \overline{v_3}\,\overline{v_4}\,\overline{v_5}\,\overline{v_6}\,\overline{v_7}\,\overline{v_8}) \vee \overline{v_1}\, v_3\, v_4\, v_5\, v_6\, v_7\, v_8$$

$$P(\mathrm{C2}) = v_1\, v_2\, (v_3\, \overline{v_4}\,\overline{v_5}\,\overline{v_6}\,\overline{v_7}\,\overline{v_8} \vee \overline{v_3}\, v_4\, v_5\, v_6\, v_7\, v_8)$$

$$P(\mathrm{C3}) = v_2\, v_3\, (\overline{v_1}\,\overline{v_4} \vee v_1\, (v_4 \vee \overline{v_4}\, v_5\, v_6\, v_7\, v_8)).$$

Building a correspondence between the morphology of lattice motion trajectories on one side and spacetime excitation dynamics on the other side allows us to give a deep insight into the intrinsic processes of excitable lattice motility. We will discuss the phenomenology of the three classes: C1, C2 and C3.

We do not consider class C0 because no long-term excitation occurs in lattices of this class as $v_1 = 0$ and $v_2 = 0$.

In examples of several lattice trajectories from each class (figure 5.71(A)) we see that all lattices of class C1 move toward the target straightforwardly, subject to probabilistic excitation of edge nodes and quasi-random blasts of excitation caused by collisions of mobile self-localized excitations.

In figure 5.72 we see that an excitation is created on lattice edges relatively close to the light target. Due to the specifics of the cell state transition functions which constitute this class, no long wavefronts are formed but swarms of mobile excitation clusters emerge. These breather-like structures travel across the lattice unchanged until they reach absorbing boundaries or collide with other mobile excitations. In our previous papers (see e.g. [26]) we have shown that two basic types of mobile self-localizations can be detected in the class C1 lattice, particularly for an excitation function represented by vector $\boldsymbol{v} = (0100000)$: 2^+-particles, which move along the rows and columns of the lattice and 3^+-particles, which run along the diagonals. In most cases the velocity vectors of localizations are collinear (subject to discreteness) to the normal from the target to the lattice edge. Thus, the mobile patterns cause actuators to generate a perfect propulsive field that pushes the lattice towards the target (figures 5.71(A) and 5.72).

The spacetime dynamics of excitation in class C2 exhibit no far-spreading clusters of excitation, because newly formed wavefronts are immediately broken due the values of the third and fourth entries of the vector $\boldsymbol{v}$: $v_3 = 0$ or $v_4 = 0$ (figure 5.73). The dynamics inside the lattice is chaotically homogeneous and therefore mostly the excitation of cells near the lattice edges contributes to the formation of the local force field. This leads to losses in lattice stability to random fluctuations, which results in long curvy trajectories (figure 5.73). Remarkably, class C3 lattices spend much time circling either near their initial positions or around the target, however not hitting it (figure 5.71(B)).

Typical excitation waves are observed in class C3 lattices. Initially the waves are generated on the lattice edges and move centripetally. Pretty soon wave generators, represented by compact groups of excited and refractory cells, are formed in the internal parts of the lattice. These generators produce excitation waves, which usually annihilate waves emitted by light-sensitive lattice edges. These wave generators are responsible for a lattice losing its stability. It is very difficult to destroy generators. Because the generators periodically emit waves, the lattice sometimes starts to rotate or circle (figures 5.71(C) and 5.74).

Eventually, the lattice either stops moving towards the target (figure 5.74) or moves in circles (figure 5.75).

If we look at excitable lattices from a practical point of view, e.g. we could think about chemical or silicon implementation of mobile excitable media, we may be particularly concerned with efficiency. The efficiency ϵ of a mobile lattice is determined by the length of the travel path from the start site to the light target and also by the average energy a lattice consumes per time step. We consider actuators as abstract entities and therefore we do not take them into consideration

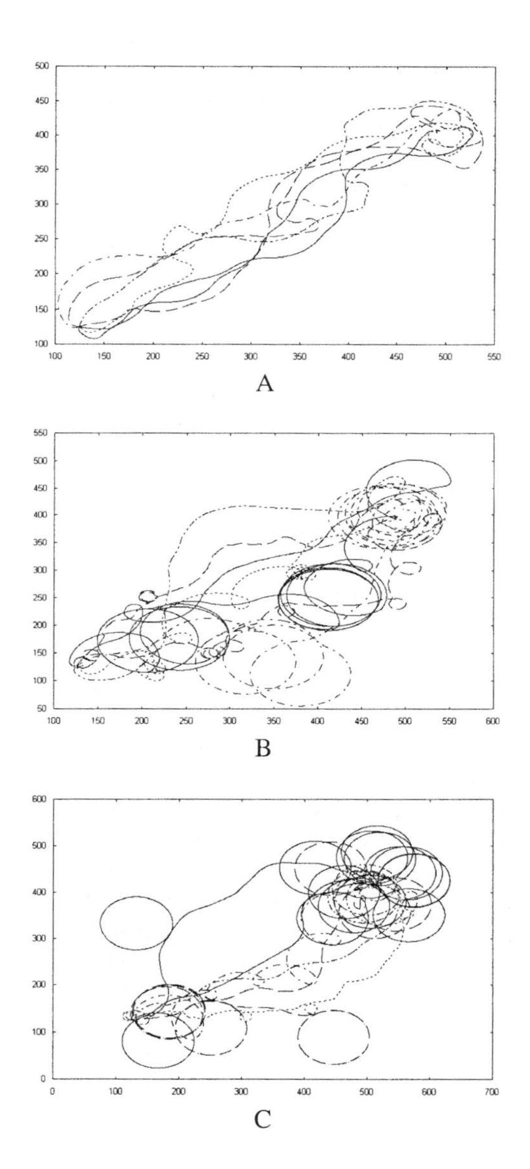

Figure 5.71. Sets of seven exemplary trajectories of robots guided by excitable lattices from different classes: (A) class C1, exemplary function $\boldsymbol{v} = (00111111)$; (B) class C2, exemplary function $\boldsymbol{v} = (11100101)$; (C) class C3, exemplary function $\boldsymbol{v} = (01110100)$. Each lattice starts its motion at the left-hand bottom corner of the arena; the light source is placed at the top right-hand corner.

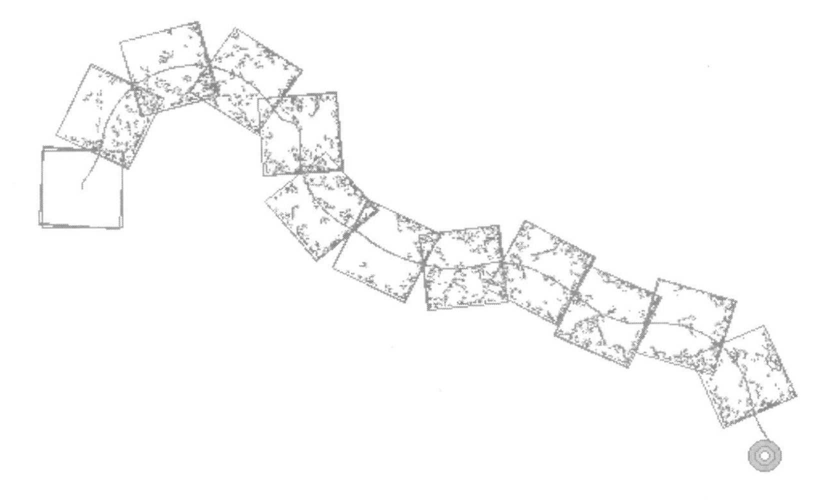

Figure 5.72. Snapshots of a lattice, governed by the function $\boldsymbol{v} = (00111111)$ of class C1, which travels towards the light target. The lattice configuration is recorded every 100th step of the experiment. The lattice travels from the left to the right.

Figure 5.73. Snapshots of a lattice, governed by the function $\boldsymbol{v} = (11101101)$ of class C2, which travels toward the light target. The lattice configuration is recorded every 100th step of the experiment. The lattice travels from the left to the right.

when analysing efficiency. So, we deal only with lattice cells.

In our model both lattice cells and actuators update their states simultaneously in discrete time, therefore we measure the travel time in discrete

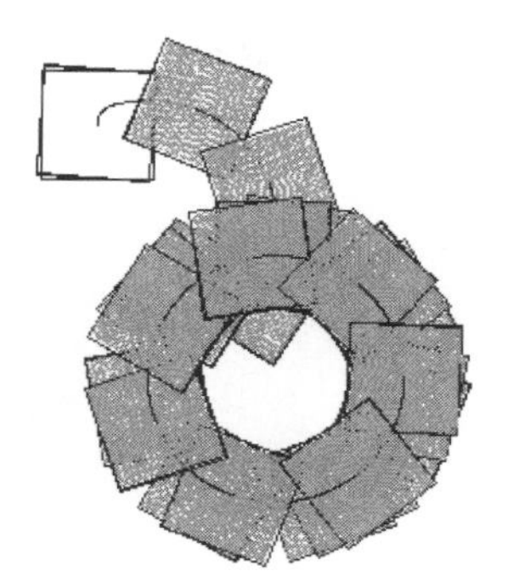

Figure 5.74. Snapshots of a lattice, governed by the function $\boldsymbol{v} = (01111101)$ of class C3, which travels toward the light target. The lattice configuration is recorded every 100th step of the experiment. The light target is shown by circles.

time steps of lattice updates. Assuming a fixed amount of energy is released when a cell is excited we can evaluate the energy as a lattice activity level. For each rule f we calculate an activity level $\alpha(f)$ as the number of excited cells summed over the whole period of a robot travelling from an initial point to the target and normalized to the travel time and number of cells in the lattice.

The distribution of travel times in successful lattice trials is shown in figure 5.76. Class C1 lattices reach the target in the shortest time. For the majority of excitation functions class C3 lattices travel 1.5 times slower than class C1 lattices. We should remember that class C1 lattices reach the target in all trials, however, only a few trials may be successful for class C3 lattices.

For any excitation function f: $\alpha(f) \leq \frac{1}{3}$.

Every lattice cell updates its state in the cycle

$$\cdots \rightarrow + \rightarrow - \rightarrow \bullet \rightarrow + \rightarrow \cdots .$$

The transitions $+ \rightarrow -$ and $- \rightarrow \bullet$ are unconditional; they happen anyway. Therefore, even in an ideal situation when all lattice cells become excited synchronously during an evolution period $[0, t]$ the maximal sum of excited cells is close to $\tilde{\alpha} = \frac{t}{3}n$, where n is the number of cells. Thus, we have a maximal value of activity level $\alpha(f) = \frac{\tilde{\alpha}}{nt} = \frac{1}{3}$. The distributions of activity levels for all possible rules are shown in figure 5.77. The distribution of activity levels of the functions from C1 has two characteristic peaks at $\alpha(f) = 0.075$ and $\alpha(f) = 0.265$; whereas almost all functions from the other two classes have near

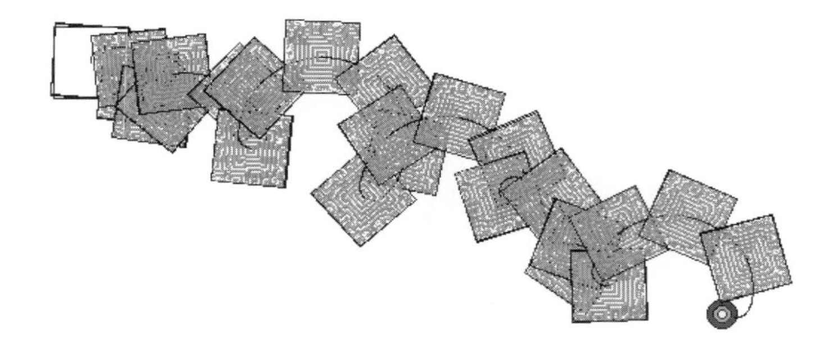

Figure 5.75. Snapshots of a lattice, governed by the function $\boldsymbol{v} = (11111001)$ of class C3, which travels toward the light target. The lattice configuration is recorded every 100th step of the experiment. The lattice travels from the left to the right.

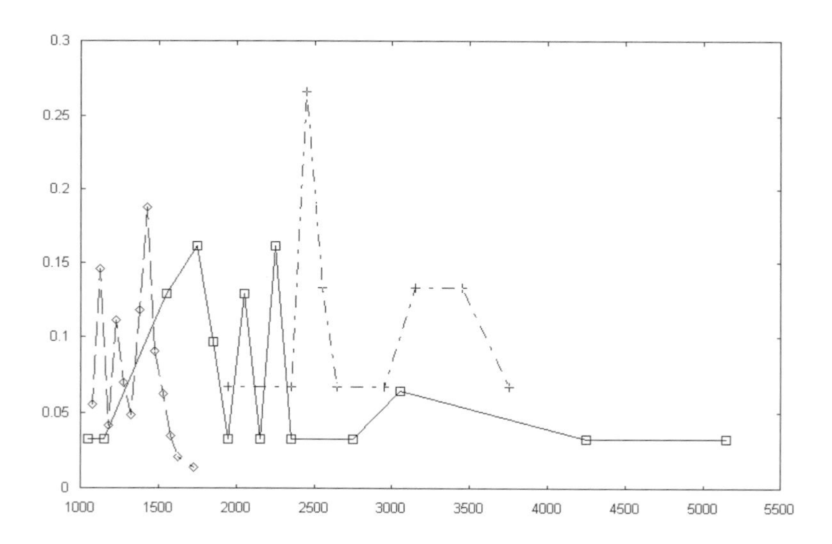

Figure 5.76. Distribution of travel times, counted only for successful trials, for lattices from classes C1 ($\diamond$), C2 ($+$) and C3 ($\square$). The horizontal axis represents travel times, the vertical axis, the normalized number of functions.

maximal activity levels: 0.315 for C2 and 0.325 for C3. So, we may speculate that

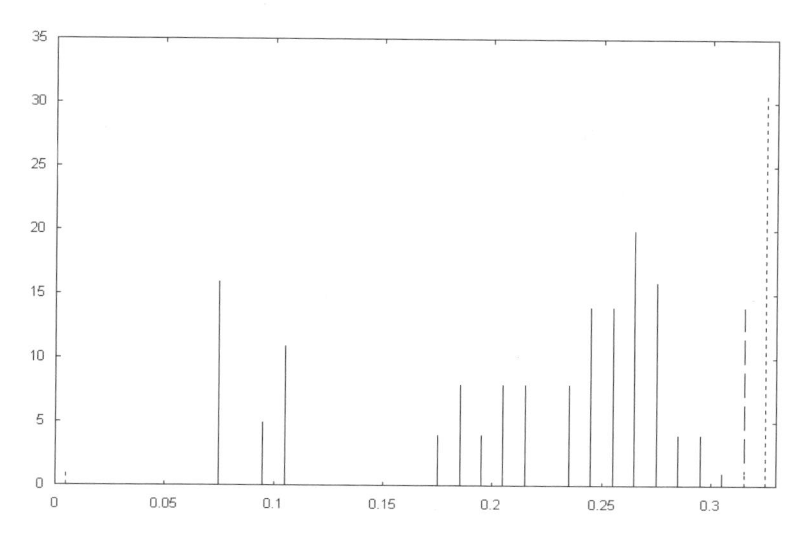

Figure 5.77. Distribution of activity levels α for the functions of classes C1 (full lines), C2 (broken lines), and C3 (dotted lines). The horizontal axis represents activity levels, the vertical axis, the number of functions.

excitable lattices with the highest levels of activity may be less suitable for the phototaxis task.

From the distributions of activity levels, travel times and success rates we can order the classes of mobile excitable lattices in the following hierarchy of efficiency:

$$\epsilon(\mathrm{C1}) > \epsilon(\mathrm{C2}) > \epsilon(\mathrm{C3}).$$

We have classified the excitation rules into several groups depending on the behaviour of mobile lattices governed by the rules and commonsense suitability of the lattices to execute phototactic behaviour. We have also given some hints on how to select 'good navigation' rules by analysing excitable-lattice activity. Obviously both the trajectory-based classification and activity-based selection require experimentation with an excitable lattice.

Is there any way to decide on the suitability of an excitation rule without experimenting with the lattice but simply by studying the rule's structure? This is generally the parametrization problem of cellular-automata rules, formulated and investigated in [630, 382] and applied to excitable lattices and lattice swarms in [25, 33]. Based on the results from our previous work (see e.g. [25, 33]) we could offer two characteristics, which may be utilized in the express design of mobile excitable lattices. The first represents some type of inhomogeneity of cell state transition rule; the second deals instead with the sensitivity of the excitation rule.

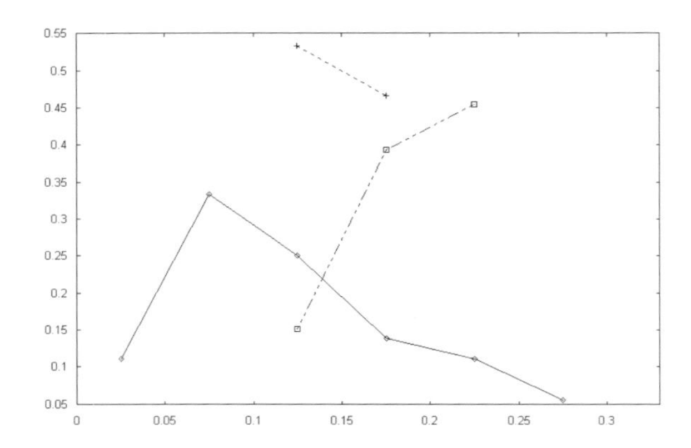

Figure 5.78. Normalized distribution of excitation functions against λ^+ values: $\diamond$ (C1), $+$ (C2), and $\square$ (C3). Only non-zero entries are shown. The horizontal axis represents values of λ^+, the vertical axis, the normalized number of functions.

For every excitation function f we can calculate the static parameter λ^+, defined in section 5.4.1. The distribution of the excitation rules with respect to their values of the parameter λ^+ is shown in figure 5.78. We see that more than one-third of all functions of class C0 have values of λ^+ concentrated in the lower part of λ^+ scale. Class C2 occupies the median of the λ^+ scale. The λ^+ values of almost half of the members of class C3 are grouped at the higher part of the λ^+ scale. Based on the results from the previous section and this λ^+ distribution we can assert the following:

> *Mobile excitable lattices governed by rules with the highest values of λ^+ are less suitable for executing phototaxis.*

The sensitivity parameter $\nu(f)$ is measured as the number of changes from 0s to 1s and from 1s to 0s in the vectors $\boldsymbol{v}$ of the local excitation function f. To appreciate the parameter ν we must recall that it is usually the excitation threshold that is employed in excitable-media models: a cell becomes excited if at least θ of its neighbours are excited. In the vector $\boldsymbol{v}$ representation of the threshold function the entries in which the indexes are less than θ are zero valued, the others take the value 1. If function f is a threshold function its ν parameter equals 1.

Recently we discussed the so-called excitation intervals (see also e.g. [23, 28, 32]), where every lattice cell is excited if the sum of its excited neighbours belongs to the interval $[\theta_1, \theta_2]$, where $1 \leq \theta_1 \leq \theta_2 \leq 8$ in the case of eight-cell neighbourhood. Obviously, function f, represented by the excitation interval $[\theta_1, \theta_2]$ has a vector $\boldsymbol{v}$ such that every entry with index $\theta_1 \leq j \leq \theta_2$ takes the value 1, and the others equal zero. Thus, we have $\nu(f) = 2$.

Generally, we can say the higher the value of $\nu(f)$ the more sensitive are the cells of an excitable lattice governed by f.

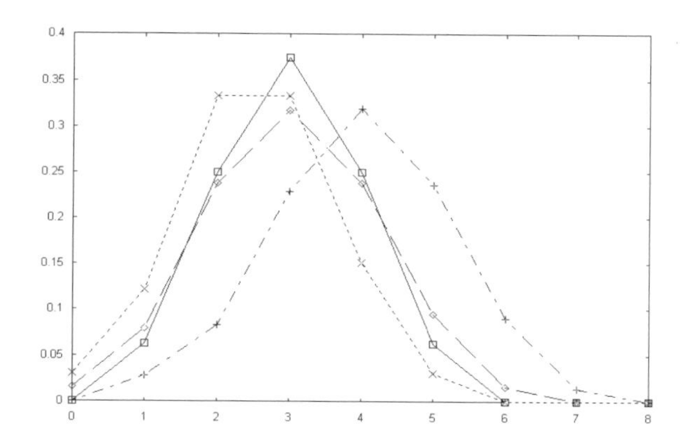

Figure 5.79. Normalized distribution of excitation functions against ν values: $\diamond$, C0; $+$, C1; $\square$, C2; $\times$, C3. The horizontal axis represents values of ν, the vertical axis, the normalized number of functions.

As we see in the distribution of functions on the parameter ν (figure 5.79) classes C3 and C0 have the lowest values of ν while classes C1 and C2 occupy the top of the ν hierarchy.

Given a choice we must prefer the excitable lattice with the most sensitive local excitation function to design mobile excitable lattices.

Both the parameters λ^+ and ν are structural in the sense that they are calculated by cell state transition rules of cellular automata and the parameters themselves do not take into account any type of dynamical computation or analysis of the spacetime configurations of the evolving systems.

As a result of computer experiments with two-dimensional excitable lattices coupled with actuators we found that the set of 256 cell excitation rules can be roughly split into four main classes. The lattices of the first class, C0, do not exhibit persistent excitation patterns; therefore they do not react to a stimulus and hence they do not move. Members of the second class, C1, show an excellent phototactic response. Their trajectories are impeccable. The lattices of the third class, C2, do well in general. However, their trajectories are lengthy and certainly not optimal. Class C2 is somewhat transitional ranging from lattices behaving straightforwardly to lattices behaving weirdly. The fourth class, C3, is constituted of lattices, which also burst into excitation when illuminated; however, their spacetime excitation dynamic is not specifically reflected in perfect locomotive activity.

The present phenomenological findings on the spacetime dynamics of mobile excitable lattices are consistent with our previous results (see e.g. [25]) on vector-based excitation rules. In our earlier paper [25] we classified lattices by their computational abilities. Does computational ability imply good phototactic behaviour?

Everyone is most interested in whether the excitable lattices which play a role in massively parallel universal computers or at least specialized image-processing devices show any signs of strong phototactic behaviour. There is no certain answer. Some C1 lattices produce self-localized excitations. Therefore, they can implement collision-based universal computing [26]. They are also move straightforwardly toward the light source. However, a significant number of the members of C1 do not do any computation (at least for known problems). Nevertheless they are good navigators. Inversely, C3 lattices can be employed in image-processing and computational geometry [31]; however, these lattices are very poor navigators.

Ideally, it would be impressive to verify our theoretical results with real-life implementations. Unfortunately, at this stage we cannot offer any artificially designed mobile arrays of actuators with cellular-automata architecture. Fabrication of the arrays, possibly based on monomolecular arrays, such as Scheibe aggregates [70, 71], may be a task of the near future.

Bibliography

[1] Abel M and Pikovsky A 1997 Parametric excitation of breathers in a nonlinear lattice *Z. Naturf. A: J. Phys. Sci.* **52** 570–2

[2] Abel M, Flach S and Pikovsky A 1997 Localization in a coupled standard map lattice *Physica* D **119** 4–21

[3] Ablowitz M J, Keiser J M and Takhtajan L A 1991 Class of stable multistate time-reversible cellular automata with rich particle content *Phys. Rev.* A **44** 6909–12

[4] Adamatzky A 1991 Neural algorithm for constructing minimum spanning tree of a finite planar set *Neural Networks World* **6** 335–9

[5] Adamatzky A 1992 Local parallel algorithms for computation of Voronoi diagram *Elektronnoe Modelirovanie* **14** 26–8

[6] Adamatzky A I 1992 Local parallel constructing of graphs with vertices on the plane *Elektronnoe Modelirovanie* **14** 24–8

[7] Adamatzky A 1993 Massively parallel algorithm for inverting Voronoi diagram *Neural Networks World* **5** 385–92

[8] Adamatzky A 1993 Complexity of identifying asynchronous nonstationary cellular automata *J. Comput. Syst. Sci.* **31** 127–36

[9] Adamatzky A 1994 *Identification of Cellular Automata* (London: Taylor and Francis)

[10] Adamatzky A 1994 Reaction–diffusion algorithm for constructing discrete generalized Voronoi diagram *Neural Networks World* **6** 635–43

[11] Adamatzky A 1994 Hierarchy of fuzzy cellular automata *Fuzzy Sets and Systems* **62** 167–79

[12] Adamatzky A 1995 Computation of discrete hull in homogeneous automata networks *Neural Networks World* **3** 241–54

[13] Adamatzky A 1995 Controllable transmission of information in excitable media: the 2^+-media *Adv. Mater. Opt. Electron.* **5** 145–55

[14] Adamatzky A 1996 On the particle like waves in the discrete model of excitable medium *Neural Networks World* **1** 3–10

[15] Adamatzky A 1996 Voronoi-like partition of lattice in cellular automata *Math. Comput. Modelling* **23** 51–66

[16] Adamatzky A 1996 Computation of shortest path in cellular automata *Math. Comput. Modelling* **23** 105–13

[17] Adamatzky A and Tolmachiev D 1997 Chemical processor for computation of skeleton of planara shape *Adv. Mater. Opt. Electron.* **7** 135–9

[18] Adamatzky A 1997 How cellular automaton plays Minesweeper *Appl. Math. Comput.* **85** 127–37

[19] Adamatzky A 1997 Universal computation in excitable media: the 2^+-media *Adv. Mater. Opt. Electron.* **7** 263–72

[20] Adamatzky A and Holland O 1997 Swarm intelligence: representations and algorithms *Proc. Int. Workshop on Distributed Artificial Intelligence and Multi-Agent Systems (St Petersburg, Russia)* (St Petersburg: Institute of Informatics, RAS) pp 13–22

[21] Adamatzky A 1997 Cellular automata labyrinths and solution findings *Comput. Graphics* **21** 519–22

[22] Adamatzky A and Holland O 1998 Voronoi-like nondeterministic partition of a lattice by collectives of finite automata *Math. Comput. Modelling* **28** 73–93

[23] Adamatzky A and Holland O 1998 Phenomenology of excitation in 2D cellular automata and lattice swarm *Chaos, Solitons and Fractals* **9** 1233–65

[24] Adamatzky A 1998 Dynamical universal computation in excitable lattices *Proc. Coll. on Universal Machines and Computation (Metz)* **2** 194–214

[25] Adamatzky A and Holland O 1998 Parametrisation of the local excitation of 2D lattices: morphology, dynamics and computation, unpublished

[26] Adamatzky A 1998 Universal dynamical computation in multidimensional excitable lattices *Int. J. Theor. Phys.* **37** 3069–108

[27] Adamatzky A, Holland O, Rambidi N G and Winfield A 1999 Wet artificial brains: Towards the chemical control of robot motion by reaction–diffusion and excitable media *Lecture Notes in Computer Science* **1674** 304–13

[28] Adamatzky A and Melhuish C 2000 Parallel controllers for decentralized robots: toward nano design *Kybernetes* **29** 733–45

[29] Adamatzky A 2000 Pathology of collective doxa. Automata models *Appl. Math. Computation* at press

[30] Adamatzky A 2001 Space-time dynamic of normalized doxatons: Automata models of pathological collective mentality *Chaos, Solitons and Fractals*

[31] Adamatzky A 2001 Reaction–diffusion and excitable processors: a sense of the uncoventional *Parallel and Distributed Computing Practices*

[32] Adamatzky A and Melhuish C 2001 Phototaxis of mobile excitable lattices *Chaos, Solitons and Fractals*

[33] Adamatzky A 2000 Choosey hot sand: reflection of grain sensitivity on pattern morphology *Int. J. Mod. Phys.* C **11** 47–68

[34] Adamatzky A and Holland O 2001 Reaction–diffusion and ant-based load balancing of communication networks *Kybernetes*

[35] Adleman L M Molecular computations of solutions to combinatorial problems *Science* **266** 1021–4

[36] Aggarwal A, Guibas L J, Saxe J and Shor P W 1989 A linear-time algorithm for computing the Voronoi diagram of a convex polygon *Discrete Comput. Geom.* **4** 591–604

[37] Agladze K, Budriene L, Ivanitsky G, Krinsky V, Shakhbazyan V and Tsyganov M 1993 Wave mechanisms of pattern formation in microbial populations *Proc. R. Soc.* B **253** 131–5

[38] Agladze K, Magome N, Aliev R, Yamaguchi T and Yoshikawa K 1997 Finding the optimal path with the aid of chemical wave *Physica* D **106** 3–4, 247–54

[39] Agladze K, Aliev R R and Yamaguchi T 1996 Chemical diode *J. Phys. Chem.* **100** 13 895–7

[40] Ahuja M and Zhu Y 1989 A distributed algorithm for minimum weight spanning tree based on echo algorithms *Proc. 9th Int. Conf. Distr. Computing Syst.* (New York: IEEE, Computer Society Press) pp 2–8

[41] Aladyev V 1980 *Mathematical Theory of Homogeneous Structures and Their Applications* (Tallinn: Valgue)

[42] Albers G, Guibas L J, Mitchell J S B and Roos T 1998 Voronoi diagram of moving points *Int. J. Comput. Geom. Appl.* **8** 365–79

[43] Aldous D J 1990 The random walk construction of uniform spanning trees and uniform labelled trees *SIAM J. Discr. Math.* **3** 450–65

[44] Aleliunas R, Karp R M, Lipton R J, Lovasz L and Rackoff C 1979 Random walks, universal traversal sequences, and complexity of maze problems *Proc. 20th IEEE Symp. Foundations of Computer Science* (New York: IEEE) pp 218–33

[45] Aleksander I (ed) 1989 *Neural Computing Architectures: The Design of Brain-Like Machines* (Cambridge, MA: MIT)

[46] Aliev R R and Panfilov A V 1996 A simple two-variable model of cardiac excitation *Chaos, Solitons and Fractals* **7** 293–301

[47] Aliev R R 1994 Oscillation phase dynamics in the Belousov–Zhabotinsky reaction. Implementation to image processing *J. Phys. Chem.* **98** 3999–4002

[48] Allouche J P and Reder C 1984 Oscillations spatio-temporelles engendrées par un automate cellulaire *Discr. Appl. Math.* **8** 215–54

[49] Alnuweiri H M and Kumar V K P 1991 Processor-time optimal parallel algorithms for digitized images on mesh-connected processor arrays *Algorithmica* **6** 698–733

[50] Alvarez L and Esclarin J 1997 Image quantization using reaction–diffusion equations *SIAM J. Appl. Math.* **57** 153–75

[51] Anderson R, Lovász L, Shor P, Spencer J, Tardós E and Winograd S 1989 Disks, balls and walls: Analysis of a combinatorial game *Am. Math. Monthly* **96** 481–93

[52] Anonymous author

[53] Appleby S and Steward S 1994 Mobile agents for control in telecommunications networks *British Telecom Technol. J.* **12** 104–13

[54] Arcelli C and Pamella G 1995 Finding grey-skeletons by iterated pixel removal *Image and Vision Comput.* **13** 159–67

[55] Atallah M J and Kosaraju S R 1984 Graph problems on a mesh-connected processor array *J. ACM* **31** 649–67

[56] Babloyantz A and Sepulchre J A 1991 Front propagation into unstable media: a computational tool *Nonlinear Wave Processes in Excitable Media* ed A V Holden, M Markus and H G Othmer (New York: Plenum) pp 343–5

[57] Baier H and Bonhoeffer F 1992 Axon guidance by gradients of a target-derived components *Science* **255** 472–5

[58] Bak P, Tang C and Wiesenfeld K 1987 Self-organized criticality: an explanation of $1/f$ noise *Phys. Rev. Lett.* **59** 381–4

[59] Bak P, Tang C and Wiesenfeld K 1988 Self-organized criticality *Phys. Rev.* A **38** 364–74

[60] Balakhovsky L S 1961 On the possibility of a modeling of simple behavioral acts by discrete homogeneous media *Probemy Kibernetiki* **5**

[61] Balakhovsky L S 1965 Certain modes of movement of an excitation in an ideal excitable tissue *Biophyzika* **10** 1175

[62] Banâtre J-P and Le Métayer D 1990 The gamma model and its discipline of programming *Sci. Comput. Programming* **15** 55–77

[63] Bang O and Peyrard M 1996 Generation of high-energy localized vibrational modes in nonlinear Klein–Gordon lattices *Phys. Rev.* E **53** 41–3

[64] Bang O, Christiansen P L, If F, Rasmussen K Ø and Gaididei Y 1994 Temperature effects in a nonlinear model of monolayer Scheibe aggregates *Phys. Rev.* E **49** 4627–35

[65] Banks E R 1971 Information and transmission in cellular automata *PhD Dissertation* (Cambridge, MA: MIT)

[66] Banzhaf W, Dittrich P and Rauhe H 1996 Emergent computation by catalytic reactions *Nanotechnology* **7** 307–14

[67] Barkley D 1991 A model for fast computer simulation of waves in excitable media *Physica* D **49** 61–70

[68] Barten W, Lücke, Kamps M and Schmitz R 1995 Convection in binary fluid mixtures. II. Localized traveling waves *Phys. Rev.* E **51** 56–62

[69] Bartnik E A, Blinowska K J and Tuszynski J A 1991 The possibility of an excitonic fast and nearly lossless energy transfer in biomolecular systems *Phys. Lett.* A **159** 67–72

[70] Bartnik E A, Blinowska K J and Tuszynski J A 1992 Analytical and numerical modelling of Scheibe aggregates *Nanobiology* **1** 239–50

[71] Bartnik E A and Tuszynski J A 1993 Theoretical models of energy transfer in two-dimensional molecular assemblies *Phys. Rev.* E **48** 1516–28

[72] Bauer G J, McCaskill J S and Otten H 1989 Traveling waves of *in vitro* evolving RNA *Proc. Natl Acad. Sci., USA* **86** 7937–41

[73] Bays C 1991 A new game of three-dimensional Life *Complex Systems* **5** 15–18

[74] Beaver D 1995 Molecular computing *Technical Report* Penn State University

[75] Beaver D 1995 A universal molecular computer *Technical Report* CSE-95-001, Penn State University

[76] Beckers R, Holland O E and Deneuborg J L 1994 From local actions to global tasks: stigmergy and collective robotics *ALIFE IV: Proc. 4th Int. Workshop on the Synthesis and Stimulation of Living Systems* ed R A Brooks and P Maes (Cambridge, MA: MIT) pp 181–9

[77] Belousov V 1984 Synergetics and biological morphogenesis *Self-Organization: Autowaves and Structures Far From Equilibrium* ed V I Krinsky (Heidelberg: Springer) pp 204–8

[78] Bennet C H 1982 The thermodynamics of computation—a review *Int. J. Theor. Phys.* **21** 905–40

[79] Bennett C H 1995 Universal computation and physical dynamics *Physica* D **86** 268–73

[80] Benioff P 1980 The computer as a physical system: a microscopic quantum mechanical Hamiltonian model of computers as represented by Turing machines *J. Stat. Phys.* **22** 563–91

[81] Berlekamp E, Conway J and Guy R 1982 *Winning Ways* vol 2 (New York: Academic)

[82] Berry G and Boudol G 1990 The chemical abstract machine *Proc. 17th Ann. ACM Symp. Principles of Programming Languages (San Francisco, CA)* (New York: ACM) pp 81–94

[83] Berry G and Boudol G 1992 The chemical abstract machine *Theor. Comput. Sci.* **96** 217–48

[84] Bhattacharyya P 1997 Transients in a one–dimensional probabilistic cellular automaton *Physica* A **245** 472–84

[85] Bitar J and Goles E 1992 Parallel chip firing games on graphs *Theor. Comput. Sci.* **92** 291–300

[86] Bjornstad O N, Ims R A and Lambin X 1999 Spatial population dynamics: analyzing patterns and processes of population synchrony *Trends in Ecology and Evolution* **14** 427–32

[87] Blagoeva A B, Dinev S G, Dreischuh A A and Naidenov A 1991 Light bullets formation in a bulk media *IEEE J. Quantum Electron.* **27** 2060

[88] Blittersdorf R, Müller and Schneider F W 1995 Chemical visualization of Boolean functions: A simple chemical computer *J. Chem. Educ.* **72** 760–3

[89] Blum H 1973 Biological shape and visual science *J. Theor. Biol.* **38** 205–87

[90] Bobenko A, Bordemannm M, Gunn C and Pinkall U 1993 On 2 integrable cellular-automata *Commun. Math. Phys.* **158** 127–34

[91] Boerlijst M C, Lamers M E and Hogeweg P 1993 Evolutionary consequences of spiral waves in a host-parasitoid system *Proc. R. Soc.* B **253** 15–18

[92] Boghnosian B M and Taylor W 1998 Quantum lattice-gas model for the many-particle Schrödinger equation in *d* dimensions *Phys. Rev.* E **57** 54–66

[93] Böhringer K-F, Donald B R and MacDonald N C 1999 Programmable force fields for distributed manipulation, with applications to MEMS actuator arrays and vibratory parts feeders *Int. J. Robotics Res.* **18** 168–200

[94] Bonabeau E 1997 From classical models of morphogenesis to agent-based models of pattern formation *Artificial Life* **3** 191–211

[95] Bonabeau E., Theraulaz G, Fourcassié V and Deneoubourg J L 1998 The phase-ordering kinetics of cemetery organization in ants *Phys. Rev.* **57** 4569–71

[96] Bonabeau E, Henaux F, Guerin S, Snyers D, Kuntz P and Theraukaz G 1998 Routing in telecommunication networks with 'smart' ant-like agents *SFI Research Paper*

[97] Bonabeau E, Dorigo M and Theraulaz G 1999 *Swarm Intelligence: From Natural to Artificial Systems* (Oxford: Oxford University Press)

[98] Boon J P, Dab D, Kapral R and Lawniczak A 1996 Lattice gas automata for reactive systems *Phys. Rep.* **273** 55–148

[99] Bosch R A 1998 Using integer programming to find still lifes and oscillators in Conway's game of life *Manuscript*

[100] Botee H M and Bonabeau E 1999 Evolving ant colony optimization *Adv. Complex Syst.* **1** 149–59

[101] Bourbakis N, Steffensen N and Saha B 1997 Design of an array processor for parallel skeletonization of images *IEEE Trans. on Circuits and Systems. II. Analog and Digital Signal Processing* **44** 284–98

[102] Bournez O and Cosnard M 1996 On the computational power of dynamical systems and hybrid systems *Theor. Comput. Sci.* **168** 417–59

[103] Branicky M S 1995 Universal computation and other capabilities of hybrid and continuous dynamical systems *Theor. Comput. Sci.* **138** 67–100

[104] Braune M and Engel H 1993 Compound rotation of spiral waves in active media with periodically modulated excitability *Chem. Phys. Lett.* **211** 534–40

[105] Broder A Z and Karlin A R 1988 Bounds on the cover time *Proc. 29th Ann. Symp. Foundations of Computer Science* (New York: IEEE) pp 479–87

[106] Broder A 1989 Generating random spanning trees *Proc. Symp. Foundations of Computer Science* (New York: IEEE) pp 442–7

[107] Brodland G W and Clausi D A 1994 Embryonic tissue morphogenesis modeled by FEM *J. Biomech. Eng.* **116** 146–55

[108] Brooks R A 1986 A robust layered control system for a mobile robot *IEEE J. Robotics Automation* **2** 14–23

[109] Brooks M 1999 *Quantum Computing and Communications* (Berlin: Springer)

[110] Brown J A and Tuszynski J A 1997 Dipole interactions in axonal microtubules as a mechanism of signal propagation *Phys. Rev.* E **56** 5834–9

[111] Brown J A and Tuszynski J A 1999 A review of the ferroelectric model of microtubules *Ferroelectrics* **220** 141–56

[112] Bu L and Chiueh T-D 1999 Solving the shortest path problem using an analog network *IEEE Trans. Circuits and Systems I: Fund. Theory Appl.* **46** 1360–3

[113] Buettner H M 1995 Computer simulation of nerve growth cone filopodial dynamics for visualization and analysis *Cell Motility and the Cytoskeleton* **32** 187–204

[114] Bureš J 1984 Leão's spreading depression, an example of diffusion-mediated propagation of excitation in the central nervous system *Self-Organization: Autowaves and Structures Far From Equilibrium* ed V I Krinsky (Heidelberg: Springer) pp 180–3

[115] Butrimenko A V 1964 On the search for optimal routes in changing graphs *Izv. Akad. Nauk SSSR. Ser. Tekhn. Kibern.* **6**

[116] Buvel R L and Ingerson T E 1984 Structure in asynchronous cellular automata *Physica* D **1** 59–68

[117] Cai D, Bishop A E and Gronbechjensen N 1995 Spatially localized, temporally quasi-periodic, discrete nonlinear excitations *Phys. Rev.* E **52** 5784–7

[118] Cai D, Bishop A R and Gronbechjensen N 1996 Discrete lattice effects on breathers in a spatially linear potential *Phys. Rev.* E **53** 1202–5

[119] Calabi L and Hartnett W E 1968 Shape recognition, prairie fires, convex deficiencies and skeletons *Am. Math. Monthly* **75** 335–42

[120] Calude C S, Casti J L and Dineen M J (ed) *Unconventional Models of Computation* (Berlin: Springer)

[121] Casati R, Costato M and Milani M 1996 Cellular automata simulation of the effects induced by an external electromagnetic field on microtubule formation dynamics *Bioelectrochem. Bioenerget.* **41** 63–9

[122] Castanho C D, Chen W and Wada K 2000 Parallel algorithm for constructing strongly convex superhulls of points *IEICE Trans. Fund. Electron., Commun. Comput. Sci.* E **83-A** 722–32

[123] Chang M S, Huang N-F and Tang C-J 1990 An optimal algorithm for constructing oriented Voronoi diagrams and geographic neighbourhood graph *Inf. Process. Lett.* **35** 255–60

[124] Chang H 1982 Magnetic-bubble conservative logic *Int. J. Theor. Phys.* **21** 955–60

[125] Chau H F and Wilczek F 1995 Simple realization of the Fredkin gate using a series of 2-body operators *Phys. Rev. Lett.* **75** 748–50

[126] Chazelle B 1984 Computational geometry on a systolic chip *IEEE Trans. Comput.* C **33** 774–85

[127] Chen D, Aubry S and Tsironis G P 1996 Breather mobility in discrete Ψ^4 nonlinear lattices *Phys. Rev. Lett.* **77** 4776–9

[128] Chen W, Nakano K and Wada K 2000 Parallel algorithms for convex hull problems and their paradigm *IEICE Trans. Inf. Syst.* E **83-D** 519–29

[129] Chen W, Wada K and Kawaguchi K 1996 Parallel robust algorithms for constructing strongly convex hulls *Proc. 12th Annual Symp. Computational Geometry* (New York: ACM) pp 133–40

[130] Chernyak Y B, Feldman A B and Cohen R J 1997 Correspondence between discrete and continuous models of excitable media: trigger waves *Phys. Rev.* E **55** 3215–33

[131] Chin F and Ting H F 1990 Improving the time complexity of message–optimal distributed algorithm for minimum-weight spanning tree *SIAM J. Comput.* **19** 612–26

[132] Chong F 1993 Analog techniques for adaptive routing on interconnection networks *MIT Transit Note* No 14

[133] Chung S G 1994 Soliton theory for classical statistical mechanics of the massive-Thirring–sine-Gordon model *Int. J. Mod. Phys.* **18** 2447–88

[134] Chopard B and Droz M 1999 *Cellular Automata Modeling of Physical Systems* (Cambridge: Cambridge University Press)

[135] Christiansen P L, Gaididei Yu B, Mezentsev V K, Musher S L, Rasmussen K O, Juul Rasmussen J, Ryzhenkova I V and Turitsyn S K 1996 Discrete localized states and localization dynamics in discrete nonlinear Schrödinger equations *Phys. Scr.* T **67** 160–6

[136] Christiansen P L, Rasmussen K Ø, Bang O and Gaididei Yu B 1995 The temperature-dependent collapse regime in a nonlinear dynamical model of Scheibe aggregates *Physica* D **87** 321–4

[137] Church A 1941 *The Calculi of Lambda Conversion* (Princeton, NJ: Princeton University Press)

[138] Cockayne E J 1970 On the efficiency of the algorithm for Steiner minimal trees *SIAM Appl. Math.* **18** 150–9

[139] Colorni A, Dorigo M and Maniezzo V 1992 Distributed optimization by ant colonies *Proc. European Conf. on Artificial Life* ed F Varela and P Bourgine (Amsterdam: Elsevier) pp 134–42

[140] Cook S and Rackoff C 1980 Space lower bounds for maze threadability on restricted machines *SIAM J. Comput.* **9** 636–52

[141] Coiras E, Santamaria J and Miravet C 1998 Hexadecagonal region growing *Pattern Recognition Lett.* **19** 1111–17

[142] Conrad M 1972 Information processing in molecular systems *Curr. Mod. Biol.* (now *Biosystems*) **5** 1–14

[143] Conrad M 1976 Complementary molecular model of learning and memory *Biosystems* **8** 119–38

[144] Conrad M and Liberman E A 1982 Molecular computing as a link between biological and physical theory *J. Theor. Biol.* **98** 239–52

[145] Conrad M 1985 On design principles for a molecular computer *Commun. ACM* **28** 464–80

[146] Conrad M 1992 Molecular computing: the lock–key paradigm *Computer* **25** 11–20

[147] Conrad M 1993 Emergent computation through self-assembly *Nanobiology* **2** 5–30

[148] Conrad M 1993 Fluctuons I. Operational analysis *Chaos, Solitons and Fractals* **3** 411–24

[149] Conrad M and Zauner K-P 1998 DNA as a vehicle for the self-assembly model of computing *Biosystems* **45** 59–66

[150] Conrad M 1999 Molecular and evolutionary computation: the tug of war between context freedom and context sensitivity *Biosystems* **52** 99–110

[151] Cook S and Rackoff C 1980 Space lower bounds for maze threadability on restricted machines *SIAM J. Comput.* **9** 636–52

[152] Cori R, Metivier Y and Zielonka W 1993 Asynchronous mappings and asynchronous cellular automata *Inf. Comput.* **106** 159–72

[153] Corne D 2000 *Applied Evolutionary Computation: Solving Industrial and Scientific Optimisation Problems* (Berlin: Springer)

[154] Cotterill R 1998 *Enchanted Looms: Conscious Networks in Brains and Computers* (Cambridge: Cambridge University Press)

[155] Cottet G H and Germain L 1993 Image processing through reaction combined with nonlinear diffusion *Math. Comput.* **61** 659–73

[156] Cramer L 1995 Animation of actin-based cell motility and cell migration (December)
`http://iccbweb.med.harvard.edu/mitchisonlab/Media/`

[157] Cramer L Molecular mechanism of cell migration
`http://www.ucl.ac.uk/lmcb/cramer.html`

[158] Crawford C and Riecke H 1999 Oscillon-type structures and their interaction in a Swift–Hohenberg model *Physica* D **129** 83–92

[159] Cronhjort M B and Nyberg A M 1996 3D hypercycles have no stable spatial structure *Physica* D **190** 79–83

[160] Cross M C and Hohenberg P C 1993 Pattern formation outside of equilibrium *Rev. Mod. Phys.* **65** 851–1112

[161] Crutchfield J P, Mitchell M and Das R 1998 The evolutionary design of collective computation in cellular automata *Preprint* adap-org/9809001

[162] Cybenko G 1989 Load balancing for distribute memory multiprocessors *J. Parallel Distrib. Comput.* **7** 279–301

[163] Dab D, Lawnizcak A, Boon J-P and Kapral R 1990 Cellular-automata models for reactive systems *Phys. Rev. Lett.* **64** 2462–5

[164] Dantzig B B 1960 On the shortest route through a network *Management Sci.* **6** 187–90

[165] Dassow J and Jurgensen H 1987 Soliton automata *Lect. Notes Comput. Sci.* **278** 95–102

[166] Dassow J and Jurgensen H 1991 Deterministic soliton automata with a single exterior node *Theor. Comput. Sci.* **84** 281–92

[167] Datta A, Pal S and Pal N R 2000 Connectionist model for convex-hull of a planar set *Neural Networks* **13** 377–84

[168] Datta A and Parui S K 1996 A dynamic neural net to compute convex hull *Neurocomputing* **10** 375–84

[169] Datta A and Parui S K 1994 A robust parallel thinning algorithm for binary images *Pattern Recognition* **27** 1181–92

[170] Davydov V A and Morozov V G 1996 Galilean transformations and evolution of autowave fronts in external field *Usp. Piz. Nauk (Physics—Uspekhi)* **39** 305–11

[171] Day A M 1991 Parallel implementation of 3D convex-hull algorithm *Computer-Aided Design* **23** 177–88

[172] Day A M and Tracey D 1998 Parallel implementations for determining the 2D convex hull *Concurrency Practice and Experience* **10** 449–66

[173] Deaton R J, Garzon M, Murphy R C, Franceschetti D R and Stevens S E 1998 On the realiability and efficiency of a DNA computation *Phys. Rev. Lett.* **80** 417–20

[174] Dehne F, Deng X, Dymond P, Fabri A and Khokhar A A 1997 A randomized parallel three-dimensional convex hull algorithm for coarse-grained multicomputers *Theor. Comput. Syst.* **30** 547–58

[175] Dehne F, Hassenklover A L, Sack J R and Santoro N 1991 Computation geometry algorithms for the systolic screen *Algorithmica* **6** 734–61

[176] Delorme M and Mazoyer J (ed) 1998 *Cellular Automata: A Parallel Model* (Dordrecht: Kluwer Academic)

[177] Dennin M, Ahlers G and Cannell D S 1996 Spatiotemporal chaos in electroconvection *Science* **272** 388–90

[178] Dennin M, Cannell D S and Ahlers G 1998 Patterns of electroconvection in a nematic liquid crystal *Phys. Rev.* E **57** 638–49

[179] Dennin M, Treiber M, Kramer L, Ahlers G and Cannell D S 1996 Origin of traveling rolls in electroconvection of nematic liquid crystals *Phys. Rev. Lett.* **76** 319–22

[180] Dennin M, Ahlers G and Cannell D S 1996 Chaotic localized states near the onset of electroconvection *Phys. Rev. Lett.* **77** 2475–8

[181] Dhar D 1990 Self-organized critical state of sandpile automaton model *Phys. Rev. Lett.* **64** 1613–16

[182] Diallo M, Ferreira A and RauChaplin A 1998 Communication-efficient deterministic parallel algorithms for planar point location and 2D Voronoi diagram *Lect. Notes Comput. Sci.* **1373** 399–409

[183] Diallo M, Ferreira A, RauChaplin A and Ubeda S 1999 Scalable 2D convex hull and triangulation algorithms for coarse grained multicomputers *J. Parallel Distrib. Comput.* **56** 47–70

[184] Dickiea J F 1995 Major crowd catastrophes *Safety Sci.* **18** 309–20

[185] Diekmann R, Muthukrishnan S and Nayakkankuppam V 1997 Engineering diffusive load balancing algorithms using experiments *Proc. IRREGULAR'97 (Lecture Notes in Computer Science 1253)* (Berlin: Springer) pp 111–22

[186] Diekmann R, Monien B and Preis R 1997 Load balancing strategies for distributed memory machines *Computer Science Technical Report Series 'SFB'* No tr-rsfb-97-050, University of Padeborn

[187] Dijkstra E A 1959 A note on two problems in connection with graphs *Numer. Math.* **1** 269–71

[188] Dilão R and Sainhas J 1997 Validation and calibration of models for reaction–diffusion systems `xxx.lanl.gov/patt-sol/9712007`

[189] Dobrushin R L, Kryukov V I and Toom A L (ed) 1990 *Stochastic Cellular Systems: Ergodicity, Memory and Morphogenesis* (Manchester: Manchester University Press)

[190] Domain C and Gutowitz H 1997 The topological skeleton of cellular automaton dynamics *Physica* D **103** 155–68

[191] Dorigo M, Maniezzo V and Colorni A 1996 The ant system: optimization by a colony of cooperating agents *IEEE Trans. Syst. Man Cybernet.* **26** 1–13

[192] Dorigo M and Gambardella L M 1997 Ant colony system: a cooperative learning approach to the traveling salesman problem *IEEE Trans. Evolutionary Comput.* **1** 53–66

[193] Dormann D, Weijer C and Siegert F 1997 Twisted scroll waves organize dictyostelium mucoroides slugs *J. Cell Sci.* **110** 1831–7

[194] Dormann D, Vasiev B and Weijer C J 1998 Propagating waves control *dictyostelium discoideum* morphogenesis *Biophys. Chem.* **72** 21–35

[195] Dowle M, Mantel R M and Barkley D 1997 Fast simulations of waves in three-dimensional excitable media *Int. J. Bifurcation Chaos* **7** 2529–46

[196] Doyle P and Snell J L 1984 *Random Walks and Electrical Networks* (Washington, DC: Mathematical Association of America)

[197] O'Dunlaing C and Yap C 1985 A 'retraction' method for planning the motion of a disc *J. Algorithms* **6** 104–11

[198] Durand-Lose J 1998 About the universality of the billiard ball model *Proc. Coll. Math. Machines Comput.* **2** 118–24

[199] Durand-Lose J O 1995 Reversible cellular automaton able to simulate any other reversible one using partitioning automata *LNCS* **911** 230–44

[200] Durret R and Griffeath D 1993 Aysmptotic behaviour of excitable cellular automata *Experimental Math.* **2** 183–208

[201] Duxois T and Peyrard M 1993 Energy localization in nonlinear lattices *Phys. Rev. Lett.* **25** 3935–8

[202] Edelsbrunner H 1987 *Algorithms in Computational Geometry* (Berlin: Springer)

[203] Edmundson D E and Enns R H 1992 Bistable light bullets *Opt. Energy* **17** 586

[204] Edmundson D E and Enns R H 1993 Fully 3-dimensional collisions of bistable light bullets *Opt. Lett.* **18** 1609

[205] Edmundson D E and Enns R H 1995 The particle-like nature of colliding light bullets *Phys. Rev.* A **51** 2491–8

[206] Edmundson D and Enns R 1996 Light bullet home page `http://www.sfu.ca/~renns/lbullets.html`

[207] Edmundson D E 1996 Unstable halo `http://www.sfu.ca/~renns/lbullets.html`

[208] Egan G 1996 *Permutation City* (Millenium)

[209] El Naschie M S 1995 A note on quantum mechanics, diffusional interference and informions *Chaos, Solitons and Fractals* **5** 881–4

[210] Elkin Yu E, Biktashev V N and Holden A V 1997 On the movement of excitation wave breaks *Chaos, Solitons and Fractals* **9** 1597–610

[211] ElGindy H and Wetherall L 1997 A simple Voronoi diagram algorithm for a reconfigurable mesh *IEEE Trans. Parallel Distrib. Syst.* **8** 1133–42

[212] Epstein I R and Showalter K 1996 Nonlinear chemical dynamics: oscillations, patterns and chaos *J. Phys. Chem.* **100** 13 132–47

[213] Epstein I R and Pojman J A 1998 *An Introduction to Nonlinear Chemical Dynamics: Oscillations, Waves, Patterns, and Chaos* (Oxford: Oxford University Press)

[214] Érdi P and Tóth T 1992 *Mathematical Models of Chemical Reactions* (New York: Wiley)

[215] Ermentrout B 1995 A heuristic description of spiral wave instability in discrete media *Physica* D **82** 154–64

[216] Evans D J and Mai S-W 1985 Two parallel algorithms for the convex hull problem in a two dimensional space *Parallel Comput.* **2** 313–26

[217] Evans D J and Stojmenovic I 1989 On parallel computation of Voronoi diagram *Parallel Comput.* **12** 121–5

[218] Fast V G and Efimov I R 1991 Stability of vortex rotation in an excitable cellular medium *Physica* D **49** 75–81

[219] Feige U 1993 A randomized time-space tradeoff of $O(mR)$ for USTCON *Proc. 34th Symp. Foundations of Computer Science* pp 238–46

[220] Feige U 1996 A fast randomized LOGSPACE algorithm for graph connectivity *Theor. Comput. Sci.* **169** 147–60

[221] Feldman A, Chernyak Y B and Cohen R J 1995 Relation between continuous and discrete models of excitable media *Proc. Ann. Int. Conf. IEEE Eng. Med. Biol.* **17** 47–8

[222] Feldman A, Yin J Z, Saxberg E H, Chernyak Y B and Cohen R J 1995 Vortex wave stability in homogeneous excitable media: simulation on a randomized discrete lattice *Proc. Ann. Int. Conf. IEEE Eng. Med. Biol.* **17** 25–6

[223] Fenwick J W and Dowell L J 1999 Electrical substation service-area estimation using cellular automata: an initial report *Proc. ACM Symp. Applied Computing* pp 560–5

[224] Feynman R P 1982 Simulating physics with computers *Int. J. Theor. Phys.* **21**

[225] Feynman R P 1999 Computing machines in the future (Nishima Memorial Lecture, 1985) *Feynman and Computation* ed A J G Hey (Reading, MA: Perseus) pp 225–30

[226] Field R J, Körös E and Noyes R M 1972 Oscillations in chemical systems. II. Thorough analysis of temporal oscillation in the bromate-cerium-malonic acid system *J. Am. Chem. Soc.* **94** 3649–65

[227] Field R J and Burger M (ed) 1985 *Oscillations and Travelling Waves in Chemical Systems* (New York: Wiley)

[228] Filatrella G and Malomed B A 1999 The ac-driven motion of dislocations in a weakly damped Frenkel–Kontorova lattice http//xxx.lanl.gov/cond--mat/9908292

[229] Fink H-W and Schoenenberg C 1999 Electrical conduction throught DNA molecules *Nature* **398** 407–10

[230] Fisher P C 1965 Generation of primes by a one-dimensional real-time iterative arrays *J. ACM* **12** 388–94

[231] FitzHugh R 1961 Impulses and physiologic states in theoretical models of nerve membrane *Biophys. J.* **11** 445–66

[232] Flach S, Kladko K and MacKay R S 1997 Energy thresholds for discrete breathers in one-, two-, and three-dimensional lattices *Phys. Rev. Lett.* **78** 1207–10

[233] Flach S and Willis C R 1997 Discrete breathers *Phys. Rep.* **295** 181–264

[234] Flocchini P, Geurts F and Santoro N 1997 CA-like error propagation in fuzzy CA *Parallel Comput.* **23** 1673–82

[235] Flocchini P, Geurts F, Mingarelli A and Santoro N 2000 Convergence and aperiodicity in fuzzy cellular automata: revisiting rule 90 *Physica* D **142** 20–8

[236] Floyd R W 1962 Algorithm 97: shortest path *Commun. ACM* **5** 345

[237] Fogel L J 1999 *Intelligence Through Simulated Evolution: Forty Years of Evolutionary Programming* (New York: Wiley)

[238] Fontata W and Buss L W 1996 The barrier of objects: from dynamical systems to bounded organizations *SFI Working Paper* 96-05-035

[239] Forinash K, Peyrard M and Malomed B 1994 Interaction of discrete breathers with impurity modes *Phys. Rev.* E **49** 3400–11

[240] Forinash K, Cretegny T and Peyrard M 1997 Local modes and localization in a multicomponent nonlinear lattices *Phys. Rev.* **55** 4740–56

[241] Frauenkorn H, Kivshar Y S and Malomed B A 1996 Multisoliton collisions in nearly integrable systems *Phys. Rev.* E **54** R2244–7

[242] Frederickson G N 1990 A distributed shortest path algorithm for a planar network *Inf. Comput.* **86** 607–25

[243] Fredkin E and Toffoli T 1982 Conservative logic *Int. J. Theor. Phys.* **21** 219–53

[244] Fredman M L and Tarjan R E 1984 Fibonacci heaps and their uses in improved network optimization algorithms *Proc. 25th Ann. Symp. Found. Comput. Sci. (Singer Island, FL; Silver Springer, NY)* (New York: IEEE) pp 338–46

[245] Freeman N C 1980 Soliton interaction in two dimensions *Adv. Appl. Mech.* **20** 1–37

[246] Frisch U, Hasslacher B and Pomeau Y 1986 Lattice-gas automata for the Navier–Stokes equation *Phys. Rev. Lett.* **56** 1505–8

[247] Frisch U, d'Humieres D, Hasslacher B, Lallemand P, Pomeau Y and Rivet J-P 1987 Lattice gas hydrodynamics in two and three dimensions *Complex Syst.* **1** 649–707

[248] Fortune S 1987 A sweepline algorithm for Voronoi diagram *Algorithmica* **2** 153–74

[249] Fu B and Beigel R 1999 Length bounded molecular computing *Biosystems* **52** 155–63

[250] Fuchssteiner B 1991 Filter automata admitting oscillating carrier waves *Appl. Math. Lett.* **4** 23–6

[251] Gallager R G, Humblet P A and Spira P M 1983 A distributed algorithm for minimum-weight spanning tree *ACM Trans. Programming Languages Syst.* **5** 66–77

[252] Gambardella L M and Dorigo M 1995 Ant-Q: a reinforcement learning approach to the traveling salesman problem *Proc. 12th Int. Conf. on Machine Learning* (San Francisco, CA: Morgan Kaufmann) pp 252–60

[253] Gardner M 1983 *Wheels, Life and Other Mathematical Amusements* (New York: Freeman)

[254] Garzon M 1995 *Models of Massive Parallelism: Analysis of Cellular Automata and Neural Networks* (Berlin: Springer)

[255] Garzon M H, Jonoska N and Karl S A 1999 The bounded complexity of DNA computing *Biosystems* **52** 63–72

[256] Garzon M H and Deaton R J 1999 Biomolecular computing and programming *IEEE Trans. Evolutionary Comput.* **3** 236–50

[257] Gaylord R J and Nishidate K 1996 *Modeling Nature. Cellular Automata Simulations with Mathematica* (Berlin: Springer)

[258] Gecseg F and Jürgensen H 1990 Automata represented by products of soliton automata *Theor. Comput. Sci.* **74** 163–81

[259] Gel'fand I M and Tsetlin M L 1960 On continuous models of control systems *Dokl. Akad. Nauk SSSR (Proc. Acad. Sci. USSR)* 131

[260] Gen M and Cheng R 2000 *Genetic Algorithms and Engineering Optimization* (New York: Wiley)

[261] Gerhardt M, Schuster H and Tyson J J 1990 A cellular automaton model of excitable media including curvature and dispersion *Science* **247** 1563–6

[262] Gerhardt M, Schuster H and Tyson J J 1990 A cellular automaton model of excitable media. II. Curvature, dispersion, rotating waves and meandering waves *Physica* **46** 392–415

[263] Gerhardt M, Schuster H and Tyson J J 1990 A cellular automaton model of excitable media inlcuding curvature and dispersion III. Fitting the Belousov–Zhabotinsky reaction *Physica* D **46** 416–25

[264] Gerhardt M, Schuster H and Tyson J J 1991 A cellular automaton model of excitable media. IV. Untwisted scroll rings *Physica* D **50** 189–206

[265] Gleiser M and Sornborger A 2000 Long-lived localized field configurations in small lattices: application to oscillons *Phys. Rev.* E **62** 1368–74

[266] Goles E and Margenstern M 1996 Sand pile as a universal computer *Int. J. Mod. Phys.* C **7** 113–22

[267] Goles E and Kiwi M A 1993 Games on line graphs and sand piles *Theor. Comput. Sci.* **115** 321–49

[268] Goodhill G J 1997 Diffusion in axon guidance *Eur. J. Neurosci.* **9** 1414–21

[269] Goodhill G J and Urbach J S 1999 Theoretical analysis of gradient detection by growth cones *J. Neurobiol.* **41** 230–41

[270] Goodman G C 1995 Mechanisms and molecules that control growth cone guidance *Ann. Rev. Neurosci.* **19** 341–77

[271] Goodman S E and Hedetniemi S T (ed) 1977 *Introduction to the Design and Analysis of Algorithms* (New York: McGraw-Hill)

[272] Goss S, Beckers R, Deneuborg J L, Aron S and Pasteels J M 1990 How trail laying and trail following can solve foraging problems of ant colonies *Behavioural Mechanisms of Food Selection (NATO ASI Series G 20)* ed R N Hughes (Berlin: Springer) pp 661–7

[273] Gravner J 1996 Cellular automata models of ring dynamics *Int. J. Mod. Phys.* C **7** 863–71

[274] Gramss T, Bornholdt S, Gross M, Mitchell M and Pellizzari T 1998 *Nonstandard Computation* (New York: Wiley–VCH)

[275] Greenberg J M and Hasting S P 1978 Spatial patterns for discrete models of diffusion in excitable media *SIAM J. Appl. Math.* **34** 515–23

[276] Griffeath D and Moore C 1997 Life without death is P-complete `http://www.santafe.edu/~Moore`

[277] Grinstein G, Jayaprakash C and He Y 1985 Statistical mechanics of probabilistic cellular autonata *Phys. Rev. Lett.* **65** 2527–30

[278] Gronewold A and Sonnenschein M 1998 Event-based modelling of ecological systems with asynchronous cellular automata *Ecological Modelling* **108** 37–52

[279] Goles E and Margenstern M 1996 Sand pile as a universal computer *Int. J. Mod. Phys.* **7** 113–22

[280] Goles E and Kiwi M 1993 Games on linear graphs and sand piles *Theor. Comput. Sci.* **115** 321–49

[281] Goles E and Matamala M 1997 Reaction–diffusion automata: three states implies universality *Theor. Comput. Syst.* **30** 223–9

[282] Gutowitz H A and Langton C 1995 Mean field theory of the edge of chaos *Proc. 3rd European Conf. On Artificial Life* (Berlin: Springer) pp 52–64

[283] Hafner G S and Tokarski T R 1998 Morphogenesis and pattern formation in the retina of the crayfish *Procambarus clarkii Cell and Tissue Res.* **293** 535–50

[284] Hagberg A A 1994 Fronts and patterns in reaction–diffusion equations *PhD Thesis* The University of Arizona

[285] Haykin S 1999 *Neural Networks: A Comprehensive Foundation* (New York: IEEE)

[286] Hameroff S and Penrose R 1996 Conscious events as orchestrated space-time selections *J. Consciousness Stud.* **1/2** 36–53

[287] Hameroff S, Smith S and Watt R 1986 Automaton model of dynamic organization in microtubules *Ann. N. Y. Acad. Sci.* **466** 949–52

[288] Hameroff S R and Watt R C 1982 Information processing in microtubules *J. Theor. Biol.* **98** 549–61

[289] Hanson J E and Crutchfield J P 1997 Computational mechanics of cellular automata: An example *Physica* D **103** 169–89

[290] Hardy J, de Pazzis O and Pomeau Y 1976 Molecular dynamics of a classical lattice gas: transport properties and time correlation functions *Phys. Rev.* A **13** 1949–61

[291] Haskins C P and Whelden R 1965 'Queenlessness', worker sibship and colony versus population structure in the formicid genus *Rhytidoponera Psyche* **72** 87–112

[292] Hasslacher B and Meyer D A 1998 Modelling dynamical geometry with lattice gas automata `http://xxx.lanl.gov`

[293] Hatayama G, Kuniba A and Takagi T 1999 Soliton cellular automata associated with finite crystals `http://xxx.lanl.gov/solv-int/9907020`

[294] Hayashi T, Nakano K and Olariu S 1998 Optimal parallel algorithms for finding proximate points, with applications *IEEE Trans. Parallel Distrib. Syst.* **9** 1153–66

[295] Heijmans H J A M and Ronse C 1990 The algebraic basis of mathematical morphology. I. Dilation and erosion *Comput. Vision Graphics Image Process.* **50** 245–95

[296] Heijmans H J A M 1994 *Morphological Image Operators* (Boston, MA: Academic)

[297] Helbing D 1992 A fluid-dynamic model for behaviour of pedestrians *Complex Syst.* **6** 391–415

[298] Helbing D, Molnar P and Schweitzer F 1998 Computer simulations of pedestrians dynamics and trail formation

[299] Hentschel H G E and Ooyen van A 1999 Models of axon guidance and bundling during development *Proc. R. Soc.* B **266** 2231–8

[300] Henze C and Tyson J J 1996 Cellular-automaton model of 3-dimensional excitable media *J. Chem. Soc.—Faraday Trans.* **92** 2883–95

[301] Hey A J G (ed) 1999 *Feynman and Computation* (Reading, MA: Perseus)

[302] Hillis D 1985 *The Connection Machine* (Cambridge, MA: MIT)

[303] Hjelmfelt A, Weinberger E D and Ross J 1991 Chemical implementation of neural networks and Turing machines *Proc. Natl Acad. Sci., USA* **88** 10 983–7

[304] Hjelmfelt A, Weinberger E D and Ross J 1992 Chemical implementation of finite-state machines *Proc. Natl Acad. Sci., USA* **89** 383–7

[305] Hjelmfelt A and Ross J 1992 Chemical implementation and thermodynamics of collective neural networks *Proc. Natl Acad. Sci., USA* **89** 388–91

[306] Hjelmfelt A and Ross J 1993 Mass-coupled chemical systems with computational properties *J. Phys. Chem.* **97** 7988–92

[307] Hjelmfelt A and Ross J 1995 Implementation of logic functions and computations by chemical kinetics *Physica* D **84** 180–93

[308] Hodgkin A L and Huxley A F 1952 A quantitative description of membrane current and its application to conduction and excitation in nerve *J. Physiol.* **117** 500–44

[309] Hohmann W, Kraus M and Schneider F W 1997 Recognition in excitable chemical reactor networks. Experiments and model simulations *J. Phys. Chem.* A **101** 7364–70

[310] Holden A V, Thompson B C and Tucker J V 1990 Can excitable media be considered as computational systems? *Nonlinear Wave Processes in Excitable Media* ed A V Holden, M Markus and H G Othmer (New York:) pp 509–16

[311] Holden A V, Tucker J V and Thompson B C 1991 Can excitable media be considered as computational system? *Physica* D **49** 240–6

[312] Holey J A and Ibarra O H 1992 Iterative algorithms for the planar convex hull problem on mesh-connected arrays *Parallel Comput.* **18** 281–96

[313] Hölldobler B and Wilson E O 1995 *Journey to the Ants* (Cambridge: The Belknap Press)

[314] Hopfield J J 1982 Neural networks and physical systems with emergent collective computational abilities *Proc. Natl Acad. Sci., USA* **79** 2554–8

[315] Hordijk W, Crutchfield J P and Mitchell M 1996 Embedded-particle computation in evolved cellular automata *Santa Fe Institute Working Paper* 96-09-073

[316] Huang S-T 1990 A fully pipelined minimum spanning tree constructor *J. Parallel Distrib. Comput.* **9** 55–62

[317] Huang C-H and Lengauer C 1989 An incremental mechanical development of systolic solutions to the algebraic path problem *Acta Informatica* **27** 97–124

[318] Ilachinski A and Halpern P 1987 Structurally dynamic cellular automata *Complex Syst.* **1** 503–27

[319] Inverardi P and Wolf A L 1995 Formal specification and analysis of software architectures using the chemical abstract machine model *IEEE Trans. Software Eng.* **21** 373–86

[320] Ishida T, Iida T, Yokoi H and Kakazu Y 2000 Development of a liquid metal mobile object field *Proc. 6th Int. Conf. on Simulation of Adaptive Behavior* (Cambridge, MA: MIT) pp 362–8

[321] Islam M N, Soccolich C E and Gordon J P 1992 Ultrafast digital soliton logic gates *Opt. Quantum Electron.* **24** S1215–35

[322] Ivanitski G R, Krinsky V I, Zaikin A N and Zhabotinsky A M 1981 Autowave processes and their role in disturbing the stability of distributed excitable systems *Sov. Sci. Rev.* **1** 79–91

[323] Ivanitsky G R, Kunisky A S and Tzyganov M A 1984 Study of 'target patterns' in a phage-bacterium system *Self–Organization: Autowaves and Structures Far From Equilibrium* ed V I Krinsky (Heidelberg: Springer) pp 214–17

[324] Jahnke W, Skaggs W E and Winfree A T 1989 Chemical vortex dynamics in the Belousov–Zhabotinsky reaction and in the two variable Oregonator model *J. Phys. Chem.* **93** 740–9

[325] Jakubowski M H, Steiglitz K and Squier R 1997 Information transfer between solitary waves in the saturable Schrödinger equation *Phys. Rev.* E **56** 7267–72

[326] Jakubowski M H, Steiglitz K and Squier R K 1998 State transformations of colliding optical solitons and possible application to computation in bulk media *Phys. Rev.* E **58** 6752–8

[327] Jakubiowski M H 1998 Using `nls` for solution and visualization of NLS (Department of Computer Science, Princeton University)

[328] Jang J-S R and Mizutani E 1996 *Neuro-Fuzzy and Soft Computing: A Computational Approach to Learning and Machine Intelligence* (Englewood Cliffs, NJ: Prentice-Hall)

[329] Jankowski S and Wanczuk R 1994 CNN model of complex pattern formation in excitable media *Proc. IEEE Int. Workshop on Cellular Neural Networks and Their Applications* (New York: IEEE) pp 333–8

[330] Jaromczyk J W and Toussaint G T 1992 Relative neighbourhood graphs and their relatives *Proc. IEEE* **80** 1502–17

[331] Jaromczyk J W and Kowaluk M 1987 A note on relative neighbourhood graphs *Proc. 3rd Ann. Symp. Computational Geometry* (New York: ACM) pp 233–41

[332] Jeong C S, Choi J J and Lee D T 1992 Parallel enclosing rectangle on SIMD-machines *Parallel Comput.* **18** 221–9

[333] Jeong C S and Lee D T 1990 Parallel geometric algorithms on mesh-connected computer *Algorithmica* **5** 1555–177

[334] Johnson D B 1977 Efficient algorithms for shortest paths in sparse networks *J. ACM* **24** 1–13

[335] Jonoska N, Karl S A and Saito M 1999 Three dimensional DNA structures in computing *Biosystems* **52** 143–53

[336] Kacperski K and Holyst J A 1999 Opinion formation model with strong leader and external impact: a mean field approach *Physica* A **269** 511–26

[337] Kalmar Z, Marczell Z, Szepesvari C and Lorincz A 1999 Parallel and robust skeletonization built on self-organizing elements *Neural Networks* **12** 163–73

[338] Kamala K and Meena M 1995 Nondeterministic, probabilistic and alternating computations on cellular array models *Theor. Comput. Sci.* **143** 23–49

[339] Kampfner R R 1995 Integrating molecular and digital computing: an information systems design perspective *Biosystems* **35** 229–32

[340] Kanaya M, Cheng G-X, Watanabe K and Tanaka M 1994 Shortest path searching for the robot walking using an analog resistive network *Proc. IEEE Int. Symp. Circuits and Systems* **6** 311–14

[341] Kapral R 1993 Chemical waves and coupled map lattices *Theory and Applications of Coupled Map Lattices* ed K Kaneko (New York: Wiley)

[342] Kari L, Gloor G and Yu S 2000 Using DNA to solve the Bounded Post Correspondence problem *Theor. Comput. Sci.* **231** 193–203

[343] Karlsson R G and Poblete P V 1983 An $O(m \log\log D)$ algorithm for shortest path *Discr. Appl. Math.* **6** 91–3

[344] Kasmacherleidinger H and Schmidschonbein H 1994 Complex dynamic order in ventricular fibrillation *J. Electrocardiography* **21** 287–99

[345] Katz J F 1998 Psychodynamique des foules *Méd. Catastrophe Urg. Collectives* **1** 155–60

[346] Kaye R 2000 Minespeeper is NP-complete *Math. Intell.* **22** 9–15

[347] Kaye R 2000 Infinite versions of minesweeper are Turing complete *Manuscript* `http://www.mat.bham.ac.uk/R.W.Kaye`

[348] Kayey R 2000 Some minesweeper configurations *Manuscript* `http://www.mat.bham.ac.uk/R.W.Kaye`

[349] Keener J P 1986 A geometrical theory for spiral waves in excitable media *SIAM J. Appl. Math.* **46** 1039–56

[350] Keener J P and Tyson J J 1986 Spiral waves in the Belousov–Zhabotinsky reaction *Physica* D **21** 300–24

[351] Keitt T H and Johnson A R 1995 Spatial heterogeneity and anomalous kinetics: emergent patterns in diffusion-limited predatory–prey interaction *J. Theor. Biol.* **172** 127–39

[352] Kenkre V M 1983 Theoretical methods for the analysis of exciton capture and annihilation *Organic Molecular Aggregates: Electronic Excitation and Interaction Processes* ed P Reineker, H Haken and H C Wolf (Berlin: Springer) pp 193–201

[353] Keynes R and Cook G M W 1995 Axon guidance molecules *Cell* **83** 161–9

[354] Kim C E and Stojmenovic I 1995 Sequential and parallel approximate convex hull algorithms *Comput. Artificial Intell.* **14** 597–610

[355] Klein R 1988 Voronoi diagram in the Moscow metric *Graphtheoretic Concepts in Computer Science: WG'88* ed J van Leeuwen (Berlin: Springer) pp 434–41

[356] Klein R 1988 Abstract Voronoi diagrams and their applications *Computational Geometry and Its Applications: CG'88* ed H Nolteimer (Berlin: Springer) pp 148–57

[357] Kock A J and Meinhardt H 1994 Biological pattern formation: from basic mechanisms to complex structures *Rev. Mod. Phys.* **66** 1481–507

[358] Koiran P, Cosnard M and Garzon M 1994 Computability with low-dimensional dynamical systems *Theor. Comput. Sci.* **132** 113–28

[359] Koiran P and Moore C 1999 Closed-form analytic maps in one and two dimensions can simulate universal Turing machines *Theor. Computer Sci.* **210** 217–2

[360] Kolmogorov A N, Petrovsky I G and Piskunov N S 1937 On the equation of diffusion related to the increasing of matter, and its application to one problem of biology *Bull. MGU. Ser. Matematika i Mekhanika. (Bull. Moscow State Univ. Ser. Mathematics and Mechanics)* **1** 1–26 (in Russian)

[361] Korohoda W, Mycielska M, Janda E and Madeja Z 2000 Immediate and long-term galvanotactic responses of *Amoeba proteus* to dc electric fields *Cell Motility and the Cytoskeleton* **45** 10–26

[362] Kostic P, Reljin I and Reljin B 1998 In-node fast optimal packet routing algorithm in Manhattan street network using cellular neural network *Proc. Mediterranean Electrotechnical Conf.—MELECON* **2** 698–700

[363] Kotropoulos C, Pitas I and Maglara A 1993 Voronoi tessellation and Delauney triangulation using Euclidean disk growing in $\mathbf{Z}^2$ *Proc. IEEE Int. Conf. Acoustics, Speech and Signal Process.* **5** 29–32

[364] Koza J R, Bennett F H III, Bennett F H, Andre D and Keane M A 1999 *Genetic Programming III: Darwinian Invention and Problem Solving* (Morgan Kaufmann)

[365] Krinski V I 1984 Autowaves: results, problems, outlooks *Self-Organization: Autowaves and Structures Far From Equilibrium* ed V I Krinsky (Heidelberg: Springer) pp 9–19

[366] Krinsky V I 1966 Spread of excitation in an inhomogeneous medium *Biofizika* **11** 676–83

[367] Krinsky V I 1984 *Self-Organization: Autowaves and Structures Far From Equilibrium* (Heidelberg: Springer)

[368] Krinsky V I, Bikhtashev V N and Efimov I R 1991 Autowave principles for parallel image processing *Physica* D **49** 247–53

[369] Krishnapuram R and Chen L F 1993 Implementation of parallel thinning algorithms using recurrent neural networks *IEEE Trans. Neural Networks* **4** 142–7

[370] Kruskal J B 1956 On the shortest subtree of a graph and the traveling problem *Proc. Am. Math. Soc.* **7** 48–50

[371] Kube C R and Bonabeau E 2000 Cooperative transport by ants and robots *Robotics and Autonomous Syst.* **30** 85–101

[372] Kuhn U 1998 Local calculation of Voronoi diagrams *Inf. Process. Lett.* **68** 307–12

[373] Kudriavcev V B, Podkolzin A S and Bolotov A A 1990 *Basics of the Theory of Homogeneous Structures* (Moscow: Nauka)

[374] Kuhnert L 1986 Photochemische manipulation von chemischen wellen *Naturwissenschaften* **76** 96–7

[375] Kuhnert L 1986 A new photochemical memory device in a light sensitive active medium *Nature* **319** 393

[376] Kuhnert L, Agladze K L and Krinsky V I 1989 Image processing using light-sensitive chemical waves *Nature* **337** 244–7

[377] Kurrer Ch and Schulten K 1991 Propagation of chemical waves in discrete excitable media: anisotropic and isotropic wave fronts *Nonlinear Wave Processes in Excitable Media (NATO ASI Series B 244)* ed A Holden *et al* (London: Plenum) pp 489–500

[378] Lam L, Lee S W and Suen C Y 1992 Thinning metholodogies—a comprehensive survey *IEEE Trans. Pattern Analysis and Machine Intell.* **14** 869–85

[379] Lambin X, Elston D A, Petty S J and MacKinnon J L 1998 Spatial asynchrony and periodic travelling waves in cyclic populations of field voles *Proc. R. Soc.* B **265** 1491–6

[380] Landauer R 1999 Information is inevitably physical *Feynman and Computation* ed A J G Hey (Reading, MA: Perseus) pp 77–92

[381] Landweber L F and Baum E B 1998 *DNA Based Computers II: Dimacs Workshop, 1996 (Dimacs Series in Discrete Mathematics and Theoretical Computer Science, V. 44)* (Providence, RI: American Mathematical Society)

[382] Langton C G 1990 Computation at the edge of chaos: phase transitions and emergent computation *Physica* D **42** 12–27

[383] Lam L and Post J (ed) 1992 *Solitons in Liquid Crystals* (New York: Springer)

[384] Laplante J P, Pemberton M, Hjelmfelt A and Ross J 1995 Experiments on pattern-recognition by chemical kinetics *J. Phys. Chem.* **99** 10 063–5

[385] Lee D T and Wong C K 1980 Voronoi diagrams in $L_1(L_\infty)$ metrics with 2-dimensional storage applications *SIAM J. Comput.* **9** 200–11

[386] Lee D T 1980 Two-dimensional Voronoi diagram in L_1 metric *J. ACM* **27** 604–18

[387] Lee K J, McCormick W D, Pearson J E and Swinney H L 1994 Experimental observation of self-replicating spots in a reaction–diffusion system *Nature* **369** 215–28

[388] Leeuwenhoek van A Arcana naturae ope et beneficic exquisitissimorum microscopicorum hugd *Batavorum 1696. Epistole physiologicae super compluribus naturae arcanis etc. ad alios d. vinos doctos Delphis.*

[389] Leung Y, Zhang J-S and Xu Z-B 1997 Neural networks for convex hull computation *IEEE Trans. Neural Networks* **8** 601–11

[390] Leven D and Sharir M 1987 Planning purely translational motion for a convex object in two-dimensional space using generalized Voronoi diagram *Discr. Comput. Geom.* **2** 9–37

[391] Levine D S 2000 *Introduction to Neural and Cognitive Modeling* (Lawrence Erlbaum Association)

[392] Liggett T M 1985 *Interacting Particle Systems* (Berlin: Springer)

[393] Light Bullets' web site `http://www.sfu.ca/~renns/lbullets.html`

[394] Lin J-C and Lin J-Y 1998 $O(1 \log N)$ parallel algorithm for detecting convex hulls on image boards *IEEE Trans. Image Process.* **7** 922–5

[395] Lin R, Olariu S, Schwing J L and Wang B F 1999 The mesh with hybrid buses: an efficient parallel architecture for digital geometry *IEEE Trans. Parallel Distr. Syst.* **10** 266–80

[396] Linde H and Engel H 1991 Autowave propagation in heterogeneous active media *Physica* D **49** 13–20

[397] Littman M and Boyan J 1993 A distributed reinforcement learning scheme for network routing *Carnegi Mellon University Computer Science Report* CMU-CS-93-165

[398] López F J and Sanz G 2000 Stochastic comparisons for general probabilistic cellular automata *Stat. Probab. Lett.* **46** 401–10

[399] Lo H-K, Spiller T and Popescu S 1998 *Introduction to Quantum Computation and Information* (Singapore: World Scientific)

[400] Luczak T and Wierman J 1988 Critical probability bounds for two-dimensional site percolation models *J. Phys. A: Math. Gen.* **21** 3131–8

[401] Lugosi E 1989 Analysis of meandering in Zykov kinetics *Physica* D **40** 331–7

[402] Lukashevich I P 1963 A computer study of continuous models of control systems *Biofizika* **8**

[403] Lumer E D and Nicolis G 1994 Synchronous versus asynchronous dynamics in spatially distributed systems *Physica* D **71** 440–52

[404] Lyons R and Peres Y 2001 *Probability on Trees and Networks* (Cambridge: Cambridge University Press)

[405] Machta J and Greenlaw R 1996 Parallel computational complexity and logical depth in statistical physics *PhysComp96: 4th Workshop on Physics and Computation* ed T Toffoli, M Biafore and J Leao (Cambridge, MA: New England Complex Systems Institute) pp 201–7

[406] Machta J and Greenlaw R 1996 The computation complexity of generating random fractals *J. Stat. Phys.* **82** 1299

[407] MacLennan B 1990 Field computation: a theoretical framework for massively parallel analog computation. parts I–IV *Computer Science Department Technical Report* CS-90-100 (Knoxville, TN: University of Tennessee)

[408] MacLennan B 2000 An overview of field computation `http://www.cs.utk.edu/~mclennan/ICCIN-FC.html`

[409] Maimistov A, Malomed B and Desyatnikov A 1998 A potential of incoherent attraction between multidimensional solitons *Preprint* patt-sol/9812009

[410] Makhankov V G 1990 *Phenomenology of Solitons* (Dordrecht: Kluwer Academic)

[411] Makowiec D 1993 Probabilistic approach to the dynamics of deterministic homogeneous and symmetric two-dimensional cellular automata *Physica* A **199** 299–311

[412] Malandain G, Bertrand G and Ayache N 1991 Topological classification in digital space *Lect. Notes Comput. Sci.* **511** 300–13

[413] Malandain G, Bertrand G and Ayache N 1993 Topological segmentation of discrete surfaces *Int. J. Computer Vision* **10** 183–97

[414] Maniatty W and Szymanski B 1997 Fine-grain discrete Voronoi diagram algorithms in L_1 and L_∞ norms *Math. Comput. Modelling* **26** 71–8

[415] Maniezzo V and Colorni A 1999 Ant system applied to the quadratic assignment problem *IEEE Trans. Knowledge Data Eng.* **11** 769–78

[416] Marangoni R, Preosti G and Colombetti G 2000 Phototactic orientation mechanism in the ciliate *Fabred salina*, as inferred from numerical simulations *J. Photochem. Photobiol. B: Biol.* **54** 185–93

[417] Marangoni R, Batistini A, Puntoni S and Colombetti G 1995 Temperature effects on motion parameters and the phototactic reaction of the marine ciliate *Fabrea salina J. Photochem. Photobiol. B: Biol.* **30** 123–7

[418] Melkonian M, Anderson R A and Schnepf E 1991 *The Cytoskeleton of Flagellate and Ciliate Protists* (Berlin: Springer)

[419] Marczell Zs, Kalmar Zs and Lorincz A 1996 Generalized skeleton formation for texture segmentation *Neural Networks World* **1** 79–87

[420] Maree A F M, Panfilov A V and Hogeweg P 1999 Migration and thermotaxis of dictyostelium discoideum slugs, a model study *J. Theor. Biol.* **199** 297–309

[421] Marek M and Ševčíková H 1984 Electrical field effects on propagating pulse and front waves *Self-Organization: Autowaves and Structures Far From Equilibrium* ed V I Krinsky (Heidelberg: Springer) pp 161–3

[422] Ševčíková H and Marek M 1986 Chemical waves in electric field: modelling *Physica* D **21** 61–77

[423] Margolus N 1984 Physics-like models of computation *Physica* D **10** 81–95

[424] Margolus N 1999 Crystalline computation *Feynman and Computation* ed A J G Hey (Reading, MA: Perseus) pp 267–305

[425] Markus M and Hess B 1990 Isotropic cellular automata for modelling excitable media *Nature* **347** 56–8

[426] Markus M, Krafczyk M and Hess B 1991 Randomized automata for isotropic modelling of two- and three-dimensional waves and spatiotemporal chaos *Nonlinear Wave Processes in Excitable Media* ed A V Holden, M Markus and H G Othmer (New York: Plenum) pp 161–81

[427] Markus M, Nagyungvarai Z and Hess D 1992 Phototaxis of spiral waves *Science* **257** 225–7

[428] Markus M and Stavridis K 1994 Observation of chemical turbulence in the Belousov–Zhabotinsky reaction *Int. J. Bifurcation Chaos* **4** 1233–43

[429] Markus M, Kloss G and Kusch I 1994 Disordered waves in a homogeneous, motionless excitable medium *Nature* **371** 402–4

[430] Markus M, Kusch I, Ribeiro A and Almeida P 1996 Class IV behaviour in cellular automata models of physical systems *Int. J. Bifurcation Chaos* **6** 1817–27

[431] Marshall G F and Tarassenko L 1993 Robot path planning using VLSI resistive grids *Proc. 2nd Int. Conf. Artificial Neural Networks (Brighton)* (Berlin: Springer) pp 163–7

[432] Martins-Ferreira H, Nedergaard M and Nicholson C 2000 Perspectives on spreading depression *Brain Res. Rev.* **32** 215–34

[433] Matsukidaira J, Satsuma J, Takahashi D, Tokihiro T and Torii M 1997 Toda-type cellular automata and its N-soliton solution *Phys. Lett.* A **225** 287–95

[434] Mase K, Fukumura T and Toriwaki J 1981 Modified digital Voronoi diagram and its applications to image processing *Systems, Computers and Controls* **12** 27–36

[435] Mazelko J and Showalter K 1991 Chemical waves in inhomogeneous media *Physica* D **49** 21–32

[436] Mazoyer J 1996 Computations on one-dimensional cellular-automata *Ann. Math. Artificial Intell.* **16** 285–309

[437] Mayer A G 1908 The cause of pulsation *The Popular Science Monthly* **LXXIII** 481–7

[438] McCullough W and Pitts W 1943 A logical calculus of the ideas immanent in nervous activity *Bull. Math. Biophys.* **5** 115–33

[439] Mead C 1989 *Analog VLSI and Neural Systems* (Reading, MA: Addison-Wesley)

[440] Meador J L 1995 Spaciotemporal neural network for shortest path optimization *Proc. IEEE Int. Symp. Circuits and Systems* **2** 801–4

[441] Meinhardt H 1982 *Models of Biological Pattern Formation* (London: Academic)

[442] Meinhardt H and Klinger M 1986 Pattern formation by coupled oscillations: The pigmentation pattern on shells of molluscs *Lect. Notes Biomath.* **71** 184–98

[443] Meinhardt H and Klinger M 1987 A model of pattern formation on the shells of moluscs *J. Theor. Biol.* **126** 63–9

[444] Meinhardt H and Klinger M 1991 Pattern formation on the shells of molluscs by travelling waves with unsual properties *Nonlinear Wave Processes in Excitable Media* ed A V Holden, M Markus and H G Othmer (New York: Plenum) pp 233–44

[445] Meinhardt H 1999 Orientation of chemotactic cells and growth cones: models and mechanisms *J. Cell Sci.* **112** 2867–74

[446] Melhuish C, Holland O and Hoddell S 1999 Convoying: using chorusing to form travelling groups of minimal agents *Robotics and Autonomous Syst.* **28** 207–16

[447] Melhuish C 2000 Autostruosis: construction without explicit planning in microrobots—a social insect approach *Proc. 7th Int. Symp. of Evolutionary Robotics (Tokyo)* (AAI Books)

[448] Melzak Z A 1961 On the problem of Steiner *Can. Math. Bull.* **4** 143–8

[449] Men'shikov M V and Pelikh K D 1989 Percolation with several defect types. An estimate of critical probability for a square lattice *Math. Notes Acad. Sci. USSR* **46** 778–85

[450] Meyer D A 1996 From quantum cellular automata to quantum lattice gases *J. Stat. Phys.* **85** 551–74

[451] Michoel T and Verbeure A 1999 Mathematical structure of magnons in quantum ferromagnets *Preprint* KUL-TF-99/10 (Insituut voor Theoretische Fysica, Katholieke Universiteit Leuven)

[452] *Microsoft*^{R} *Minesweeper, Help.* ©1981–1997, Microsoft Corp., by R Donner and C Johnson

[453] Mikhailov A S 1991 Kinematics of wave patterns in excitable media *Nonlinear Wave Processes in Excitable Media* ed A V Holden, M Markus and H G Othmer (New York: Plenum) pp 127–60

[454] Mikhailov A S 1993 Collective dynamics in models of communicating populations *Interdisciplinary Approaches to Nonlinear Complex Systems* ed H Haken and A Mikhailov (Berlin: Springer) pp 77–88

[455] Michalewicz Z 1996 *Genetic Algorithms + Data Structures = Evolution Programs* (Berlin: Springer)

[456] Mills J 1995 The continuous retina: image processing with a single-sensor artificial neural field network *Technical Report* 443, Department of Computer Science, Indiana University

[457] Minsky M L 1982 Cellular vacuum *Int. J. Theor. Phys.* **21** 537

[458] Minsky M 1999 Richard Feynman and cellular vacuum *Feynman and Computation* ed A J G Hey (Reading, MA: Perseus) pp 117–30

[459] Mitchell M 1996 Computation in cellular automata: a selected review *Santa Fe Institute Working Paper* 96-09-074

[460] Mitchell M, Hraber P T and Crutchfield J P 1993 Revisiting the edge of chaos: evolving cellular automata to perform computation *Complex Syst.* **7** 89–130

[461] Mitchell M, Crutchfield J P and Hraber P T 1994 Evolving cellular automata to perform computation: mechanisms and impediments *Physica* D **75** 361–91

[462] Mitchell M 1996 *An Introduction to Genetic Algorithms* (Cambridge, MA: MIT)

[463] Möbius D and Kuhn H 1988 Energy transfer in monolayers with cyanine dye Scheibe aggregates *J. Appl. Phys.* **64** 5138–41

[464] Moffat A M and Takaoka T 1987 An all pairs shortest path algorithm with expected time $O(n \log n)$ *SIAM J. Comput.* **6** 1023–31

[465] Molofsky J 1994 Population dynamics and pattern formation in theoretical populations *Ecology* **75** 30–9

[466] Moller R 1999 Path planning using hardware time delays *IEEE Trans. Robotics and Automation* **15** 588–92

[467] Moore C and Nilsson M 1998 The computational complexity of sandpiles *SFI Working Paper* 98-08-071

[468] Moriwaki S, Nagai A, Satsuma J, Tokihiro T, Torii M, Takahashi D and Matsukidaira J 1999 $2 + 1$ dimensional soliton cellular automaton *London Math. Soc. Lect. Notes Ser.* **255** 224–35

[469] Morita K and Ueno S 1992 Computation universal model of 2D 16-state reversible partitioned cellular automaton *IEICE Trans. Inf. Syst.* E **75–D** 141–7

[470] Mornev O A, Aslanidi O V, Aliev R R and Chailakhian L M 1996 Soliton-like regime in FitzHugh–Nagumo equations: reflection of colliding pulses of excitation *Proc. Russ. Acad. Sci.* **347** 123–5

[471] Müler S C, Mair T and Steinbock O 1998 Traveling waves in yeast extract and in cultures of *Dictyostelium discoideum Biophys. Chem.* **72** 37–47

[472] Müller S C and Plesser Th 1991 Dynamics of spiral centers in the ferroin-catalyzed Belousov–Zhabotinsky reaction *Nonlinear Wave Processes in Excitable Media* ed A V Holden, M Markus and H G Othmer (New York: Plenum) pp 15–22

[473] Münster A F, Hasal F, Snita D and Marek M 1994 Charge distribution and electric field effects on spatiotemporal patterns *Phys. Rev.* E **50** 546–51

[474] Muratov C B 1999 Traveling wave solutions in the Burridge–Knopoff model `http://xxx.lanl.gov/patt-sol/9901003`

[475] Nagy-Ungvarai Z and Hess B 1991 Control of dynamic pattern formation in the Belousov–Zhabotinsky reaction *Physica* D **49** 33–9

[476] Nakagaki T, Yamada H and Tóth Á 2000 Intelligence: maze-solving by an amoeboid organism *Nature* **407**

[477] Nakamura K 1974 Asynchronous cellular automata and their computational ability *Systems, Computer, Control* **5** 58–66

[478] Nakamura K 1997 Parallel universal simulation and self–reproduction in cellular spaces *IEICE Trans. Inf. Syst.* E **80D** 547–52

[479] Neumann von J 1966 *Theory of Self-Reproducing Automata* (Chicago, IL: University of Illinois Press)

[480] Neusius C, Olszewski J and Scheerer D 1992 Efficient distributed thinning algorithm *Parallel Comput.* **18** 47–55

[481] Nieuwkoop P D, Gordon R and Bjorklund N K 1999 The neural induction process; its morphogenetic aspects *Int. J. Developmental Biol.* **43** 615–23

[482] Oficjalski J and Bialynicki-Birula I 1978 Collision of gaussons *Acta Phys. Polon.* B **9** 759–75

[483] Olariu S, Schwing J L and Zhang J Y 1992 Efficient image computations on reconfigurable meshes *Lect. Notes Comput. Sci.* **634** 589–94

[484] Omohundro S 1984 Modelling cellular automata with partial differential equations *Physica* D **10** 128–34

[485] Ooyen van A and Willshaw D J 1999 Competition for neurotrophic factor in the development of nerve connections *Proc. R. Soc.* B **266** 883–92

[486] Orda A and Rom R 1990 Shortest path and minimum delay algorithms in networks with time-dependent edge-length *J. ACM* **37** 607–25

[487] Orponen P and Matamala M 1996 Universal computation by finite two-dimensional coupled map lattices *PhysComp96: Proc. Workshop on Physics and Computation* extended abstract

[488] Orponen P and Matamala M 1996 Universal computation by finite two-dimensional coupled map lattices *Proc. Physics and Computation (Boston, MA)* (Cambridge, MA: New England Complex Systems Institute)

[489] Ortoleva P J 1992 *Nonlinear Chemical Waves* (Chichester: Wiley)

[490] Ostrovskaya E A, Kivshar Yu S, Chen Z and Segev M 1998 Solitonic gluons `http://xxx.lanl.gov/patt-sol/9808005`

[491] Packard N H and Wolfram S 1985 Two-dimensional cellular automata *J. Stat. Phys.* **38** 901–46

[492] Park J K, Steiglitz K and Thurston W P 1986 Soliton-like behaviour in automata *Physica* D **19** 423–32

[493] Papatheodorou T S, Ablowitz M J and Saridakis Y G 1988 A rule for fast computation and analysis of soliton automata *Stud. Appl. Math.* **79** 173–84

[494] Paulett J E and Ermentrout G B 1994 Stable rotating waves in two-dimensional discrete active media *SIAM J. Appl. Math.* **54** 1720–44

[495] Paun G, Rozenberg G, Salomaa A and Brauer W 1998 *DNA Computing: New Computing Paradigms* (Berlin: Springer)

[496] Perez-Munuzuri V, Perez-Villar V and Chua L O 1993 Autowaves for image processing on a two-dimensional CNN array of excitable nonlinear circuits: flat and wrinkled labyrinths *IEEE Trans. Circuits and Systems I: Fund. Theor. Appl.* **40** 174–81

[497] Petrov V, Quyang Q and Swinney H L 1997 Resonant pattern formation in a chemical system *Nature* **338** 655–7

[498] Pfaffmann J O and Conrad M 2000 Adaptive information processing in microtubule network *Biosystems* **55** 47–58

[499] Pighizzini G 1994 Asynchronous automata versus asynchronous cellular automata *Theor. Comput. Sci.* **132** 179–207

[500] Plath P J, Plath J and Schwietering J 1997 Collision patterns of mollusc shells *Discr. Dynam. Nature Soc.* **1** 57–76

[501] Plesser Th, Kingdon R D and Winters K H 1991 The simulation of chemical waves in excitable reaction–diffusion–convection systems by finite differences and finite element methods *Nonlinear Wave Processes in Excitable Media* ed A V Holden, M Markus and H G Othmer (New York: Plenum) pp 451–67

[502] Plewczyński D 1998 Landau theory of social clustering *Physica* A **261** 608–17

[503] Prajer M, Fleury A and Laurent M 1997 Dynamics of calcium regulation in *Paramecium* and possible morphogenetic implication *J. Cell Sci.* **110** 529–35

[504] Prassler E and Milios E 1991 Parallel path planning in unknown terrains *Proc. SPIE—The Int. Soc. Opt. Eng.* **1388** 2–13

[505] Preparata F P and Shamos M I 1985 *Computational Geometry: An Introduction* (Berlin: Springer)

[506] Preston K and Duff M 1984 *Modern Cellular Automata. Theory and Applications* (New York: Plenum)

[507] Price C B, Wambacq P and Oosterlinck A 1990 Image enhancement and analysis with reaction–diffusion paradigm *IEE Proc. Commun. Speech and Vision* **137** 136–45

[508] Price C B, Wambacq P and Oosterlinck A 1993 The plastic coupled map lattice: a novel image processing paradigm *Chaos* **2** 351–63

[509] Prim R C 1957 Shortest connection networks and some generalizations *Bell Syst. Tech. J.* **36** 1389–401

[510] Propoi A I 1995 Excitable media and nonlocal search *Automation and Remote Control* **56** 1042–50

[511] Propoi A I 1996 Problems of optimization and learning for nonlocal search in excitable media. 1. Problem of optimization *Automation and Remote Control* **57** 45–52

[512] Pour-El M B 1974 Abstract computability and its relation to the general purpose analog computer (some connections between logic, differential equations and analog computers) *Trans. Am. Math. Soc.* **199** 1–28

[513] Quinn J and Deo N 1984 Parallel graph algorithms *ACM Comput. Surv.* **16** 319–415

[514] Rambidi N G, Chernavskii D S and Sandler Yu M 1991 Towards a biomolecular computer. I. Ways, means, objectives *J. Mol. Electron.* **7** 105–14

[515] Rambidi N G and Chernavskii D S 1991 Towards a biomolecular computer. II. Information processing and computing devices based on biochemical nonlinear dynamic system *J. Mol. Electron.* **7** 115–25

[516] Rambidi N G 1992 Towards a biomolecular computer *Biosystems* **27** 219–22

[517] Rambidi N G 1992 An approach to computational complexity: nondiscrete biomolecular computing *Computer* **25** 51–4

[518] Rambidi N G, Maximychev A V and Usatov A V 1994 Molecular neural network devices based on nonlinear dynamic media *Biosystems* **33** 125–37

[519] Rambidi N G, Maximychev A V and Usatov A V 1994 Molecular image processing devices based on chemical reaction systems. I. General principles for implementation *Adv. Mater. Opt. Electron.* **4** 179–90

[520] Rambidi N G, Maximychev A V and Usatov A V 1994 Molecular image processing devices based on chemical reaction systems. II. Implementation of Blum–type algorithms *Adv. Mater. Opt. Electron.* **4** 191–201

[521] Rambidi N G and Maximychev A V 1995 Molecular neural network devices based on non-linear dynamic model: basic primitive information processing operations *Biosystems* **36** 87–99

[522] Rambidi N G and Maximichev A V 1995 Molecular image-processing devices based on chemical reaction systems. III. Some operational characteristics of excitable light-sensitive media used for image processing *Adv. Mater. Opt. Electron.* **5** 223–31

[523] Rambidi N G and Maximychev A V 1995 Molecular image processing devices based on chemical reaction systems. IV. Image-processing operations performed by active media functioning in oscillating mode *Adv. Mater. Opt. Electron.* **5** 233–41

[524] Rambidi N G and Maximychev A V 1997 Molecular image processing devices based on chemical reaction systems. V. Processing images with several levels of brightness and some application potentialities *Adv. Mater. Opt. Electron.* **7** 161–70

[525] Rambidi N G and Maximychev A V 1997 Molecular image processing devices based on chemical reaction systems. VI. Processing half-tone images and neural network architecture of excitable media *Adv. Mater. Opt. Electron.* **7** 171–82

[526] Rambidi N G 1997 Biomolecular computer: roots and promises *Biosystems* **44** 1–15

[527] Rambidi N G and Yakovenchuck D 1999 Finding path in a labyrinth based on reaction–diffusion media *Adv. Mater. Opt. Electron.* **9** 67–72

[528] Rambidi N G 2000 Image processing using light-sensitive chemical waves *Personal Communication*

[529] Rasmussen K Ø, Aubry S, Bishop A R and Tsironis G P 1999 Discrete nonlinear Schrödinger breathers in a phonon bath `http://xxx.lanl.gov/patt-sol/9901002 v2`

[530] Rasmussen K Ø, Gaididei Yu B, Bang O and Christiansen P L 1996 Nonlinear and stochastic modelling of energy transfer in Scheibe aggregates *Math. Comput. Simul.* **40** 339–58

[531] Ram Reddy M K, Dahlem M, Zykov V S and Müller S C 1995 The effect of an illumination jump on wave propagation in the Ru-catalyzed Belousov–Zhabotinsky reaction *Chem. Phys. Lett.* **236** 111–16

[532] Reif J H 1998 Paradigms for biomolecular computing *Proc. 1st Int. Conf. on Unconventional Models of Computation* (Singapore: Springer) pp 72–93

[533] Reimen N and Mazoyer J A 1992 Linear speed-up theorem for cellular automata *Theor. Comput. Sci.* **101** 59–98

[534] Requardt M 1996 Emergence of space-time on the Planck scale within the scheme of dynamic cellular networks and random graphs `http://xxx.soton.ac.uk/ps/hep-th/9612185`

[535] Requardt M 1996 Discrete mathematics and physics on the Planck-scale exemplified by means of a class of cellular network models and their dynamics `http://xxx.soton.ac.uk/ps/hep-th/9605103`

[536] Reshod'ko L V and Bures J 1975 Computer simulations of reverberating spreading depression in a network of cell automata *Biol. Cybernet.* **18** 181–9

[537] Riecke H 1995 Attractive interaction between pulses in a model for binary-mixture convection `http://xxx.lanl.gov/patt-sol/9502005`

[538] Riecke H and Granzow G D 1998 Localization of waves without bistability: worms in nematic electroconvection `http://xxx.lanl.gov/patt-sol/9802003`

[539] Riecke H 1999 Worms in electroconvection of nematic liquid crystals `http://www.esam.northwestern.edu/riecke/research/Worm/worm.html`

[540] Ripoll A, Senar M A, Cortes A and Luque E 1988 Mapping and dynamic load-balancing strategies for parallel programming *Comput. Artificial Intell.* **17** 481–91

[541] Riecke H and Granzow G D 1998 Localization of supercritical waves: worms in nematic electroconvection `http://xxx.lanl.gov/patt-sol/9802003`

[542] Rocha A F, Rebello M P and Miura K 1998 Toward a theory of molecular computing *Inf. Sci.* **106** 123–57

[543] Ross J, Müller S C and Vidal Ch 1988 Chemical waves *Science* **240** 460–5

[544] Rubin H and Wood D H (ed) 1999 *DNA Based Computers III: Dimacs Workshop, 1997 (Dimacs Series in Discrete Mathematics and Theoretical Computer Science 48)* (Providence, RI: American Mathematical Society)

[545] Rushton W A H 1937 Initiation of the propagated disturbance *Proc. R. Soc.* B **124** 210–43

[546] Russell F M 1995 Evidence for energetic lattice excitations and practical applications *Abstracts of Coherent Structures in Physics and Biology (Edinburgh)* `http://www.ma.hw.ac.uk/~chris/sca`

[547] Sager B M 1996 Propagation of traveling waves in excitable media *Genes and Development* **10** 2237–50

[548] Saha P K, Chaudhuri B B, Dutta D and Majumder D D 1997 A new shape preserving parallel thinning algorithm for 3D digital images *Pattern Recognition* **12** 1939–55

[549] Sala D M and Cios K J 1999 Solving graph algorithms with networks of spiking neurons *IEEE Trans. Neural Networks* **10** 953–7

[550] Schönfisch B 1997 Anisotropy in cellular automata *Biosystems* **41** 29–41

[551] Schönfisch B and de Roos A 1999 Synchronous and asynchronous updating in cellular automata *Biosystems* **51** 123–43

[552] Schoonderwoerd R, Holland O, Bruten J and Rothkratz L 1996 Ant-based load balancing in telecommunication networks *Adaptive Behaviour* **5** 169–207

[553] Schoonderwoerd R, Holland O and Bruten J 1997 Ant-like agents for load balancing in telecommunication networks *Proc. 1st Int. Conf. On Autonomous Agents (Marina del Rey, USA)* (New York: ACM) pp 209–16

[554] Sherratt J A 1996 Periodic travelling waves in a family of deterministic cellular-automata *Physica* D **95** 319–35

[555] Shi H, Gader P and Li H 1998 Parallel mesh algorithms for grid graph shortest paths with application to separation of touching chromosomes *J. Supercomputing* **12** 69–83

[556] Shin F Y and Wong W T 1994 Fully parallel thinning with tolerance to boundary noise *Pattern Recognition* **27** 1677–95

[557] Schimansky-Geier L, Mieth M, Rose H and Malchow H 1995 Structure formation by active Brownian particles *Phys. Lett.* A **207** 140–8

[558] Showalter K and Tyson J J 1987 Luther's 1906 discovery and analysis of chemical laws *J. Chem. Educ.* **64** 742–4

[559] Showalter K 1995 Quadratic and cubic reaction–diffusion fronts *Nonlin. Sci. Today* **4** 3–10

[560] Shu C and Buzton H 1995 Parallel path planning on the distributed array processor *Parallel Comput.* **21** 1749–67

[561] Siegelman H T 1996 The simple dynamics of super Turing theories *Theor. Comput. Sci.* **168** 461–72

[562] Siegelman H T and Sontag E D 1994 Analog computation via neural networks *Theor. Comput. Sci.* **131** 331–60

[563] Siegert F and Weijer C J 1989 Digital image processing of optical density wave propagation in *Dictyostelium discoideum* and analysis of the effects of caffeine and ammonia *J. Cell Sci.* **93** 325–35

[564] Sime J D 1995 Crowd psychology and engineering *Safety Sci.* **21** 1–14

[565] Sinha S and Ditto W L 1998 Dynamics based computation *Phys. Rev. Lett.* **81** 2156–9

[566] Sipper M 1997 *Evolving of Parallel Cellular Machines* (Berlin: Springer)

[567] Sperry R W 1963 Chemoaffinity in the orderly growth of nerve fiber patterns and connections *Proc. Natl Acad. Sci., USA* **50** 703–10

[568] Squier R K and Steiglitz K 1993 2D FHP lattice-gases are computation universal *Complex Syst.* **7** 297–307

[569] Squier R K and Steiglitz K 1994 Programmable parallel arithmetic in cellular automata using a particle model *Complex Syst.* **8** 311–23

[570] Skolnick M M and Marineau P 1992 Pebble Pond: a morphological wave propagation algorithm for representing spatial point patterns *Proc. SPIE: Image Algebra and Morphological Image Process.* **1769** 344–55

[571] Sóle R V, Miramontes O and Goodwin B C 1993 Emergent behavior of insect societies: global oscillations, chaos and computation *Interdisciplinary Approaches to Nonlinear Complex Systems* ed H Haken and A Mikhailov (Berlin: Springer) pp 77–88

[572] Stauffer D and Aharony A 1992 *Introduction to Percolation Theory* (London: Taylor and Francis)

[573] Starmer C F 1998 A biased history of the physics of neuro- and cardiac electrophysiology with a reading list `http://www.musc.edu/~tarmerf/biblio.html`

[574] Stark W R and Hughes W H 2000 Asynchronous, irregular automata nets: the path not taken *Biosystems* **55** 107–17

[575] Staunton R C 1996 An analysis of hexagonal thinning algorithms and skeletal shape representation *Pattern Recognition* **29** 1131–46

[576] Stefanyuk V L 1971 On mutual assistance in the collective of radio stations *Inf. Transmission Prob.* **7** 103–7

[577] Stefanuk V L 1997 From multi-agent systems to collective behaviour *Proc. Int. Workshop Distributed Artificial Intelligence and Multi-Agent Systems (St Petersburg)* (St Petersburg: Institute of Informatics, RAS) pp 223–4, 327–38

[578] Steiglitz K, Kamal I and Watson A 1988 Embedded computation in one-dimensional automata by phase coding solitons *IEEE Trans. Comput.* **37** 138–45

[579] Steinbock O, Zykov V and Muller S C 1993 Control of spiral-wave dynamics in active media by periodic modulation of excitability *Nature* **366** 322–4

[580] Steinbock O, Siegert F, Müller S and Weijer C J 1993 Three-dimensional waves of excitation during *Dictyostelium* morphogenesis *Proc. Natl Acad. Sci., USA* **90** 7332–5

[581] Steinbock O, Kettunen P and Showalter K 1995 Anisotropy and spiral organizing centers in patterned excitable media *Science* **269** 1857–60

[582] Steinbock O, Tóth A and Showalter K 1995 Navigating complex labyrinths: optimal paths from chemical waves *Science* **267** 868–71

[583] Steinbock O, Kettunen P and Showalter K 1996 Chemical wave logic gates *J. Phys. Chem.* **100** 18 970–5

[584] Stifter S 1991 An axiomatic approach to Voronoi diagrams in 3D *J. Comput. Syst. Sci.* **43** 361–79

[585] Subramanian D, Druschel P and Chen J 1997 Ants and reinforcement learning: a case study in routing in dynamic networks *Proc. IJCAI—97: Int. Joint Conf. AI* (Morgan Kaufmann) pp 832–8

[586] Sudha N, Nandi S and Sridharan K 1999 Parallel algorithm to construct Voronoi diagram and its VLSI architecture *Prof. IEEE Int. Conf. on Robotics and Automation* **3** 1683–8

[587] Supowit K J 1988 The relative neighbourhood graph, with application to minimum spanning tree *J. ACM* **30** 428–48

[588] Sutner K 1997 Linear cellular automata and Fischer automata *Parallel Comput.* **23** 1613–34

[589] Takahashi D and Matsukidaira J 1995 On discrete soliton-equations related to cellular-automata *Phys. Lett.* A **209** 184–8

[590] Takaoka T 1988 An efficient parallel algorithm for the all pairs shortest path problem *Proc. Int. Workshop on Graph-Theoretic Concepts in Computer Science (Lecture Notes in Computer Science 344) (Amsterdam)* (Berlin: Springer) pp 276–87

[591] Toffoli T and Margolus N 1985 *Cellular Automata Machines* (Cambridge, MA: MIT)

[592] Toffoli T and Margolus N 1991 Programmable matter *Physica* D **47** 263–72

[593] Toffoli T 1998 Programmable matter methods *Future Generation Comput. Syst.* **16** 187–201

[594] Toffoli T 1999 Action, or the fungibility of computation *Feynman and Computation* ed A J G Hey (Reading, MA: Perseus) pp 349–92

[595] Toffoli T 1998 Non-conventional computers *Encyclopedia of Electrical and Electronic Engineering* ed J Webster (New York: Wiley)

[596] Toussaint G T 1980 The relative neighbourhood graph of a finite planar set *Pattern Recognition* **12** 261–8

[597] Toussaint G T 1989 Computational geometry: recent developments *New Advances in Computational Geometry* ed R A Earnshow and B Wyvill (Berlin: Springer) pp 23–51

[598] Toussaint G T 1991 Some unsolved problems on proximity graphs *Proc. 1st Workshop on Proximity Graphs* (New Mexico: Computing Research Laboratory, New Mexico State University)

[599] Tokihiro T, Takahashi D, Mattsukidaira J and Satsuma J 1996 From soliton-equations to integrable cellular-automata through a limiting procedure *Phys. Rev. Lett.* **76** 191–6

[600] Tolmachev D and Adamatzky A 1996 Chemical processor for computation of Voronoi diagram *Adv. Mater. Opt. Electron.* **6** 191–6

[601] Thomson W (Lord Kelvin) 1876 On an instrument for calculating integral of the product of two giden functions *Proc. R. Soc.* **24** 266–75

[602] Tóth A and Showalter K 1995 Logic gates in excitable media *J. Chem. Phys.* **103** 2058–66

[603] Toyozawa Y 1983 Localization and delocalization of an exciton in the phonon field *Organic Molecular Aggregates: Electronic Excitation and Interaction Processes* ed P Reineker, H Haken and H C Wolf (Berlin: Springer) pp 90–106

[604] Tsetlin M L 1973 *Automaton Theory and Modeling of Biological Systems* (New York: Academic) (translated from the Russian edition 1969 *Issledovaniya po Teorii Avtomatov i Modelirovaniyu Biologicheskikh Sistem* (Moscow: Nauka))

[605] Tsimring L S and Aranson I S 1997 Localized and cellular patterns in a vibrated granular layer *Phys. Rev. Lett.* **79** 213–22

[606] Tu Y 1997 Worm structure in modified Swift–Hohenberg equation for electroconvection *Phys. Rev.* E **56** 3765–8

[607] Turing A 1952 The chemical basis of morphogenesis *Phil Trans. R. Soc.* B **237** 37–72

[608] Tuszynski J A, Trpisova B, Sept B and Brown J A 1997 Physical issues in the structure and function of microtubules *J. Struct. Biol.* **118** 94–106

[609] Trpisova B and Tuszynski J A 1997 Possible link between 5'-triphosphate hydrolysis and solitary waves in microtubules *Phys. Rev.* E **55** 3288–305

[610] Tsyganov I M, Aliev R R and Ivanitsky G R 1997 Dissipative pulsars in excitable media *Proc. Russ. Acad. Sci.* **352** 699–703

[611] Turing A 1936 On computable numbers, with an application to the Entscheidungsporblem *Proc. London Math. Soc.* **42** 230–65

[612] Tyson J J, Alexander K A, Manoranjan V S and Murray J D 1989 Spiral waves of cyclis AMP in a model of slime mold aggregation *Physica* D **34** 193–207

[613] Tzionas P G, Tsalides P G and Thanailakis A 1994 New, cellular automaton-based, nearest neighbor pattern classifier and its VLSI implementation *IEEE Trans. VLSI Syst.* **2** 343–53

[614] Tzionas O, Thanailakis A and Tsalides Ph 1997 An efficient algorithm for the largest empty figure problem based on a 2D cellular automaton architecture *Image and Vision Comput.* **15** 35–45

[615] Tzionas P G, Thanailakis A and Tsalides P G 1997 Collision-free path planning for a diamond-shaped robot using two-dimensional cellular automata *IEEE Trans. Robotics and Automation* **13** 237–50

[616] Ulam S M 1960 *A Collection of Mathematical Problems* (New York: Interscience) p 30

[617] Ubeda S 1995 Pyramidal thinning algorithm for SIMD parallel machines *Pattern Recognition* **28** 1993–2000

[618] Valentine F A 1964 *Convex Sets* (New York: McGraw-Hill)

[619] Vanag V K 1997 Investigation of the stochastic Oregonator by the probability cellular automaton: frequency-multiplying bifurcation *J. Phys. Chem.* A **101** 7074–84

[620] Varela F J, Maturana H R and Uribe R 1974 Autopoiesis: the organization of living systems, its characterization and a model *Biosystems* **5** 187–96

[621] Venetianer P L, Szolgay P C, Kenneth R, Roska T and Chua L O 1994 Analog combinatorics and cellular automata—key algorithms and layout design *Proc. IEEE Int. Workshop on Cellular Neural Networks and their Applications* (New York: IEEE) pp 249–54

[622] Venkataramani S C and Ott E 1998 Spatio-temporal bifurcation phenomena with temporal period doubling: patterns in vibrated sand *Phys. Rev. Lett.* **80** 3495

[623] Venkatasubramanian S, Krithivasan K and Pangan C P 1989 Algorithms for weighted graph automaton *Theor. Inf. Appl.* **23** 251–79

[624] Vergis A, Steiglitz K and Dickinson B 1986 The complexity of analog computation *Math. Comput. Simul.* **28** 91–113

[625] Vicsek T, Czirók A, Farkas I J and Helbing D 1999 Application of statistical mechanics to collective motion in biology *Physica* A **274** 182–9

[626] Vleugels J and Overmars M 1998 Approximation of Voronoi diagram of convex sites in any dimension *Int. J. Comput. Geom. Appl.* **8** 201–21

[627] Voorhees B 1996 *Computational Analysis of One-Dimensional Cellular Automata* (Singapore: World Scientific)

[628] Voronoi G 1908 Nouvelles applications des parametres continus à la theorie des formes quadratiques; deuxième memoire: recherches sur les paralleloedres primitifs *J. R. Angew. Math.* **134** 198–287

[629] Vyalyi M N, Gordeyev E N and Tarasov S P 1996 The stability of the Voronoi diagram *Comput. Math. Math. Phys.* **36** 405–14

[630] Walker C C and Ashby W R 1966 On temporal characteristics of behavior in certain complex systems *Kybernetik* **3** 100–8

[631] Warman J M, de Haas M P and Rupprecht A 1996 DNA: a molecule wire? *Chem. Phys. Lett.* **249** 319–22

[632] Weimar J R, Tyson J J and Watson L T 1992 Third generation cellular automata for modeling excitable media *Physica* D **55** 328–39

[633] Weimar J R, Tyson J J and Watson L T 1992 Diffusion and wave propagation in cellular automaton models of excitable media *Physica* D **55** 309–27

[634] Weimar J R 1996 Cellular automata for reaction–diffusion systems

[635] Weimar J 1998 *Simulation with Cellular Automata* (Logos)

[636] Weinstein M I 1999 Excitation thresholds for nonlinear localized modes on lattices `http://xxx.lanl.gov/patt-sol/9903001`

[637] Li W 1992 Phenomenology of non-local cellular automata *J. Stat. Phys.* **68** 829–82

[638] White T, Pagurek B and Oppacher F 1998 Connection management using adaptive mobile agents *Proc. Int. Conf. Parallel Distributed Processing Techniques and Applications (CSREA)* (New York: IEEE) pp 802–9

[639] Wichterman R 1986 *The Biology of Paramecium* (New York: Plenum)

[640] Wiener N and Rosenblueth A 1946 The mathematical formulation of the problem of conduction of impulses in a network of connected excitable elements, specifically in cardiac muscle *Arch. Inst. Cardiol. Mex.* **16** 205–65

[641] Wiener N 1961 *Cybernetics, or Control and Communication in the Animal and the Machine* (Cambridge, MA: MIT)

[642] Williams C P and Clearwater S H 1999 *Ultimate Zero and One: Computing at the Quantum Frontier* (New York: Copernicus)

[643] Winfree A T, Winfree E M and Seifert H 1985 Organizing centers in a cellular excitable medium *Physica* D **17** 109–15

[644] Winfree A T 1987 *When Time Breaks Down* (Princeton, NJ: Princeton University Press)

[645] Wolfram S 1984 Universality and complexity in cellular automata *Physica* D **10** 1–35

[646] Wolfram S 1986 Cellular automaton fluids. 1: Basic theory *J. Stat. Phys.* **45** 471–526

[647] Wolfram S 1996 *Cellular Automata and Complexity* (Reading, MA: Addison-Wesley)

[648] Wuensche A and Lesser M 1992 *The Global Dynamics of Cellular Automata* (Reading, MA: Addison-Wesley)

[649] Wuensche A 1997 *Attractors Basins of Discrete Networks (CSRP 461)* University of Sussex at Brighton

[650] Wuensche A 1999 Classifying cellular automata automatically: Finding gliders, filtering, and relating space-time patterns, attractor basins, and the Z parameter *Complexity* **4** 47–66

[651] Wuensche A 2000 DDLab: Discrete Dynamics Lab `http://www.ddlab.com/`

[652] Zu Z B, Zhang J S and Leung Y W 1998 An approximate algorithm for computing multidimensional convex hulls *Appl. Math. Comput.* **94** 193–226

[653] Xu C-Z and Lau F C 1994 Optimal parameters for load balancing with the diffusion method in mesh networks *Parallel Process. Lett.* **4** 139–47

[654] Xu C, Monien B, Löling R and Lau F 1995 Nearest-neighbour algorithms for load-balancing in parallel computers *Concurrency—Practice and Experience* **7** 707–36

[655] Yamada K, Yokoi H and Kakazu Y 1996 Adaptive distributed system of artificial cells using vibrating potential method. An application to nTSP *Proc. Adaptive Distributed Parallel Computing Symp. (Dayton, OH)* (Cleveland, OH: Case Western Reserve University) pp 73–87

[656] Yang T and Yang L-B 1997 Fuzzy cellular neural network: a new paradigm for image processing *Int. J. Circuit Theor. Appl.* **25** 469–81

[657] Yang T, Yang C-M and Yang L-B 1998 Differences between cellular neural network based and fuzzy cellular neural network based mathematical morphological operations *Int. J. Circuit Theor. Appl.* **26** 13–25

[658] Yap C K 1987 An $O(n \log n)$ algorithm for the Voronoi diagram of a set of simple curve segments *Discr. Comput. Geom.* **2** 365–93

[659] Yokoi H and Katazu Y 1992 An approach to the traveling salesman problem by a bionic model *HEURISTIC, J. Knowledge Eng.* **32** 13–27

[660] Yokoi H and Kakazu Y 1992 Theories and applications of autonomous machines based on the vibrating potential method *Proc. Int. Symp. Distributed Autonomous Robotic Systems* (New York: IEEE) pp 31–8

[661] Yokoi H, Mizumo T, Takita M and Kakazu Y Amoebae model using vibrating potential field *Proc. Adaptive Distributed Parallel Computing Symp. (Dayton, OH)* (Cleveland, OH: Case Western Reserve University) pp 292–314

[662] Yokoi H, Yu W, Hakura J and Kakazu Y 1998 A morpho-functional machine: An artificial amoeba based on the vibrating potential method *Proc. 6th Int. Conf. Artificial Life* ed C Adami *et al* (Cambridge, MA: MIT) pp 477–82

[663] Zabusky N J and Kruskal M D 1965 Interaction of 'solitons' in a collisionless plasma and the recurrence of initial states *Phys. Rev. Lett.* **15** 240–3

[664] Zaikin A N and Zhabotinsky A M 1970 Concentration wave propagation in two-dimensional liquid-phase self-oscillating system *Nature* **225** 535

[665] Zakharov A A 1972 *Vnutrividovyie Otnoshenia Muraviov* (*Interspecies Relations in Ants*) (Moscow: Nauka)

[666] Zakharov A A 1978 *Ant, Family, Colony* (Moscow: Nauka)

[667] Zakharov A A 1991 *Organization of Ant Communities* (Moscow: Nauka)

[668] Zauner K P and Conrad M 1996 Parallel computing with DNA: toward antiuniversal machine *4th Int. Conf. On Parallel Problem Solving From Nature* (Berlin: Springer)

[669] Zhabotinsky A M and Zaikin A N 1973 Autowave processes in a distributed chemical system *J. Theor. Biol.* **40** 45–61

[670] Zhabotinsky A M and Rovinsky A B 1984 Mathematical model of chemical active media *Self–Organization: Autowaves and Structures Far From Equilibrium* ed V I Krinsky (Heidelberg: Springer) pp 140–6

[671] Zhabotinsky A M, Eager M D and Epstein I R 1993 Refraction and reflection of chemical waves *Phys. Rev. Lett.* **71** 1526–9

[672] Ziegler J, Dittrich P and Banzhaf W 1997 Towards a metabolic robot controller *Information Processing in Cells and Tissues* (New York: Plenum) pp 305–18

[673] Zuev S A 1987 Bounds for the percolation threshold for a square lattice *Theor. Prob. Appl.* **32** 551–3

[674] Zukicke Ch, Ebeling W and Schimansky-Geier L 1989 Dynamic pattern processing with adaptive excitable media *Bioinformatics* **22** 261– 75

[675] Zuse K 1969 *Rechnender Raum* (Braubschweig: Vieweg)

[676] Zuse K 1982 The computing universe *Int. J. Theor. Phys.* **21** 589–600

[677] Zykov V S 1980 Analytical evaluation of the dependence of the speed of an excitation wave in a two-dimensional excitable medium on the curvature of its front *Biophysics* **25** 906–11

[678] Zykov V S 1987 *Simulation of Wave Processes in Excitable Media* (Manchester: Manchester University Press)

[679] Zykov V S 1986 Cycloidal circulation of spiral waves in excitable medium *Biofizika* **31** 862–5

Index